ENGINEERING CHEMISTRY

HEMANT KUMAR JOSHI, Ph.D.
Assistant Professor, Department of Chemistry
Nanhi Pari Seemant Engineering Institute, Pithoragarh
Campus Institute of Veer Madho Singh Bhandari
Uttarakhand Technical University, Dehradun, Uttarakhand

NARENDRA SINGH BHANDARI, Ph.D.
Professor, Department of Chemistry
Soban Singh Jeena University, Almora (Ex. VC), Uttarakhand

SUSHIL KUMAR JOSHI, Ph.D.
Professor, Department of Chemistry
Soban Singh Jeena University, Almora, Uttarakhand

PHI Learning Private Limited
Delhi-110092
2025

*In fond memory of **Shri Asoke K. Ghosh** (October 1942 – February 2024), Founder Chairman and Managing Director of PHI Learning, whose vision endlessly inspires.*

The Legacy Continues....

Published by Pushpita Ghosh, PHI Learning Private Limited, Rimjhim House, 111, Patparganj Industrial Estate, Delhi-110092 and Printed by Syndicate Binders, A-20, Hosiery Complex, Noida, Phase-II Extension, Noida-201305 (N.C.R. Delhi).

₹825.00

ENGINEERING CHEMISTRY
Hemant Kumar Joshi, Narendra Singh Bhandari and Sushil Kumar Joshi

ISBN-978-81-19364-33-6 (Print Book)
ISBN-978-81-19364-32-9 (e-Book)

श्रीकृष्णार्पणमस्तु

Contents

UNIT 3: WATER CHEMISTRY AND CORROSION

UNIT 4(B): CHEMISTRY OF ENGINEERING MATERIALS FUEL AND LUBRICANTS

Preface

Welcome to the interesting and fascinating world of engineering chemistry. This subject serves as a connection between chemistry and engineering applications. As technology advances and industry demands increase, there is an increasing need for budding engineers to have a comprehensive understanding of the principles of chemistry.

This book has been designed to meet the needs of B.Tech. students of Veer Madho Singh Bhandari Uttarakhand Technical University, Dehradun. It provides an organized way of learning the fundamentals of engineering chemistry and is designed to fit well with the university curriculum. In this book, students will learn the fundamentals of chemistry and their practical applications in the field of engineering. This book covers the fundamentals of atomic structure and chemical bonding to the intricacies of thermodynamics, electrochemistry, organic reactions and spectroscopy. Each chapter presents a clear and comprehensive overview of the subject matter.

The lecture video for each chapter has been prepared so that viewers can easily understand the concept. This will be an invaluable learning tool for students. By scanning the QR codes that are placed in the opening page of each chapter, the readers can access the videos. We really hope that the students and instructors will find this book informative and interesting.

Finally, we would like to express our heartfelt gratitude to Almighty, the Director of the institute, the faculty members, experts, reviewers and the publisher, PHI Learning Private Limited who have contributed their invaluable insight and expertise in the development of this book. We would like to mention here that we have taken every precaution to make the book error free even then we do not claim to be infallible. Therefore, recommendations and helpful critique from the readers are requested and will be greatly appreciated for improvement in the future editions.

May this book help you succeed in both your academic and professional endeavours!

Best wishes,

Hemant Kumar Joshi
Narendra Singh Bhandari
Sushil Kumar Joshi

UNIT 1

Atomic and Molecular Structure

CHAPTER 1

Atomic Structure and Chemical Bonding

1.1 HISTORY OF THE ATOM

In 460 BC, **Democritus**, Leucippus, and Epicurus advance the concept of the atom and put forth the earliest perspectives regarding its configuration and interconnections. Utilizing a pestle and mortar, he pulverized substances and reduced them to ever-smaller particles, which he termed ATMA (Greek for "indivisible"). He considered an atom to be an inert solid, excluding any entities that engage in interactions with other atoms from its volume. In contrast, contemporary atoms governed by quantum mechanics exhibit significant reactivity and engage in interactions through electric and magnetic fields. Democritus said, "The more any indivisible exceeds, the heavier it is".

Democritus

John Dalton

John Joseph Thomson

Ernest Rutherford

In 1808 **John Dalton** said that he thought everything was made up of small spheres. They were very flexible and could bounce around. They named them "atoms." He noticed that the atom is the smallest object that can't be broken down into smaller ones. But research showed that the atom is not the smallest piece of mass. An atom consists of three fundamental components, namely electrons, protons, and neutrons. Electrons are elementary subatomic particles characterized by a negative charge and a relatively low mass. Protons and neutrons are positively charged and neutrally charged particles, with almost the same mass, that are found in the middle of an atom.

John Joseph Thomson did an experiment in 1897 that showed a strong electric field can "pull" negatively charged particles out of any metal. In 1898, he discovered that atoms could sometimes send out an electron, which was a much smaller negative particle. Because of this, atoms can't be made up of only one part. If they are, they must have some kind of internal structure that controls their physical and chemical traits. Thompson comes up with the idea that an atom is made up of electrons that are spread out unevenly inside an elastic sphere. The electrons are encompassed by a surrounding medium of positive charge, which serves to neutralize their inherent negative charge, analogous to plums enveloped by a pudding-like substance.

1.2 PLUM-PUDDING MODEL OF ATOM

Thomson's atomic model, sometimes referred to as the "plum-pudding" model, postulated that the positive charges within an atom were uniformly distributed throughout a spherical structure encompassing the atom's dimensions. The electrons were concealed within the homogeneous backdrop, as depicted in Figure 1.1. Thomson postulated that under conditions of increased thermal energy, the electrons within an atom would exhibit mobility within their respective energy levels, hence giving rise to the phenomenon of electromagnetic radiation.

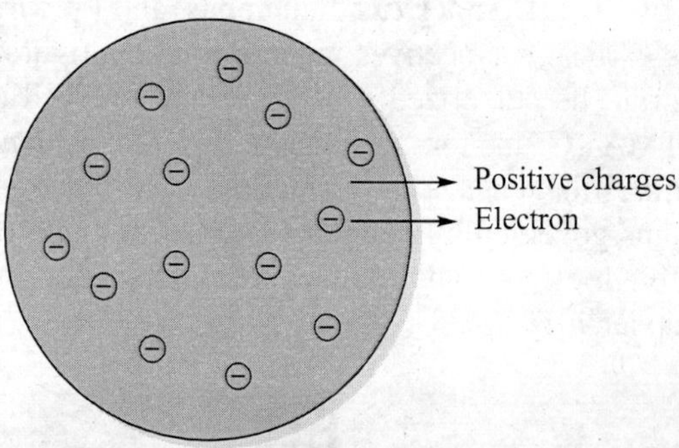

Figure 1.1 Plum-pudding model.

Ernest Rutherford and Frederic Soddy won the Nobel Prize in science in 1908 for their work on scattering experiments, which are also known as radioactive transmutation of chemical elements. They hit a thin piece of gold paper with alpha particles that came from a radioactive source. It was found that most of the helium nuclei went through the foil, but a few were pushed back. As you can see in Figure 1.2, though, some helium atoms bounced right back. Rutherford's new data let him come up with a more detailed model that had a nucleus in the middle. He said that the positive charge was all in a core in the middle, and negative electrons were all around it.

1.2.1 Experiments of Geiger and Marsden

The gold foil that Geiger and Marsden were shooting at was only a few microns thick. They found that even though most of them went through. About 1 out of every 10,000 hits. It was shown by Marsden and Geiger that some particles were thrown off of thin gold-leaf objects at angles greater than 90°.

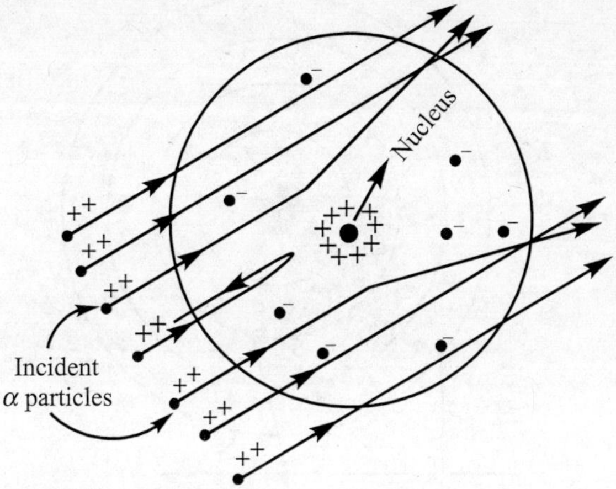

Figure 1.2 Scattering of alpha particles by the atom.

1.3 NIELS BOHR AND HYDROGEN ATOM MODEL

Niels Bohr

Theories that were popular at the time, before the work of Danish scientist **Niels Bohr**, could not explain why atoms were stable. At the Victoria University in Manchester, **Niels Bohr** learned from Rutherford. It was in 1913 that he used the work of Einstein and Planck to apply a new theory to hydrogen, the smallest atom. Bohr built on Rutherford's idea by keeping the idea that the nucleus held all the protons and neutrons and adding the idea that the electrons were moving around the nucleus in orbits, like planets moving around the sun. There are a certain amount of electrons in each orbit.

Niels Bohr's general assumptions:

1. A hydrogen atom consist stationary states i.e. orbits with revolving electrons, which does not radiate energy, around dense nucleus.
2. In an atom, the energy levels available to an electron are discrete and limited.
3. The energy levels of an electron within an atom can be altered through a process known as a "transition," wherein the electron moves from one energy level to another. The change of energy ΔE can be explained as follows, considering E_1 and E_2 as first orbit and second orbit energy,

$$\Delta E = E_1 - E_2$$

   Transitions between stationary states in physics are not governed by classical laws; instead, electrons undergo a phenomenon known as a "quantum leap" where they abruptly change from one state to another.

4. The formula for the energy levels of the electron in the hydrogen atom was derived by Bohr.

$$E = -\frac{R_h}{n^2} \quad n = 1, 2, 3, ..., \infty \quad \text{(for H atom)}$$

   R_h is a constant with a value of 2.18×10^{-18} J.

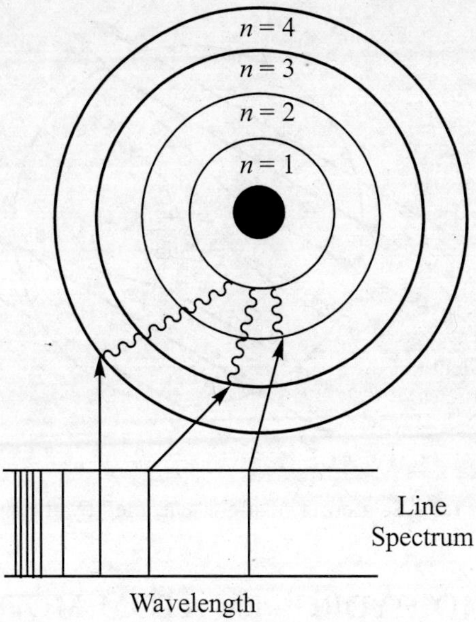

Figure 1.3 Representation of various energy levels and line spectrum.

5. The quantization of angular momentum occurs in discrete units of $h/2\pi$, where h represents Planck's constant and h-bar denotes the reduced Planck's constant. When quantum numbers reach extremely large values, the binding energies decrease to such an extent that transitions between stationary states can occur without the discrete quantum nature of these transitions being readily apparent.
6. When an electron experiences a transition from a state of higher energy to a state of lower energy, it releases its excess energy in the form of a photon.

$$\text{Energy of emitted photon} = h\nu = E_i - E_j$$

$$E_i = -\left(\frac{R_h}{n_i^2}\right); \quad E_j = -\left(\frac{R_h}{n_j^2}\right)$$

If a substitute is made into the above equation that represents the energy of the emitted photon,

$$h\nu = E_f - E_i = \left(-\frac{R_h}{n_f^2}\right) - \left(-\frac{R_h}{n_i^2}\right)$$

$$h\nu = -R_h\left(\frac{1}{n_f^2} - \frac{1}{n_i^2}\right)$$

where R_h is Rydberg constant with value 1,09,677 cm^{-1}, n_i and n_j are initial and final orbit numbers.

A hydrogen atom is produced when hydrogen molecule undergo through bombardment of electrons, during this process transition of hydrogen atom takes place with absorption and emission of light. After analyzing the light hydrogen spectrum containing number of lines is obtained. These lines are categorized into different series named after their discoverers shown in the Figure 1.4.

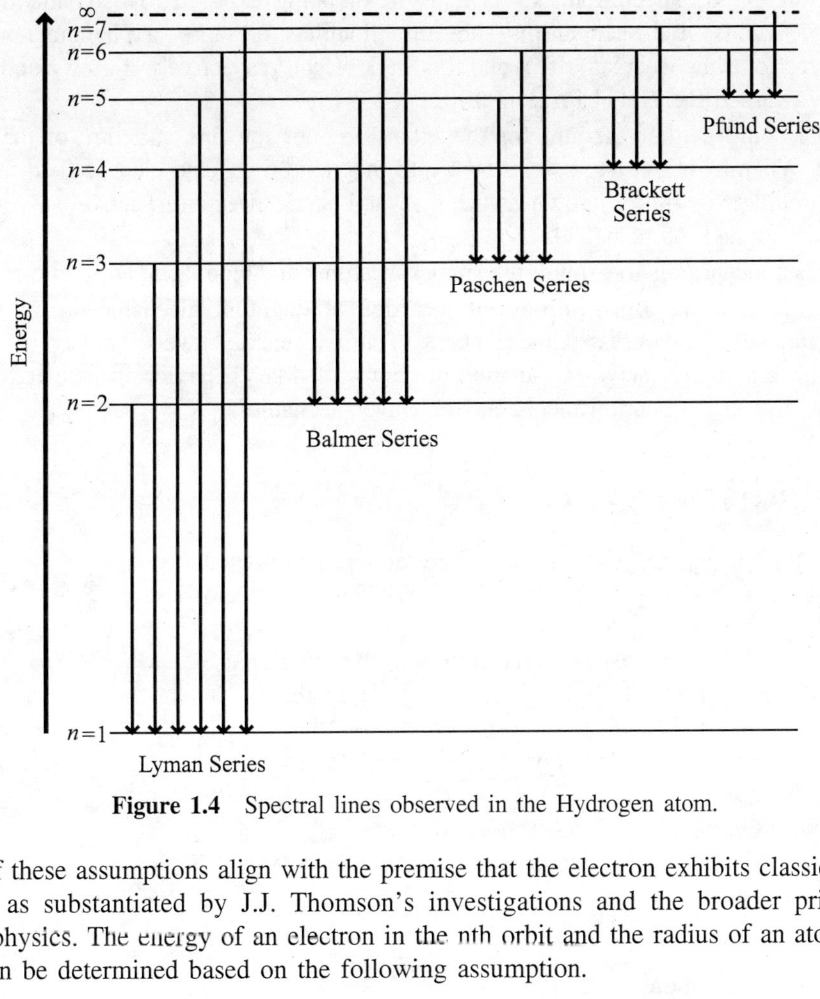

Figure 1.4 Spectral lines observed in the Hydrogen atom.

All of these assumptions align with the premise that the electron exhibits classical particle behavior, as substantiated by J.J. Thomson's investigations and the broader principles of classical physics. The energy of an electron in the nth orbit and the radius of an atom with an n orbit can be determined based on the following assumption.

$$E_n = -\frac{2\pi me^4}{n^2h^2}; \quad r_n = \frac{4\pi\varepsilon_0 n^2 h^2}{me^2} \equiv n^2 a_0$$

1.3.1 Limitations of the Bohr's Model

The Bohr model was a significant step toward developing a quantum theory of matter, and it was responsible for explaining the spectral lines of hydrogen. However, but it had its limitations.

1. In the first place, Bohr's model is only applicable to atoms and ions with a single electron (H^+ and Li^+). The only change that needs to be made to all of the equations is to substitute Ze for e, where Z represents the total number of positive charges in the nucleus.
2. The model was unable to take into consideration either the intensities or the fine structure of the spectral lines. As a result of being exposed to magnetic fields and electric fields, hydrogen atoms split into doublets or more, a phenomenon that is referred to as the Zeeman effect and the Stark effect, respectively (Lorentz and Zeeman were awarded the Nobel Prize in 1902).
3. Model could neither account for the intensities nor the fine structure of the spectral lines. When hydrogen atoms were put into magnetic fields and electric field they splits into doublets or more, known as Zeeman and Stark effects respectively (Nobel prize to Lorentz and Zeeman 1902).
4. Bohr's concept fails to explain the shapes of atoms and binding of atom into molecules.
5. Bohr's theory not only introduced the idea of quantum mechanics in an artificial manner, but it also created the concept of atomic energy levels. However, it did not provide a comprehensive explanation of the "wave-like" behavior of the electron. Next, we will discuss quantum mechanics or wave mechanics.

1.4 de BROGLIE RELATION

In 1802, **T. Young** suggested that light is a wave and **J.J. Thomson** proposed that electron is a particle in 1899. In 1905, **A. Einstein** proved that light is also a particle.

The discovery of the **de Broglie relation** was the initial piece of information that led to the development of quantum theory. In the year 1923, **Louis de Broglie** came to the conclusion that if light have particle properties, then it is possible that particles of matter possess wave qualities.

He hypothesized that a particle having a mass of ***m*** *and a velocity of* ***v*** *would have a wavelength λ that is related with it.*

Louis de Broglie

$$\lambda = \frac{h}{mv}$$

This particular equation is referred to as the de Broglie relation.

The wavelengths of electrons are on the order of a few picometers (1 pm = 10^{-12} m). Observation of the wave nature of electrons ought to be possible under the appropriate conditions. The wave nature of electrons has been proved experimentally by Davisson and Germer (Figure 1.5).

We are no longer able to consider an electron to have a certain orbit within an atom where it belongs. Knowing the precise position and velocity of such an orbit is necessary in order to provide a description of it. In the year 1927, Werner Heisenberg demonstrated, using the principles of quantum physics, that it is no longer feasible to know both simultaneously.

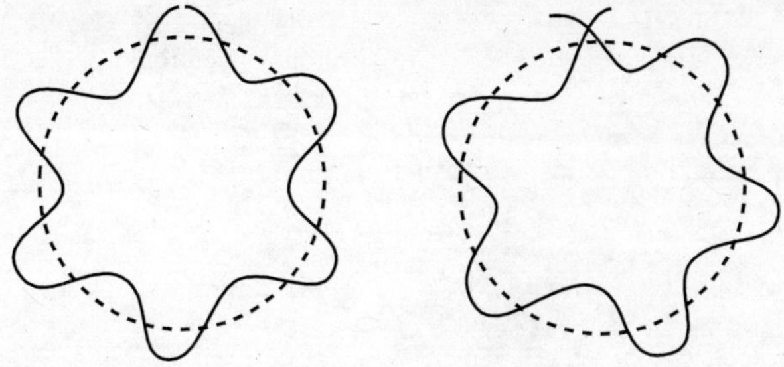

(a) In phase (b) Out of phase

Figure 1.5 Illustration of wave motion of electrons (a) in phase (b) out of phase.

1.5 HEISENBERG AND UNCERTAINTY PRINCIPLE

The product of uncertainty in position (Δx) and the uncertainty in momentum ($m\Delta v_x$) of a particle cannot be smaller than $h/4\pi$, whether it is equal or greater. From the standpoint of this principle, it is not possible to ascertain simultaneously the position and momentum of a tiny particle such as an electron.

$$(\Delta x)(m\Delta v_x) \geq \frac{h}{4\pi}$$

When the value of m is big, such as in the case of a baseball, the uncertainties are low. However, when it comes to electrons, significant uncertainties make it impossible to define an exact orbit.

The uncertainty principle can also be stated in terms of energy and time. Therefore,

$$(\Delta E)(\Delta t) \geq \frac{h}{4\pi}$$

In order to accurately estimate the energy of a system with a precision of ΔE, the measurement must be conducted over a time period of magnitude $h/\Delta E$. This equation is utilized to approximate the level of sharpness exhibited by spectral lines. Typically, lines that result from the transition of an atom's ground state are narrow. This is because the electron responsible for the transition spends a significant amount of time, Δt, in the ground state, resulting in a minimal uncertainty in the energy level. The width of the line is directly connected to the difference in energy levels, ΔE by $\Delta \nu = \frac{\Delta E}{h}$. The duration of excited states can occasionally be quite brief, and the transition between these states can generate diffuse lines due to the uncertainty of the energy levels.

In spite of the fact that we are unable to exactly define the orbit of an electron, we are able to determine the ***probability*** of finding an electron at a particular location around the nucleus. A mathematical equation known as a wave function, which is symbolized by the symbol (ψ), was used by **Erwin Schrödinger** to define this as **the state of the system**. The product of wave function (ψ^2) defines the probability of finding a particle in a particular region of space.

Erwin Schrödinger

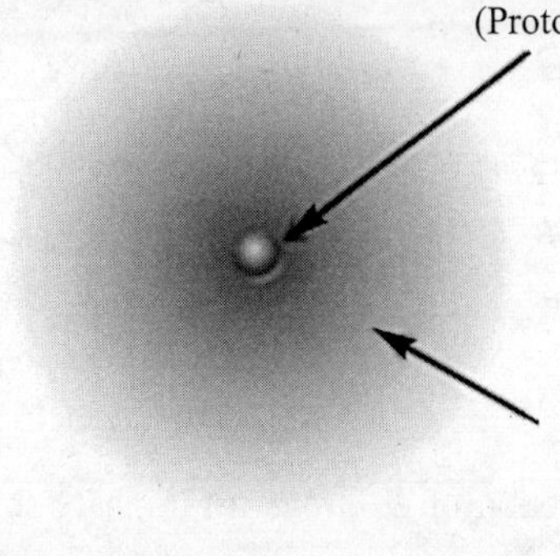

Figure 1.6 Illustration of atom with electron cloud.

There must be a mathematical function that is the solution to a differential equation that characterizes electrons, atoms, and molecules if matter possesses wave-like qualities. This indicates that there must be a mathematical function. The differential equation is referred to as the Schrodinger equation, and the solution to this equation is referred to as the wave function, abbreviated as $\Psi(x, t)$. An example of a "wave function" that is related with a matter wave is $\Psi(x, t)$, which represents **a state of a matter**. This was not a derivation; rather, it was an argument using plausibility. This argument is not the reason that we trust the Schrodinger equation; rather, we believe it because its predictions are consistent with experiment. There are restrictions on the validity of it. It is only applicable to a single non-relativistic particle in this form, which means that it must have a rest mass that is not zero and a speed that is significantly lower than c.

As is the case with the classical wave equation, the Schrodinger wave equation is a partial differential equation in x (one dimension) and t (time). It is of the first order in time, in contrast to the classical wave equation for waves. The complex integer i is included in the equation that describes the Schrodinger wave. Because of this, its solutions are fundamentally complicated, in contrast to classical waves, in which the utilization of complex numbers is merely a mathematical convenience.

1.6 THE CLASSICAL WAVE EQUATION

In the past, we have observed that the equation for the wave in one dimension, where wave is propagating in x axis at time t (Figure 1.7), is represented as follows:

$$\frac{\partial^2 \psi}{\partial x^2} = \frac{1}{v^2}\frac{\partial^2 \psi}{\partial t^2}$$

Where v represents the velocity of the wave and $\psi(x, t)$ represents the displacement of the wave in one dimension. The concept of matter waves in free space can be utilized with this. A potential solution is something that we could try: i.e.

$$\psi(x,t) = e^{i(kx-\omega t)}$$

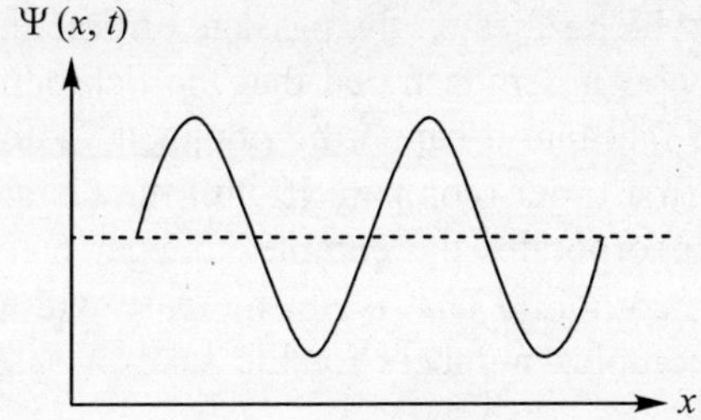

Figure 1.7 Wave propagation in x direction.

Try a modified wave equation of the following type:

$$\frac{\partial^2 \psi}{\partial x^2} = \alpha \frac{\partial \psi}{\partial t} \quad \text{(Where } \alpha \text{ is a constant)}$$

We get same solution as before: e.g.

$$\psi(x, t) = e^{i(kx - \omega t)}$$

Hence, the equation for matter waves in *free space* is:

$$-\frac{\hbar}{2m}\frac{\partial^2 \psi}{\partial x^2} = i\hbar \frac{\partial \psi}{\partial t}$$

For

$$\psi(x, t) = e^{i(kx - \omega t)}$$

Then we have

$$-\frac{k^2\hbar^2}{2m}\psi(x, t) = \hbar\omega\psi(x, t)$$

We can write this equation in this form:

(Kinetic energy) × wave function = (Total energy) × wave function

1.7 THE TIME-DEPENDENT SCHRÖDINGER EQUATION

If a particle has a potential $V(x, t)$ in one dimension, x, and time t then

$$E = \frac{p^2}{2m} + V(x, t)$$

We can also write

(Kinetic Energy + Potential Energy) × wave function = (Total energy) × wave function

$$-\frac{\hbar^2}{2m}\frac{\partial^2 \Psi}{\partial x^2} + V(x, t)\Psi = i\hbar \frac{\partial \Psi}{\partial t}$$

The Schrodinger equation, which relies on the passage of time, is one of the postulates that support quantum physics. It has been demonstrated that the Schrodinger equation is consistent with all of the tests, despite the fact that it cannot be obtained. Within the context of time, the Schrodinger equation is of the first order (compare it with the classical wave equation). Given that the Schrodinger equation incorporates the complex integer *i*, it follows that the solutions to this equation are, in essence, complex. That is not the case with classical waves, which are characterized by the usage of complex numbers for the sake of ease.

1.8 THE HAMILTONIAN OPERATOR

The left-hand side of the Schrodinger equation that is dependent on time can be written as follows:

$$-\frac{\hbar^2}{2m}\frac{\partial^2\Psi}{\partial x^2}+V(x,t)\Psi=\left(-\frac{\hbar^2}{2m}\frac{\partial^2}{\partial x^2}+V(x,t)\right)\Psi=\hat{H}\Psi$$

In this context, the term $\hat{H}$ **"Hamiltonian operator"** refers to the differential operator that acts as a representation of the **total energy of the particle**.
Thus

$$\hat{H}=\left(-\frac{\hbar^2}{2m}\frac{\partial^2}{\partial x^2}+V(x)\right)=\frac{\hat{p}_x^2}{2m}+\hat{V}(x)$$

Where the mathematical expression of the momentum operator is

$$\hat{p}_x=-i\hbar\frac{\partial}{\partial x}$$

Thus, time dependent Schrödinger equation is expressed as:

$$\hat{H}\Psi=i\hbar\frac{\partial\Psi}{\partial t}$$

1.9 SOLVING THE TIME-DEPENDENT SCHRÖDINGER EQUATION

Assuming that the potential is not dependent on the passage of time, so that *V(x, t)* = *V(x)*, the Schrodinger equation that is time dependent is as follows:

$$i\hbar\frac{\partial\Psi}{\partial t}=-\frac{\hbar^2}{2m}\frac{\partial^2\Psi}{\partial x^2}+V(x)\Psi$$

Left hand side involves variation of ψ with t while right hand side involves variation of ψ with x. Hence look for a separated solution:

On the left-hand side, there is a variation of ψ with t, and on the right-hand side, there is a variation of ψ with x. Now, we have a solution that is separated as a result:

$$\Psi(x,t)=\psi(x)T(t)$$

$$-\frac{\hbar^2}{2m}T\frac{\partial^2\psi}{\partial x^2}+V(x)\psi T=i\hbar\psi\frac{\partial T}{\partial t}$$

then

$$-\frac{\hbar^2}{2m}T\frac{\partial^2\psi}{\partial x^2}+V(x)\psi T=i\hbar\psi\frac{\partial T}{\partial t}$$

Dividing by ψT:

$$-\frac{\hbar^2}{2m}\frac{1}{\psi}\frac{\partial^2\psi}{\partial x^2}+V(x)=i\hbar\frac{1}{T}\frac{\partial T}{\partial t}$$

Left-hand side is only dependent on x, while right-hand side is solely dependent on t. As a result of the fact that this is true for every x and t, both sides must equal a constant, denoted by the letter E(E = separation constant).

Thus, the expression becomes:

$$i\hbar\frac{1}{T}\frac{\partial T}{\partial t}=E$$

$$-\frac{\hbar^2}{2m}\frac{1}{\psi}\frac{\partial^2\psi}{\partial x^2}+V(x)=E$$

1.10 TIME-INDEPENDENT SCHRÖDINGER EQUATION

If we solve the time dependent equation, then we will get the following expression.

$$i\hbar\frac{1}{T}\frac{dT}{dt}=E \quad\Rightarrow\quad \frac{dT}{T}=-\frac{iE}{\hbar}\,dt \quad\Rightarrow\quad T(t)=Ae^{-iEt/\hbar}$$

This is precisely the same as a wave $e^{-i\omega t}$, where $E = \hbar\omega$. Since this is the case, $T(t)$ is dependent on the energy E. We need to discover a solution to the space component of the problem in order to determine what the energy actually is.

At this point, the space equation is

$$-\frac{\hbar^2}{2m}\frac{\partial^2\psi}{\partial x^2}+V(x)\psi=E\psi \quad \text{or} \quad \hat{H}\psi=E\psi$$

The time independent Schrodinger equation, sometimes known as TISE, is presented here.

1.11 EIGENVALUE EQUATIONS

Schrödinger's equation is an example of an eigenvalue equation, which is written as follows:

$$\hat{H}\psi=E\psi$$

Where the Hamiltonian operator is denoted by the symbol $\hat{H}$. This can be written as follows:

$$\hat{H} = \hat{T} + \hat{V} = -\frac{\hbar^2}{2m}\frac{d^2}{dx^2} + V(x)$$

A wave function, denoted by ψ, is an eigen function of $\hat{H}$. E, on the other hand, represents the total energy ($T + V$) and is an eigenvalue of $\hat{H}$. In this case, E is a constant.

1.12 TIME-INDEPENDENT SCHRÖDINGER EQUATION FOR A FREE-PARTICLE

In the case of a free particle with the value $V(x)$ equal to zero, the Schrodinger equation is as follows:

$$-\frac{\hbar^2}{2m}\frac{\partial^2\psi}{\partial x^2} = E\psi$$

and has solutions

$$\psi = e^{ikx} \quad \text{or} \quad e^{-ikx} \quad \text{where} \quad E = \frac{k^2\hbar^2}{2m}$$

Consequently, the complete answer to the Schrodinger equation that is independent of time is as follows:

$$\Psi(x,t) = \psi(x)\,T(t) = e^{i(\pm kx - Et/\hbar)}$$

Identifies waves that are moving in either the x direction or the y direction ($\pm\, x$ direction) with the following characteristics: (i) An angular frequency, $\omega = E/\hbar \Rightarrow E = \hbar\omega$ (ii) The equation for a wave vector is $k = (2mE)^{1/2}/\hbar = p/\hbar \Rightarrow p = h/\lambda$.

1.13 INTERPRETATION OF $\Psi(x, t)$

As was indicated earlier, the solutions to the time-dependent Schrodinger equation are inherently complex. There is no possibility that $\Psi(x, t)$ is a physical wave, such as electromagnetic waves. The relationship between $\Psi(x, t)$ and actual physical measurements on a system can therefore be established.

1.13.1 The Born Interpretation

At position x and time t, the probability of finding a particle in a short length dx is equal to

$$\Psi^*(x,t)\Psi(x,t)\,dx = |\Psi(x,t)|^2\,dx = P(x,t)dx$$

Since Ψ is a complex quantity, it is difficult to understand how it may be compatible with actual physical measurements on a system.

1.13.2 Interpretation of $|\Psi|^2$

There is a real interpretation of the expression $\Psi\Psi^*$, which is the probability per unit length (or volume in three dimensions). This interpretation is necessary for a probability distribution. One way to understand the value of the quantity $|\Psi|^2$ is to consider it as the probability that the particle can be located at a specific location x and at a specific time t. The wave function is transformed into a particle as a result of the process of measurement, which causes it to "collapse."

"According to the Born interpretation, the *probability amplitude* is denoted by the symbol Ψ, the *probability density* is denoted by $\Psi^*\Psi$ (= $P(x, t)$), and the *probability* is denoted by $\Psi^*\Psi$ dx."

1.14 EXPECTATION VALUES

Consequently, if we are aware of the value of $\Psi(x, t)$, then the calculation of $\Psi^*\Psi\, dx$ enables us to determine the average position:

$$\bar{x} = \sum_i x_i P(x_i)\,\delta x$$

If the limit of x is $\delta x \to 0$, then the summation is as follows:

$$\bar{x} = \langle x \rangle = \int_{-\infty}^{\infty} xP(x)dx = \int_{-\infty}^{\infty} x|\Psi(x,t)|^2\, dx$$

The average, which is sometimes referred to as the *expectation value*, is of great significance in the field of quantum mechanics. Its primary function is to supply us with the average values of physical properties. This is because, in many instances, precise values cannot be determined, even in hypothetical situations.

In similar way

$$\langle x^2 \rangle = \int_{-\infty}^{\infty} x^2 P(x)dx = \int_{-\infty}^{\infty} x^2 |\Psi(x,t)|^2\, dx$$

1.14.1 Momentum and Energy Expectation Values

In order to calculate the expected value of *momentum*, momentum must first be represented as a quantum mechanical operator by the following steps:

$$\langle p_x \rangle = \int_{-\infty}^{\infty} \Psi^*(x,t)\frac{\hbar}{i}\frac{\partial}{\partial x}\Psi(x,t)dx$$

where

$$\hat{p}_x = \frac{\hbar}{i}\frac{\partial}{\partial x}.$$

$\hat{p}_x$ is the operator for the x component of momentum.

This is the equation for the expectation value of the *average energy* of a free particle. According to the fact that V equals zero, the expected amount of energy for a particle that is traveling in one dimension is

$$\langle E \rangle = \int_{-\infty}^{\infty} \Psi^*(x,t)\left(-\frac{\hbar^2}{2m}\frac{\partial^2}{\partial x^2}\right)\Psi(x,t)dx$$

where

$$\langle E \rangle = \frac{\langle p^2 \rangle}{2m}$$

1.15 NORMALISATION

In order to discover a particle anywhere, the total probability of finding it must be one:

$$\int_{-\infty}^{\infty} P(x)dx = \int_{-\infty}^{\infty} |\Psi(x,t)|^2 \, dx = 1$$

The *Normalisation condition* is the name given to this particular need. The reason for this requirement is that the Schrodinger equation is linear inΨ, and hence, if Ψ is a solution of the time-dependent Schrodinger equation, then $c\Psi$ is also a solution of the Schrodinger equation, where c is a constant.

If the original wave function that has not been normalized is $\Psi(x, t)$, then the normalization integral can be expressed as follows:

$$N^2 = \int_{-\infty}^{\infty} |\Psi(x,t)|^2 \, dx$$

The normalized wave function Ψ_{norm}, which has been rescaled, is equal to that of $(1/N)\ \Psi$.

EXAMPLE 1.1 What value of N normalises the function $N\, x(x - L)$ of $0 \le x \le L$?

EXAMPLE 1.2 Find the probability that a system described by the function $2^{1/2}\sin(\pi x)$ where $0 \le x \le 1$ is found anywhere in the interval $0 \le x \le 0.25$.

1.16 BOUNDARY CONDITIONS FOR Ψ

In order for ψ to be considered a solution of the Schrödinger equation and to provide a representation of a system that can be observed physically, it is necessary for ψ to fulfil certain requirements.

First Condition: In order to ensure that the probability density is specified in a manner that is unique, the function must be continuous and have a single value from both x and t. Illustrations of wave functions that are not suitable (Figure 1.8). Figure 1.9 (a) requires that the function be a single-valued function of x and t.

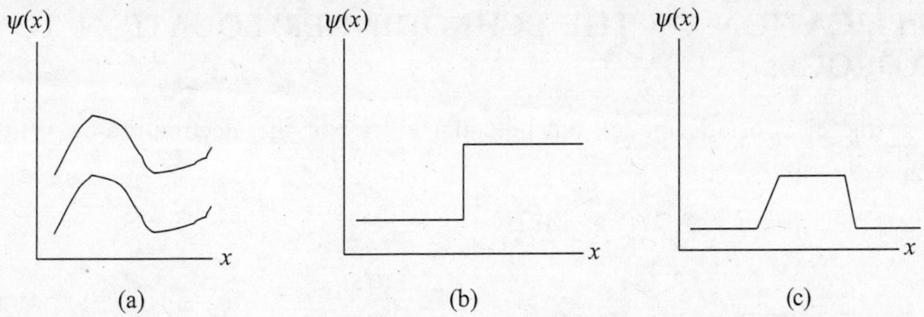

Figure 1.8 Wave functions that are (a) no single valued (b) discontinuous (c) gradient discontinuous.

Second Condition: We must possess a finite normalization integral in order to be able to compute a normalized probability. The wave function must be able to be normalized which indicates that the $\psi \rightarrow 0$ as $x \rightarrow \infty$ (Figure 1.9(b)).

Third Condition: The first derivative must be continuous, with the exception of places where the potential is always infinite. Figure 1.9(c) depicts that the function $\psi(x)$ be a continuous function of x.

Fourth Condition: The *slope* of ψ must be continuous; more specifically, the slope of $d\psi(x)/dx$ must be continuous, with the exception of sites where potential is infinite (Figure 1.9(d)).

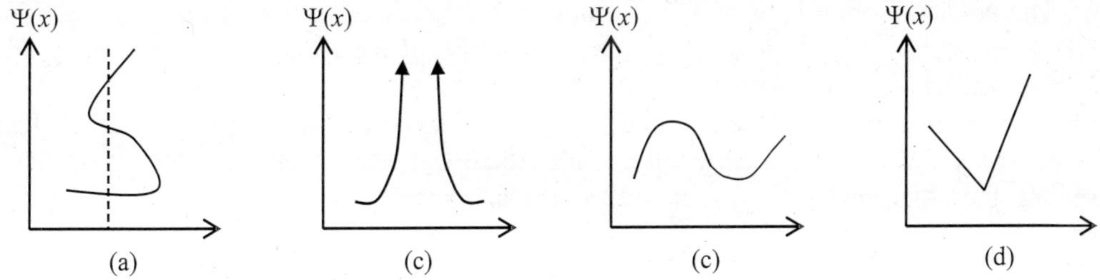

Figure 1.9 Boundary conditions for wave function.

1.17 STATIONARY STATES

According to what we saw, the wave function continucs to oscillate in time even when the potential is not reliant on the passage of time. This is the solution to the Schrodinger equation that is dependent on the complete time.

$$\Psi(x,t) = \psi(x)\,T(t) = \psi(x)e^{-iEt/\hbar}$$

In contrast, the probability distribution is *static* (unchanging):

$$P(x,t) = |\Psi(x,t)|^2 = \psi^*(x)e^{+iEt/\hbar}\;\psi(x)e^{-iEt/\hbar} = |\psi(x)|^2$$

Therefore, a solution to the Schrodinger equation that is independent of time is sometimes referred to as a stationary state.

1.18 APPLICATION OF THE SCHRÖDINGER EQUATION TO THE HYDROGEN ATOM

According to the electrostatic model, the potential energy of the electron-proton system can be approximated as follows:

$$V(r) \propto \frac{1}{r}$$

$$V(r) = -\frac{e^2}{4\pi\varepsilon_0 r}$$

Due to the presence of a "spherically symmetric" potential, the electron is "confined" to the atom.

Rewrite the three-dimensional time-independent Schrödinger Equation.

$$-\frac{\hbar}{2m}\frac{1}{\psi(x,y,z)}\left[\frac{\partial^2\psi(x,y,z)}{\partial^2 x}+\frac{\partial^2\psi(x,y,z)}{\partial^2 y}+\frac{\partial^2\psi(x,y,z)}{\partial^2 z}\right] = E - V(r)$$

1.18.1 Hydrogen-like Atoms (He^+ or Li^{++})

Replacing e^2 with Ze^2 (the atomic number is Z) in the equation. Using appropriate reduced mass μ. The potential, often known as the central force $V(r)$ is a function of the distance r that separates the electron and the proton. In order to take use of the radial symmetry, we need to transform to spherical polar coordinates.

When examining these "planetary like" models, it is more convenient to examine them using spherically polar coordinates (r, θ, ϕ) rather than Cartesian coordinates (x, y, z) with θ representing the polar angle and ϕ representing the azimuthal angle.

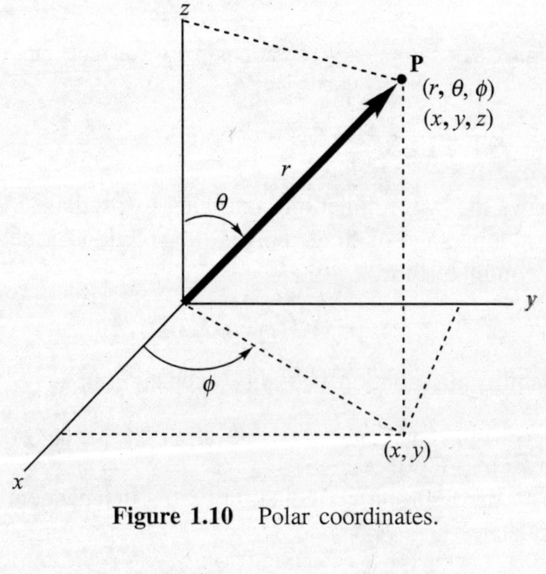

Figure 1.10 Polar coordinates.

The value of θ can range from 0 to π. The value of ϕ can range from 0 to 2π. Where r can range from 0 to ∞.

Cartesians to Polar

$$x = r\sin(\theta)\cos(\phi);\ y = r\sin(\theta)\sin(\phi);\ z = r\cos(\theta)$$

Polar to Cartesians

$$r = \sqrt{x^2 + y^2 + z^2} \quad \cos(\theta) = \frac{z}{r} \quad \tan(\phi) = \frac{x}{y}$$

The Coulomb potential should be inserted into the Schrodinger equation after it has been converted. For the hydrogen atom, the Schrodinger equation is as follows:

$$\hat{H}\psi = E\psi$$

$$(\hat{T} + \hat{V})\psi = E\psi$$

$$\left(-\frac{\hbar^2}{2m_e}\nabla^2 - \frac{e^2}{4\pi\varepsilon_0}\frac{1}{r}\right)\psi = E\psi$$

where

$$\nabla^2 = \frac{1}{r^2}\frac{\partial}{\partial r}\left(r^2\frac{\partial}{\partial r}\right) + \frac{1}{r^2}\frac{1}{\sin\theta}\frac{\partial}{\partial\theta}\left(\sin\theta\frac{\partial}{\partial\theta}\right) + \frac{1}{r^2}\frac{1}{\sin^2\theta}\frac{\partial^2}{\partial\phi^2}$$

The Schrodinger equation for the Hydrogen atom:

$$(\hat{T} + \hat{V})\psi = E\psi$$

Putting the value of T and V

$$\left[-\frac{\hbar^2}{2m_e}\left(\frac{1}{r^2}\frac{\partial}{\partial r}\left(r^2\frac{\partial}{\partial r}\right) + \frac{1}{r^2}\frac{1}{\sin\theta}\frac{\partial}{\partial\theta}\left(\sin\theta\frac{\partial}{\partial\theta}\right) + \frac{1}{r^2}\frac{1}{\sin^2\theta}\frac{\partial^2}{\partial\phi^2}\right) - \frac{e^2}{4\pi\varepsilon_0}\frac{1}{r}\right]\psi = E\psi$$

This, can be broken down into the following categories:

$$\psi_{n,l,m_l}(r,\theta,\phi) = R(r)_{n,l}Y_{l,m_l}(\theta,\phi) = R(r)_{n,l}\Theta_{l,m_l}(\theta)\Phi_{m_l}(\phi)$$

As a function of r, θ, and ϕ, the wave function ψ is a function. The equation can be separated. The solution could be the result of three different functions.

$$\psi(r,\theta,\phi) = R(r)f(\theta)g(\phi)$$

We are able to divide the equation shown above into three distinct differential equations, each of which is dependent on a single coordinate, namely r, θ, or ϕ. In order to solve the Schrodinger equation for the radial component:

$$\left.-\frac{\hbar^2}{2m_e}\frac{1}{r^2}\frac{d}{dr}\left(r^2\frac{dR}{dr}\right)+\left(\frac{\hbar^2}{2m_e}\frac{l(l+1)}{r^2}-\frac{e^2}{4\pi\varepsilon_0}\frac{1}{r}\right)R(r)=E\,R(r)\right\}\text{gives } R_{n,l}(r)$$

When it comes to the quadratic components of the Schrodinger equation:

$$\left.\begin{array}{l}(1-x^2)\dfrac{d^2P}{dx^2}-2x\dfrac{dP}{dx}+\left(l(l+1)-\dfrac{m^2}{1-x^2}\right)P(x)\\ x=\cos\theta\end{array}\right\}\text{gives}\,\Theta(\theta)$$

$$\left.\frac{d^2\Phi}{d\phi^2}=-m^2\Phi(\phi)\right\}\text{ gives }\Phi(\phi)$$

Wave functions: $y_{n,l,m}(r, \theta, \phi) = R_{n,\,l}(r)\ Y_{l,\,m}\ (\theta, \phi)$ are called orbitals

1.18.2 Orthogonally Modified Laguerre Functions; $R_{n,l}(r)$

We obtained them by the usage of a Numerov solution. Defines the events that take place "inside" the orbital. A total of *n-l*-1 radial nodes are present. Associated Laguerre polynomials, often known as $G_{n,\,1}\,(Zr/a_0)$, are terms that are dependent on the values of *n* and l.

In the following, we will find a number of radial wavefunctions ($R_{n,\,l}(r)$)) that have been derived for the hydrogen atom.

n	*l*	$R_{n,\,l}\,(r)$
1	0	$\sqrt[3]{(Z/a_0)}\,2e^{(-Zr/a_0)}$
2	0	$\sqrt[3]{(Z/2a_0)}\,2\left(1-\dfrac{Zr}{2a_0}\right)e^{(-Zr/2a_0)}$
2	1	$\sqrt[3]{(Z/2a_0)}\dfrac{2}{\sqrt{3}}\left(\dfrac{Zr}{2a_0}\right)e^{(-Zr/2a_0)}$
3	0	$\sqrt[3]{(Z/3a_0)}\,2\left[1-(2Zr/3a_0)+\dfrac{2}{3}\left(\dfrac{Zr}{3a_0}\right)^2\right]e^{(-Zr/3a_0)}$
3	1	$\sqrt[3]{(Z/3a_0)}\left(4\sqrt{2}/3\right)\left(\dfrac{Zr}{3a_0}\right)\left(1-\dfrac{1}{2}\dfrac{Zr}{3a_0}\right)e^{(-Zr/3a_0)}$
3	2	$\sqrt[3]{(Z/3a_0)}\left(2\sqrt{2}/3\sqrt{5}\right)\left(\dfrac{Zr}{3a_0}\right)^2e^{(-Zr/3a_0)}$

The probabilities that the electron is to be located between *r* and *r* + *dr* is the radial probability function, which is denoted by the symbol $P_{n,\,l}(r)$.

$$P_{n,l}(r)dr=\int\left|R_{n,l}(r)Y_{l,m_l}(\theta,\varphi)\right|^2r^2dV$$

$$=4\pi r^2\left|R_{n,l}\right|^2r^2dr$$

The following are some examples of radial probability functions that are representative:

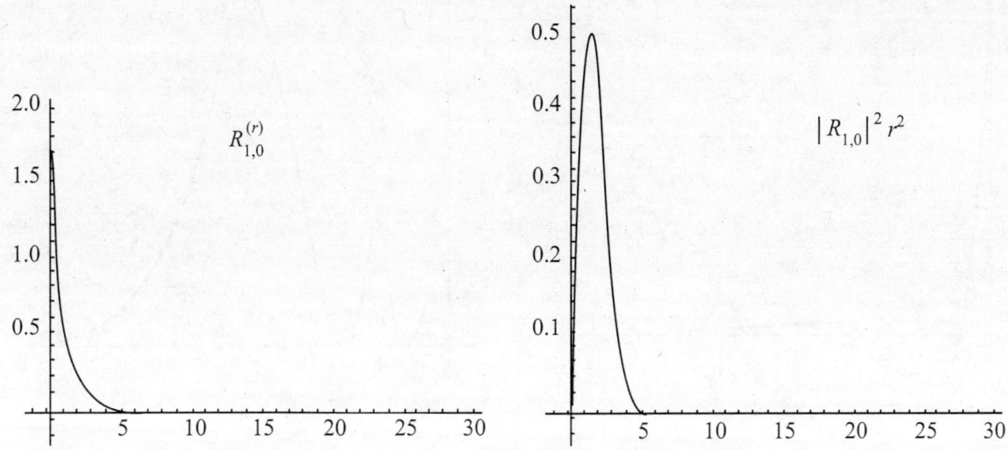

Figure 1.11 Radial wave functions and radial probability functions for $n = 1$ and $l = 0$.

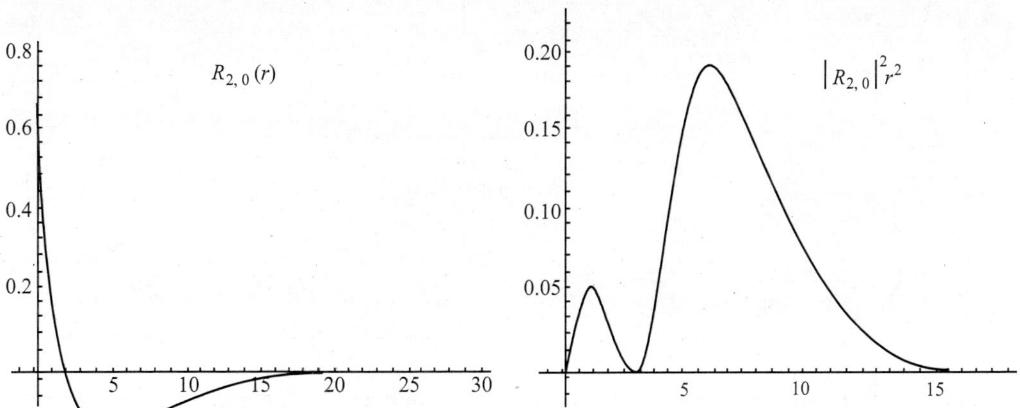

Figure 1.12 Radial wave functions and radial probability functions for $n = 2$ and $l = 0$.

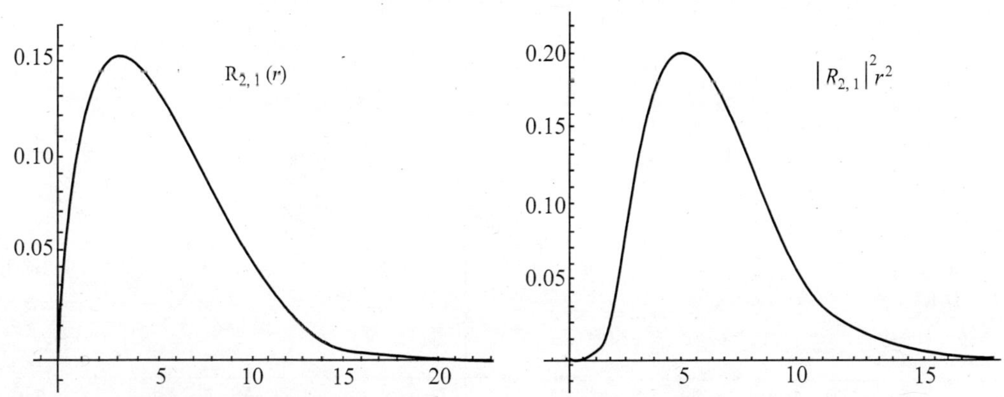

Figure 1.13 Radial and radial probability functions for $n = 2$ and $l = 1$.

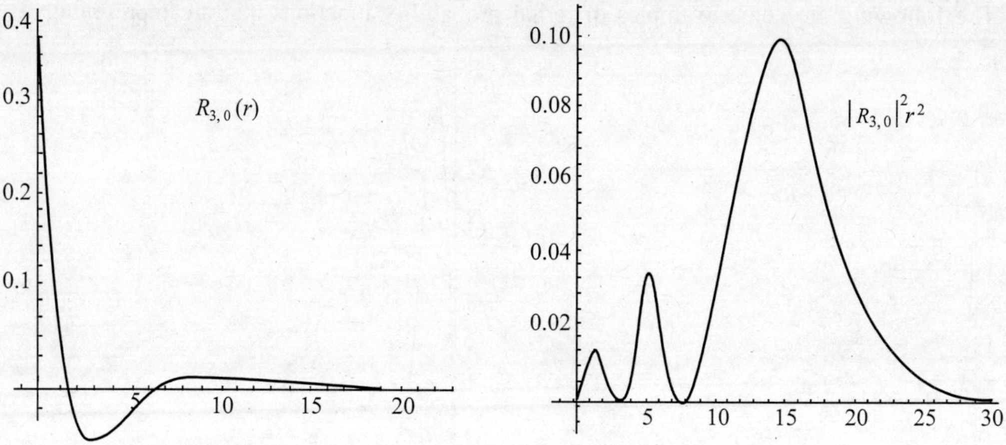

Figure 1.14 Radial and radial probability functions for $n = 3$ and $l = 0$.

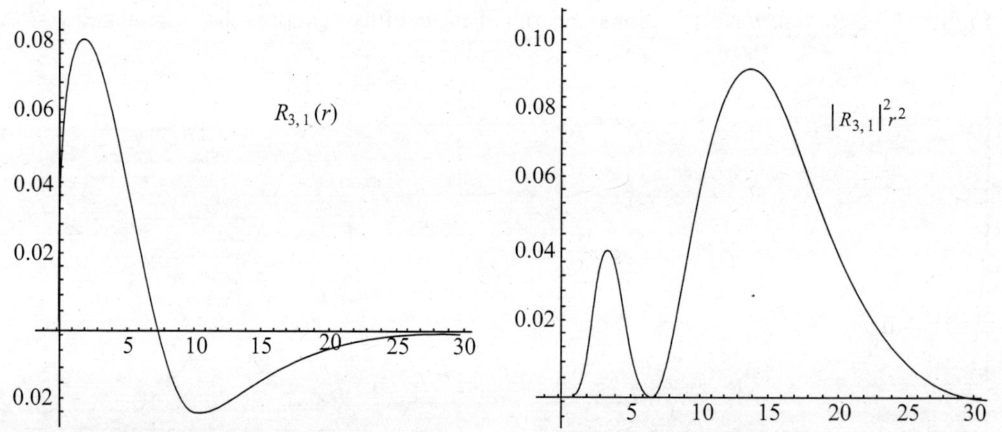

Figure 1.15 Radial and radial probability functions for $n = 3$ and $l = 1$.

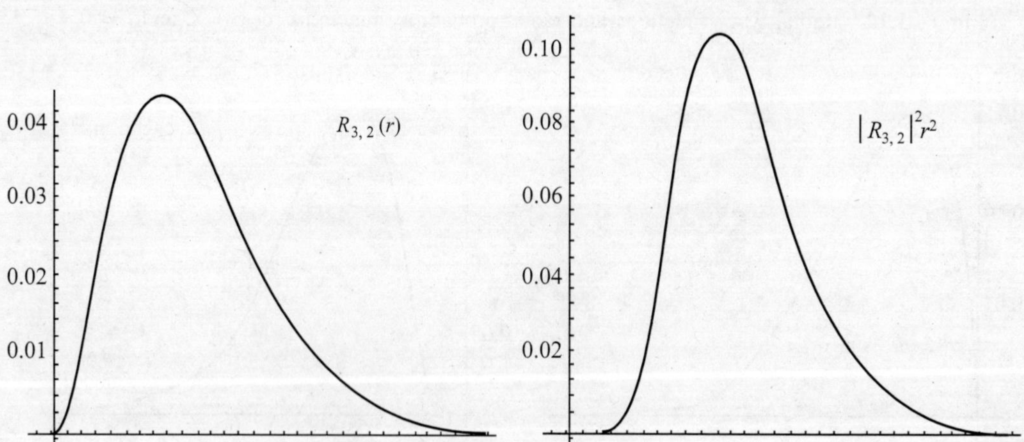

Figure 1.16 Radial and radial probability functions for $n = 3$ and $l = 2$.

At the origin, the radial probability density disappears due to the r^2 factor, and this occurs even for situations in which $l = 0$. There are $n - l - 1$ nodes in the distribution for each state, given that their values are n and l. There are n maxima in the distribution of states with $l = 0$, and the amplitude of these maxima increases as the distance from the origin increases.

1.19 QUANTUM NUMBERS

In accordance with the principles of quantum mechanics, each electron is characterized by four quantum numbers.

1. Principal quantum number (n)
2. Angular momentum quantum number (l)
3. Magnetic quantum number (m_l)
4. Spin quantum number (m_s)

The first three elements are responsible for defining the wave function of a certain electron. The fourth quantum number represent the magnetic property of electrons.

1.19.1 Principal Quantum Number (n)

One of the shell numbers that an electron can accommodate is denoted by the primary quantum number, which is denoted by the letter n. When the value of n is low, the orbital that represents the size of the atom also decreases. The energy of the electron decreases in proportion to the reduced value of n. A higher value of n results in an increase in the amount of energy that an electron possesses. The formula $2n^2$ is used to determine the maximum number of electrons that can "reside" on the shell.

1.19.2 Angular Momentum or Azimuthal Quantum Number (l)

There are subshells within a particular shell that have varied shapes, and the angular momentum quantum number (l) is what differentiates them from one another. Subshells are the divisions that are made within each main shell. There are n subshells that are contained within each shell of the quantum number n, and each of these subshells has a unique shape. When the value of the principle quantum number (n) is known, the value of the variable l can be any integer between 0 and (n – 1), inclusive.

The various subshells are represented by the letters *s*, *p*, *d*, *f*, and so on. These letters are taken from the initial letter of **s**harp, **p**rimary, **d**iffused, and **f**undamental, respectively. All the possible values are tabulated in Table 1.1.

Letter	*s*	*p*	*d*	*f*	*g*	...
l	0	1	2	3	4	...

1.19.3 Magnetic Quantum Number (m_l)

It is possible to differentiate between orbitals inside a particular sub-shell that have various shapes and orientations in space by using the magnetic quantum number, abbreviated as m_l.

The electrons of the atom position themselves in specific regions of space around the nucleus when the atom is subjected to an external magnetic field. Orbitals are the name given to this. Every subshell is further subdivided into "orbitals," and each orbital has the capacity to accommodate a different pair of electrons. It is possible for the value of m_l to take on any integer value between $-l$ and $+l$; hence, the values of m could be $(2l+1)$. This means that the energy of each orbital within a certain subshell is the same.

1.19.4 Spin Quantum Number (m_s)

The spin quantum number, denoted by the symbol m_s, is a value that describes the two possible spin orientations of electrons that are included within a particular orbital. Only two electrons can be contained within each orbital, and their spins must be in opposition to one another. The clockwise spin (+1/2) and the anti-clockwise spin (–1/2) are the two potential values for the m_s parameter.

Table 1.1 The quantum numbers of atomic orbitals that are allowed to have certain values.

n	l	m_l	**Subshell Notation**	**Number of Orbitals in the Subshell**
1	0	0	1s	1
2	0	0	2s	1
2	1	–1, 0, +1	2p	3
3	0	0	3s	1
3	1	–1, 0, +1	3p	3
3	2	–2, –1, 0, +1, +2	3d	5
4	0	0	4s	1
4	1	–1, 0, +1	4p	3
4	2	–2, –1, 0, +1, +2	4d	5
4	3	–3, –2, –1, 0, +1, +2, +3	4f	7

1.20 SHAPES OF ORBITALS

The term "orbitals" refers to the region of space that surrounds a nucleus and contains an electrons or pair of electrons spends between 90 and 95 percent of its time. The *s*-orbitals have the shape of a symmetrically spherical sphere, with the nucleus being surrounded by an electron cloud and the chance of finding electrons surrounding the nucleus being the same in all directions. The difference between the *1s* orbital and the *2s* orbital is due to the presence of one node, which is a location where the chance of finding electrons is zero. Two nodes are present in the *3s* orbital, while the number of nodes in the *ns* orbital is $(n-1)$ (Figure 1.17).

The *p*-orbital is composed of dumb-bell shape, which are lobes, and they are separated by a plane with zero electron density, which is known as the nodal plane. There are two lobes that make up a *p* orbital, and they are separated by a gap. They have the same amount of energy along the *x*, *y*, and *z* axes, and their names are p_x, p_y, and p_z (Figure 1.18).

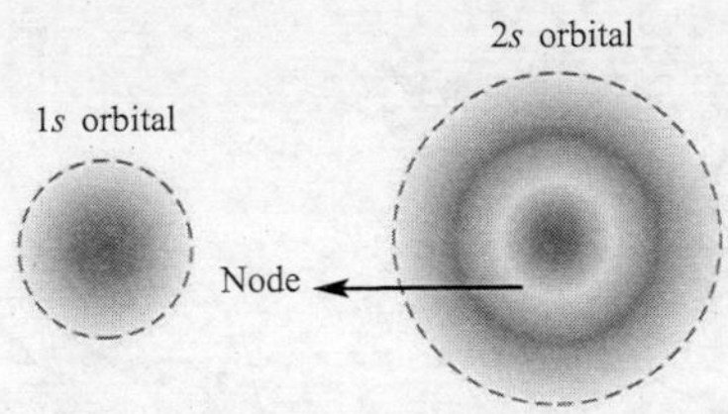

Figure 1.17 Shapes of 1*s* and 2*s* orbitals in space

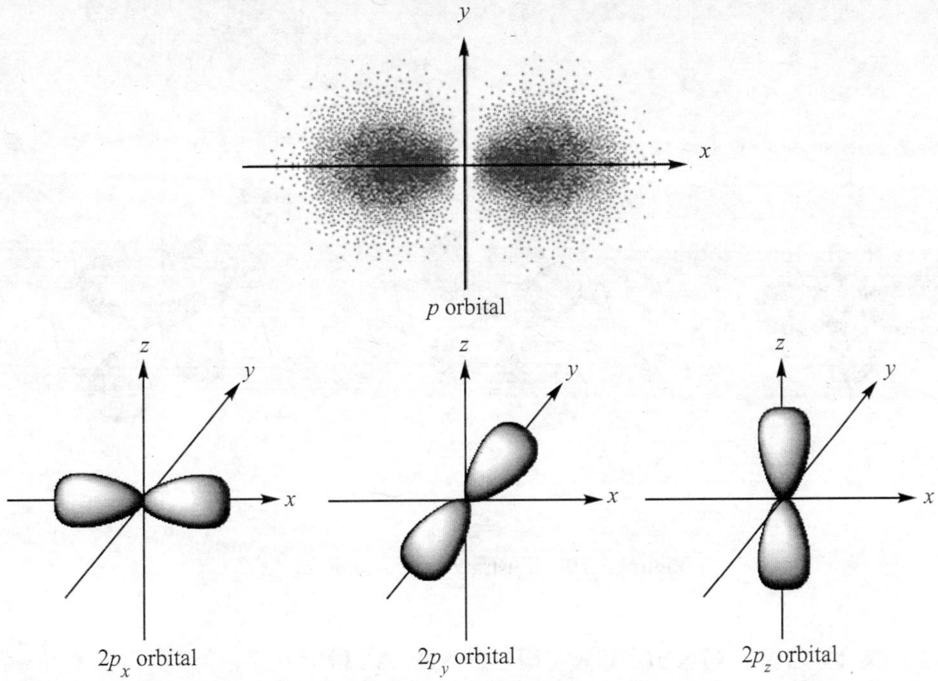

Figure 1.18 Different orientations of *p*-orbitals.

The forms of *d*-orbitals are double dumbbells that are separated into five sections: d_{xy}, d_{yz}, and d_{xz} (locations where lobes are laying between the axes), and $d_{x^2} - d_{y^2}$ and d_{z^2} (locations where lobes are laying on axes) (Figure 1.19).

1.21 PAULI AND EXCLUSION PRINCIPLE

In 1925, Wolfgang Pauli presented the principle, which stated that "no two electrons in an atom can have all four quantum numbers the same." In accordance with this concept, it is possible for two electrons to have the same quantum number in an orbital of an atom; but, the spin quantum number must be different, specifically $s = +1/2$ and $s = -1/2$. This means that an orbital can contain two electrons that have the opposite spin.

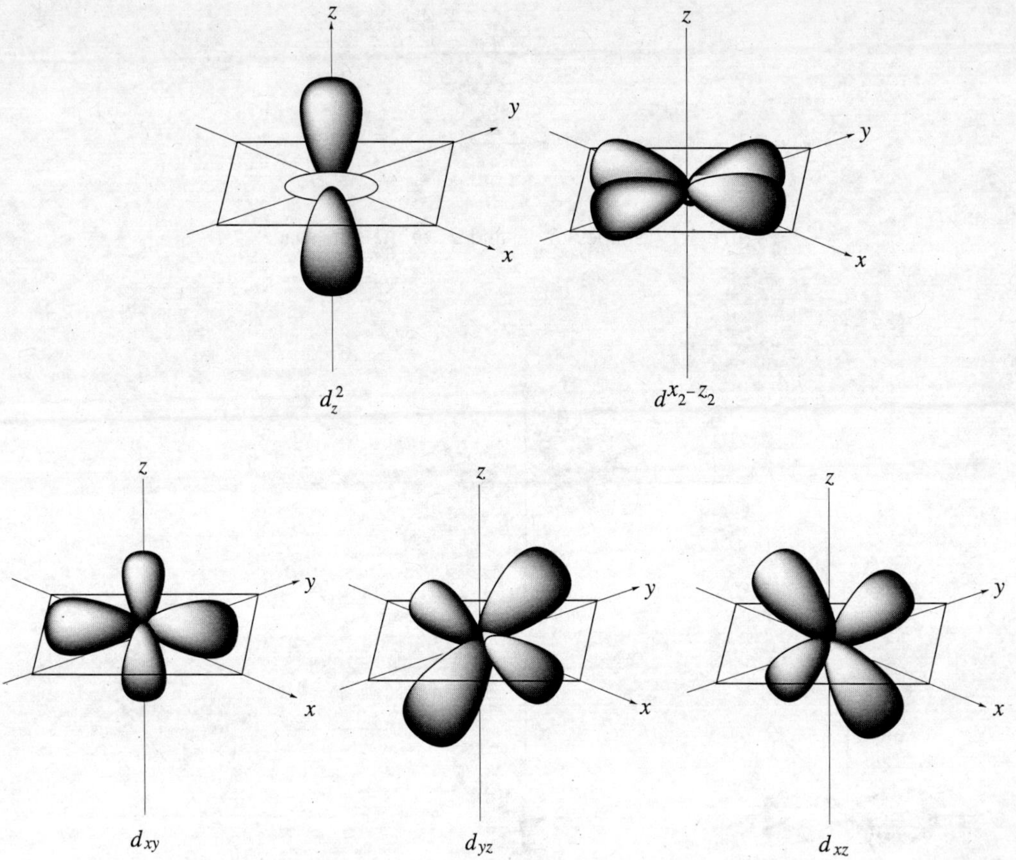

Figure 1.19 Illustration of *d*-orbitals.

1.22 ELECTRON CONFIGURATION OF ATOMS

The ground-state electron configuration of an atom is a description of the orbitals that the electrons of the atom occupy. This configuration is also known as the lowest-energy arrangement possible. For the purpose of filling electrons in various orbitals, the following norms are recognized. Table 1.2 contains a tabular representation of the ground-state electrical configuration of the elements.

Rule 1 states that the orbitals with the lowest energy are filled first, in the order of $1s \rightarrow 2s \rightarrow 2p \rightarrow 3s \rightarrow 3p \rightarrow 4s \rightarrow 3d \rightarrow 4p \rightarrow 5s \rightarrow 4d \rightarrow 5p$ and more.

Rule 2 states that an orbital can only be occupied by two electrons, and those electrons must have opposite spins.

Rule 3 states that if there are two or more vacant orbitals of equal energy, then one electron will occupy each of them with the spins parallel until all of the orbitals are half full.

Table 1.2 The electronic configurations of elements 1–18.

Elements	Atomic Number	Orbitals								
		1*s*	2*s*	$2p_x$	$2p_y$	$2p_z$	3*s*	$3p_x$	$3p_y$	$3p_z$
H	1	1								
He	2	2								
Li	3	2	1							
Be	4	2	2							
B	5	2	2	1						
C	6	2	2	1	1					
N	7	2	2	1	1	1				
O	8	2	2	2	1	1				
F	9	2	2	2	2	1				
Ne	10	2	2	2	2	2				
Na	11	2	2	2	2	2	1			
Mg	12	2	2	2	2	2	2			
Al	13	2	2	2	2	2	2	1		
Si	14	2	2	2	2	2	2	1	1	
P	15	2	2	2	2	2	2	1	1	1
S	16	2	2	2	2	2	2	2	1	1
Cl	17	2	2	2	2	2	2	2	2	1
Ar	18	2	2	2	2	2	2	2	2	2

1.23 HYBRIDIZATION

The term "mixing" is denoted by the phrase "hybridization." Within the framework of the valence bond approach to covalent bonding, it is the combination of pure atomic orbitals that results in the generation of hybrid orbitals. In order to create hybrid orbitals, it is possible to combine the characteristics of atomic orbitals that are relatively near in energy. Depending on the atomic orbitals that are involved and the percentage contributions that they make, the characteristics of a hybrid orbital can be determined. An example of a hybrid orbital is a group of identical orbitals that have been reformed from pure atomic orbitals. These orbitals are used to represent specific types of covalent bonds.

Linus Pauling

1.23.1 Hybrid Orbital Properties

It is possible to calculate the total number of hybrid orbitals by adding up all of the atomic orbitals that have been mixed. Hybridization is not a real physical event; rather, it is a rationalization of the shape that was discovered by experimentation. In the process of hybridization, the energy of atomic orbitals is preserved.

EXAMPLE: In the case of tetrahedral C, the *p* orbitals undergo a downward movement of ¼ of the energy difference that exists between the *s* orbitals and the *p* orbitals, while the *s* orbitals undergo an upward movement of ¾.

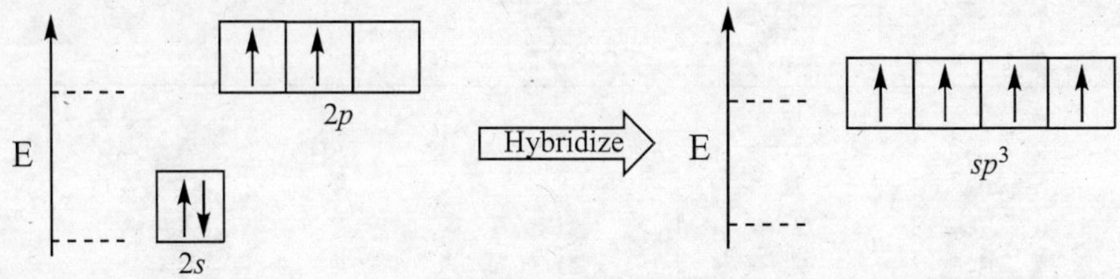

Note: **Linus Pauling** was the one who first proposed the idea of orbital hybridization. In 1954, he was awarded the Nobel Prize in Chemistry for his study, which was published under the title "Nature of the Chemical Bond."

1.23.2 *sp* Hybridization

During the process of *sp* hybridization, an *s* orbital will only join with one of the three *p* orbitals in order to produce three atomic orbitals that are comparable to one another. The two unhybridized *2p* orbitals have not been altered in any way (Figure 1.20). This hybridization strategy is utilized by the majority of compounds that include beryllium by nature. Like BeF_2 and $HgCl_2$, among other examples.

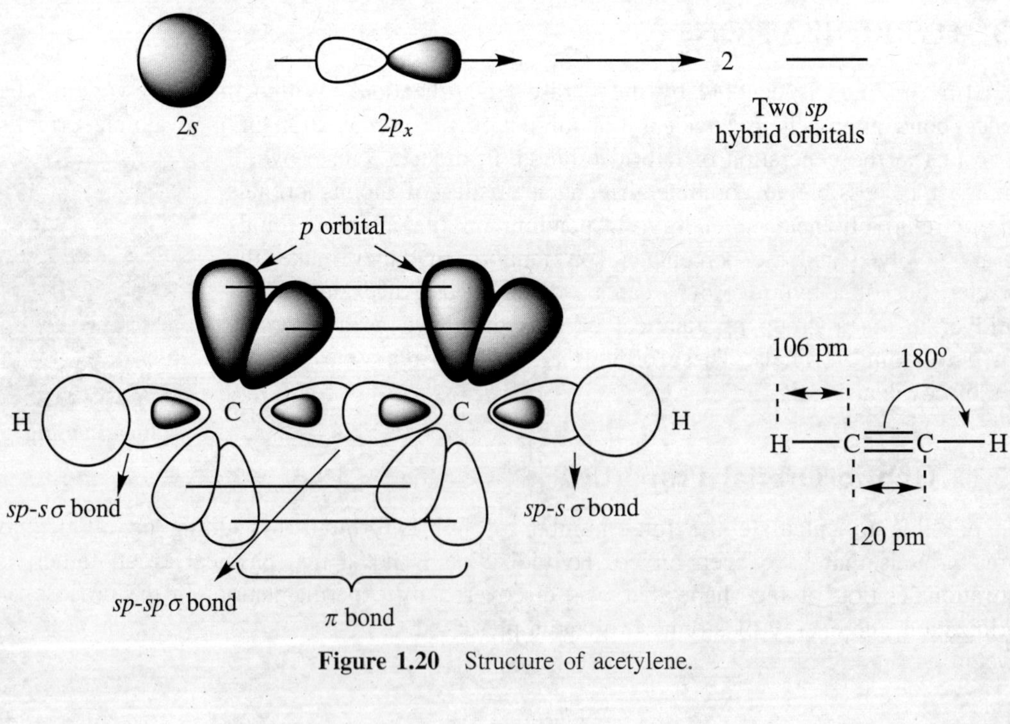

Figure 1.20 Structure of acetylene.

1.23.3 sp^2 Hybridization

To create three atomic orbitals that are equal to one another, sp^2 hybridization involves the combination of an *s*-orbital with only two of the three *p*-orbitals. During hybridization, one of the 2*p* orbitals that has not been hybridized is left unaltered. The three sp^2 hybrid orbitals are all planar and lie at an angle of 120 degrees to each other; as a result, they are organized in a trigonal configuration. The hybridization strategy described here is typical of the majority of boron-containing compounds. These include C_2H_6, BF_3, and SO_3, amongst others.

The Structure of Ethylene

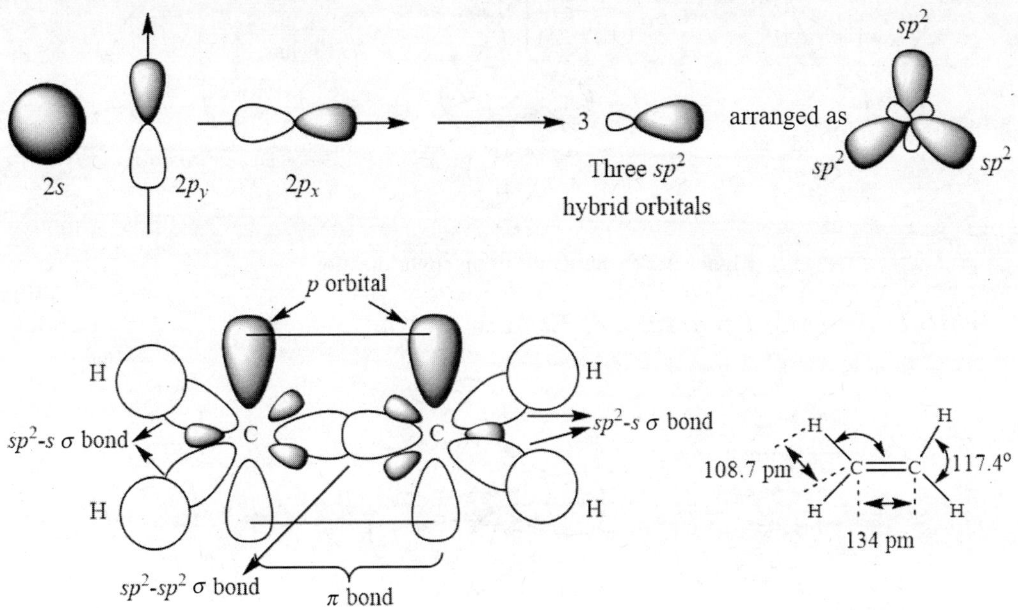

Figure 1.21 Structure of ethylene.

1.23.4 sp^3 Hybridization

In order to create four comparable atomic orbitals, sp^3 hybrid orbitals, it is possible for a *s* orbital and three *p* orbitals to join, also known as hybridize. These hybrid orbitals have a tetrahedral orientation and a difference of 109.5 degrees between them. Such is CH_4, NH_3, H_2O and NH_4^+, amongst other examples.

The Structure of Methane

There are four electrons in the valence shell of methane carbon, which is denoted by the $2s^2$ $2p_x^1 2p_y^1$, and electrons in the excited state of methane carbon occur as a result of hybridization between the hybridization of $2s^1$ $2p_x^1 2p_y^1 2p_z^1$.

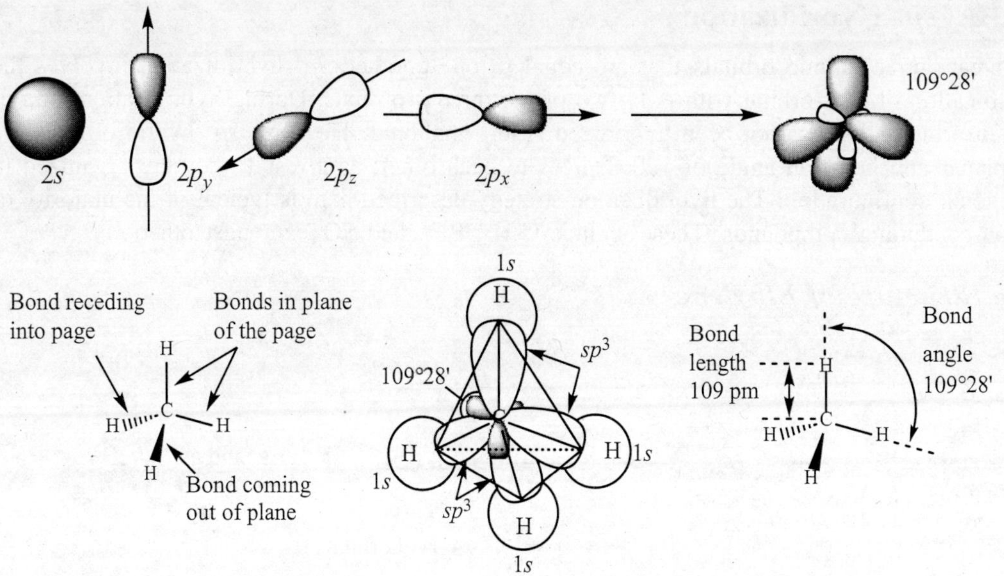

Figure 1.22 Structure of sp^3 hybridization.

sp^3 *Orbitals and the Structure of Ethane*

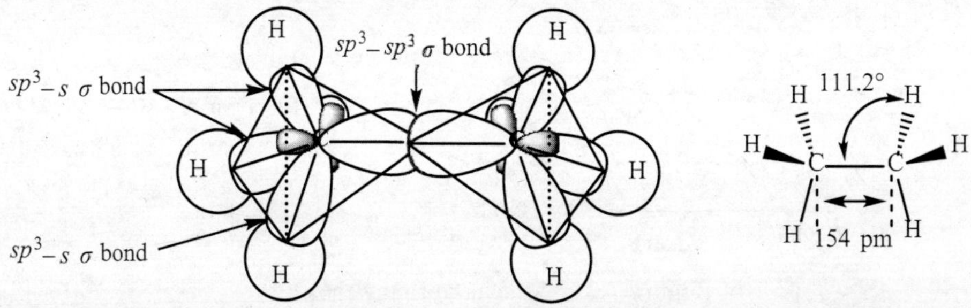

Figure 1.23 Structure of ethane.

1.23.5 sp^3d Hybridization

One of the five orbitals that are created in the process of sp^3d hybridization is formed by the hybridization of one *s* orbital, three *p* orbitals, and one *d* orbital. Within a trigonal bipyramid, the five orbitals are oriented in the direction of the four corners. The hybridization of phosphorus, denoted by the letter P. It is also possible to explain the shapes of seesaw molecules, t-shaped molecules, and some linear molecules using this hybridization strategy. For example PF_5, SF_4 and BrF_3, and so on.

1.23.6 sp^3d^2 Hybridization

One of the six orbitals that are generated in the process of sp^3d^2 hybridization consists of the hybridization of one *s* orbital, three *p* orbitals, and two *d* orbitals. At the corners of a regular octahedron, the six orbitals are oriented in a specific direction. In addition, the forms of square pyramidal and square planar molecules can be explained by this hybridization scheme. For instance, SF_6, ClF_5, XeF_4 and PF_6^-, amongst others.

1.24 THE NATURE OF CHEMICAL BONDS

Atoms form bonds with one another in such a way that for each atom in the bond, the electron configuration of the noble gas that is closest to it in atomic number is acquired. By adhering to the octet rule, which is the tendency to react in ways that result in an outer shell consisting of eight valence electrons, any bong structure that is formed should satisfy this requirement.

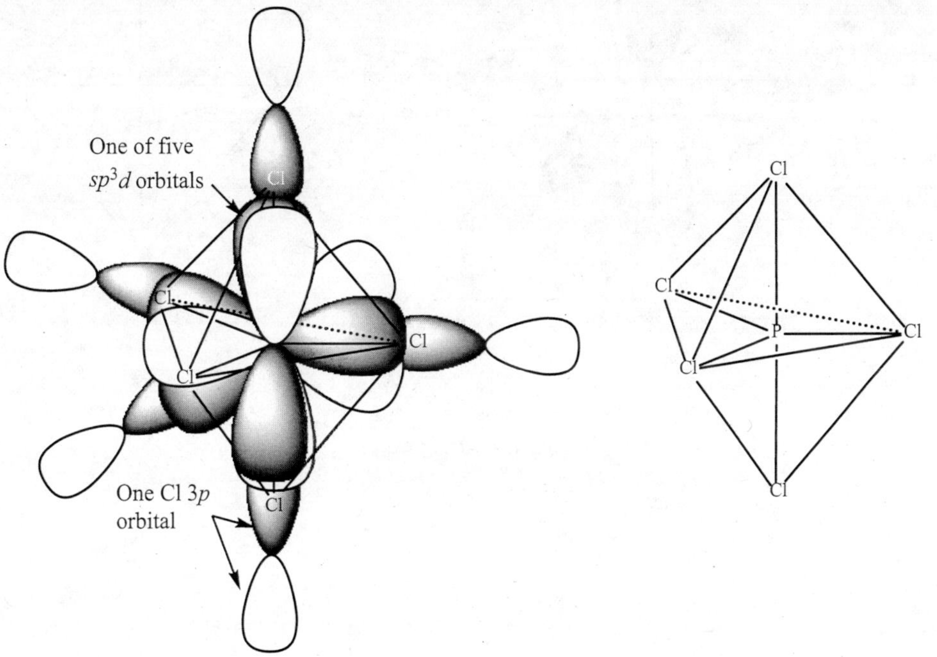

Figure 1.24 Bonding and structure of trigonal bipyramidal PCl_5.

1.24.1 Ionic Bond

Ionic bond is a term that will be used to describe a chemical bond that is formed as a result of the electrostatic attraction between an anion and a cation (like Na^+ Cl^-). An anion is formed when an atom acquires electrons, while a cation is formed when an atom reduces the number of electrons it possesses.

1.24.2 Covalent Bond

Whenever two atoms share one or more pairs of electrons, this results in the formation of a chemical bond. When two atoms come into close proximity to one another and a singly occupied orbital on one atom overlaps a singly occupied orbital on the other atom, a cooperative connection is formed between the two atoms. As a result of the electrons being coupled in the overlapping orbitals, they are naturally drawn to the nuclei of both atoms, which ultimately results in the formation of a bond between the atoms.

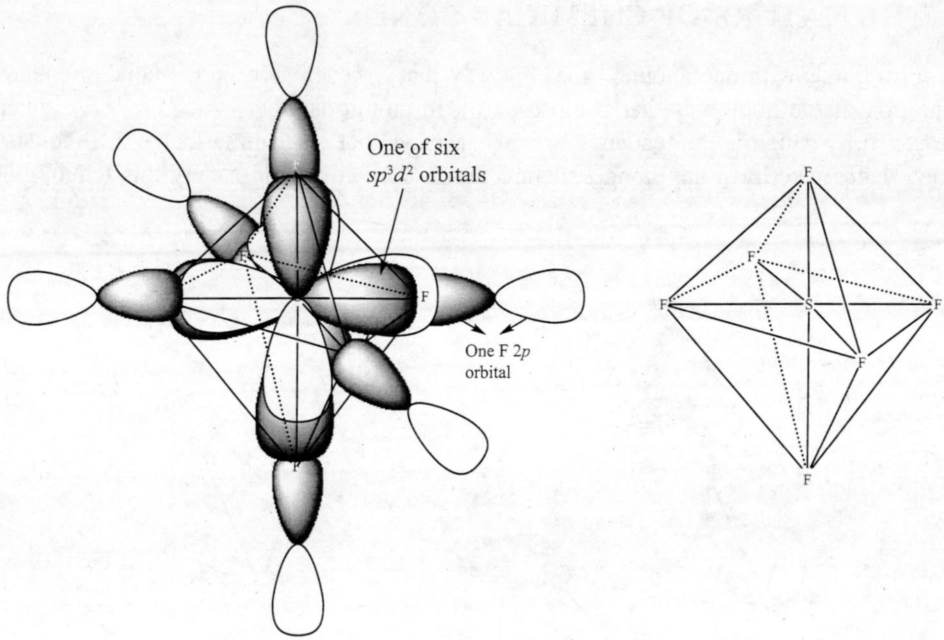

Figure 1.25 Bonding and structure of octahedral SF_6.

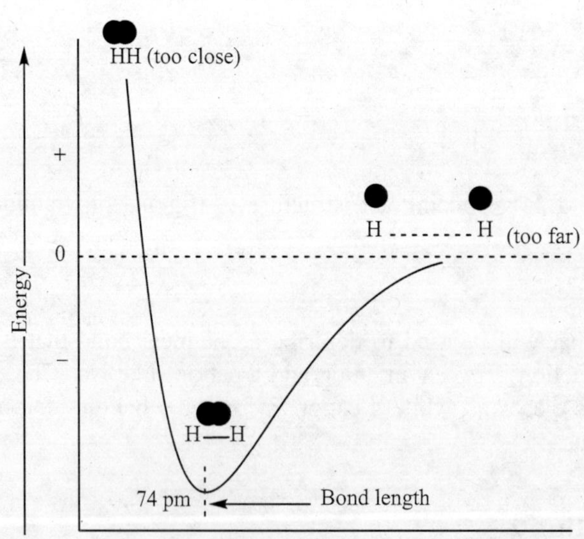

Figure 1.26 Covalent bond formation of Hydrogen molecule.

1.25 VALENCE BOND THEORY

The Lewis theory of electron pair bonds is expanded upon by the valence bond theory, which also includes the concepts of orbital overlap and hybridization. In order to effectively account for the molecular geometry that is predicted by VSEPR, hybridization is required. Creating hybrid atomic orbitals involves making linear combinations (sums and differences) between hydrogen-like orbitals in the valence shell of each atom. This allows for the construction of intermediate atomic orbitals. **Heitler and London** introduced VBT in 1927, and **Pauling and Slater** extended it in 1931. The following assumption was the one that they postulated.

1. A covalent bond is formed through the overlapping of atomic orbitals belonging to two atoms situated in their respective valence shells.
2. The process of overlapping is only possible for atomic orbitals that are partially filled. The formation of a bond resulted in the acquisition of a pair of electrons with opposite spins.
3. The formation of the newly established orbital, characterized by enhanced stability compared to previous orbital, arises from the phenomenon of atomic orbital overlap.
4. The strength of the chemical relationship is enhanced when there is a higher degree of overlap. The recently formed molecule exhibits a reduced energy state and an increased level of stability in comparison to the individual atoms that were separated.
5. For the initiation of bond formation, it is necessary for electrons to undergo a process of unpairing and subsequent relocation to vacant orbitals within the same valence shell, which possess relatively higher energy levels.

1.25.1 VBT Limitations

The oxygen molecule has a bond length and strength consistent with a double bond, and it contains two unpaired electrons.

1. Valence bond theory predicts the double bond, but not the paramagnetism of oxygen.
2. Resonance is another example of the limitations of valence bond theory. Bond lengths and strengths are intermediate between single, double or triple bonds.
3. Molecular orbital theory is often a better approach to use with molecules that have extended π systems.
4. In many basic compounds, the valence bond hypothesis is unable to adequately describe the bonding that occurs. VBT was unable to provide an explanation for a few of the following points.

1.26 MOLECULAR ORBITAL THEORY

According to the valence bond theory, a bond between two atoms should be seen as an overlap of orbitals obtained from those two atoms. Or, to put it another way, the formation of the bond in HCl occurs when an H (1*s*) orbital overlaps with a Cl (3*p*) orbital. After the limitations of the valence bond hypothesis were determined, a new concept regarding molecular orbitals was presented. **I.E. Lennard-Jones** and **Charles Coulson** were the ones who eventually developed it after it was initially proposed by **F. Hund** and **R.S. Milliken** in the year 1930.

1. The atomic orbitals (AOs) of two atoms overlap to form additional orbitals that are referred to as molecular orbitals (MOs).
2. The total number of MOs that are created will be equal to the total number of AOs that overlap.
3. The overlapping of two AOs results in the formation of two MOs: a bonding molecular orbital (BMO) with a lower energy and an anti-bonding molecular orbital (ABMO) with a higher energy.
4. Every MO-like AO has the capacity to accommodate two electrons that have two different spins.
5. In accordance with the different forms of overlapping MOs, the design of MOs can be categorized as σ, σ^*, π, π^*, δ and δ^*.
6. The filling of electrons in different MOs operates in accordance with the rule of Aufbau and Hund as well as Pauli's Principle.
7. If Ψ_1 and Ψ_2 are wave functions of two AOs that overlap, then the process of their overlapping is referred to as Linear Combination of Atomic Orbitals (LCAO) methods.
8. The paired and unpaired electrons on an orbital of a molecule are responsible for the magnetic characteristics of the molecule. Due to the presence of both unpaired and paired electrons on the MOs and ABMOs, the molecule will exhibit both paramagnetic and diamagnetic properties in its natural state.
9. With the assistance of the following formula, one is able to determine the bond order, which stands for the magnitude of the strength of the bond that exists between the AOs.

$$\text{Bond Order} = \frac{(\text{No. of e}^-\text{'s in bonding MO's}) - (\text{No. of e}^-\text{'s in antibonding MO's})}{2}$$

1.26.1 Bonding and Antibonding Orbitals

When two atomic orbitals are put together, a pair of bonding orbitals with lower energy is produced as a result. The majority of the electron density, which is negative, is found in bonding orbitals, which are located between two positive nuclei. Through the process of subtracting two atomic orbitals from one another, a collection of anti-bonding orbitals with a higher energy level is produced (Figure 1.27 and 1.28). When it comes to antibonding orbitals, the majority of the electron density is located on the side that is opposite the region where the bond needs to be established.

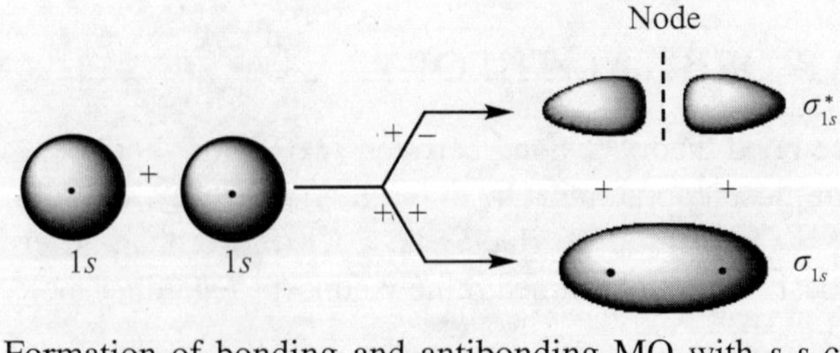

Figure 1.27 Formation of bonding and antibonding MO with *s-s* combination.

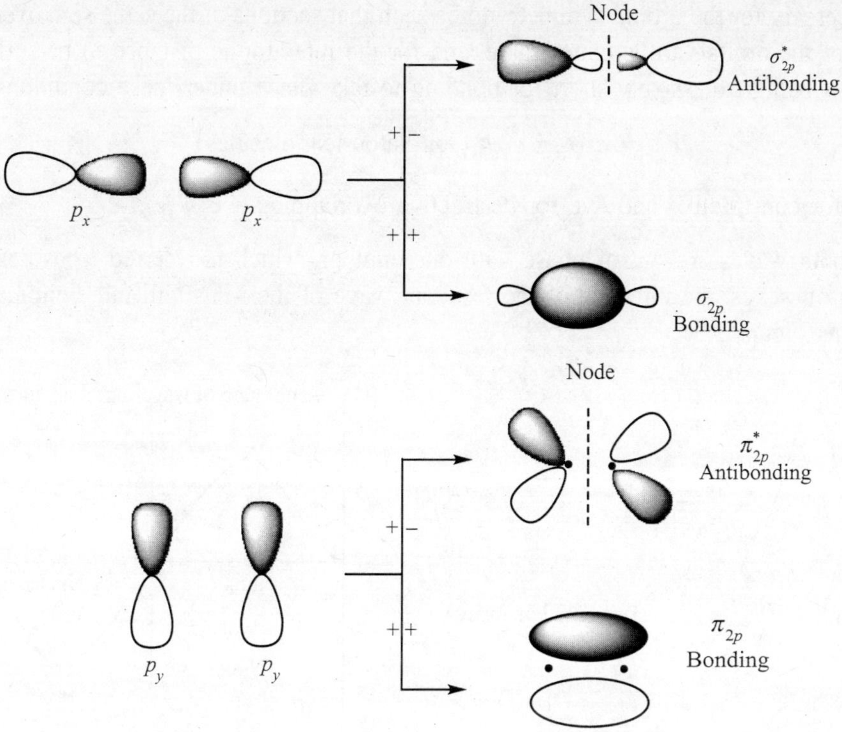

Figure 1.28 Formation of bonding and antibonding MO with *p-p* combination.

1.26.2 Linear Combination of Atomic Orbitals (LCAO)

The construction of a molecular orbital is possible by the linear combination of molecular orbitals. Either the symmetry of two atomic orbitals must remain unchanged when rotated about the internuclear line in order to produce bonding and antibonding MOs, or both atomic orbitals must change symmetry in the same manner in order to produce bonding and antibonding MOs. The linear combinations of atomic orbitals that make up molecular orbitals are referred to as LCAO. MOs behave in a manner similar to that of a wave. The intensity of the resultant wave of the MOs will significantly rise (waves will reinforce) when both waves are in phase with one another, which is when constructive interference of waves occurs. This is seen in Figure 1.29.

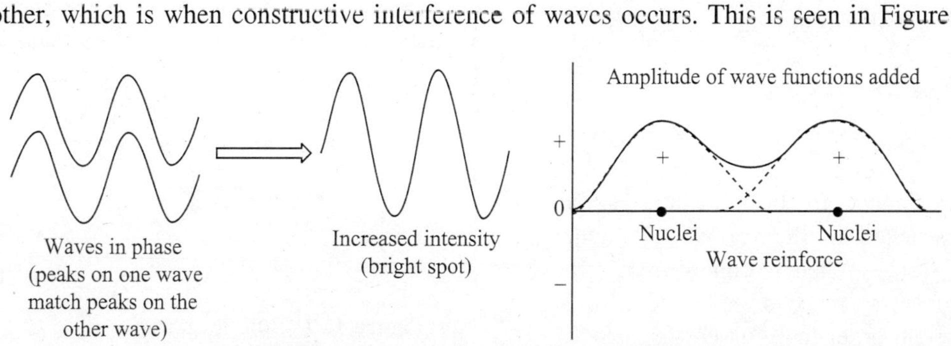

Figure 1.29 Constructive interference of waves.

It is necessary for the orbital symmetry to be such that sections of the same sign overlap, for the energy of the orbitals to be comparable, and for the interatomic distance to be sufficiently short without being excessively short for bonding to take place under these conditions.

$$\Psi_{ab} = c_a\psi_a \pm c_b\psi_b \text{ (for diatomic molecules)}$$

where +ve for constructive and –ve for destructive overlapping.

When both waves are out of phase with one another, which is referred to as destructive interference of waves, the intensity of the resultant wave of the MOs will fall significantly, as shown in this Figure 1.30.

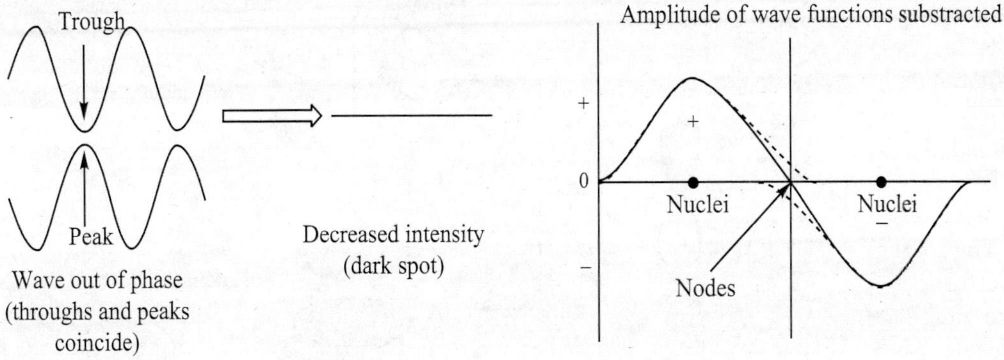

Figure 1.30 Destructive interference of waves.

1.26.3 Molecular Orbital (MO) Energy Level Diagram for Homo-nuclear Diatomic Molecule

Atomic orbitals of the same atoms that have the same energy and symmetry overlap with one another in the highest possible order to produce molecular orbitals. Several examples of this category are going to be presented in the following paragraphs.

The Bonding in H_2

As a result of the fact that H possesses only a single electron and a single 1*s* AO, the MO diagram for the molecule H_2 can be constructed in the following manner (Figure 1.32). When it comes to H_2, the order of MOs is $\sigma_{1s}^2\ \sigma_{1s}^*$. In the case of H_2 the bond order in H_2 is (2 – 0)/2 = 1.

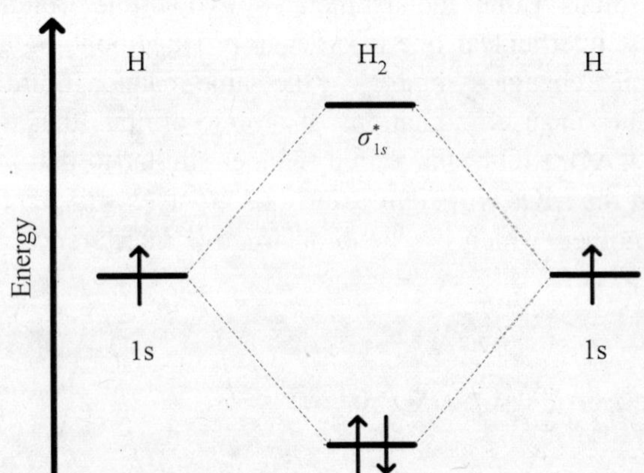

Figure 1.31 MO diagram for H_2.

Due to the fact that electrons are paired, it possesses a diamagnetic property.

The Bonding in He_2

In addition, He possesses only 1*s* AO, which means that the MO diagram for the molecule He_2 may be constructed in the same manner. The only difference is that the 1*s* AO on He contains two electrons, as seen in Figure 1.32. The bond order in He_2 is (2 – 2)/2 = 0, so it is impossible for the molecule to exist.

The cation He_2^+, which is characterized by the removal of one of the electrons in the σ^* MO, would possess a bond order of (2 – 1)/2 = 1/5, indicating that it is possible to anticipate the existence of such a cation (Figure 1.33).

It is possible to write the electron configuration of this cation in the same manner that we write the configurations of atoms, with the exception that the MO labels are used instead of the AO labels: He_2^+ is equal to $\sigma^2_{1S}\sigma_{1S}{*}^1$.

The molecular orbital theory is a powerful theory because it enables us to make predictions about whether or not molecules should exist. Additionally, it provides us with a precise picture of the electronic structure of any hypothetical molecule that we can envision.

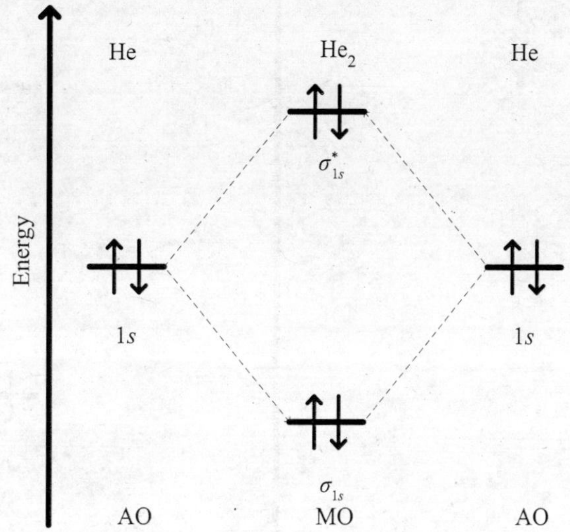

Figure 1.32 MO diagram for He_2.

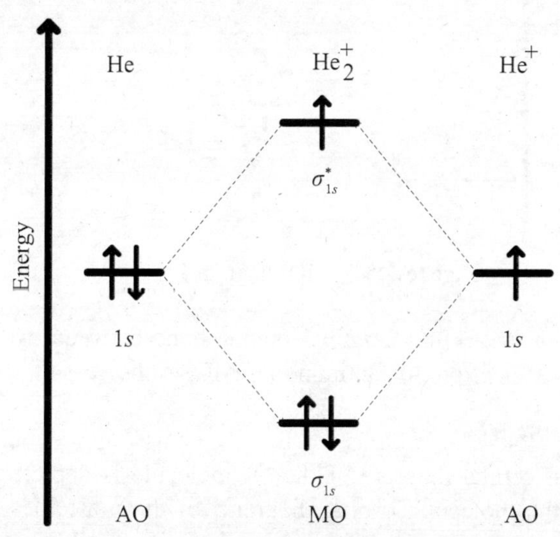

Figure 1.33 MO diagram for He_2^+.

The Bonding in Li_2

The order of MOs for Li_2 is $\sigma_{1s}^2\sigma_{1s}^{*2}\sigma_{2s}^2\sigma_{2s}^{*0}$ The MO diagram is shown in Figure 1.34.

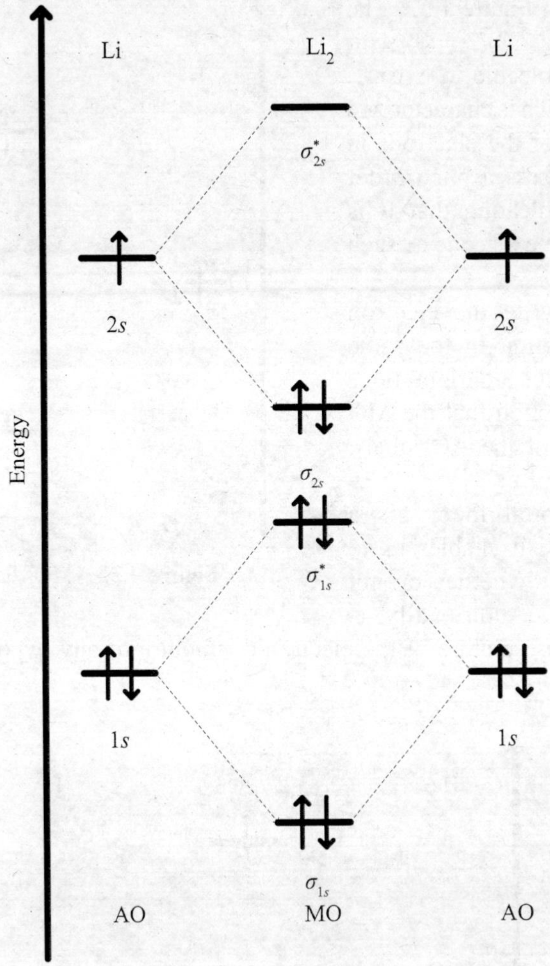

Figure 1.34 MO diagram for Li_2.

Bond order = $[N_b - N_a]/2$ = [4 – 2]/2 = 1 (single bond between two atoms). The fact that it possesses two unpaired electrons in antibonding MO is the reason why it is paramagnetic in nature.

The diagram of the molecular orbitals for B_2, C_2, and N_2 is identical to the one depicted in Figure 1.36 molecular orbital diagram of the O_2 molecule is depicted in Figure 1.37. This Figure is distinct from the molecular orbital diagrams of the molecules O_2, F_2, and Ne_2. The order of energy of MO $\sigma_{2_{pz}}$ is the sole fundamental distinction between the two. The most essential thing to note is that the ordering of the orbitals is altered as a result of mixing between the 2*s* and $2p_z$ orbitals. According to quantum physics, the amount of mixing that will occur between a given set of orbitals of the same symmetry will increase in proportion to the degree to which they are closer in energy to one another. The energy of the MOs that are created are altered as a result of these mixing processes. When the atomic number grows, we are aware that the distance that separates the *ns* orbital and the *np* orbital also increases.

This indicates that the amount of mixing reduces as we move over the second row of the periodic table until there is no longer any s-p_z mixing, which occurs at the O_2 level. Increasing the atomic number results in an increase in both the effective nuclear charge and the electronegativity of the atoms. The drop in energy of the corresponding orbitals from Li_2 to F_2 can be attributed to this particular reason (Figure 1.35). It is possible to gain an understanding of the patterns of bond lengths and energy by analyzing the size of each atom (Table 1.3).

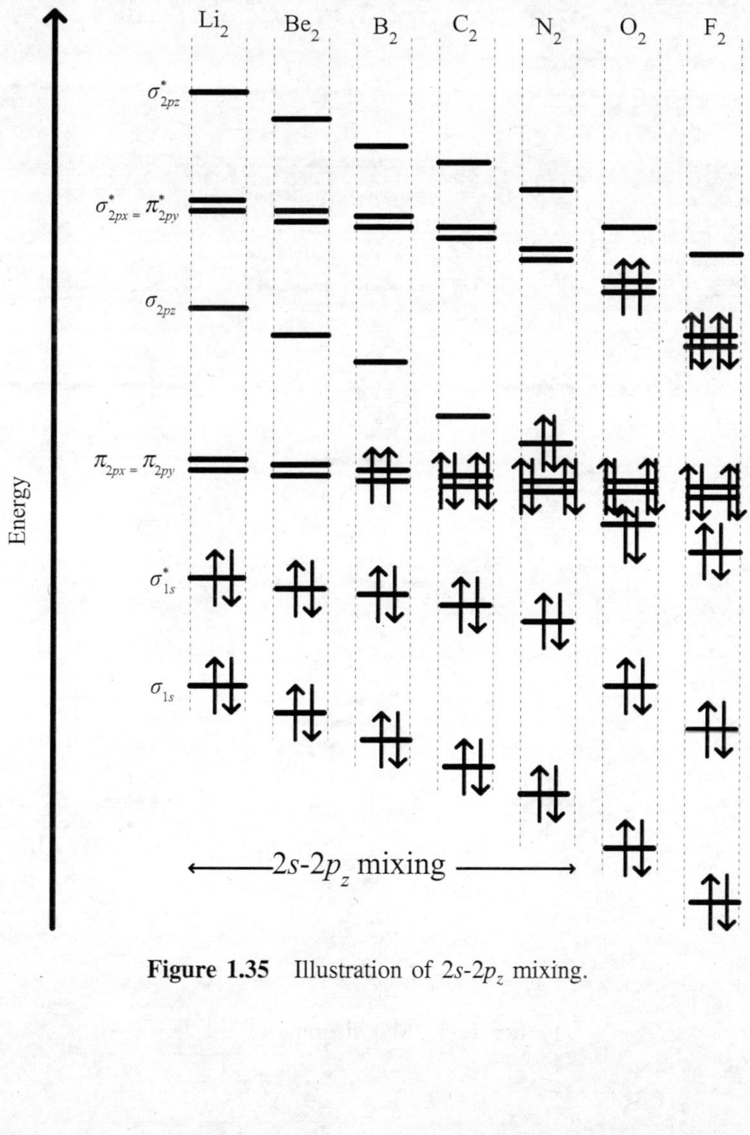

Figure 1.35 Illustration of 2*s*-$2p_z$ mixing.

Table 1.3 Different properties of molecule from Li_2 to Ne_2.

Molecule	Li_2	Be_2	B_2	C_2	N_2	O_2	F_2	Ne_2
Bond Order	1	0	1	2	3	2	1	0
Bond Length (Å)	2.67	n/a	1.59	1.24	1.01	1.21	1.42	n/a
Bond Energy (kJ/mol)	105	n/a	289	609	941	494	155	n/a
Diamagnetic (d)/ Paramagnetic (p)	d	n/a	p	d	d	p	d	n/a

The Bonding in B_2

In the formation of MO of B_2 molecule, five electrons ($1s^2\ 2s^2\ 2p^1$) from each boron atom is contributed. The electronic configuration for B_2 molecule is as follows (Figure 1.36).

$$\sigma_{1s}^2, \sigma_{1s}^{*2}, \sigma_{2s}^2, \sigma_{2s}^{*2}, \pi_{2p_x}^1 = \pi_{2p_y}^1, \sigma_{2p_z}^0, \pi_{2p_x}^{*0} = \pi_{2p_y}^{*0}, \sigma_{2p_z}^{*0}$$

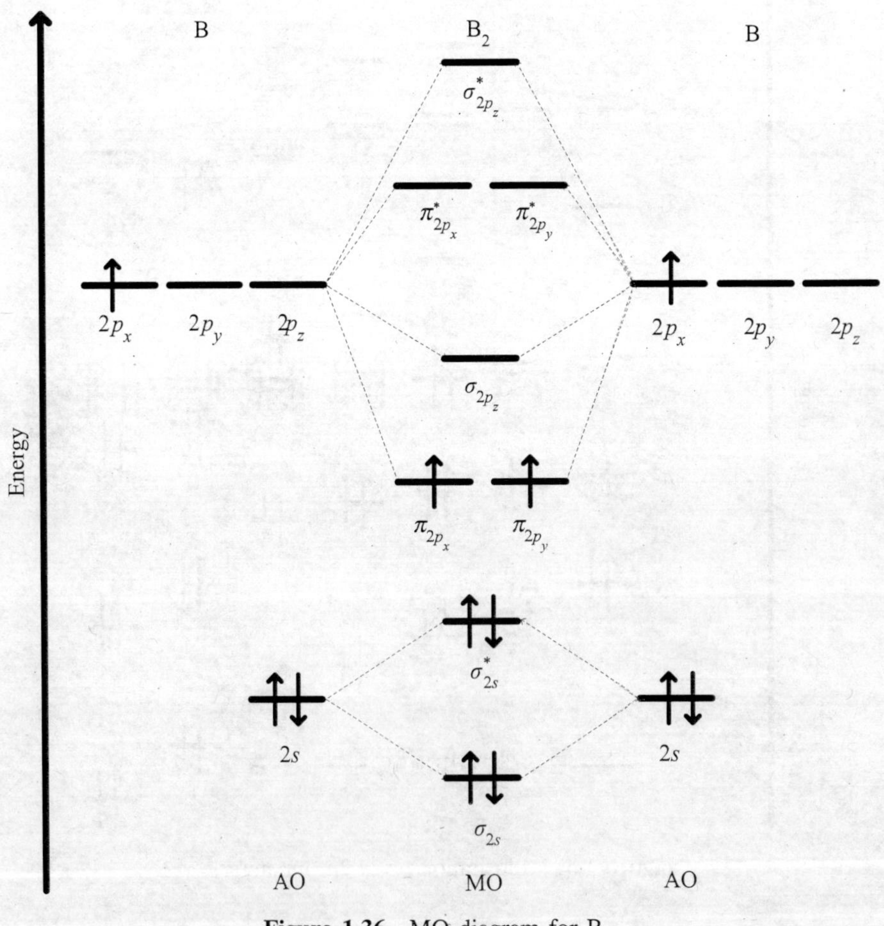

Figure 1.36 MO diagram for B_2.

The Bonding in O_2

There are total eight electrons in the oxygen atom and has sixteen electron in O_2. Electrons are arranged in molecular orbitals as the following manner.

$$\sigma_{1s}^2, \sigma_{1s}^{*2}, \sigma_{2s}^2, \sigma_{2s}^{*2}, \sigma_{2p_z}^2, \pi_{2p_x}^2 = \pi_{2p_y}^2, \pi_{2p_x}^{*1} = \pi_{2p_y}^{*1}, \sigma_{2p_z}^{*0}$$

The MO diagram is shown in Figure 1.37.

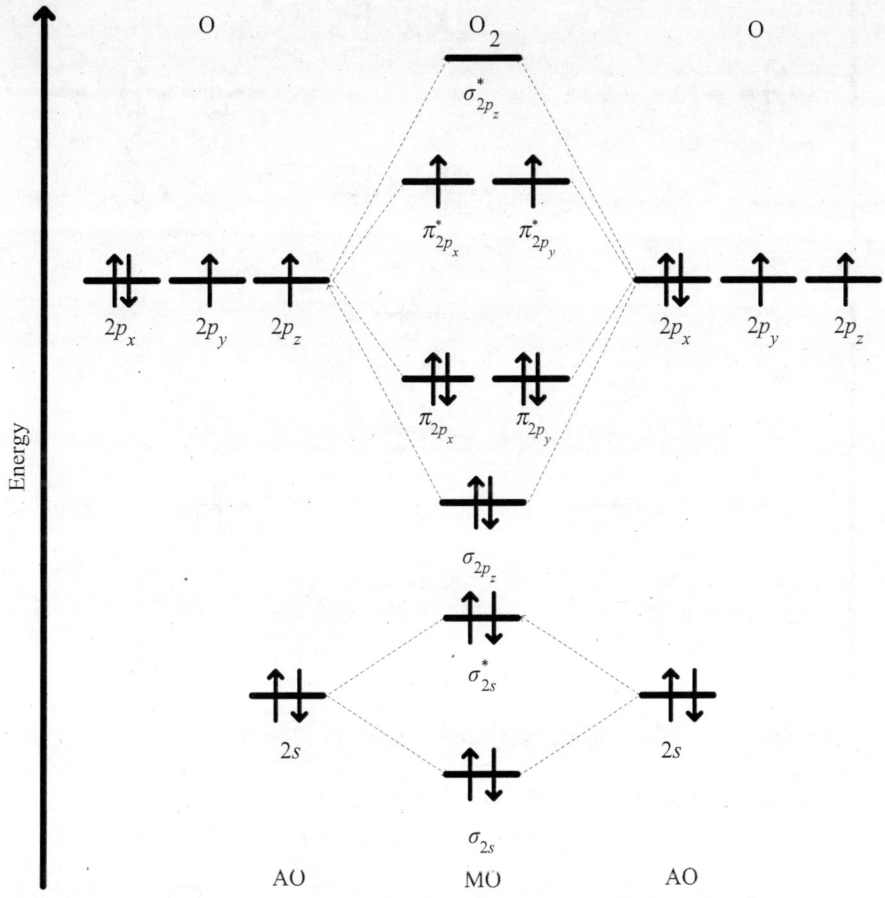

Figure 1.37 MO diagram for O_2.

Bond order = $[N_b - N_a]/2$ = [10 – 6]/2 = 2 (double bond between two atoms). It is paramagnetic in nature because it has two unpaired electron in antibonding MO.

The Bonding in F_2

A Fluorine atom has nine electrons. Thus it has eighteen electrons in F_2. Electrons are arranged in molecular orbitals as the following manner:

$$\sigma_{1s}^2, \sigma_{1s}^{*2}, \sigma_{2s}^2, \sigma_{2s}^{*2}, \sigma_{2p_z}^2, \pi_{2p_x}^2 = \pi_{2p_y}^2, \pi_{2p_x}^{*2} = \pi_{2p_y}^{*2}, \sigma_{2p_z}^{*0}$$

The MO diagram is shown in Figure 1.38.

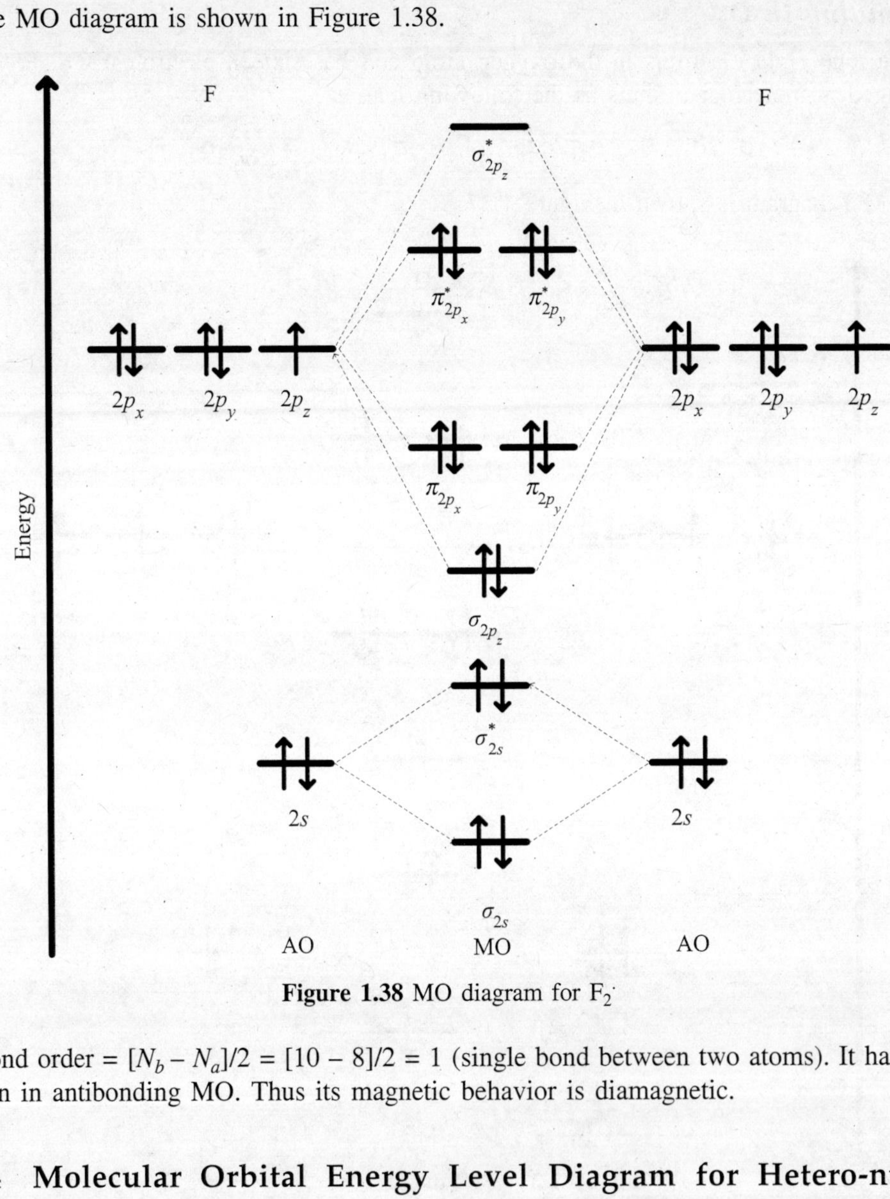

Figure 1.38 MO diagram for F_2.

Bond order = $[N_b - N_a]/2 = [10 - 8]/2 = 1$ (single bond between two atoms). It has paired electron in antibonding MO. Thus its magnetic behavior is diamagnetic.

1.26.4 Molecular Orbital Energy Level Diagram for Hetero-nuclear Diatomic Molecule

Atomic orbitals of different atom overlap together in maximum order with same energy, same symmetry to give molecular orbitals. There are some examples of this category is discussed.

The Bonding in NO

Nitrogen has seven and Oxygen has eight electrons. Thus, Nitric oxide contain fifteen electrons. Because the contributions are not equal, the MO diagram of NO will be skewed (Figure 1.39).

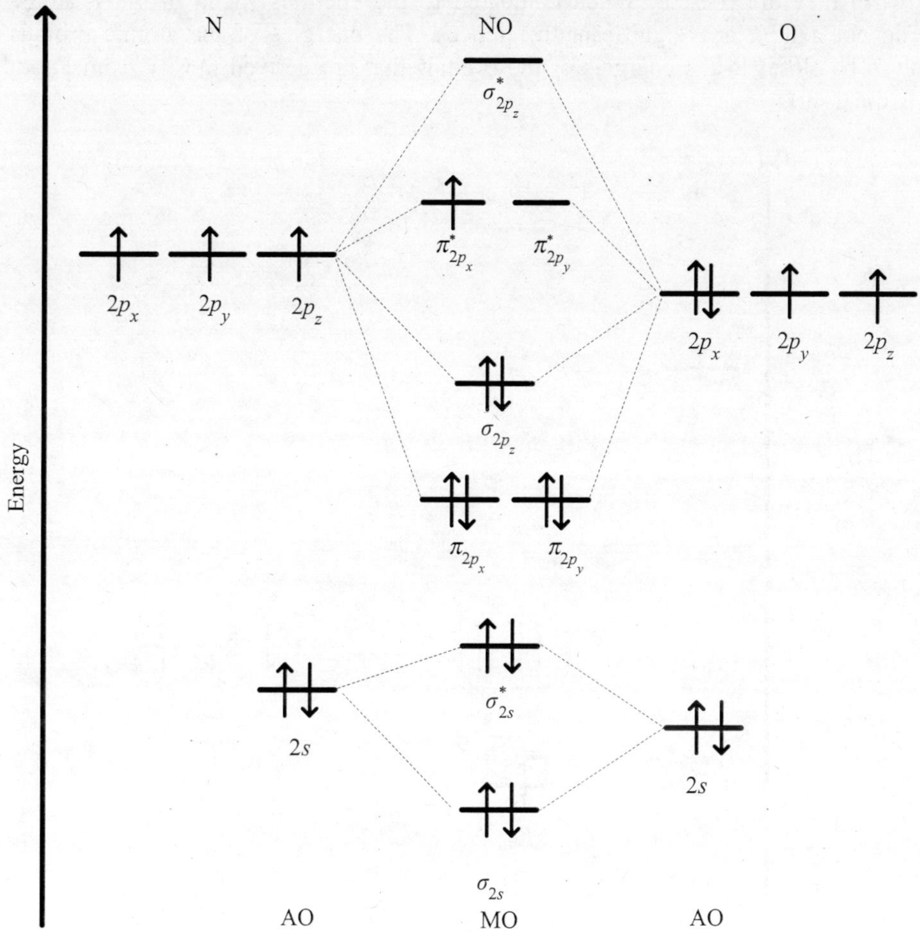

Figure 1.39 MO diagram for NO.

Electronic configuration of NO molecule is as follows:

$$\sigma_{1s}^2, \sigma_{1s}^{*2}, \sigma_{2s}^2, \sigma_{2s}^{*2}, \sigma_{2p_z}^2, \pi_{2p_x}^2 = \pi_{2p_y}^2, \pi_{2p_x}^{*1} = \pi_{2p_y}^{*0}, \sigma_{2p_z}^{*0}$$

Bond order = $[N_b - N_a]/2$ = [8 – 3]/2 = 2.5. It has unpaired electron in antibonding MO. Thus its magnetic behavior is paramagnetic.

The Bonding in HF

Although there is some mixing between the H 1*s* orbital and the F 2*s* orbital in the case of HF, the $2p_z$ orbital is the one with which it interacts the majority of the time. Additionally, F possesses the $2p_x$ and $2p_y$ orbitals, which are unable to interact with the H 1*s* orbital due to the fact that they possess the incorrect symmetry! We shall discover that the overlap integral, denoted by the letter S, is equal to zero if we attempt to merge these orbitals with the ones on H. These orbitals are referred to be non-bonding orbitals since they are only found on the

F atom when they are present. When compared to the energies found in the F atom, these orbitals do not exhibit any significant differences. The energies of the atomic orbitals of F are going to be closer to the energies of the orbitals that are derived mostly from F, and vice versa (Figure 1.40).

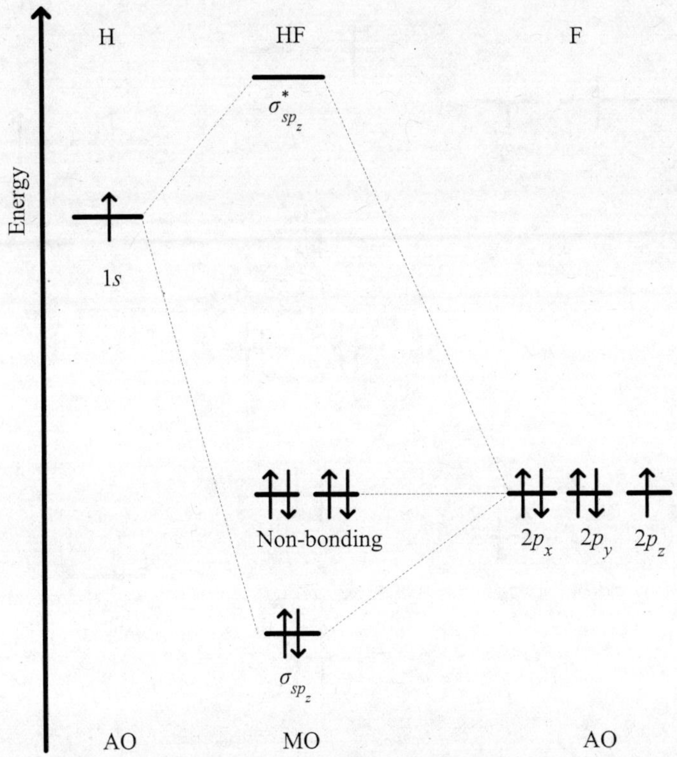

Figure 1.40 MO diagram for HF.

SOLVED PROBLEMS

PROBLEM 1.1 Calculate the de Broglie wavelength of electron moving with the velocity 1.2×10^5 m/sec.

Solution We know that the de Broglie equation is $\lambda = \frac{h}{mv}$

So

$$\lambda = \frac{h}{mv} = \frac{6.626 \times 10^{-35}\text{ Js}}{(9.1 \times 10^{-31}\text{ kg})(1.2 \times 10^5\text{ m/s})}$$

$$= 6.068 \times 10^{-9}\text{ m}$$

$$= 6.068\text{ Å}$$

PROBLEM 1.2 Find the uncertainty in the position of an electron when the mass of an electron is 9.1×10^{-28} g and the uncertainty in velocity is equal to 2×10^{-3} cm/sec.

Solution we know Heisenberg Uncertainty equation: $\Delta x \times \Delta P \geq \frac{h}{4\pi}$.

We can calculate uncertainty in the momentum, $\Delta P = m\Delta V = 9.1 \times 10^{-28}$ gm $\times 2 \times 10^{-3}$ cm/sec $= 1.82 \times 10^{-30}$ g-cm/sec $= 1.82 \times 10^{-35}$ kg-m/sec.

Therefore, $\Delta x = \frac{h}{4\pi\,\Delta P} = 2.9$ m $= 2.9 \times 10^{2}$ cm.

QUESTIONS

1.1 Draw the molecular orbital diagram for CO molecule. Write the electronic configuration, calculate the bond order and predict the magnetic behavior of CO.

1.2 Derive Schrodinger wave equation and write down its applications.

1.3 Derive an wave equation for particle in a box when $0>l>1$.

1.4 How do you find the degeneracy of a particle in a box?

1.5 With the help of molecular orbital diagram, calculate the bond order of O_2^-, O_2^{2-}, He_2^+, NO^-. Also write their magnetic character.

1.6 On the basis of molecular orbital theory explain the paramegnetic nature of O_2 molecule.

1.7 Explain Hamiltonian and Hermitian Operator with examples.

1.8 Discuss the postulate of valance bond theory and its limitation. Also give structure and hybridization of XeF_4.

1.9 Draw the MO diagram and calculate bond order for CO molecule.

1.10 Explain the formation of bonding and antibonding molecular orbital on the basis of wave functions.

1.11 Which species out of H_2, H_2^{2+} and H_2^{2-} are paramagnetic and why?

1.12 Explain Linear Combination of Atomic Orbitals (LCAO).

CHAPTER 2

Bonding in Coordination Compounds: Crystal Field Theory

2.1 INTRODUCTION

When an orbital of a ligand containing lone pairs coincides with an unoccupied orbital of a metal. Referred to be a coordinate covalent bond on occasion. Ligands are required to possess unshared electron pairs. A Metal-Ligand bond is established when a Lewis acid and a Lewis base interact. The ligands possess non-bonding electrons, which function as Lewis bases. The metal, acting as a Lewis acid, possesses unoccupied orbitals. The characteristics of the metal can be significantly modified by the coordination ligands and geometry, resulting in changes in color and susceptibility to oxidation. Silver metal atom can easily form coordination with NH_3, illustrated below:

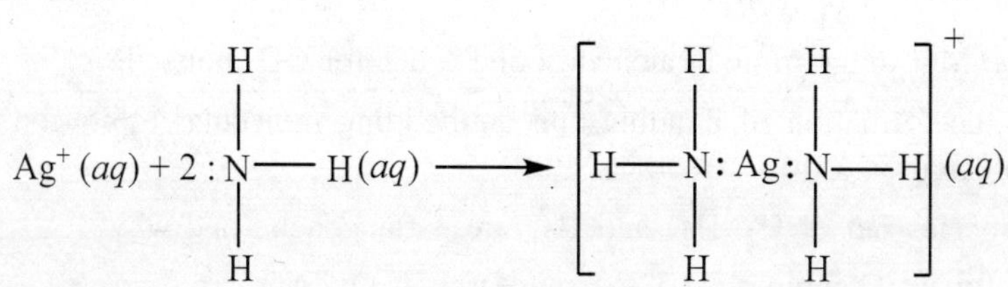

2.2 CRYSTAL FIELD THEORY

Valence bond theory is a helpful tool for visualizing the bonding in complexes. However, it falls short in explaining the characteristics of ligands, such as color and electronic spectra, as well as the influence of temperature on magnetic moment and magnetic susceptibilities. It also does not account for why certain complexes exhibit high spin while others exhibit low spin, or the stability of complexes. Bethe and van Vleck introduced the crystal field theory to elucidate these characteristics. The proposition was put up by the scientist Hans Bethe in 1929. J.H. van Vleck offered subsequent adjustments in 1935 to incorporate a certain degree of covalency in the interactions. The alterations are often known as Ligand Field Theory. Originally, this theory was specifically used to analyze ionic crystals, which is why it is referred to as crystal field theory.

This Theory is founded upon the subsequent assumptions:

The ligands which are ionic, such as Cl^-, OH^-, and CN^-, are considered to be negative point charges, while ligands which are neutral, such as H_2O, NH_3, and Py, are considered to be dipoles due to their dipolar nature. When the ligand is a neutral molecule, the negative end of the dipole is oriented towards the metal ion. The metal-ligand link is non-covalent, meaning that there is no orbital overlap. The bonding in complexes is solely of an electrical nature. In complexes, there are two types of electrostatic forces to consider:

1. The first type is the attraction between a metal cation and a negatively charged ligand or the negative end of a polar ligand (i.e., dipole).
2. The second form of electrostatic interaction involves the repulsion between the unpaired electrons on the ligands and the electrons in the *d*-orbital of the metal cation, as well as the repulsion between the nuclei of the metal cation and the ligands to a lesser degree.

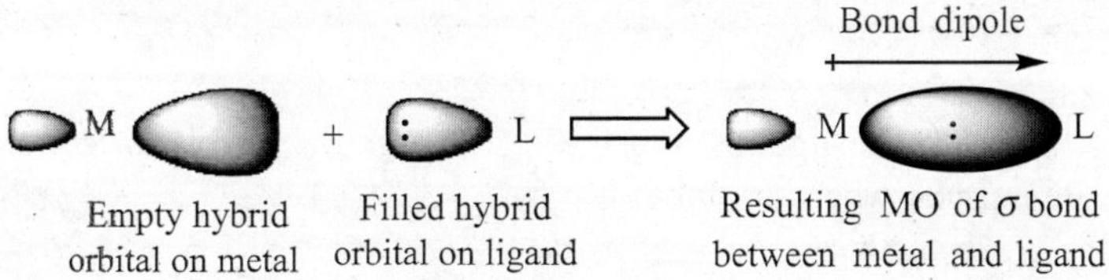

Another factor to consider is the repulsion that arises between the ligands. The five *d*-orbitals of a free metal ion exhibit degeneracy, meaning they possess identical energy levels. Upon formation of a complex, the presence of ligands disrupts the degeneracy of the *d*-orbitals by means of their electrostatic field, resulting in the equalization of orbital energies. The energy of the orbitals aligned with the ligands is increased more than those not aligned with the ligands due to the repulsion between the *d*-electrons and the ligands. Certain ligands exhibit a greater affinity for specific *d*-orbitals, such as d_{z^2} and $d_{x^2-y^2}$, compared to other orbitals. The outcome is contingent upon the relative alignment of the orbital and ligand.

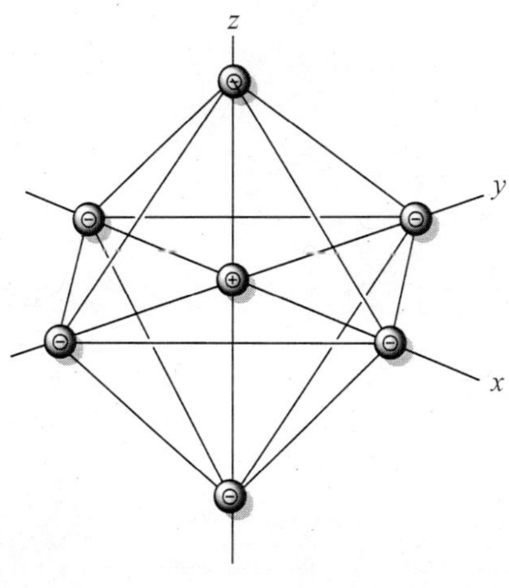

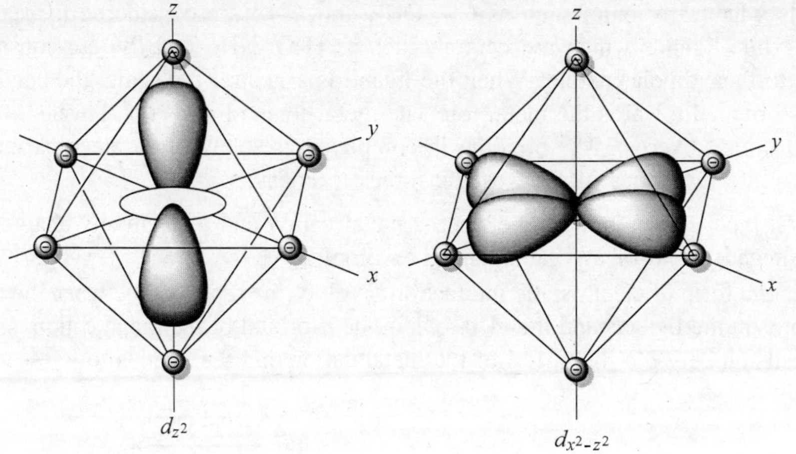

d_{z^2} $d_{x^2-z^2}$

The interaction between the ligands and the d_{xy}, d_{yz}, and d_{zx} orbitals is weaker due to the positioning of the ligands between the lobes.

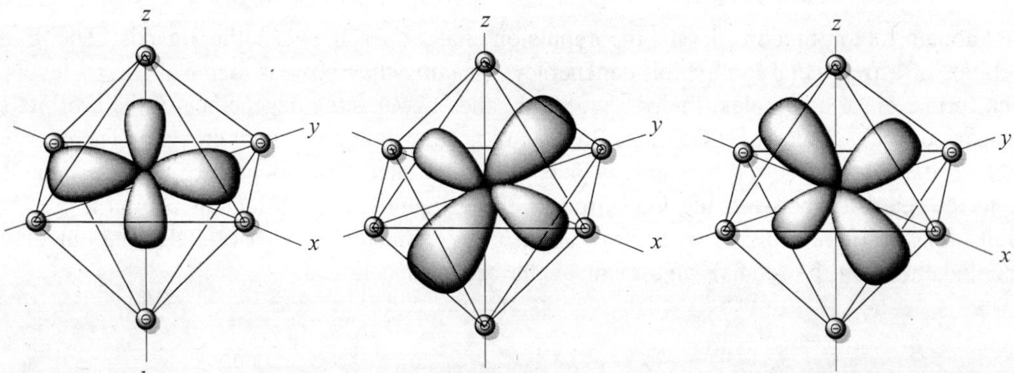

d_{xy} d_{yz} d_{xz}

To comprehend CFT, one must possess knowledge of the geometric characteristics and alignment of the five d-orbitals.

2.3 SPLITTING OF *d*-ORBITAL IN OCTAHEDRAL COMPLEX

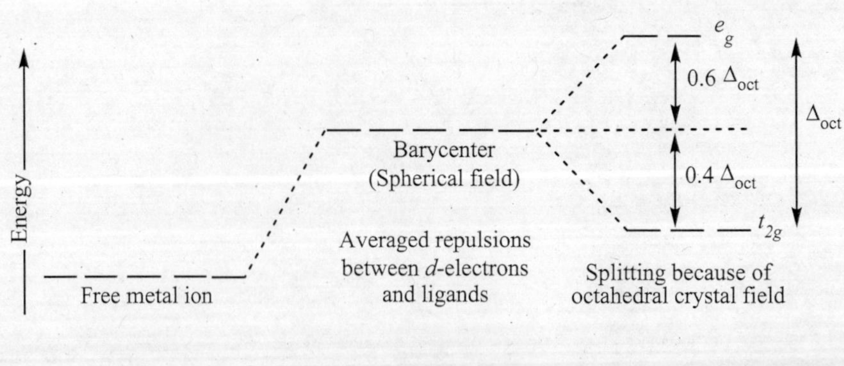

Crystal Field Splitting in Octahedral Complex

All five *d*-orbitals in a free metal cation are degenerate, meaning they have equal energy. In an octahedral complex, denoted as $[ML_6]^{n+}$, the metal cation is positioned at the central point of the octahedron, while the six ligands are located at the six corners. The six corners are oriented along the Cartesian coordinates, specifically the *x*, *y*, and *z*-axis. When the ligands are positioned infinitely far from the metal cation, the electrostatic field of the ligands will not have any effect on the degeneracy of the metal cation's five *d*-orbitals. As the ligands approach the metal cation, two electrostatic forces come into play:

1. The attractive force between the metal cation and the ligand, and
2. The repulsive force between the *d*-electrons of the metal cation and the lone pairs of the ligands.

The stronger the repulsion between the metal cation and ligands, the closer the ligands will be to the metal cation. Consequently, there will be more repulsion between the metal *d*-electrons and the lone pair of electrons on the ligand. When the ligands are in close proximity to the metal cation, an electrostatic repulsive force also arises between the ligands. These two repulsive forces result in the adoption of an octahedral geometry, which positions the ligand at the internuclear distance from the metal cation and maximizes the distance between them.

The repulsive interaction between the *d*-electrons of the metal and the electrons of the ligand leads to a rise in the potential energy of the metal's *d*-electrons. Keep in mind that the potential energy increases as the force of repulsion becomes stronger. If all six ligands uniformly surround the metal cation, they are arranged in a spherically symmetric manner, meaning that each of the six ligands is equidistant from each of the *d*-orbitals.

The energy of each *d*-orbital will be increased uniformly, resulting in the preservation of degeneracy among all five *d*-orbitals. This scenario is purely fictional and possesses the mean energy level of a group of *d*-orbitals. A really symmetrical field is never achieved in a real octahedral complex. Consequently, the five *d*-orbitals are not equally affected.

The two *d*-orbitals ($d_{x^2-y^2}$ and d_{z^2}) directly face the ligands, whereas the three *d*-orbitals (d_{xy}, d_{yz} and d_{zx}) are positioned between the paths of the incoming ligand. Consequently, $d_{x^2-y^2}$ and d_{z^2} orbitals will experience greater repulsion compared to the d_{xy}, d_{yz} and d_{zx} orbitals. Consequently, the energy levels of the $d_{x^2-y^2}$ and d_{z^2} orbitals will be elevated, while the energy levels of the other three orbitals, which are located farther from the ligand, will be reduced compared to the hypothetical energy state.

In a free metal cation, the five degenerate *d*-orbitals are now divided into two sets of *d*-orbitals with differing energies. The higher energy level consists of two orbitals ($d_{x^2-y^2}$ and d_{z^2}) with the same energy, while the lower energy level consists of three orbitals (d_{xy}, d_{yz}, and d_{zx}) with equal energy. The $d_{x^2-y^2}$ and d_{z^2} orbitals form the e_g set, which is characterized by double degeneracy. The d_{xy}, d_{yz}, and d_{zx} orbitals form the t_{2g} set, which is characterized by triple degeneracy.

Assuming that the distance between the metal cation and the ligands has remained unchanged, the total potential energy (or average energy) of the system must likewise remain constant, resembling that of the spherical field before it split. The state of average energy is commonly known as the barycenter.

Crystal field splitting refers to the division of five *d*-orbitals of a metal cation into two distinct sets with varying energy levels. The energy disparity between two sets of orbitals resulting from an octahedral field is quantified using the parameter Δ_0 or $10D_q$ (Differential of Quanta), with the "o" in Δ_0 representing octahedral.

Since the energy of barycentre remains constant, the total energy decrease of the t_{2g} set must be equal to the total energy increase of the e_g set. Therefore since there are two e_g orbitals, they must increase by $0.6\Delta_o$ or $6D_q$ and the three t_{2g} orbitals must decrease by $0.4\Delta_o$ or $4D_q$ per electron. The decreased energy t_{2g} of or e_g.

As the energy of the barycenter is constant, the total energy decrease of the t_{2g} set must be equivalent to the total energy increase of the e_g set. Hence, because to the presence of two e_g orbitals, they will experience an increment of $0.6\Delta_o$ or $6D_q$, whereas the three t_{2g} orbitals will see a reduction of $0.4\Delta_o$ or $4Dq$ per electron. The energy levels t_{2g} and e_g have been reduced.

$$\text{CFSE} = [-0.4 \times n(t_{2g}) + 0.6 \times n(e_g)]\ \Delta_o + nP$$

The variables $n(t_{2g})$ and $n(e_g)$ represent the number of electrons present in the corresponding energy levels. $n = 0, 1, 2, 3, \ldots$, P stands for Pairing. If the crystal field splitting energy (CFSE) is significantly greater than the pairing energy (P), electron pairing takes place. If the crystal field stabilization energy (CFSE) is relatively small, there is no pairing of electrons, meaning that the energy of the pairing (P) is greater than the CFSE.

2.4 FACTORS AFFECTING THE MAGNITUDE OF SPLITTING

The magnitude of splitting (Δ_o) of *d*-orbitals is influenced by many parameters related to the surrounding ligands.

1. Metal's oxidation state: The magnitude of splitting increases with the higher positive charge of the metal cation. As the oxidation state of the metal increases, the ligand moves closer to it, resulting in greater splitting of the metal's *d*-orbitals. For example, magnitude of splitting for $[Co(H_2O)_6]^{2+}$ complex is 9200 cm^{-1} whereas it is 20760 cm^{-1} for $[Co(H_2O)_6]^{3+}$

2. Same Oxidation state but Different *d*-electrons: Typically, for a complexes of 3*d* or 4*d* series of transition metals where the metal cation has the same oxidation state but a different number of *d*-orbitals electrons, the value of Δ_o drops as the number of *d*-electrons increases. The increased amount of *d*-electrons hinders the approach of ligands to the metal cation. E.g. magnitude of splitting for ($3d^7$) $[Co\ (H_2O)_6]^{2+}$ complex is 9200 cm^{-1} whereas it is 8500 cm^{-1} ($3d^8$) for $[Ni(H_2O)_6]^{2+}$.

3. Principal Quantum Number (*n*) of the *d*-orbital: For complexes with metal cations having identical oxidation states and the same number of *d*-electrons, the magnitude of the increase in size for homologous complexes within a specific group is approximately 30% to 50% when transitioning from 3*d* to 4*d*, and a similar rise occurs when transitioning from 4*d* to 5*d*. The reason for this is that

1. When transitioning from 3*d* to 4*d* and 4*d* to 5*d*, the size of the *d*-orbital expands while the electron density within them drops. Consequently, the ligands are able to come into closer proximity with the metal cation that possesses a bigger *d*-orbital.

2. A larger metal cation experiences reduced steric hindrance.

For example Δ_o for $[Co(NH_3)_6]^{2+}$ = 24800 cm^{-1} 163 kJ/mol

Δ_o for $[Rh(NH_3)_6]^{2+}$= 34100 cm^{-1} 213 kJ/mol

Δ_o for $[Ir(NH_3)_6]^{2+}$= 41200 cm^{-1} 314 kJ/mol

4. Behavior of Ligands: The ligands can be categorized as either weak or strong ligands. Ligands that induce a minor degree of splitting in *d*-orbitals are referred to as weak ligands, while those that induce a significant splitting are referred to as strong ligands. The ligands have been organized in ascending order based on their crystal field splitting strength, which induces the splitting of *d*-orbitals. This arrangement was determined by studying the impact of these ligands on the spectra of transition metal ions. Weak field ligands, which have a crystal field splitting energy (CFSE) in the range of 7000–30000 cm^{-1}, create a tiny splitting. On the other hand, ligands that have a high splitting energy cause a significant splitting. Powerful ligands with high crystal field splitting energy (usually greater than 30000 cm^{-1}).

(Weak end) O_2^{2-} < I^- < Br^- < S^{2-} < SCN^- (S–bonded) < Cl^- < N_3^- < F^- < NCO^- < OH^- < $C_2O_4^{2-}$ < H_2O < NCS^- (N–bonded) < CH_3CN < gly (glycine) < py (pyridine) < NH_3 < en (ethylenediamine) < bipy (2,2'-bipyridine) < phen (1,10-phenanthroline) < NO_2^- < PPh_3 < CN^- < CO < CH_2 (strong end)

This order is usually called as **Spectrochemical series.** The relative field strength of common ligands remains consistent regardless of the specific metal cation or complex shape. The impact of ligands on complexes of certain metal is depicted below with the value of magnitude of splitting.

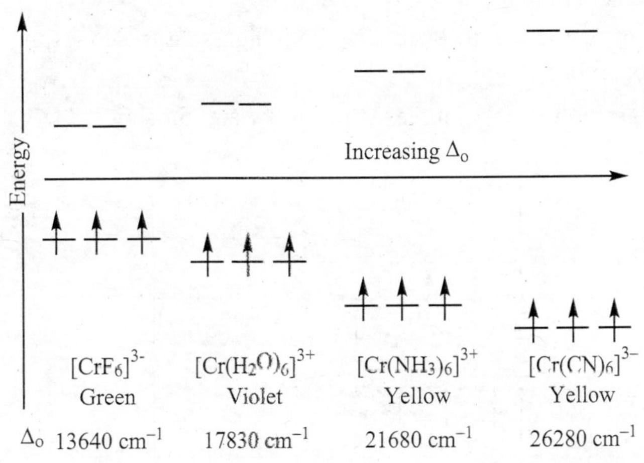

5. Number of Ligands: The crystal field splitting (Δ_o) grows proportionally with the number of ligands. As an illustration, the value of CFSE for octahedral is greater than the value of CFSE for tetrahedral ($\Delta_o > \Delta_t$). While square planar complexes have less ligands compared to octahedral complexes, the value of Δ_{sp} is larger than Δ_o. Square planar complexes are generated by strong ligands with d8 metal cations from the 3*d*-series transition metals, as well as the 4*d* and 5*d* series d^8 metal cations, which can have either weak or strong ligands. The heightened crystal field splitting is attributed to the more potent ligands and transition metal cations from the $4d$ or $5d$ series. In square planar complexes of d^8 metal cations, the d_{z^2} orbital, which contains two electrons, is stabilized, while the unoccupied $d_{x^2-y^2}$ orbital is destabilized.

The increasing order of CFSE value for different complexes.

$$\Delta_t < \Delta_{sp} < \Delta_o$$

The Δ_t and Δ_o is realted as

$$\Delta_t = \frac{4}{9}\Delta_o$$

and Δ_{sp} and Δ_t is realted as

$$\Delta_{sp} = \Delta_t + \Delta_t + \Delta_t$$

Now

$$\Delta_{sp} = 1.33\Delta_o = 1.33 \times \frac{9}{4}\Delta_t$$

2.5 PAIRING ENERGY

Pairing energy refers to the amount of energy needed to compel the two unpaired electrons within a single orbital to pair up. The mean pairing energy, P, is obtained when several electrons are linked together. It can be acquired by the examination of electronic spectra.

1. If the value of Δ_o is greater than $P(\Delta_o > P)$, it indicates a preference for the low spin complex.
2. If the value of Δ_o is less than $P(\Delta_o < P)$, it indicates a preference for the formation of the high spin complex.
3. If the value of Δ_o is equal to $P(\Delta_o = P)$, both high spin and low spin complexes coexist in equal proportions.

Typically, for transition metal complexes in the $4d$ and $5d$ series, the value of Δ_o is larger than that of P.

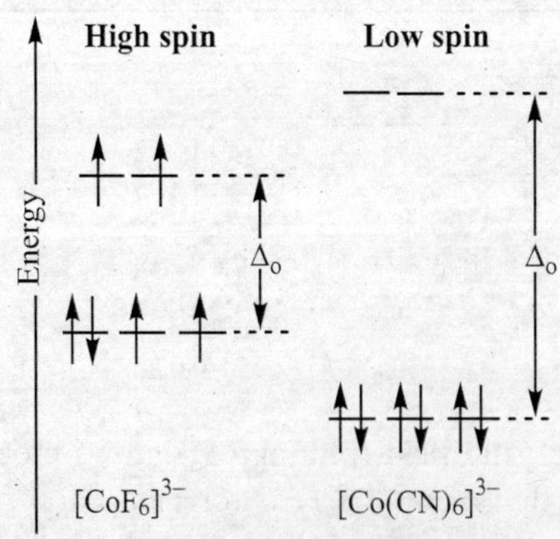

In a weak field octahedral complex of a transition metal from the $3d$-series with an oxidation number less than or equal to +3, the value of Δ_o is negligible, resulting in the absence of electron pairing. Thus, in complexes with d^4, d^5, d^6, and d^7 configurations under weak fields, electron pairing does not occur. Complexes that possess the highest possible number of unpaired electrons are referred to as high spin or spin free complexes. The term "high spin" or "spin-free" is employed to describe these complexes, as they possess the same spin state as the *d*-orbitals of unbound metal cations. The value of Δ_o is significant in octahedral complexes of transition metals from the 3*d*-series with an oxidation number larger than or equal to +2. In complexes with strong magnetic fields and d^4, d^5, d^6, and d^7 configurations, the *d*-electrons will couple up according to Hund's rule.

Complexes that possess the highest possible number of paired electrons are referred to as low spin or spin paired complexes. The term "low spin" or "spin paired" is employed to describe complexes that possess a greater number of paired electrons (or spin) compared to the free metal cation. It should be noted that octahedral complexes with weak fields are always low spin complexes. The metal cations in the 3*d*-transition series, with an oxidation number larger than or equal to +4, as well as the cations in the 4*d* and 5*d* transition metal series, always tend to form low spin complexes when paired with weak ligands. As an illustration, the $[NiF_6]^{2-}$ ion, with a nickel oxidation state of +4, exhibits low spin and diamagnetism, despite the weak ligand properties of F^-. The complex $[Rh(H_2O)_6]^{3+}$ exhibits low spin and diamagnetic properties, while being a weak ligand. An anomaly is noticed in the case of transition metals belonging to the 3*d*-series, where Co^{3+} ions tend to form low spin complexes with H_2O and O^{2-} ions, despite the fact that H_2O and O^{2-} are considered weak ligands.

2.6 PROPERTIES OF TRANSITION METAL COORDINATION COMPLEXES

2.6.1 Magnetic Moment

The spin-only magnetic moment, denoted as μ, of a complex having a total spin quantum number, can be expressed as:

$$\mu = \sqrt{S(S+1)}\mu_B \qquad \mu_B = \text{Bohr magneton}$$

This method is employed to ascertain the quantity of unpaired spins present in a complex, hence enabling the identification of its ground-state structure. Compounds are categorized as diamagnetic if they experience a repulsive force when exposed to a magnetic field, and paramagnetic if they experience an attractive force when exposed to a magnetic field.

Tbale calculated spin-only magnetic moments.

Ion	**N**	**S**	**μ/μ_B**	
			Cal.	**Exp.**
Ti^{3+}	1	½	1.73	1.7–1.8
V^{3+}	2	1	2.83	2.7–2.9
Cr^{3+}	3	1½	3.87	3.8
Mn^{3+}	4	2	4.90	4.8–4.9
Fe^{3+}	5	2½	5.92	5.9

2.6.2 Color

The color of a metal complex is influenced by its ligands. "Charge transfer" can result in vibrant coloration. Electrons from the ligand transition to the orbitals of the metal. Chlorine does not possess *d*-orbitals. Instead, it has orbitals that are of higher energy. Light can undergo reflection or absorption when interacting with an object. When no light is absorbed and all light is reflected, the resulting color is white. An object appears black when it absorbs all light and reflects none. What happens if only one hue is absorbed? Here are some example of colored compounds.

$KMnO_4$ $KCrO_4$ $KClO_4$

Complimentary Color Wheel

When a color is absorbed, the opposite color is perceived. Assimilate the color orange, perceive the color blue. Assimilate the color red, perceive the color green. The chemical formula $Ti(H_2O)$ exhibits absorption in the green regions and resulting in a perceived purple color. A d^1 complex is formed when there is just one electron (e^-) in the system. This electron occupies the lowest energy orbital, specifically one of the three degenerate t_{2g} orbitals. The purple color is caused by the absorption of light, which leads to the elevation of the t_{2g} electron to the e_g level $t_{2g}^1 e_g^0 \Rightarrow t_{2g}^0 e_g^1$. Which is illustrated as follows:

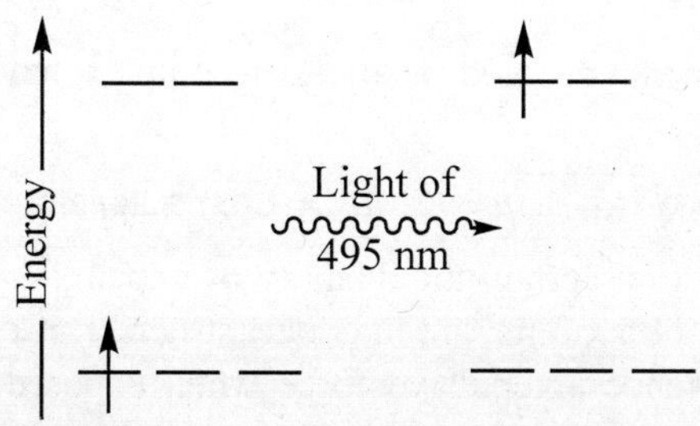

The UV-Vis absorption spectra indicates that this transition takes place at a peak of 20300 cm^{-1}, which corresponds to an energy difference (Δ_o) of 243 kJ/mol. The conversion factor for 1000 cm^{-1} is 11.96 kJ/mol or 2.86 kcal/mol or 0.124.

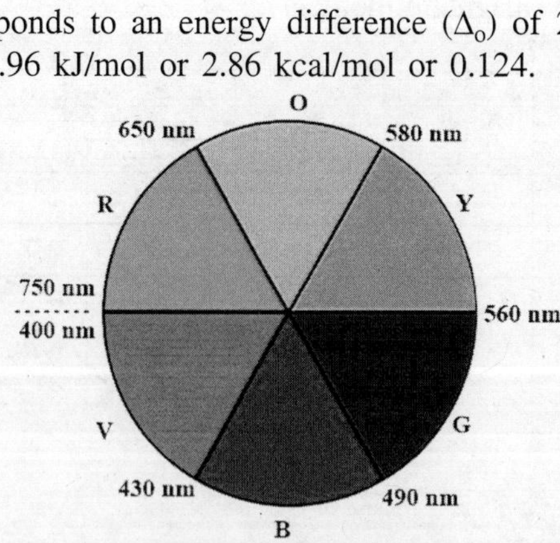

2.7 APPLICATION OF CFSE

Enthalpy of Hydration

Hydration refers to the process in which water molecules gather from the ion when one mole of an ionic crystal is dissolved in water. During this process, a certain quantity of energy is liberated, known as hydration energy. Let us examine the changes in enthalpy for M^{2+} ions.

$$M^{2+}(g) + 6\,H_2O(l) = \left[M(H_2O)_6\right]^{2+}(aq)$$

Calcium ions (Ca^{2+}), manganese ions (Mn^{2+}), and zinc ions (Zn^{2+}) have electronic configurations of d^0, d^5, and d^{10}, respectively. As a result, the crystal field splitting energy (CFSE) for these ions is zero. Additional crystal field splitting energy (CFSE) causes other metal ions to depart from the anticipated linear trend.

The hydration energy of a metal cation rises when the effective nuclear charge increases and the ionic radii drop. These two conditions induce the water molecules to approach the metal cation more closely, leading to an enhanced electrostatic attraction between the metal cation and the water molecule. For dipositive transition metal cations of the 3d-series, the effective nuclear charge increases and the ionic radii drop as you move across a period. The hydration energy should exhibit a consistent increase as we move from Ca^{2+} to Zn^{2+}.

So,
$$\text{Hydration energy} \propto \frac{\text{charge of the cation}}{\text{size of the cation}}$$

For instance, the hydration energy of Co^{2+} is less than that of Co^{3+} ($Co^{2+} < Co^{3+}$). Due to the equal size of both ions, the ion with a larger charge exhibits greater hydration energy.

Lattice Energy

Lattice energy refers to the energy generated when a mole of an ionic crystal is created from its constituent gaseous ions. Lattice energy refers to the amount of energy needed to separate one mole of an ionic crystal into its constituent gaseous ions. As per Born-Lande's equation, the lattice energy of an ionic crystal rises as the product of Z^+ and Z^- grows, and as the interionic distance (r_0) decreases. The lattice energy of the halides of dipositive metal ions in the 3*d*-series transition element is expected to grow in a linear fashion from Ca^{2+} to Zn^{2+} ions.

Ionic Radii of Divalent Metal Ions of 3d-series Transition Element

The ionic radii of divalent and trivalent metal cations belonging to the 3*d*-series transition metals are likely to decrease in a consistent manner from Ca^{2+} to Zn^{2+} when placed in a low spin or high-spin octahedral field. The cause for this phenomenon is the enhanced electrostatic attraction between metal cations and ligands, resulting from the augmented effective nuclear charge and reduced shielding effect of *d*-electrons. Consequently, the ligands and metal cations are brought into closer proximity.

Evidence of Metal-ligand Covalent Bonding in Complexes

$Cr(CO)_6$ is an easily evaporating molecule, while $Ni(CO)_4$ is in a liquid state. This suggests the presence of a covalent link between the metal and ligand, rather than an ionic bond. If an ionic bond were present, $Cr(CO)_6$ would exhibit non-volatility, while $Ni(CO)_4$ would exist as a solid.

Electron Spin Resonance (ESR)

The ESR spectrum of $[IrCl_6]^{2-}$ indicates that the solitary unpaired electron is predominantly localized on the metal atom, with 70% of its localization, while the remaining 30% is localized on the chloride ion. This suggests that there is electron sharing and therefore some degree of covalent bonding between the metal and ligands.

The Nuclear Magnetic Resonance (NMR)

The fluoride NMR has identified the electron delocalization in the fluoro complexes of the paramagnetic metal ion. It is only conceivable when an unpaired electron spends a significant amount of time on the ^{19}F nucleus.

The Nephelauxetic Effect

The presence of electronic repulsion within the *d*-orbitals of transition metal cations results in the formation of several energy levels, which are determined by the specific arrangement of electrons within the *d*-orbitals. The energy discrepancy between two energy levels can be mathematically represented by interelectronic repulsion parameters, specifically known as Racah parameters B and C. The energy difference between two levels with identical spin multiplicities can be solely represented in terms of B, while the energy difference between two levels with differing spin multiplicities can be expressed in terms of both B and C. Experimental observations, specifically from electronic spectra of complexes, indicate that the magnitude of B and C decreases during complex formation. The decreased values of B and C suggest a decrease in electron density on the metal cation, indicating the delocalization of the electron cloud over both the metal cation and the ligands. This implies the presence of covalent bonding between the metal cation and ligands. As the values of B and C decrease, the electron cloud becomes more spread out and the covalency increases. The phenomenon in which the electron cloud is spread out over both the metal cation and the ligand is referred to as the nephelauxetic effect.

Nuclear Quadrupole Resonance (NQR)

The NQR spectrum of some complexes with halide ions as ligands, such as $[PtCl_4]^{2-}$, $[PdCl_4]^{2-}$, indicates that the metal-ligand connection exhibits a combination of ionic and covalent character.

2.8 LIMITATIONS OF CRYSTAL FIELD THEORY

The theory of crystal field was formulated by Hans Bethe and Van Vleck in 1935. This idea is founded on the premise that the interaction between a metal ion and ligands is only of an electrostatic character. As the ligands approach the center metal atom or ion, the five degenerated *d*-orbitals of the central atom undergo a process called splitting, where they separate into distinct energy levels due to the electrostatic field created by the ligands.

1. Crystal field theory effectively elucidates the structural, magnetic, color, electronic, thermodynamic, and kinetic characteristics of complexes. Nevertheless, this theory possesses significant constraints.
2. In CFT, only the *d*-electrons of the metal ion are addressed, whereas other orbitals such as *s* are disregarded.

3. This idea has overlooked the covalent nature present in transition metal complexes. It considers the metal-ligand interaction to be solely ionic.
4. The crystal field theory (CFT) is unable to account for the varying strengths of ligands as presented in the spectrochemical series. Specifically, it fails to explain why H_2O is a more potent ligand than the OH^- ion.
5. Furthermore, CFT has not taken into account the Pi bonding in complexes, regardless of whether it is metal to ligand or ligand to metal.
6. This theory is irrelevant to the ligands' orbits. As a result, it is incapable of elucidating any characteristics pertaining to ligand orbitals and their interaction with metal orbitals.
7. The explanation of the influence of π bond on Δ_o is lacking.
8. Compounds in which the metal is in a zero oxidation state and the ligand is neutral lack electrostatic interaction between the metal and the ligands.

2.9 TETRAHEDRAL COMPLEXES

Tetrahedral compounds exhibit orbital inversion. The reason for this is once again due to the orientation of orbitals and ligands. The value of Δ is consistently modest and the spin state is consistently low, indicating a lower number of ligands. Additionally, there is a division between tetrahedral and square planar arrangements.

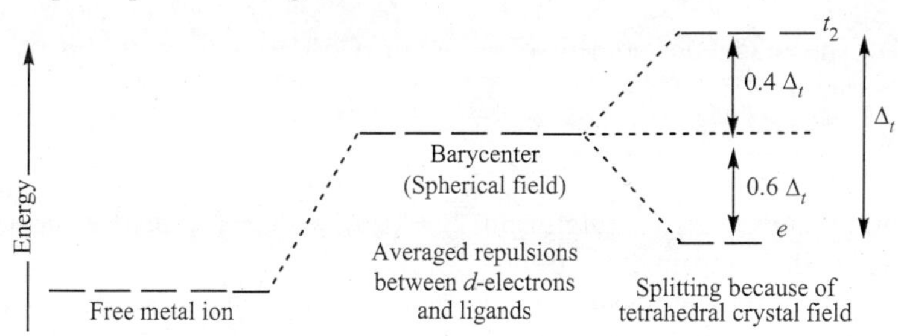

Imagine a tetrahedral molecule surrounded by a cube, with metal ions located in the center of the cube. The ligands are situated in the four alternate corners of the cube, while the remaining four corners are empty. The two 'e' orbitals are aligned towards the centroid of the cube's face, whereas the three 't_2' orbitals are oriented towards the centroid of the cube's edges. Consequently, the angle created by the e-orbitals, metal, and ligand is exactly half of the tetrahedral angle, which is equal to 54°44'. The angle formed between the t_2-orbitals, metal, and ligand is precisely one-third of the tetrahedral angle, measuring 35°16'. This is equivalent to 109°28'/3. The t_2 orbitals are located in closer proximity to the direction of ligand approach in comparison to the e orbitals. Thus, t_2 orbitals have more energy than e-orbitals.

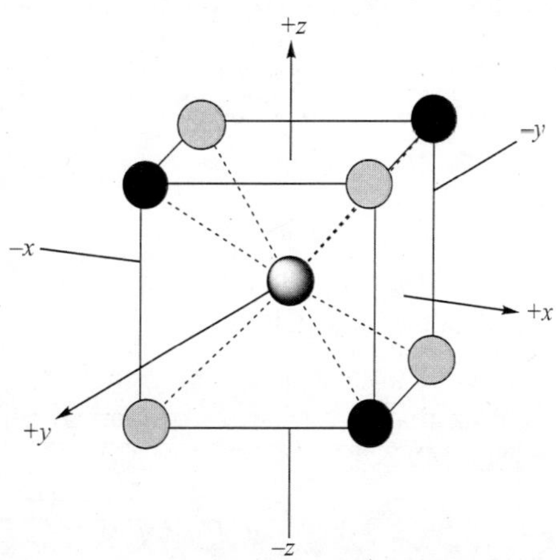

$\Delta_t < \Delta_o$; $\Delta_t = 4/9\ \Delta_o$

The tetrahedral complex contains a total of four ligands, resulting in a ligand field that is approximately two-third the size of the octahedral field. The orientation of ligand approach in a tetrahedral compound does not align with the *d*-orbitals. This decreases the field by a ratio of 2/3. Thus, the value of Δ_t is approximately equal to 4/9 of Δ_o, calculated by multiplying 2/3 by 2/3. Consequently, all tetrahedral complexes exhibit high-spin behavior due to the fact that the crystal field stabilization energy (CFSE) is typically lower than the pairing energy. Therefore, it is uncommon to observe low spin combinations. The square planar form is typically preferred when a highly potent field ligand is present.

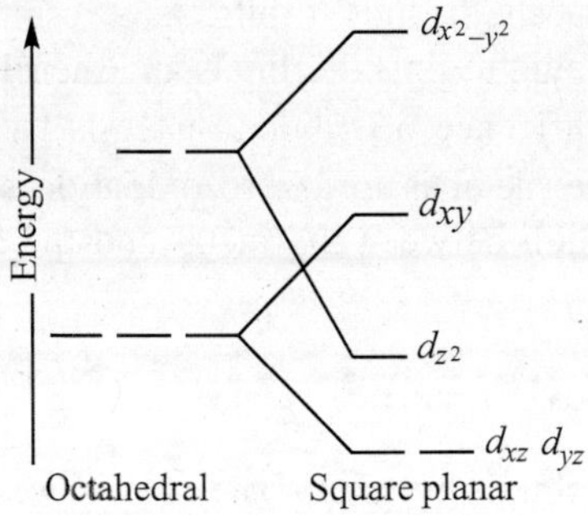

SOLVED PROBLEMS

PROBLEM 2.1 The magnetic moment of a certain Co(II) complex is 4.0 μ_B. What is its *d*-electron configuration?

Solution

A Co(II) complex is d^7.

Two possible configurations: $t_{2g}^5 e_g^2$ (high-spin, S = 1½) with 3 unpaired electrons or $t_{2g}^6 e_g^1$ (Low-spin, S = ½) with 1 unpaired electron.

The spin-only magnetic moments are 3.87 μ_B and 1.73 μ_B. Therefore, the only consistent assignment is the high-spin configuration $t_{2g}^5 e_g^2$.

PROBLEM 2.2 $[Fe(CN)_6]^{3-}$ should be an inner orbital complex. Ignoring the pairing energy, the value of crystal field stabilization energy for this complex is (–) ____________ Δ_o.

Solution Fe is present in Fe(III) (+3) oxidation state in $[Fe(CN)_6]^{3-}$. This complex is inner orbital complex with d^5 (with pairing). So its configuration is t^5_{2g}. So

$$\text{CFSE} = 5 \times \frac{-2}{5}\Delta_o = -2\Delta_o$$

PROBLEM 2.3 What is the difference between spin only magnetic moment values of $[Co(H_2O)_6]Cl_2$ and $[Cr(H_2O)]Cl_3$.

Solution The electronic configuration of Co is: $4s^2 3d^7$. So Co^{2+} has $3d^7$ electronic configuration. We know that as per spectrochemical series H_2O is weak field ligand. Number of unpaired electrons (n) = 3

$$\therefore \quad \mu_1 = \sqrt{n(n+2)} = \sqrt{15}\ \text{BM}$$

The electronic configuration of Cr is: $4s^1 3d^5$. So Cr^{3+} has $3d^3$ electronic configuration.

$$\text{Number of unpaired electrons } (n) = 3,\ \mu_2 = \sqrt{n(n+2)} = \sqrt{15}\,\text{BM}$$

So, the difference between spin only magnetic moment values = $\mu_2 - \mu_2 = 0$

PROBLEM 2.4 What is the spin only magnetic moment (μ_s) of $K_3|Fe(CN)_6]$.

Solution In $K_3|Fe(CN)_6]$ CN have –1 oxidation state. Let oxidation state of Fe in the complex is *x*. So $x - 6 = -3 \Rightarrow x = +3$. The electronic configuration of Fe is: $4s^2 3d^6$. So Fe^{3+} has $3d^5$ electronic configuration. We know that as per spectrochemical series CN is strong field ligand. Number of unpaired electrons $(n) = 1$

$$\therefore \qquad \mu_s = \sqrt{n(n+2)} = \sqrt{1(1+2)} = 1.7\,\text{BM}$$

PROBLEM 2.5 If $[Cu(H_2O)_4]^{2+}$ absorbs a light of wavelength 600 nm for d-d transition, then the value of octahedral crystal field splitting energy for $[Cu(H_2O)_6]^{2+}$ will be ...... $\times 10^{-21}$ J. [Nearest Integer] (Given: $h = 6.63 \times 10^{-34}$ Js and $c = 3.08 \times 10^8$ ms^{-1})

Solution $[Cu(H_2O)_4]^{2+}$ is tetrahedral complex and $[Cu(H_2O)_6]^{2+}$ is octahedral complex.

$$\therefore \qquad \Delta_t = \frac{4}{9} \times \Delta_O$$

$$\Delta_t = \frac{6.63 \times 10^{-34} \times 3.08 \times 10^8}{600 \times 10^{-9}}$$

$$\Delta_o = \frac{9}{4} \times \frac{6.63 \times 10^{-34} \times 3.08 \times 10^8}{600 \times 10^{-9}} \approx 765 \times 10^{-21}\,\text{J}$$

PROBLEM 2.6 Amongst $[Fe(H_2O)_3Cl_3]$, $K_3[Fe(CN)_6]$ and $[Co(NH_3)_6]Cl_3$, the spin-only magnetic moment value of the inner-orbital complex that absorbs light at shortest wavelength is ............B.M. [nearest integer]

Solution The complex $[Fe(H_2O)_3Cl_3]$, $K_3[Fe(CN)_6]$ and $[Co(NH_3)_6]Cl_3$ is outer-orbital complex, inner-orbital, and inner-orbital complex. Since CN^- is a strong field ligand than NH_3. Hence $K_3[Fe(CN)_6]$ is the inner-orbital complex that absorbs light at shortest wavelength. The electronic configuration of Fe is: $4s^2 3d^6$. So Fe^{3+} has $3d^5$ electronic configuration. Since CN^- will do pairing, so number of unpaired electron $(n) = 1$.

$$\therefore \qquad \mu_s = \sqrt{n(n+2)} = \sqrt{1(1+2)} = 1.7\,\text{BM} \approx 2\,\text{BM}$$

PROBLEM 2.7 Reaction of $[Co(H_2O)_6]^{2+}$ with excess ammonia and in the presence of oxygen results into a diamagnetic product. Number of electrons present in t_{2g}-orbitals of the product is.........

Solution The following reaction is involved when reaction of $[Co(H_2O)_6]^{2+}$ with excess ammonia and in the presence of oxygen results into a diamagnetic product.

$$[Co(H_2O)_6]^{2+} \xrightarrow[O_2]{NH_3} [Co(NH_3)_6]^{3+} + e^-$$

The electronic configuration of Co is: $4s^2 3d^7$. So Co^{3+} has $3d^6$ electronic configuration. Since NH_3 is a strong field ligand, so number of unpaired electron $(n) = 0$. $3d^6 \rightarrow t_{2g}^6 e_g^0$

PROBLEM 2.8 The spin-only magnetic moment value of an octahedral complex among $CoCl_3.4NH_3$, $NiCl_2.6H_2O$ and $PtCl_4.2HCl$, which upon reaction with excess of $AgNO_3$ gives 2 moles of AgCl is.........

Solution

$$CoCl_3.4NH_3 \xrightarrow[\text{excess}]{AgNO_3} [Co(NH_3)_4 \cdot Cl_2] + AgCl$$

$$NiCl_2.6H_2O \xrightarrow[\text{excess}]{AgNO_3} [Ni(H_2O)]^{2+} + 2AgCl$$

$$PtCl_4.2HCl \longrightarrow [PtCl_6]^{4-} + \text{No AgCl ppt}$$

$$[Ni(H_2O)]^{2+} = d^8 = t_{2g}^6 e_g^2 = 2 \text{ unpaired electrons}$$

$$\text{Magnetic moment} = \sqrt{2(2+2)} = 2\sqrt{8}»3$$

PROBLEM 2.9 Acidified potassium permanganate solution oxidises oxalic acid. The spin-only magnetic moment of the manganese product formed from the above reaction is.......... B.M. (Nearest integer)

Solution

$$KMnO_4(\text{acidic medium}) + H_2C_2O_4 \longrightarrow CO_2 + Mn^{2+}$$

Mn^{2+} has 5 unpaired electrons.

$$\text{Spin only magnetic moment} = \sqrt{5(5+2)} = \sqrt{5 \times 7} = \sqrt{35} \approx 5.92\,BM \cong 6\,\text{BM}$$

PROBLEM 2.10 Spin only magnetic moment of $[MnBr_6]^{4-}$ is...... B.M. (round off to the closest integer

Solution

$$Mn^{2+} \longrightarrow t_{2g}^3 e_g^2 \text{ (five unpaired electrons)}$$

$$\mu_s = \sqrt{5(5+2)} = \sqrt{5 \times 7} = \sqrt{35} = 5.91 \approx 6$$

QUESTIONS

2.1 Explain $[Co(NH_3)_6]^{3+}$ is an inner orbital complex while $[Ni(NH_3)_6]^{2+}$ is an outer orbital complex. Also comment on the magnetic properties of both complexes.

2.2 What are the magnetic properties of transition metals? How does CFT explain magnetic properties?

2.3 What are the main postulates of crystal field theory (CFT)? Discuss crystal field splitting in $[Co(NH_3)_6]^{2+}$ complex ion.

2.4 What is crystal field theory? How does this theory account for the fact that $[CoF_6]^{3-}$ is paramagnetic but $[Co(NH_3)_6]^{3+}$ is diamagnetic though both are octahedral.

2.5 The magnetic moment of the complex $[Mn(NCS)_6]^{4-}$ is 6.06 μ_B. What is its electron configuration?

CHAPTER 3

Bonding in Metals (Band Theory)

3.1 INTRODUCTION

An effective method for illustrating the distinction between conductors, insulators, and semiconductors is to graph the accessible energy levels for electrons within these materials. The energy levels in the case of unbound atoms are discrete, whereas in this scenario, they form bands.

3.1.1 Electronic Properties

Properties related to the behaviour and characteristics of electronic systems and devices. The final band that is entirely occupied (at a minimum temperature of 0 K) is referred to as the Valence Band. The subsequent band possessing greater energy is referred to as the Conduction Band. The Conduction Band can either be unoccupied or partially occupied. The Band Gap, also known as the Forbidden Gap, refers to the energy differential between the bottom of the Conduction Band (CB) and the top of the Valence Band (VB).

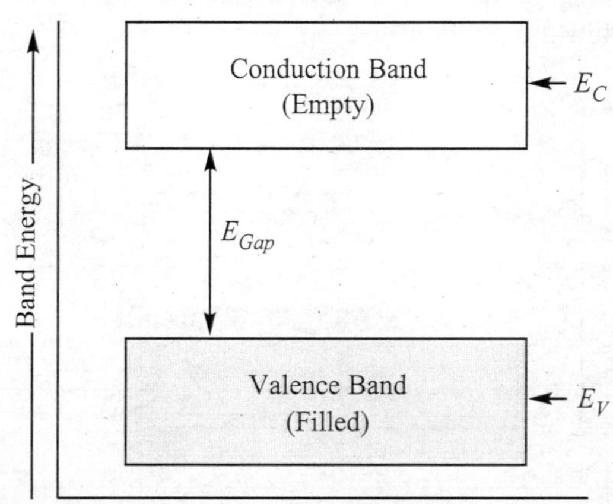

3.2 ENERGY BANDS FOR SOLIDS

The Fermi level, which represents the highest energy level of accessible electrons at low temperatures, is a crucial parameter in the band theory. The relative position of the Fermi level with respect to the conduction band is a critical determinant of electrical characteristics.

3.3 ENERGY BANDS IN INSULATORS

The majority of solid substances exhibit insulating properties, which can be explained by the band theory of solids. According to this theory, there exists a significant prohibited gap between the energy levels of the valence electrons and the energy level at which electrons can easily travel within the material (known as the conduction band).

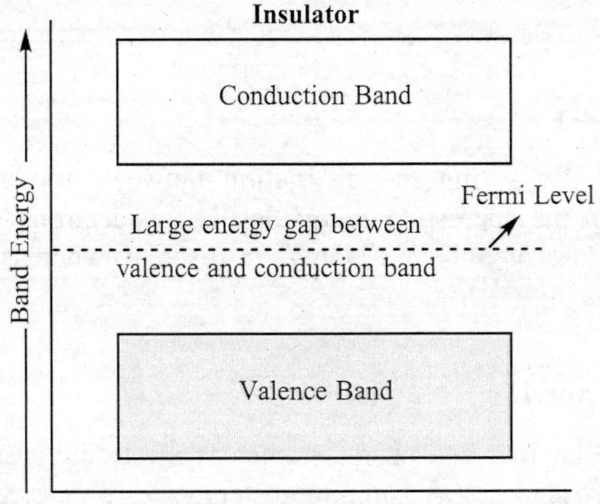

Metals possess a distinctive characteristic among solids according to the band theory, as they exhibit excellent electrical conductivity. This phenomenon can be attributed to the fact that their valence electrons are basically unbound. In the band theory, the depiction involves an overlap between the valence band and the conduction band, allowing a portion of the valence electrons to be mobile within the material.

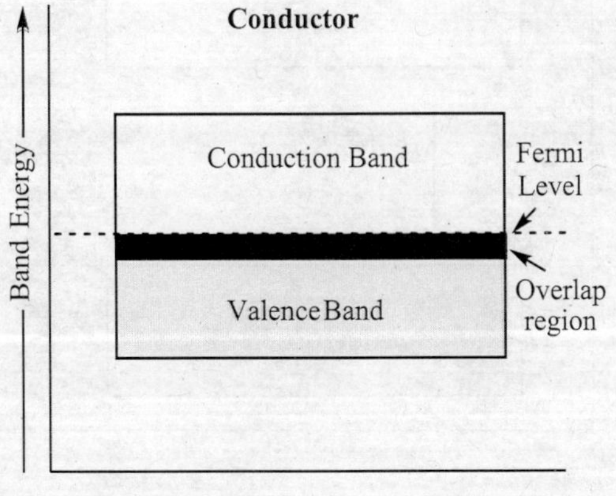

3.4 SEMICONDUCTOR

A semiconductor is a substance that has electrical properties that fall between those of insulators and conductors. Two examples of elements that can be used are germanium and silicon. Semiconductors are materials characterized by an almost empty conduction band and an almost filled valence band, with a very narrow energy gap (about 1 eV) between the two.

3.4.1 Types of Semiconductors

The different types of semiconductors are illustrated below:

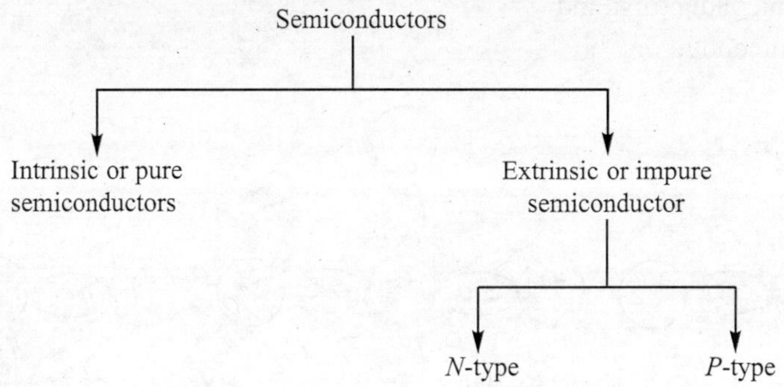

3.4.1.1 Intrinsic Semiconductors

An intrinsic semiconductor is composed of semiconductor material in a highly purified state. Two examples of semiconductors with specific energy gaps are pure germanium and silicon, which have forbidden energy gaps of 0.72 eV and 1.1 eV, respectively. At typical ambient temperature, the energy gap is sufficiently narrow that numerous electrons have enough energy to traverse the short gap between the valence and conduction bands. An intrinsic semiconductor can be described as a semiconductor where the quantity of conduction electrons is equivalent to the quantity of holes. The figure below displays the energy band diagram of an intrinsic semiconductor at ambient temperature.

Semiconductor

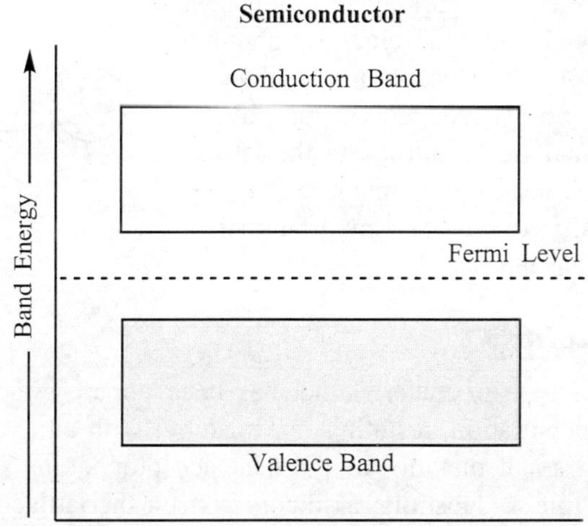

The "Fermi level" refers to the highest energy level of the electron collection at a temperature of absolute zero. The maximum energy level that an electron can occupy in the valence band at absolute zero temperature is referred to as the Fermi energy.

3.4.1.2 Extrinsic Semiconductors

Extrinsic or impurity semiconductors refer to intrinsic semiconductors that have been modified by adding a very small amount (about 1 part in 10^8) of an appropriate impurity or doping substance. Extrinsic semiconductors can be classified into two categories based on the type of doping material employed:

1. *N*-type semiconductors, and
2. *P*-type semiconductors.

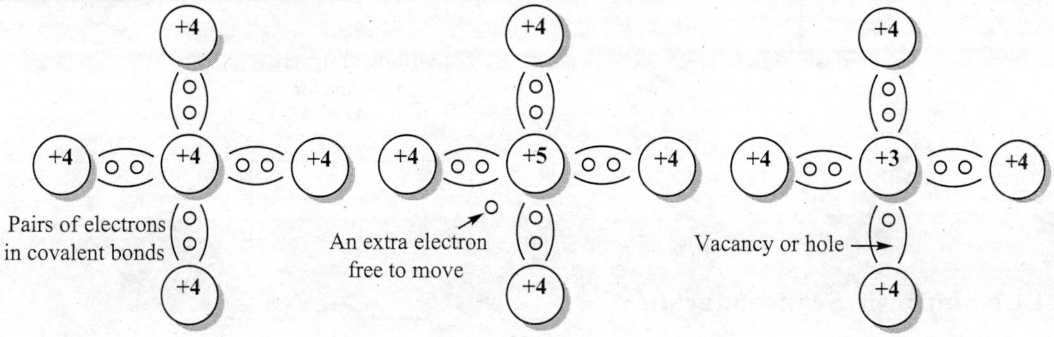

N-type *P*-type

1. *N-type Semiconductor*

N-type semiconductor refers to a type of semiconductor material that has an excess of negatively charged electrons, allowing for the flow of electric current. By introducing a small quantity of pentavalent donor atoms such as phosphorus (P) and arsenic (As), it is possible to substitute a silicon atom in the lattice with a donor atom. In this substitution, four of the donor atom's valence electrons participate in covalent bonds, while one additional electron remains unbound. This is an *N*-type semiconductor, which exhibits enhanced conductivity compared to intrinsic semiconductors, owing to the presence of additional free electrons in the lattice. These electrons are referred to as predominant or majority current carriers. Additionally, there are a small number of holes known as minority carriers.

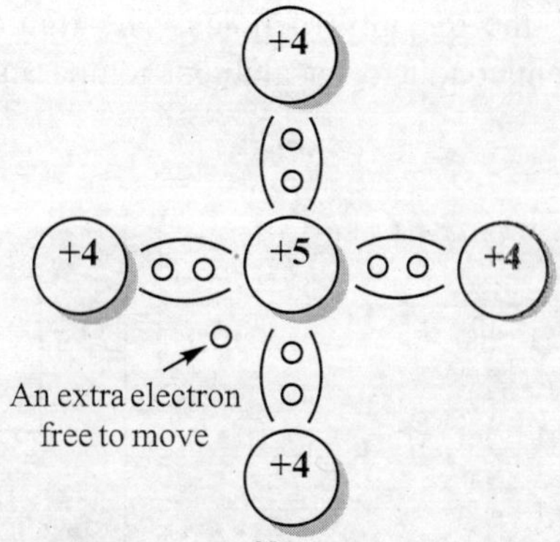

N-type

2. *P-type Semiconductor*

P-type semiconductor is a type of material that has been doped with impurities to create a positive charge carrier concentration, resulting in a material with an excess of positive charge carriers. By introducing a small quantity of trivalent acceptor atoms such as boron (B) and aluminium (Al), it is possible to substitute a silicon atom in the lattice with an acceptor atom that possesses only three valence electrons. This substitution results in the formation of three covalent bonds and creates a void in the lattice structure, commonly referred to as a hole. This is a *P*-type semiconductor with enhanced conductivity compared to intrinsic semiconductors, attributed to the presence of prominent or majority current carriers known as holes in the lattice. Additionally, there is a small quantity of free electrons known as minority carriers.

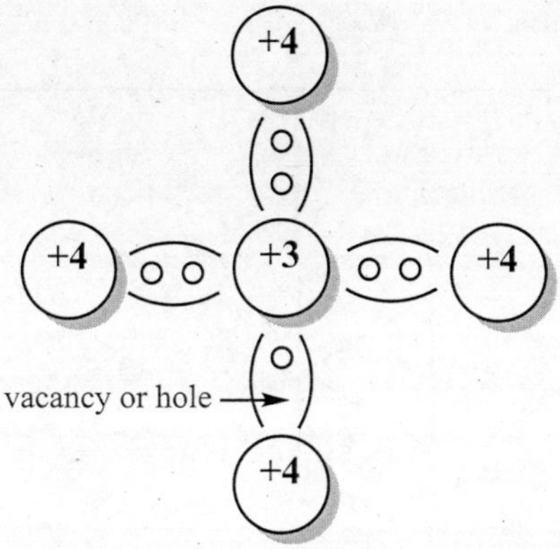

P-type

Comparison of *P*- and *N*-type Semiconductors

N-type Semiconductor	*P*-type Semiconductor
1. Containing pentavalent atoms as impurities. Electrons constitute the predominant charge carriers. Holes are charge carriers that are present in lower numbers compared to other types of charge carriers.	1. Containing trivalent atoms as impurities. Majority charge carriers in a material are referred to as holes. Electrons function as charge carriers with a minority status.
2. It gives out electrons, therefore being referred to as donor atoms.	2. It has the ability to accept electrons, so it is referred to as acceptor atoms.
3. The elements utilized for doping are Bismuth, Antimony, Arsenic, and Phosphorus.	3. The elements utilized for doping are Aluminium, Gallium, Indium, and Boron.

3.5 P-N JUNCTION DIODE

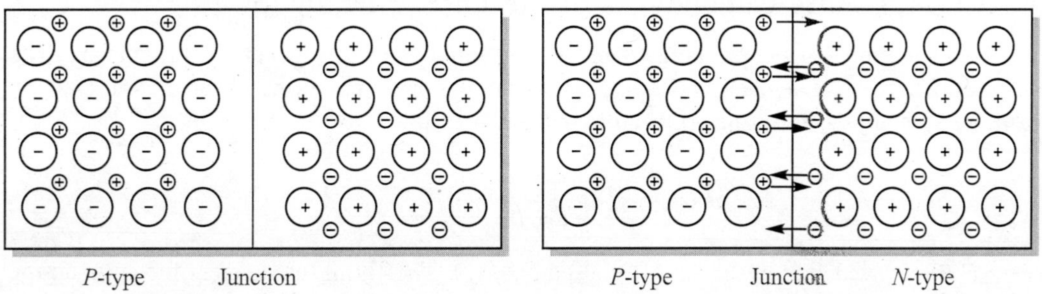

Construction

These devices are comprised of a P-N junction that is created in either a Germanium (Ge) or Silicon (Si) crystal. The circuit sign is depicted in Figure below. The regions designated as *P* and *N* types are commonly known as the anode and cathode, respectively. In figure below, the arrowhead denotes the typical direction of current flow under forward biasing. The hole flow occurs in the same direction.

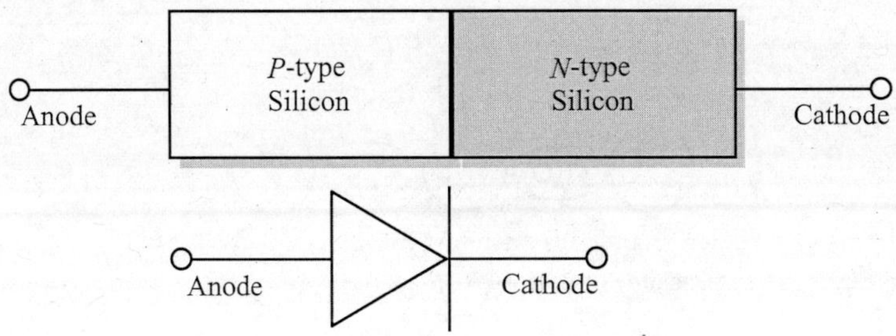

Working

A P-N junction diode is a unidirectional component that exhibits low resistance when biased in the forward direction and acts as an insulator when biased in the reverse direction. Therefore, these diodes are primarily utilized as rectifiers to convert alternating current into direct current.

3.6 SEMICONDUCTOR ENERGY BANDS

In intrinsic semiconductors such as silicon and germanium, the Fermi level is positioned about at the midpoint between the valence and conduction bands. At absolute zero (0 K), there is no conduction. However, at elevated temperatures, a limited number of electrons can access the conduction band and generate a certain amount of electrical current. Additional energy levels are introduced in doped semiconductors. The relationship between conductivity and temperature can be mathematically described using the Fermi function, which enables the determination of the number of electrons in the conduction band. There are two scenarios; the material has a direct band gap and an indirect gap.

3.6.1 Temperature Dependence of the Energy Bandgap

As the temperature rises, the energy bandgap of semiconductors generally diminishes. The relationship between the energy bandgap, E_g, and temperature, T, has been determined by experimental analysis, resulting in the following expression:

$$E_g(T) = E_g(0) - \frac{\alpha T^2}{T + \beta}$$

Where $E_g(0)$, α, and β represent the fitting parameters. The fitting parameters for germanium, silicon, and gallium arsenide are mentioned in the Table below.

	Germanium	Silicon	GaAs
$E_g(0)$ (ev)	0.7437	1.166	1.519
α (meV/K)	0.477	0.473	0.541
β(K)	235	636	204

SOLVED PROBLEMS

PROBLEM 3.1 Explain *n*-doping of silicon by diffusion technique.

Answer The diffusion approach is used to achieve *n*-doping of silicon. This process involves introducing impurity atoms, such as phosphorus or arsenic, into the silicon crystal lattice. These impurity atoms have one more valence electron than silicon, which creates an excess of electrons in the crystal structure. As a result, the silicon becomes n-doped, meaning it has an abundance of negatively charged electrons available

QUESTIONS

1.1 A small amount of phosphorus is added to a sample of silicon to improve its ability to conduct electricity.
 (a) Find out the type of semiconductor?
 (b) Explain the increasing conductivity of semiconductor after the addition of phosphorus.

1.2. Explain the following terms:
 (a) Valence band
 (b) Conduction band
 (c) Insulator
 (d) Extrinsic semiconductor
 (e) *N*-type semiconductor

1.3 What are the tunnel and Zener diode?

1.4 What is called doping?

1.5 Define Fermi level and Fermi energy.

1.6 What is a transistor and what are the two types?

1.7 Explain the characteristics of insulators, conductors and semiconductors.

1.8 What is the result of adding a small amount of indium to germanium metal through doping?

1.9 Provide two examples of extrinsic semiconductors that result from the process of doping.

1.10 Provide two instances of intrinsic semiconductors that can also be classified as examples of extrinsic semiconductors.

1.11 What is the difference between (i) a conductor and an insulator (ii) a conductor and a semiconductor?

1.12 Explain conductors, semiconductors and insulators with the help of Band theory.

Unit 2

Thermodynamic Functions

CHAPTER 4

Thermodynamics

4.1 INTRODUCTION

The term "thermo-dynamic," initially introduced by Thomson, also known as Lord Kelvin, originates from Greek and may be broken down into two components: "*therme*" meaning heat, and "*dynamis*" meaning power. In a more formal context, thermodynamics can be defined as the scientific discipline that encompasses the study of heat, work, and the properties of matter that are interconnected with heat and work.

Or

Thermodynamics is commonly defined as the scientific discipline concerned with the conversion of heat into mechanical work, as well as the reverse process of converting mechanical work into heat.

Or

The field of scientific study concerned with the examination of energy levels and the exchange of energy among systems and various states of matter.

A primary objective of thermodynamics is to establish a formalized correlation among heat, work, and energy. In addition to the establishment of the three fundamental laws, several state functions characteristic of thermodynamics have been introduced over time. For instance, the concept of entropy was first presented by **Rudolf Clausius** in 1855 as part of the reformulation of the enunciation of the second law. The concept of enthalpy was introduced by **Germain Henri Hess** and **Hermann von Helmholtz** between 1840 and 1869 as a measure of heat content under constant pressure in the field of thermochemistry. The free energies were ultimately designated as a tribute to the contributions of **Josiah Willard Gibbs** and **Hermann von Helmholtz**, as documented in the History of Chemistry and Energy Balance of Reacting Systems. Several scientists mentioned before are depicted in Figure 4.1.

4.2 THERMODYNAMIC SYSTEM

The subject of study is a thermodynamic system, which refers to a fixed mass. A "thermodynamic system" refers to a segment of the physical realm that consists of a substantial quantity of

James Watt **Sadi Carnot** **James Joule** **Germain Henri Hess**

Hermann von Helmholtz **Rudolf Clausius** **Walther Nernst** **Josiah Willard Gibbs**

Figure 4.1 Scientist responsible for laws of thermodynamics.

particles, such as atoms or molecules. Illustrative instances include the states of matter, namely gases, liquids, and solids, as well as their amalgamations. The system is partitioned into an "external ambient" or "surroundings" through the establishment of a clearly defined boundary surface. To clarify, the term "surroundings" refers to everything that exists outside of the system, while the interface that separates the system from its surroundings is referred to as the "system boundary."

The permeability of these limits to streams of mass, heat, and work may vary. If the presence of certain streams is prohibited, the system is referred to as "isolated," indicating that it remains unaffected by external factors. The boundary surfaces allow for the transfer of mass, heat, and work between the system and its surroundings. This type of system is commonly referred to as "open".

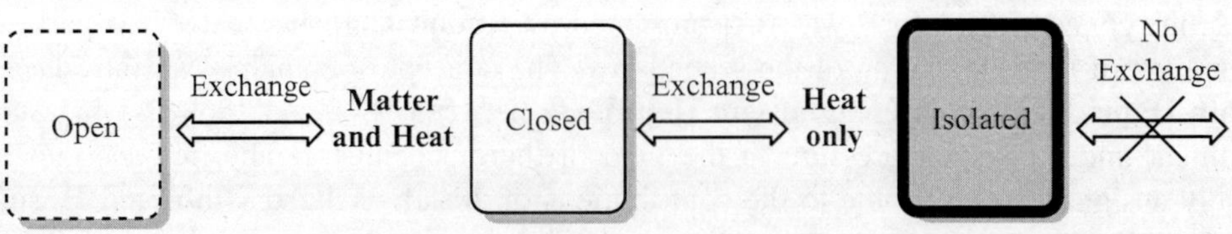

The composite entity including both the system and its surrounding environment is commonly referred to as the universe. In general, the characteristics of the walls encompassing the systems establish the limitations that are present within the system. Hence, a partition that restricts or permits the transfer of thermal energy is referred to as "adiabatic" or "diathermal" correspondingly, but a partition that restricts the transfer of mechanical work is termed as "rigid." Figure 4.2 illustrates the representation of the system, surrounds, and system boundary within the context of a universe. Ice is an illustrative example of a single phase. Another phase that exists is liquid water. A glass containing ice water can be described as a two-phase mixture, where the phase boundaries are located at the periphery of each individual ice cube.

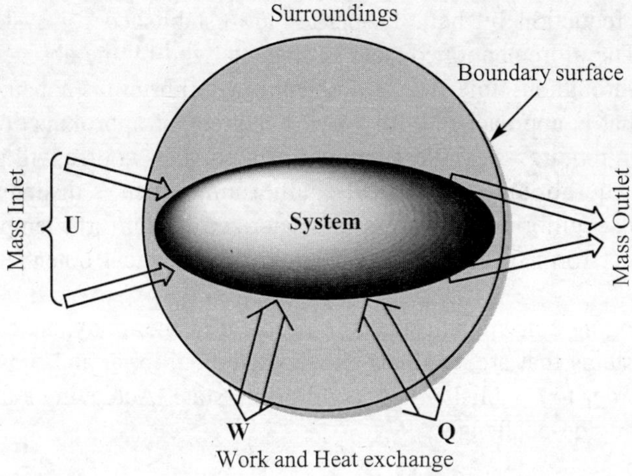

Figure 4.2 Representation of the system, surrounds, and system boundary.

The illustration given below provides a comprehensive overview of the categorization of thermodynamic systems based on their respective interactions with the surrounding environment.

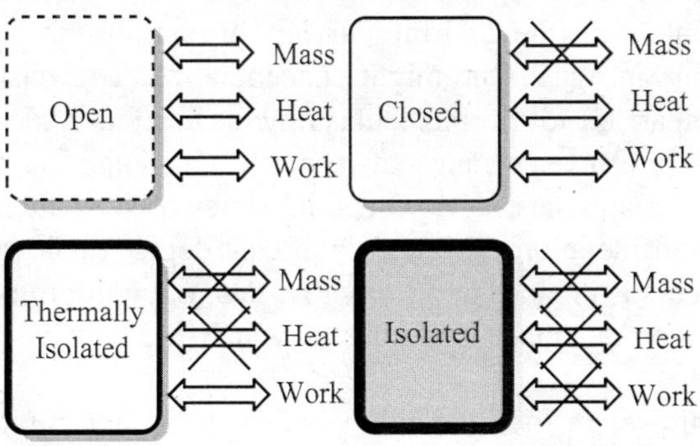

A state is a particular condition characterized by macroscopic properties that can be observed. On the other hand, a property is a measurable quantity that solely relies on the state of the system and remains unaffected by the history of system. Illustrative instances of qualities encompass temperature and pressure. Two states can be considered equal if they possess identical properties. If state 1 is characterized by temperature T_1 and pressure P_1, and state 2 is characterized by temperature T_2 and pressure P_2, state 1 is considered identical to state 2 if and only if T_1 is equal to T_2 and P_1 is equal to P_2.

In the field of thermodynamics, there exist two significant categories of properties that are given due consideration. An extensive property refers to a characteristic of a system that is directly proportional to the mass or extent of the system. Examples of extensive characteristics include mass, total volume, and total energy. Furthermore, an intensive property is a characteristic of a system that remains constant regardless of the system's mass. Intensive properties, such as temperature and pressure, serve as illustrative examples.

In a general context, when a system is divided into two equal parts and its qualities are re-evaluated, it is seen that intense properties exhibit no alteration, whereas extended properties undergo a reduction by half. Properties are established for systems that are in a state of equilibrium. The aforementioned state is characterized by the absence of any observed spontaneous changes throughout time. The attainment of equilibrium is not an absolute outcome, but rather a process that is approached with varying degrees of approximation. The attainment of ultimate equilibrium requires an endless amount of time. This course will primarily focus on two distinct forms of equilibrium: Mechanical equilibrium, which is distinguished by uniform pressure, and Thermal equilibrium, which is characterized by uniform temperature. Chemical equilibrium is a distinct form of equilibrium wherein the chemical potentials are observed to be equal.

One of the challenging conceptual aspects in the field of thermodynamics involves the task of reconciling two systems that are originally in separate equilibria, and subsequently bringing them into contact in order to establish a new equilibrium state. Achieving a task without taking time into mind can provide challenges.

4.3 PROCESSES AND CYCLES

Frequently, systems experience a transition in their state, indicating a modification in one or more characteristics of the system. The investigation of how these qualities would undergo changes in a temporal context lies beyond the purview of equilibrium thermodynamics, which is a subject of considerable intrigue. One might conceptualize these changes as occurring at a very gradual pace in comparison to the molecular time scales that underlie them. Fortunately, this will enable us to tackle a diverse range of technical challenges.

In addition, we establish a process as a sequential series of state alterations. It is postulated that the processes under consideration exhibit a significant degree of slowness, thereby ensuring that each stage of the process is in close proximity to a state of equilibrium. Special designations are assigned to specific commonplace procedures, which are derived from the Greek term "*isos*" denoting "equal."

The term "isothermal" refers to a condition in which the temperature remains constant. Similarly, "isobaric" describes a situation where the pressure remains constant, allowing for changes in the system's boundaries, such as variations in volume. Lastly, "isochoric" pertains to a state in which the volume remains constant.

The adiabatic process is characterized by the absence of heat transfer across the boundaries of a system. A fundamental concept in the field of thermodynamics is the notion of a cycle, which refers to a sequence of actions that ultimately brings a system back to its initial state. The cycle can be described as a thermodynamic process that completes a whole circuit.

4.4 ZEROTH LAW OF THERMODYNAMICS

The axiomatic method is being employed in this course. It is important to note that an axiom is inherently unprovable. This statement can only be verified through empirical experimentation. The validity of the axiom can be refuted through the conduct of a singular experiment that yields a negative outcome. The assumption known as the zeroth law of thermodynamics is widely regarded as being of utmost importance. The zeroth law was designated afterwards to the establishment of the first and second laws, hence its nomenclature. In the event that a more foundational assumption were to be identified, it may potentially be designated as the "First law of thermodynamics."

The zeroth law of thermodynamics states that if two bodies are in thermal equilibrium with a third body, then they are in thermal equilibrium with each other.

The precise origins of the zeroth law remain uncertain. The attribution of this notion is ascribed by Sommerfeld to R.H. Fowler in a review of a thermodynamics book published in 1931. Later on, Fowler and Guggenheim expressly introduce the term "zeroth law of thermodynamics." The corresponding proposition in mathematical logic asserts that if x is equal to y and x is equal to z, then y is equal to z. This proposition is, in reality, comparable to the initial common notion stated by Euclid, which states that objects that are equal to the same object are likewise equal to each other.

The inclusion of the zeroth law in thermodynamics theory allows for the utilization of a thermometer as a means of quantifying temperature. However, it is necessary to provide a clear definition of a scale. The Celsius (°C) temperature scale, which was historically employed, established 0°C as the temperature at which water freezes and 100°C as the temperature at which water boils. The aforementioned quantities exhibited variations in response to changes in pressure. Consequently, distinct measurements would be obtained at higher elevations compared to lower elevations, rendering it an unsuitable benchmark. Prior to May 2019, the Celsius scale was characterized by a roughly identical definition, wherein the triple point of water was designated as 0.01°C and absolute zero in Kelvin was denoted as –273.15°C.

The triple point of water is characterized as the thermodynamic state in which all three phases of water (solid, liquid, and gas) are simultaneously present and in equilibrium. The current definition of temperature is based on the Boltzmann constant, comparable to the Kelvin scale. This definition ensures that the triple point of water remains at a temperature of 0.01°C, with consideration for experimental uncertainties. The conversion formula from the Celsius scale to the absolute Kelvin scale is expressed as K = °C + 273.15.

In the context of temperature measurement, the English counterparts are degrees Fahrenheit (°F) for relative temperature and degrees Rankine (°R) for absolute temperature. The temperature conversions can be expressed as follows: $T(°R) = 1.8T(K)$, $T(°F) = 1.8T(°C) + 32$, and $T(°F) = T(°R) - 459.67$.

4.4.1 The Historical Perspective

In 1798, Benjamin Thompson postulated a correlation between mechanical activity and the production of heat based on his observations of the drilling process employed in the construction of cannons. The concept of reversibility was initially articulated by Nicolas Carnot in the year 1824. During the period from 1840 to 1849, James Joule, a brewmaster by profession, conducted experiments to determine the increase in temperature resulting from mechanical stirring. These experiments aimed to establish a quantitative relationship between work and heat.

Heat

The field of thermodynamics emerged to study the phenomenon of heat. Thermodynamics encompasses the study of heat and energy, treating them as an undividable fluid that perpetually moves from regions of higher energy to those of lower energy. Therefore, the signs of the symbols (+) and (–) in the context of energy transfer indicate the direction of energy flow, with the positive sign denoting energy flowing from the surroundings to the system, and the negative sign denoting energy flowing from the system to the surroundings.

Work

Work can be classified as a distinct form of energy. In contrast to heat, the movement of energy can occur both into and out of a system, hence causing alterations. Consider the hypothetical scenario of exerting energy to elevate a book. In this context, the effort performed results in a transformation of the book's potential energy, without any association to thermal energy.

Internal Energy, *U*

Changes in the internal energy of a system, denoted as ΔU or dU, are positive when the energy of the system increases. The transfer of energy in the form of heat results in an increase in the amount of heat added. The work performed on the system is positive. There have been instances where it has been postulated that the energy change resulting from work performed by a system is positive. This differs with the formulation presented earlier, indicating a difference in perspective. It is advisable to use caution when perusing alternative sources, as the potential for misinterpretation over sign changes can permeate subsequent discussions within the field of thermodynamics.

Mechanical Work Description

Work = force required to move a body x distance that body is moved

$$w = \int_{x_0}^{x} F dx$$

Because $F = PA$, force = pressure per unit area

$$w = \int_{x_0}^{x} F dx = -(PA)(x - x_0) = -PdV$$

This describes a piston's movement and potential 'energy'.

4.5 FIRST LAW OF THERMODYNAMICS

In 1873, Gibbs postulated the Law of Conservation of Energy, which asserts that energy is conserved in every given operation, meaning it cannot be created or destroyed, but can only be moved. Hence, it can be inferred that the overall energy of the universe remains constant. Energy, nonetheless, has the capacity to undergo conversion from one form to another or to be exchanged between a system and its surroundings, or vice versa. The net energy change can be expressed as the difference between the heat transferred across a boundary and the work performed by the system.

The change in internal energy (ΔU) of a system can be calculated by summing the heat (q) added to the system and the work (w) done on or by the system.

$$\Delta U = q + w$$

In the context of thermodynamics, the variable q represents heat, while the variable w represents work. A portion of the thermal energy entering a system is transformed into mechanical work, resulting in no net increase in the system's internal energy.

4.5.1 Work and the First Law

Work can be conceptualized as a mathematical function that depends on both pressure and volume. The change in work (dw) can be expressed as the product of pressure (P) and the change in volume (dV).

$$dw = PdV$$

The symbol PdV represents the infinitesimal change in volume that occurs at a given pressure due to the force applied over a specific distance, which has the dimensions of work.

Restate the first law as:

$$dU = dq - PdV$$

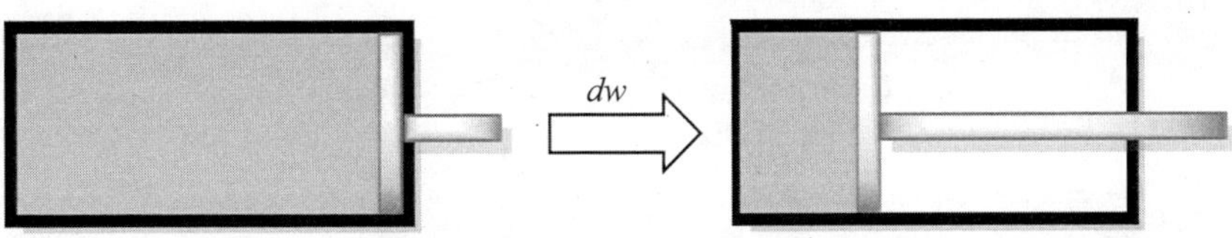

4.5.2 Energy Change with Volume and Heat

When transitioning from state 1 to state 2, the change in internal energy (dU) can be calculated using the equation $dU = dq - PdV$.

$$\int_1^2 dU = \int_1^2 dq - P\int_1^2 dV$$

The equation for yields can be expressed as $U_2 - U_1 = (q_2 - q_1) - P(V_2 - V_1)$.

Let q_p can be defined as the difference between q_2 and q_1. By multiplying the PV terms and rearranging the equation, we can express it as: The equation q_p represents the difference between the sum of the energy (E) and the product.

$$q_p = (E_2 + PV_2) - (E_1 - PV_1)$$

The measurability of q_p can be ascertained through the observation of temperature variations that arise from alterations in energy levels, such as those occurring during a chemical reaction.

4.5.3 Enthalpy (H)

The equation $H = U + PV$ represents the relationship between the enthalpy (H), internal energy (U), and pressure-volume work (PV). The total differential for enthalpy can be expressed as follows:

$$dH = dU + PdV + VdP.$$

In relation to our earlier discussion on integrated change in state:

The first equation can be expressed as $H_1 = U_1 - PV_1$, while the second equation is $H_2 = U_2 - PV_2$. The change in enthalpy, denoted as ΔH, can be calculated as $H_2 - H_1$, which is equivalent to the heat transfer at constant pressure and volume, represented as q_p.

It is important to note that in the context of energy, its absolute value is not ascertainable; rather, the focus lies on the significance of its change. Hence, the measurement of change is conducted with respect to a reference state, wherein pure elements at a temperature of 25ºC and a pressure of 1 bar possess an enthalpy value of zero, denoted as H°_f.

4.5.4 Spontaneous Processes

Spontaneous processes are characterized by their ability to occur without the need for external intervention. The gas contained in vessel B will undergo spontaneous effusion into vessel A. However, after the gas has diffused into both vessels, it will no longer exhibit spontaneous effusion. Phenomena that exhibit spontaneity in a certain direction are characterized by non-spontaneity when occurring in the opposite direction.

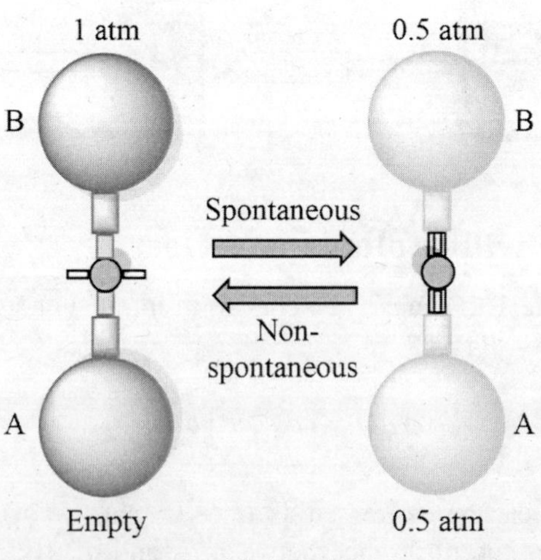

It is observed that certain processes, which exhibit spontaneity at a specific temperature, may exhibit non-spontaneity at different temperatures. At temperatures over 0°C, the process of ice melting is considered spontaneous. The opposite process becomes non-spontaneous when the temperature falls below zero degrees Celsius.

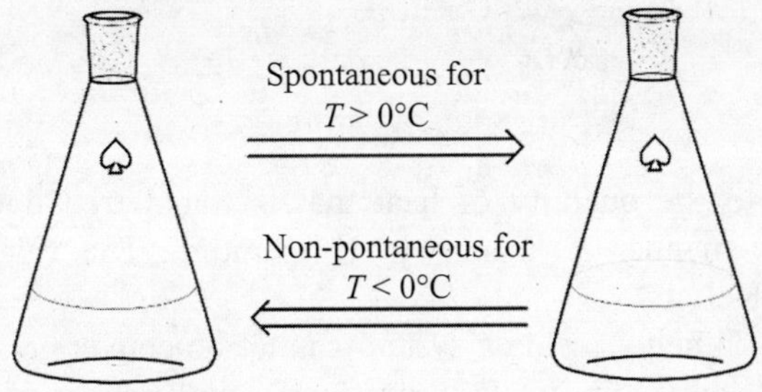

4.5.5 Reversible Processes

In the context of thermodynamics, a reversible process refers to a system undergoing a transformation wherein both the system and its surroundings may be restored to their initial conditions by precisely reversing the aforementioned process. In a reversible process, the magnitude of changes is exceedingly minute.

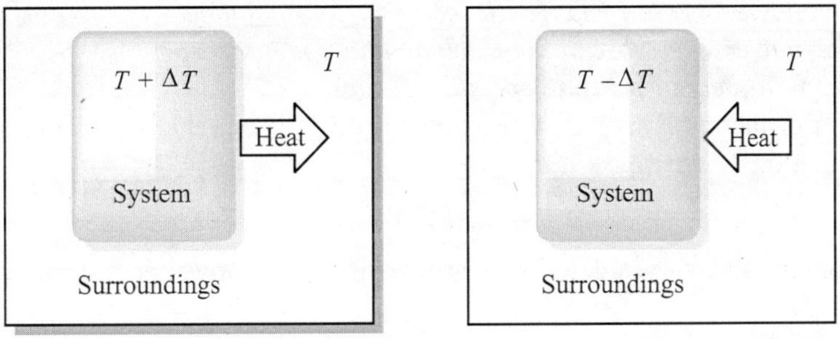

Irreversible processes are characterized by their inability to be completely reversed through an exact reversal of the changes made to the system. All spontaneous processes exhibit irreversibility. All processes in the physical world are inherently irreversible.

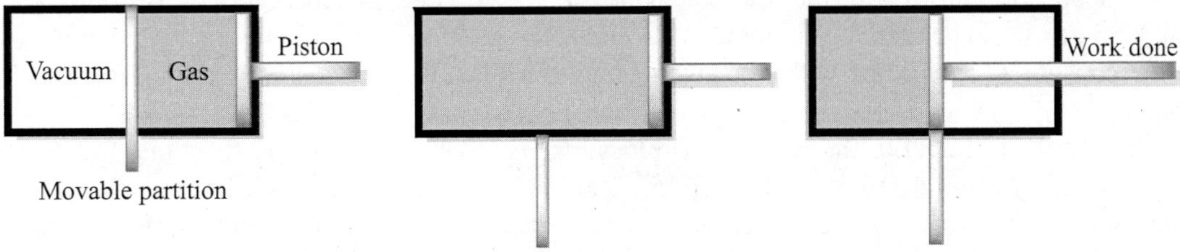

4.6 SECOND LAW OF THERMODYNAMICS

The second law of thermodynamics introduces the concept of entropy, denoted by the symbol S. The concept of entropy (S) was introduced by Rudolph Clausius during the 19th century. Clausius held a firm belief in the importance of the ratio between the amount of heat transferred and the corresponding temperature at which it is transferred. In the context of a thermodynamic system, an isothermal process refers to a process that takes place at a constant temperature.

$$dS = \frac{dq_{rev}}{T} \quad dS > \frac{dq}{T}$$
(reversible) (irreversible)

The term q_{rev} refers to the quantity of heat that is transferred during a process that is conducted in a reversible manner at a constant temperature. The variable T represents the temperature measured in Kelvin.

A portion of the total enthalpy inside a system cannot be converted into useful work, such as pressure-volume (PdV) work, due to its utilization in the process of increasing entropy. It might be reformulated that the process of increasing entropy necessitates a certain degree of effort. Entropy can be conceptualized as a quantitative indicator of the degree of disorder or unpredictability within a given system. This pertains to the diverse ways of molecular mobility.

Similar to total energy (E) and enthalpy (H), entropy (S) is also a state function. Hence, the change in entropy (ΔS) can be expressed as the difference between the final entropy (S_{final}) and the initial entropy ($S_{initial}$). Therefore, $\Delta S = S_{final} - S_{initial}$

The second law of thermodynamics states that the total entropy of the universe remains constant for reversible processes, whereas it increases for spontaneous processes. The increase in entropy of the universe is observed in actual and spontaneous processes. However, it is possible for entropy to decrease inside specific systems.

In the context of thermodynamics, a reversible process is one that may be reversed without any loss or dissipation of energy.

$$\Delta S_{universe} = \Delta S_{system} + \Delta S_{surroundings} = 0$$

The phenomenon under consideration is irreversible, meaning that it occurs naturally and cannot be reversed.

$$\Delta S_{universe} = \Delta S_{system} + \Delta S_{surroundings} > 0$$

4.6.1 Entropy on the Molecular Scale

The concept of entropy at the molecular level was elucidated by Ludwig Boltzmann. Temperature is a quantitative parameter that characterizes the mean kinetic energy exhibited by the constituent molecules within a given sample. Molecules demonstrate various forms of motion:

Translational refers to the process of relocating the complete molecule from one location to another. **Vibrational** refers to the periodic motion exhibited by atoms within a molecule. **Rotational** motion refers to the movement of a molecule around an axis or the rotation around sigma (σ) bonds.

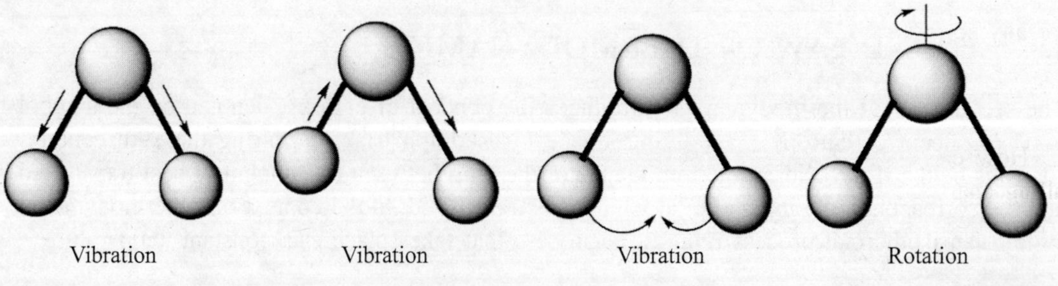

Boltzmann conceptualized the dynamic movements of a collection of molecules at a specific moment in time. This would be like to capturing a momentary representation of all the molecules. The author characterized this sampling as a microstate within the context of the thermodynamic system. Every thermodynamic state is characterized by a distinct quantity of microstates, denoted as *W*. The equation for entropy is given by $S = k \ln W$, where *k* represents the Boltzmann constant with a value of 1.38×10^{-23} J/K.

Implications

The implications of the subject matter are significant and warrant further investigation. An increase in the number of particles leads to an increase in the number of states, which in turn results in an increase in entropy. An increase in temperature leads to a greater number of energy levels, resulting in an increase in entropy. The presence of less structural organization, such as in the comparison between a gas and a solid, leads to an increase in the number of states available, thus resulting in a higher level of entropy. The quantity of microstates and, thus, the entropy exhibits a tendency to rise as temperature, volume (in the case of gases), and the count of molecules with independent motion increase.

4.6.2 The Second Law

The principle of energy conservation, known as the first law of Thermodynamics, states that in every given process, the total energy remains constant. Nevertheless, the First Law fails to provide any insight into the spontaneous direction in which a process will occur. The violation of the First Law of Motion would not occur if a book were to abruptly displace from a table and sustain a position at an elevated distance above the table's surface. The First Law of Thermodynamics would not be violated in the event that all oxygen molecules present in the ambient air within this enclosed space were to instantaneously migrate into a gas cylinder and remain confined therein, with the valve being in an open state.

Nevertheless, empirical evidence suggests that specific processes exhibit spontaneous directionality in either one direction or the other. This enables us to infer the Second Law.

According to the Kelvin formulation, it is not feasible to design an engine that operates in cycles and only accomplishes the transfer of heat from a reservoir while generating work as its sole outcome.

According to the Clausius formulation, it is deemed unattainable to create an engine that, when operating in cycles, solely accomplishes the transfer of heat from a colder body to a hotter body without any other effects.

Directionality from the Second Law

For any spontaneous irreversible process, entropy is always increasing

$$dS > \frac{dq}{T}$$

How can a reaction ever proceed if order increases? Why are minerals in the earth not falling apart as we speak?

4.6.3 Meaning of Entropy and the Second Law

Entropy is a quantitative metric that characterizes the level of disorder or unpredictability shown by a given system. The level of disorder (or randomness) in a system is directly proportional to its entropy. According to the second law of thermodynamics, it is observed that the level of disorder, or entropy, in the universe tends to increase in the course of any actual physical process. The decrease in entropy, which represents the level of order in a system, is contingent upon a simultaneous increase in the entropy, or level of disorder, of the surrounds. This increase must be greater in magnitude, ensuring that the overall entropy of the universe consistently rises.

Entropy quotes...

"There is a great difference between energy and availability of energy...The availability of energy is always decreasing. This is... what is called the entropy law, which says the entropy is always increasing."

—Richard Feynman

"The thermodynamic sense of order decrease that is enshrined in the second law is at first sight in conflict with many of the complicated things that we see going on around us. We see complexity and order increasing with time in many situations: when we tidy up our office,... the evolution of complex life-forms from the simpler ones...

"In many of these cases, we must be careful to pay attention to all the order and disorder that is present in the problem. Thus the process of tidying the office requires physical effort on someone's part. This causes ordered biochemical energy stored in starches and sugars to be degraded into heat. If one counts this into the entropy budget, then the decrease in entropy or disorder associated with the tidied desk is more than compensated for by the other increases."

Barrow (1990)

4.6.4 Entropy and Physical States

The phenomenon of entropy is directly proportional to the degree of molecular freedom of motion. Consequently, the entropy of the gaseous state ($S(g)$) is greater than the entropy of the liquid state ($S(l)$), which in turn is greater than the entropy of the solid state ($S(s)$). Therefore, $S(g) > S(l) > S(s)$.

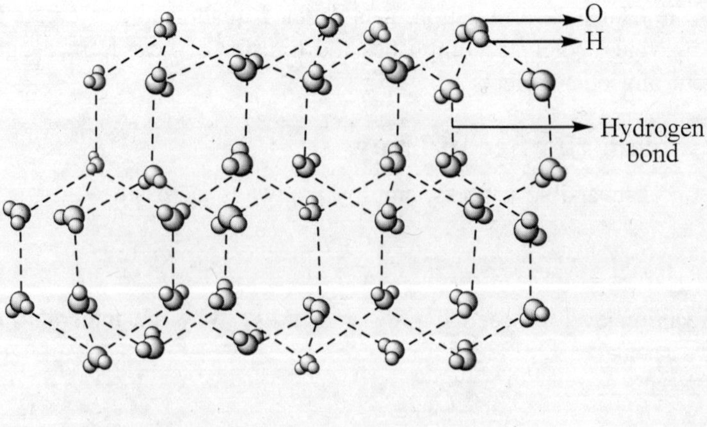

Dissolution of a Solid

Ions possess a greater degree of entropy, characterized by an increased number of available states. However, certain water molecules exhibit lower entropy due to their arrangement around ions. Typically, there is a general upward trend observed in variable *S*. An exception to this phenomenon occurs with strongly charged ions, which have a strong ability to induce alignment of several water molecules in their vicinity.

Entropy Changes

In a general context, the phenomenon of entropy tends to exhibit an increase when the transition from liquid and solid states to gaseous states occurs. Liquids or solutions can be derived from solid substances. The quantity of gas molecules experiences an increase. The number of moles increases.

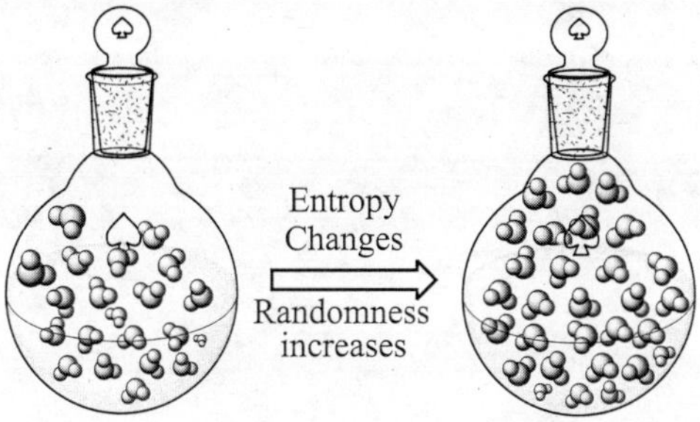

The equation $dU = dq_{rev} + dw$ can be expressed as $TdS - PdV$, which combines the first and second laws of thermodynamics. In the case of a process occurring at a constant volume, denoted as *V*, and entropy, denoted as *S*, the change in internal energy, represented as *dU*, is equal to zero. Consequently, no alterations occur within the system, and the energy remains unchanged. The concept being discussed is equilibrium.

If the change in internal energy (ΔU) is greater than zero, it indicates a spontaneous reaction in which the products have a higher energy than the reactants.

When the change in internal energy, denoted as ΔU, is negative ($\Delta U < 0$), it indicates that the reaction is spontaneous and proceeds from reactants to products.

4.7 THE FUNDAMENTAL EQUATION

The combination of the first and second laws of thermodynamics results in the equation: Change in internal energy (*dU*) is equal to the product of temperature (*T*) and change in entropy (*dS*), minus the product of pressure (*P*) and change in volume (*dV*).

$$dU = TdS - PdV$$

This step is of paramount importance, as it serves as the fundamental basis for the majority of thermodynamic computations.

QUESTIONS

4.1 For an ideal gas $C_p = 2.5R$, calculate the change in entropy when 3 moles of this gas is heated from 300 K to 600 K at:
(a) Constant pressure
(b) Constant volume.

4.2 Explain second law of thermodynamics with examples.

4.3 Define entropy of a system.

4.4 Define entropy and Gibbs free energy.

4.5 Calculate the standard free energy change (ΔG^o) of the reaction: $CO(g) + 2O_2(g) \rightarrow CO_2(g)$ $\Delta H^\circ = -282.84$kJ. The standard entropy of $CO_2(g)$, $CO(g)$ and $O_2(g)$ are 213.80, 197.90 and 205.01 J K mol^{-1}, respectively. Is this reaction feasible at standard state?

CHAPTER 5

Electrochemistry

5.1 INTRODUCTION

Electrochemistry is a specialized field within the realm of chemistry that centers on the utilization of electrical energy derived from chemical processes that occur spontaneously, as well as the application of electrical energy to induce chemical reactions that are non-spontaneous in nature. In essence, the phenomena occurring at the interfaces between phases characterized by different conductivities, along with the concurrent conversion of chemical energy into electrical energy.

The passage of an electric current through a substance has the potential to induce a chemical reaction, while conversely, a chemical reaction can generate an electric current within an external circuit. This phenomena encompasses the examination of electrolysis, conductance, and the quantification of electromotive force (EMF).

5.2 SOLUTIONS OF ELECTROLYTES

Electrolytes are chemical compounds that dissociate into ions when dissolved in a solution, hence enabling the conduction of electric current. Electrolytes refer to chemicals that exhibit electrical conductivity either in aqueous solutions or in a molten form. Electrolytes can be classified into two main categories: strong electrolytes, such as sodium chloride, hydrochloric acid, and sodium sulphate, and weak electrolytes, including ephedrine and phenobarbital. Non-electrolytes are substances that do not exhibit electrical conductivity under the given conditions.

5.2.1 Types of Electrolytes

In accordance with the Degree of dissociation (α), electrolytes can be categorized into the following categories:

1. **Strong electrolytes** are substances that undergo a significant ionization process, with a degree of ionization greater than thirty percent ($\alpha > 30\%$), when they are dissolved

in water. For example, consider HCl, HBr, $HClO_4$, H_2SO_4, HNO_3, HI, NaOH, KOH, KCl, KBr, NaCl, in addition to other ionic compounds that are soluble.

2. **Medium strong electrolytes** are substances that undergo ionization with degree of ionization (α) 2–30%. For example, H_3PO_4, H_3PO_3, etc.
3. **Weak electrolyte** is a chemical that is only weakly ionized to a small extent, specifically less than 2%. For example, H_2S, HCN, H_2CO_3, HF, CH_3COOH, NH_4OH, and so on as well.

5.2.2 Properties of Electrolyte Solutions

When a flow of direct electric current is applied to an electrolytic cell, a chemical reaction is created within the cell. Electrolysis is the term that is most generally used to refer to the method presented above. The phrase "electrolysis" may be broken down into two parts: "electro," which is a term that describes electricity and the movement of electrons, and "-lysis," which is a term that indicates a link or interaction. It is possible to provide a thorough definition of the term "electrolysis" by combining these components.

The cathode is the point at which electrons are injected into the cell. These electrons first come from either a battery or a generator (Figure 5.1). Immediately after entering the cell, these electrons get involved in a process that involves joining with positively charged ions, which are also referred to as cations that are present in the solution.

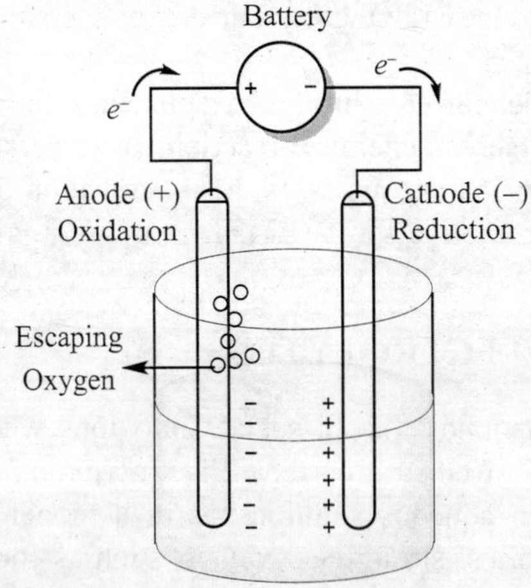

Figure 5.1 Representation of cell.

There is a reduction reaction that takes place as a consequence of this event. The anions, which are also known as negatively charged ions, play a role in the transportation of electrons throughout the solution. They then release these electrons at the anode, which ultimately leads to the oxidation of these anions. It is important to note that the process of oxidation involves the removal of electrons from a chemical species, whereas the process of reduction includes the addition of electrons to a chemical species. Electrolytes are characterized by the simultaneous flow of electric current and the physical transfer or migration of matter across their surfaces. As a result of the transfer of matter in question, the breakdown of matter is facilitated at the precise areas where electricity enters or exits the electrolyte, which in turn helps to facilitate the passage of electrons.

5.3 RESISTANCE

Opposition of conductivity is referred to as resistance, and it is a measurement of the impediment that is placed in the path of current. When a conductor is measured in terms of its length (l), its resistance is proportional to the length, and it is inversely proportional to the area of the cross-section (Figure 5.2).

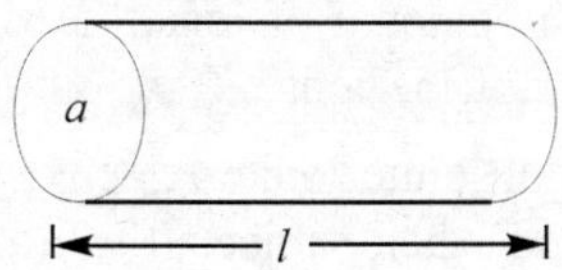

Figure 5.2 Measurement of resistance of a conductor.

For a conductor of uniform cross section (a) and length (l); Resistance, R, is represented as follows:

$$R \propto l \quad \text{and} \quad R \propto \frac{l}{a} \quad \therefore \quad R = \rho \frac{l}{a}$$

Here, ρ (rho) is the constant of proportionality, which is sometimes referred to as specific resistance or resistivity of the system. There is a correlation between the composition of the material and the resistance. Ohms (Ω) are the units that it uses.

$$R = \rho, \quad \text{if} \quad l = 1 \text{ cm}, a = 1$$

To put it another way, specific resistance is the resistance that exists between the two faces of a conductor that is one centimeter squared opposite one another.

5.4 CONDUCTANCE

Conductance refers to the inverse of the resistance shown by an electrolyte in impeding the passage of electric current. Hence, it serves as an indicator of the conductor's (or solution's) electrical conductivity. The unit of conductance is commonly expressed as ohm^{-1} or mho.

$$\text{Conductance} = \frac{1}{\text{Resistance}}$$

The electrolyte region situated between the parallel and identical electrodes, as depicted in the diagram, should be taken into consideration. Each electrode has an area of 'a' square centimeters, and the distance between them is denoted as 'l' centimeters. The application of

Ohm's law to electrolytic conductors is possible due to the fact that electrolytes, similar to metallic conductors, exhibit resistance to the flow of electric current.

Ohm's law establishes a relationship between the resistance denoted by R in a given medium, the applied voltage represented by V, and the current flowing through the medium, denoted by I.

$$V = IR$$

Georg Simon Ohm, born on 16 March 1789 and passing away on 6 July 1854, was a German physicist. He was the son of Johann Wolfgang Ohm, who worked as a locksmith. The unit of electrical resistance, known as the ohm, is named in his honor.

The conductance, which is the inverse of resistance, is principally determined by three major factors: the amount of ions, the size of charge carried by each ion, and the mobility of the ions.

Note: Ionic mobility, also known as absolute ionic velocity, is quantified as the displacement of an ion over a period of one second when subjected to a potential gradient of one volt per centimeter.

5.5 SPECIFIC CONDUCTANCE

The reciprocal of the specific resistance of an electrolytic solution is referred to as specific conductance. The symbol 'κ' 'kappa' is used to represent it. The resistance, denoted as 'R', is directly influenced by the size of the conductor.

$$R \propto l \quad \text{and} \quad R \propto \frac{1}{a}$$

$$\therefore \quad R \propto \frac{l}{a}; \quad R = \rho \frac{l}{a}$$

Where R is resistance, ρ is specific resistance

Or

$$\frac{1}{\rho} = \frac{1}{R} \times \frac{l}{a}$$

$$\kappa = \frac{1}{\rho}$$

$$\therefore \quad \text{Specific conductance} = \text{Conductance} \times \frac{l}{a}$$

Similarly to conductance $1/R$, specific conductance, which is represented by the symbol 'κ'(Kappa), is also referred to as $1/\rho$. Mho per cm is the unit that is used to measure specific conductance. The siemens (S) is the unit of conductivity that is used in the International System of Units (SI), and it is equivalent to either ohm^{-1} or mho.

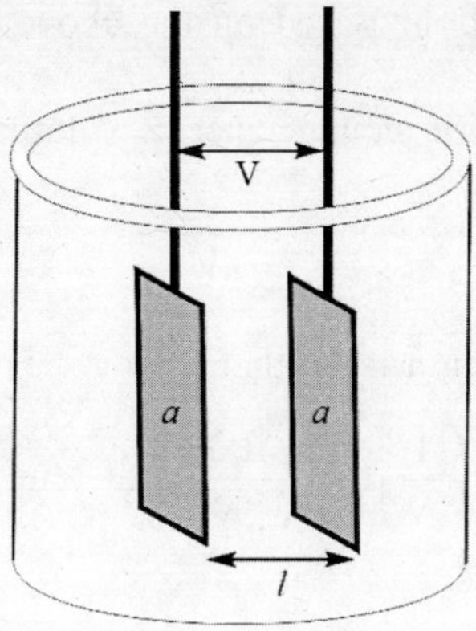

Figure 5.3 Measurement of Specific conductivity of cell.

The solution that is being investigated is placed inside of a conductivity cell that is constructed out of Pyrex glass and has two platinum electrodes that are established in parallel to one another (Figure 5.3). *l*/*a* is referred to as the cell constant for a particular cell. Specific conductance is equal to conductance multiplied by the cell constant. Thus,

$$\text{Specific conductance} = \text{Conductance} \times \text{Cell constant}$$

5.6 MOLAR CONDUCTANCE

The ability of all of the ions that are included within one mole of electrolyte to conduct electricity in the solution is what is meant by the term "molar conductivity," which can be defined as the measurement of the strength of the solution. With regard to the definition that has been established,

Molar conductance = sp. conductance × vol. in cubic centimeter (cc) containing 1 mole of the electrolyte

or

$$\Lambda_m = \kappa V$$

If M is the molar concentration in mol L^{-1}, then;

$$\Lambda_m = \frac{1000}{M}\kappa$$

The unit of equivalent conductance and molar conductance is mho cm^2 mol^{-1} or S cm^2 mol^{-1}.

If a volume of 1 cubic centimeter of an electrolytic solution is considered, it may be theoretically assumed that this volume contains one equivalent of the electrolyte. The equivalent weight of the electrolyte is equivalent to its molar weight.

5.7 EQUIVALENT CONDUCTANCE

The equivalent conductance of an electrolytic solution can be defined as the measure of the conductivity shown by all the ions included within 1 equivalent of the electrolyte in the solution, considering a specific level of dilution. Based on the above described concept of equivalent conductance, it can be inferred that

Equivalent conductance = specific conductance × volume in cubic centimeter containing 1 eq. of the electrolyte

or
$$\Lambda_{eq} = \kappa V$$

If normality of electrolytic solution is N, then:

$$V = \left(\frac{1}{N}\right) \text{L} = \left(\frac{1000}{N}\right) \text{mL} \,(\text{cm}^3)$$

$\therefore$
$$\Lambda_{eq} = \frac{1000}{N} \kappa$$

The unit of equivalent conductance is mho cm^2 eq^{-1} or S cm^2 eq^{-1}.

5.7.1 Equivalent Conductance at Infinite (∞) Dilution

The equivalent conductance exhibits an upward trend as dilution rises; however, it eventually reaches a saturation point beyond which it remains constant and does not exhibit additional increments. The equivalent conductance at infinite dilution, given as Λ_∞, refers to the greatest value of equivalent conductance. If the observed rise in equivalent conductance upon dilution can be solely attributed to the increase in the degree of dissociation of the electrolyte, then we can express it as follows:

$$\text{Degree of dissociation } \alpha = \frac{\Lambda_c}{\Lambda_\infty}$$

where Λ_c is equivalent conductance and α is the degree of dissociation at a concentration c of the weak electrolyte.

5.7.1.1 Arrhenius Theory of Electrolytic Dissociation

In the theory of dissociation, which was initially proposed by Arrhenius, it is postulated that when electrolytes are dissolved in a solution, the solute takes on an ionic form inside the solution.

$$NaCl + H_2O \longrightarrow Na^+ + Cl^- + H_2O$$

$$HCl + H_2O \longrightarrow H_3O^+ + Cl^-$$

$$CH_3COOH + H_2O \rightleftharpoons CH_3COO^- + H_3O^+$$

Particularly in much diluted solutions, Arrhenius did not believe that strong electrolytes were totally ionized. The level of molecular ionization, which is quantified by the degree of dissociation α, served as the basis for differentiation between strong and weak electrolytes. On the other hand, a weak electrolyte is characterized by a low degree of ion dissociation, and a strong electrolyte is characterized by a high degree of ion dissociation.

There are two approaches available for determining the degree of dissociation:

Conductance measurements are **one method** that can be utilized to get a determination regarding the degree of separation. To be more specific, the concept that is being discussed is the analogous conductance at infinite dilution. The symbol Λ is used to represent a metric that measures the degree to which a solute is dissociated into its constituent ions. On the other hand, the symbol Λ_c is used to designate the number of solute particles that exist as ions at a specific concentration c.

Svante August Arrhenius, a Swedish physicist and physical chemist, born on February 19, 1859, in Vik, Sweden, gained recognition for his contributions to the field of science. Notably, he formulated the theory of electrolytic dissociation and developed a model to explain the phenomenon of the greenhouse effect. The Nobel Prize for Chemistry was bestowed upon him in 1903.

Therefore, the equation can be used to mathematically express the fraction of solute molecules that go through the process of ionization, which is also referred to as the degree of dissociation.

$$\alpha = \frac{\Lambda_c}{\Lambda_\infty}$$

The following is an **additional method** that can be utilized to ascertain the value of α: In the following method, the van't Hoff factor, which is represented by the symbol i, can be associated with the degree of dissociation, denoted by the symbol α.

$$\alpha = \frac{i-1}{v-1}$$

where v is the number of ions that are created as a result of the ionization of the electrolyte, such as for sodium chloride, v equals two ($v = 2$), and for calcium chloride, v equals three ($v = 3$).

The Arrhenius theory is widely acknowledged as the framework that is recognized for describing the behavior of only weak electrolytes. We can accurately determine the degree of dissociation of a weak electrolyte by employing the conductance ratio or by acquiring the van't Hoff factor, which is represented by the letter i. Both of these methods are effective in achieving the desired result. When it comes to strong electrolytes, the demonstration is complete dissociation in solutions that are either moderately concentrated or dilute.

Furthermore, it is worth noting that there exists a significant discrepancy between the calculated value of α acquired from the i ratio and the calculated value of α obtained from the conductivity ratio when it comes to strong electrolyte solutions with concentrations that surpass about 0.5 M.

According to the Arrhenius hypothesis, the degree of dissociation does not provide an acceptable explanation for the behavior of strong electrolytes. As an alternative, it is more convenient to think of a strong electrolyte as going through the process of complete ionization and to include a variable that quantifies the degree to which the solute is moving away from complete ionization. Within the context of this purpose, the exploitation of the activity and activity coefficient is beneficial.

5.8 ACTIVITY AND ACTIVITY COEFFICIENT

The presence of a significant quantity of ions with opposite charges in solutions of strong electrolytes leads to mutual interactions between them via interionic attractive forces. In the case of nonelectrolyte solutions, irrespective of their concentration, the quantity of ions present is minimal and the interionic attractive forces are negligible. At the context of strong electrolytes, it is observed that ions have the tendency to form ion pairs when present at high concentrations. As a result, it is possible for a strong electrolyte to undergo complete ionization while exhibiting incomplete dissociation into free ions. The solution is characterized by '*a*' **"effective concentration,"** also known as **activity**. The initial introduction of this concept can be attributed to Lewis and Randall.

The activity (a) of a solute (m) is typically lower than its actual or stoichiometric concentration. This discrepancy is not primarily due to partial ionization of strong electrolytes, but rather arises from the electrostatic forces of contact that effectively remove some ions from participating in the solution. In the scenario of infinite dilution, where the ions are sufficiently enough apart to prevent any interaction between them, the activity (a) of an ion is equivalent to its concentration.

$$a = m$$

Because of the decreased effective concentration or activity of ions, which falls below the molal concentration, the ratio decreases below unity as the concentration of the solution increases. This is because the molal concentration is lower than the effective concentration. The practical activity coefficient (γ) is the term used to refer to this ratio.

$$a = \gamma m$$

The ionic activity of a cation and an anion in an aqueous solution can vary. The mean ionic activity $a_{\pm}$ characterizes the behavior of an electrolyte.

$$a_{\pm} = (a_{+}^{m} a_{-}^{n})^{1/(m+n)}$$

The stoichiometric coefficients of the corresponding ions that are present in the solution are denoted by the exponents m and n according to the context in which a_{+} represents the activity of the cation and a_{-} represents the activity of the anion.

Therefore, a sodium chloride (NaCl) solution has a mean ionic activity of $a_{\pm} = (a_{Na^+} a_{Cl^-})^{1/2}$

In contrast, it can be shown that a solution of $FeCl_3$ exhibits a mean ionic activity of $a_{\pm} = (a_{Fe^{3+}} a_{Cl^-}^{3})^{1/4}$.

Variation of Specific Conductance with Dilution

The specific conductivity (κ) is influenced by two factors: (i) the quantity of ions, and (ii) the quantity of water. Consequently, as the dilution increases (i.e., the amount of water increases), the conductivity of 1 cm^3 of the electrolytic solution decreases. This is due to the decrease in the number of ions and the increase in the number of water molecules present in 1 cm^3 of the solution.

Variation of Equivalent Conductance with Dilution

The electrical conductivity of an electrolyte is attributed to the presence of ions, which becomes more pronounced as the electrolyte is further diluted. This is because the extent of dissociation of the electrolyte increases as the dilution level rises. Put simply, the molar conductivity, also known as the equivalent conductivity, exhibits an upward trend as the quantity of ions in a solution, which contains 1 equivalent or 1 mole of electrolyte, rises due to dilution. Kohlrausch conducted an investigation into the impact of dilution on the molar conductivity (Λ_m) of both strong and weak electrolytes, revealing a notable disparity in their respective behaviors.

Nevertheless, it is important to note that there are two distinct patterns in which equivalent conductivity fluctuates in relation to concentration, as depicted in Figure 5.4. The graph illustrates the relationship between equivalent conductivity and the square root of concentration (c). The Figure 5.4 illustrates the presence of two distinct types of curves.

(i) The equivalent conductance (Λ_{eq}) of strong electrolytes, such as hydrochloric acid (HCl), sodium chloride (NaCl), barium chloride (BaCl), and sodium acetate (CH_3COONa), exhibit a linear increase when the square root of concentration decreases. When the square root of concentration approaches zero, the plot of Λ against $\sqrt{C}$ can be extended to estimate the value of Λ^{∞}, which represents the conductance at infinite dilution. Λ^{∞} is determined by the intercept on the vertical axis of the Λ vs $\sqrt{C}$ in Figure below.

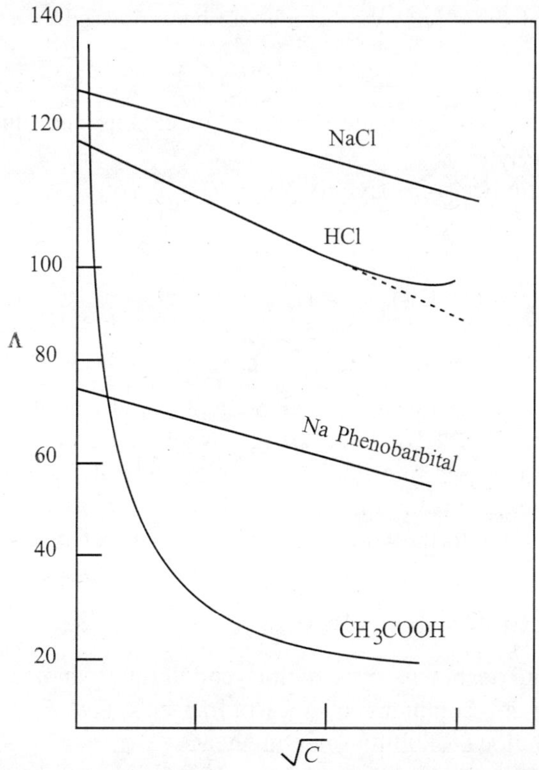

Figure 5.4 Variation of equivalent conductivity with concentration.

(ii) A distinct variation in behavior is noticed when examining weak electrolytes, such as acetic acid (CH_3COOH). A significant rise in the value of Λ is observed when concentrations are low, hence impeding the ability to obtain a precise estimation of Λ^{∞}. The rapid upward trajectory observed for acetic acid can be attributed to the phenomenon wherein the dissociation of weak electrolytes intensifies with dilution, leading to a significant augmentation in the quantity of ions capable of conducting electrical current. The value of Λ for a weak electrolyte cannot be determined from extrapolation of the curve to zero concentration.

5.9 KOHLRAUSCH'S LAW

An ionic species, regardless of whether it is a cation or an anion, contributes a constant value to the equivalent conductance of the electrolyte at a certain temperature when it is in a state of infinite dilution. This characteristic holds true regardless of whether or not other ionic species are present in conjunction with the ionic species. The contributions indicated above are usually known as the equivalent ionic conductance at infinite (∞) dilution, which is represented by the symbols λ_c^{∞} and λ_a^{∞} for cations and anions, respectively.

In accordance with Kohlrausch's law, the equivalent conductance of an electrolyte at infinite dilution can be calculated by adding up the equivalent ionic conductances of the electrolyte's cations and anions. This occurs due to the fact that every ion makes a specific contribution to the overall conductance of the electrolyte, which can be expressed as

$$\Lambda_{eq}^{\infty} = v_c\lambda_c^{\infty} + v_a\lambda_a^{\infty}$$

The variables v_c and v_a represent the quantities of cations and anions, respectively, per formula unit of the electrolyte.

For example, for acetic acid (CH_3COOH), $v_c = v_a = 1$, so:

$$\Lambda_{eq,\, CH_3COOH}^{\infty} = \lambda_{H^+}^{\infty} + \lambda_{CH_3COO^-}^{\infty}$$

And for magnesium chloride ($MgCl_2$), $v_c = 1$, $v_a = 2$, so:

$$\Lambda_{eq,\, MgCl_2}^{\infty} = \lambda_{Mg^{2+}}^{\infty} + 2\lambda_{Cl^-}^{\infty}$$

Friedrich Wilhelm Georg Kohlrausch was a German physicist who was born in Rinteln, Germany, on October 14, 1840, and passed away on January 17, 1910. He was successful in his research into the properties of electrolytes, which are substances that conduct electricity in solutions by the transfer of ions, and he made significant contributions to the knowledge of how they behave.

5.9.1 Applications of Kohlrausch's Law

The dissociation constant, degree of dissociation, and limiting molar conductivity of weak electrolyte solutions are all computed using Kohlrausch's law. It is also utilized in the computation of a salt solution's solubility measurement.

5.9.1.1 Determination of Molar Conductivity of a Weak Electrolyte

As previously stated, the determination of Λ_{eq}^{∞} values for weak electrolytes such as CH_3COOH and NH_4OH cannot be achieved through extrapolation of equivalent conductivity values to zero concentration. This is due to two reasons: (a) even at extremely high dilutions, these electrolytes do not undergo complete ionization, and (b) the resulting curve does not exhibit a linear relationship.

Kohlrausch's law offers a means of determining the Λ_{eq}^{∞} value of weak electrolytes based on Λ_{eq}^{∞} measurements obtained from strong electrolytes. Consider the objective of determining the Λ_{eq}^{∞} value for acetic acid. The value of Λ_{eq}^{∞} can be derived from the Λ_{eq}^{∞} values of strong electrolytes such as HCl, NaCl, and CH_3COONa. According to Kohlrausch's law, it may be stated that:

$$\Lambda_{eq,\,HCl}^{\infty} + \Lambda_{eq,\,CH_3COONa}^{\infty} - \Lambda_{eq,\,NaCl}^{\infty}$$

$$= (\lambda_{H^+}^{\infty} + \lambda_{Cl^-}^{\infty}) + (\lambda_{Na^+}^{\infty} + \lambda_{CH_3COO^-}^{\infty}) - (\lambda_{Na^+}^{\infty} + \lambda_{Cl^-}^{\infty})$$

$$= \lambda_{H^+}^{\infty} + \lambda_{CH_3COO^-}^{\infty}$$

$$= \Lambda_{eq,\,CH_3COOH}^{\infty}$$

The values of Λ_{eq}^{∞} for strong electrolytes, namely HCl, CH_3COONa, and NaCl, can be determined using the process of extrapolation of Λ vs. $\sqrt{C}$ graphs. Therefore, the Λ_{eq}^{∞} value of any weak electrolyte can be determined by utilizing the Λ_{eq}^{∞} values of specific strong electrolytes.

5.9.1.2 Determination of Degree of Dissociation

The term "degree of dissociation (or ionization)" refers to the proportion of the total number of molecules that have undergone dissociation into ions.

$$\alpha = \frac{\text{No. of molecules dissociated into ions}}{\text{Total number of molecules taken}}$$

It is also understood that the conductivity of a solution is attributed to the existence of ions inside the solution. The conductivity increases in direct proportion to the number of ions present.

$\therefore$ Λ_{eq}^{c}, The equivalent conductivity observed at a specific level of dilution

$$\propto \text{No. of molecules dissociated into ions at this dilution} \qquad \text{(i)}$$

However, when the concentration approaches infinity, it may be observed that all molecules exist in their ionic state.

$$\therefore \qquad \Lambda_{eq}^{\infty} \propto \text{Total number of molecules taken} \qquad \text{(ii)}$$

When (i) is divided by (ii), the result is obtained.

$$= \frac{\Lambda_{eq}^{c}}{\Lambda_{eq}^{\infty}} = \frac{\text{No. of molecules dissociated into ions at a particular dilution}}{\text{Total number of molecules taken}} \qquad \text{(iii)}$$

Therefore, the degree of dissociation at any given dilution can be defined as the quotient of the equivalent conductivity at that particular dilution and the molar conductivity at infinite dilution.

The determination of the value of Λ_{eq}^{c} can be achieved through direct measurement, whereas the determination of Λ_{eq}^{∞} can be facilitated through the utilization of Kohlrausch's law, i.e.,

$$\Lambda_{eq}^{\infty} = v_c \lambda_c^{\infty} + v_a \lambda_a^{\infty}$$

Therefore, the value of α may be computed.

5.9.1.3 Determination of Solubility of Sparingly Soluble Salts

In everyday contexts, ionic salts such as AgCl, $BaSO_4$, $CaCO_3$, Ag_2CrO_4, $PbSO_4$, PbS, $Fe(OH)_3$, etc., are commonly considered to be insoluble. However, it is important to note that these salts do exhibit minimal, yet measurable solubilities in aqueous solutions. The solubility of a sparingly soluble salt can be determined by measuring the specific conductivity (κ) of a saturated solution of the salt. Given the low concentration of salt in the solution, it is reasonable to consider the comparable conductivity at high dilution as Λ_{eq}^{∞}.

In the case of salts with very sparing solubility:

$$\Lambda_{eq}^{c} \cong \Lambda_{eq}^{\infty} = \kappa v$$

The determination of the value of Λ_{eq}^{∞} can be accomplished by the utilization of Kohlrausch's law.

$$v = \frac{\Lambda_{eq}^{c}}{\kappa} = \frac{\Lambda_{eq}^{\infty}}{\kappa} = \frac{\lambda_c^{\infty} + \lambda_a^{\infty}}{\kappa}$$

But v cm^3 of saturated solution contains = 1 eq.

∴ 1000 cm^3 or 1 L of saturated solution contains = (1000/v) eq.

Hence, solubility of salt,

$$S = \frac{1000}{v}\ \text{eq/L} = \frac{1000 \times E}{v}\ \text{g/L} = \frac{1000 \times E \times \kappa}{\lambda_c^{\infty} + \lambda_a^{\infty}}\ \text{g/L}$$

The variable E represents the equivalent weight of the salt.

5.9.1.4 Determination of Ionic Mobility

The equivalent ionic conductance (λ) and ionic mobility (μ) are distinct concepts, although they exhibit a proportional relationship.

$$\lambda_c^{\infty} \propto U_c^{\infty} \quad \text{or} \quad \lambda_c^{\infty} = KU_c^{\infty}$$
$$\lambda_a^{\infty} \propto U_a^{\infty} \quad \text{or} \quad \lambda_a^{\infty} = KU_a^{\infty}$$

It is possible to demonstrate that the proportionality constant K is equivalent to 96,500 coulombs. If there is a field of one volt per centimeter, the ionic mobility is measured in centimeters per second.

From a mathematical standpoint,

$$\text{Ionic mobility} = \frac{\text{Ionic velocity}}{\text{Potential gradient}}$$
$$= \frac{\text{Ionic velocity (cm/s)}}{\text{Potential difference (volt)/Distance between the electrodes (cm)}}$$

The unit of ionic mobility is thus $cm^2volt^{-1}s^{-1}$.

The relationship between chemical energy and electrical energy is the focus of the scientific discipline of electrochemistry. It is composed of two distinct kinds of cells:

1. The electrolytic cell, and
2. Electrochemical cell.

5.10 ELECTROLYTIC CELL

An electrolytic cell is a device that either can transform electrical energy into chemical energy or can perform electrolysis, which is a chemical reaction that involves oxidation and reduction. Both of these processes are examples of electrolysis. The process of electrolysis is accomplished by utilizing electricity within a cell that is electrolytic. Through the utilization of an applied voltage, a chemical reaction that does not take place naturally can be carried out effectively. The source of the electric current is a third person or entity that is not directly involved. There must be a difference between the potential that is provided by the external source and the potential that is provided by the naturally occurring reverse reaction. The process of electrolysis involves the utilization of electrical energy in order to compel a chemical reaction that does not occur spontaneously to take place.

5.11 ELECTROCHEMICAL CELL

The term "electrochemical cell" refers to a cell that employs electrochemistry in order to convert chemical energy into electrical energy. The chemical energy that is produced as a byproduct of a redox process is converted into electrical energy by means of this apparatus, which is employed to accomplish this transformation. These devices are frequently referred to as galvanic cells or voltaic cells, after the Italian scientists Luigi Galvani and Alessandro Volta, who were the pioneers in the field of converting chemical energy into electrical energy and called their cells after themselves. Galvani and Volta were the first people to transform chemical energy into electrical energy. A spontaneous redox reaction is carried out in an indirect manner in an electrochemical cell, and the drop in free energy that occurs during a chemical reaction is turned into electrical energy. This transformation takes place in the electrochemical cell. The reduction and oxidation processes that take place in an indirect redox reaction are carried out in two distinct vessels that are referred to as half-cells from one another. This kind of reaction is more complicated than a redox reaction that is carried out directly.

The Italian physician and scientist **Luigi Galvani** was known for his work. His birthday is September 9th, 1737, and he was born in Bologna, which was a part of the Papal States at the time. Bologna, which was a member of the Cisalpine Republic at the time of his death, was the location where he passed away on December 4, 1798. It was ultimately the result of his research that led to the creation of a form of battery that is known as the voltaic pile. Having a reliable supply of electrical current is now within reach thanks to the existence of this battery.

Luigi Galvani

Italian physicist **Alessandro Volta** is most known for his invention of the electric battery, which was the first source of continuous current. Volta is also noted for his contributions to the field of electrochemistry. Volta was born on February 18, 1745, in the city of Como, which is located in the region of Lombardy in Italy. He resided there until his death on March 5, 1827.

Alessandro Volta

An electrochemical cell contains a conductor that is submerged in an electrolyte solution. This conductor is referred to as an electrode. Electrodes are the fundamental components that can be found in an electrochemical cell. The process of electricity being transferred from one electrolyte solution to another is influenced by a number of factors, including, for instance, the migration of potassium ions in the salt bridge in one direction and the migration of chloride ions in the opposite direction. A few examples of cathodic reactions that are common. A cathode is a specific kind of electrode that is utilized in the process of reduction.

$$Ag^{+}(aq) + e^{-} \longrightarrow Ag(s)$$

$$Fe^{3+}(aq) + e^{-} \longrightarrow Fe^{2+}(aq)$$

$$NO_3^{-}(aq) + 10H^{+} + 8e^{-} \longrightarrow NH_4^{+} + 3H_2O$$

The following is an example of a typical anodic reaction. Oxidation takes place at an anode, which is a type of electrode.

$$Cu(s) \longrightarrow Cu^{2+}(aq) + 2e^{-}$$

$$2Cl^{-}(aq) \longrightarrow Cl_2(g) + 2e^{-}$$

$$Fe^{2+}(aq) \longrightarrow Fe^{3+}(aq) + e^{-}$$

5.11.1 Daniel Cell

It is composed of two half-cells that make up the Daniel Cell (Fig. 5.5). There is a possibility that the half-cell on the left has a zinc metal electrode that has been submerged in a solution of zinc sulphate. The electrode of the right-hand half-cell is formed of copper metal, and it is immersed in a solution of copper sulphate. The two sections of the cell are connected to one another by a salt bridge, which prevents the solution from being mixed mechanically.

The following observation can be made once the zinc and copper electrodes have been linked together by a wire:

1. The external circuit does indeed have a flow of electric current going through it.
2. The mass of the copper rod increases while the mass of the zinc rod decreases.
3. The concentration of the $ZnSO_4$ solution becomes more concentrated, whereas the concentration of the $CuSO_4$ solution becomes less concentrated.
4. The solutions in both of the compartments do not provide any electrical charge to the overall system.

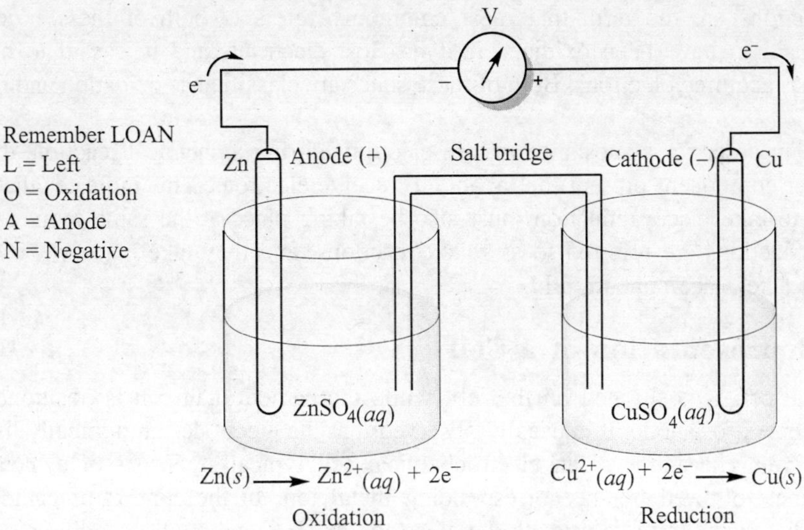

Figure 5.5 Daniel Cell

As the electric current moves through the external circuit, electrons move from the zinc electrode to the copper electrode. This journey takes place in the opposite direction. The zinc electrode is where the oxidation process that is required to produce zinc ions takes place. These zinc ions are then released into the solution after they have been produced. To arrive at the copper electrode, the electrons that are emitted at the electrode travel across the external circuit until they reach their destination. After arriving at their destination, they are utilized in the process of reducing Cu^{2+} ions to metallic copper, which is subsequently deposited on the electrode via the method.

5.11.1.1 Cell Reaction

The Zn metal that is located at the anode loses electrons, which causes it to dissolve, while the metal that is located at the cathode obtains electrons, which causes it to grow.

The following chemical reaction can only be finished if the flow of electrons in the cell is from the Zn-electrode to the Cu-electrode.

At the anode, oxidation takes place, half-cell reaction is

$$Zn(s) \longrightarrow Zn^{2+}(aq) + 2e^-$$

At the cathode, reduction takes place, half-cell reaction is

$$Cu^{2+}(aq) + 2e^- \longrightarrow Cu(s)$$

Net reaction is $Zn(s) + Cu^{2+}(s) \longrightarrow Zn^{2+}(aq) + Cu(s)$ (Cell reaction)

5.11.1.2 Redox Reactions

During the process of oxidation, a substance loses electrons, whereas during the process of reduction, a substance acquires electrons. Oxidation and reduction are two different processes.

Chemical reactions are the term that most commonly refers to both of these processes. A substance is said to have been oxidized if it has lost electrons, and it is said to have been reduced if it has acquired electrons. Both of these states are classified as oxidation and reduction, respectively.

However, in order for a substance to lose electrons during a chemical reaction, there must first be another component present that is able to receive electrons. This indicates that in order for oxidation to take place, reduction must also be taking place at the same moment, or vice versa. These reactions are referred to as redox reactions, and they take place when oxidation and reduction take place concurrently.

5.11.1.3 Representation of a Cell

The anode half-cell is positioned on the left, while the cathode half-cell is positioned on the right. In the context of anode, it is usually observed that the electrodes are typically listed prior to the electrolyte. The anode in an electrochemical cell typically consists of a metal, which is depicted first, followed by the corresponding metal ion. In the context of cathodes, it is customary to list the electrolyte prior to the electrode. In the context of cathodic processes, the metal ion is typically denoted first, followed by the corresponding metal, with a vertical line separating the two. The two half-cells are demarcated by a double vertical line. A salt bridge is represented by a double vertical line.

For example

1. A cell constructed by coupling Zn electrode dipped in 0.5M $ZnSO_4$ and Ni electrode dipped in 0.05M $NiSO_4$ can be represented by

$$Zn/ZnSO_4(0.5\,m)\,||\,NiSO_4(0.05\,m)/Ni$$

2. Daniell cell can be represented as

Anode half-cell Phase boundary | Salt Bridge | Cathode half-cell Phase boundary

$$Zn(s)\,|\,Zn^{2+}(aq)\;\;||\;\;Cu^{2+}(aq)\,|\,Cu(s)$$

Electrons flow this way →

3. Representation of a cell with one Cu electrode and Ag electrode

Cell reaction:

$$Cu(s) + 2Ag^{+}(aq) \longrightarrow Cu^{2+}(aq) + 2Ag(s)$$

Half-cell reactions:

Cathode (reduction): $2Ag^{+}(aq) + 2e \longrightarrow 2Ag(s)$

Anode (oxidation): $Cu(s) \longrightarrow Cu^{2+}(aq) + 2e$

The cell can be represented as:

$$Cu(s)\,|Cu^{2+}(aq)||Ag^{+}(aq)|\,Ag(s)$$

5.11.1.4 Salt Bridge and its Function

In most cases, the salt bridge is made out of a tube that is formed like an upside-down U and is filled with a concentrated solution of inert electrolytes. An inert electrolyte is distinguished by the fact that its ions do not take part in any electrochemical processes, nor do they undergo any chemical reactions with the electrolytes that are present in the two half-cells. Examples of chemicals that are typically utilized include potassium chloride (KCl), potassium nitrate (KNO_3), ammonium nitrate (NH_4NO_3), and potassium sulphate (K_2SO_4). This particular electrolyte is distinguished by the simultaneous movement of both the cation and the anion, which is the characteristic that sets it apart from other options.

At a high temperature, gelatin or agar-agar is dissolved in a concentrated aqueous solution of a benign electrolyte. This process is done in order to prepare a salt bridge. Following that, the solution that was produced is poured into a U-tube. By putting the solution through a cooling process, it is possible to turn it into a substance that resembles gel while it is contained within the U-tube. Cotton wool is used to seal the ends of the U-tube in order to reduce the damage that is caused by diffusion effects.

Functions of Salt Bridge

1. The basic purpose of an electrical circuit is to provide continuity by facilitating the unbroken flow of ions between its various components. This is the circuit's principal function.
2. In order to make it easier for ions to move from one solution to another while simultaneously preventing the two solutions from mixing together.
3. The purpose of the connecting wire is to create a connection between the solutions of two half cells, so completing the circuit of the cell. This is accomplished by establishing a connection of the wire.
4. It has been discovered that the diffusion of ions that are in conflict with one another between compartments is prevented.
5. The solutions that are contained within the two half-cells are constantly kept in a state that is electrically neutral. The anodic half-cell is responsible for the migration of cations into the solution, which leads to the accumulation of extra positive charge in the region of the anode. Due to the presence of this high concentration of positive charge, the migration of electrons away from the anode is more difficult. Additionally, in the cathodic half-cell, the buildup of positive ions takes place around the cathode as a consequence of their reduction and subsequent deposition. This occurs in a manner that is analogous to the previous example. The salt bridge is responsible for supplying an enough quantity of both positive and negative ions in order to accomplish the goal of ion neutralization. This is accomplished by the salt bridge, which serves to maintain the solution's electrical neutrality.
6. It is one of its purposes to prevent the occurrence of liquid-liquid junction potential, which is the potential difference that occurs when two solutions come into contact with one other. Other functions include preventing the occurrence of the potential difference.
7. The manufacture of salt bridges requires the utilization of salts such as potassium chloride (KCl) and ammonium nitrate (NH_4NO_3). Ions such as K^+ and Cl^- have

mobility that is comparable to one another and migrate in the direction of the cathode and anode compartments, respectively, in response to the difference in potential that is applied.

5.12 ELECTRODE POTENTIAL

A metal goes through a process in which it acquires either a positive or negative charge in relation to the solution when it is submerged in a solution that contains its respective ions. This process occurs when the metal is immersed in the solution. The tendency of the metal to either lose or receive electrons upon contact with its corresponding solution is what causes the establishment of a potential differential between the electrode/metal and the electrolyte/solution. This potential differential manifests itself as a potential difference between the two. Electrode potential is the measure of the difference in electric potential that is produced when a metal and a solution are brought into contact with one another. When oxidation takes place at the electrode in comparison to the typical hydrogen electrode, the potential of the electrode can be referred to as the oxidation potential. The electrode potential is referred to as the reduction potential.

The standard electrode potential, which is symbolized by the symbol E^0, is the electrode potential in a half-cell in which a metal rod is submerged in a solution with a concentration of one molar while the temperature is kept at 298 degrees Kelvin. The electrode potential that is referred to as the "standard electrode potential" is the potential that is present in a half-cell when the concentrations of all of the constituent species are equal to one. In accordance with the regulation established by the International Union of Pure and Applied Chemistry (IUPAC), the word "standard reduction potentials" has been changed to "standard electrode potentials."

For example, a zinc plate that is submerged in a solution that contains Zn^{2+} ions will acquire a negative charge in comparison to the solution when it is introduced into the solution. Consequently, a potential difference is established between the zinc plate and the solution. Specifically, this potential difference is referred to as the "electrode potential of zinc" in the scientific community. Copper acquires a positive charge in relation to the solution when it is introduced into a solution that contains Cu^{2+} ions. This occurs in a manner that is analogous to the previous example. It is determined that there is a potential difference between the solution and the copper plate. The formation of an electrical double layer at the interface between a metal and a solution is responsible for the establishment of a potential difference. This layer is responsible for the potential difference.

In the following manner, the phenomena of negative charge development, which can be observed on a zinc plate, or positive charge development, which can be noticed on a copper plate, can be explained in greater detail. There are two different charges that make themselves known when a metal rod is submerged in the salt solution that corresponds to it.

1. As a result of the transport of metal ions from the electrode to the solution, an accumulation of excess electrons occurs, which ultimately results in the formation of a negative charge on the electrode.
2. The reduction process occurs when metal ions that are present in a solution receive electrons from the electrode. This causes the electrode to accumulate a positive charge, which is the consequence of the reduction process.

5.12.1 Measurement of Electrode Potential

It is not possible to determine the absolute value of the electrode potential of a single electrode, which is also referred to as the single electrode potential. This is because there is neither an oxidation nor a reduction half-reaction present. It is only possible to carry out the measurement by making use of a reference electrode, which is typically in the form of an electrode by itself. Commonly referred to as the standard or normal hydrogen electrode (SHE or NHE), the electrode is utilized for the purpose of reference or reference reasons. The formation of a link between a particular electrode and a standard hydrogen electrode, which is made possible through the utilization of a salt bridge, is required in order to ascertain the electrode potential of a particular electrode. The following is a concise summary of one of the most significant characteristics of the typical electrode potential, which is symbolized by the symbol E°:

1. The standard electrode potential is a relative number since it reflects the potential of an electrochemical cell with the standard hydrogen electrode serving as the reference electrode. This type of electrode has been assigned a value of 0.000 V.
2. A reduction reaction is the only type of reaction that is relevant to the standard electrode potential for a half-reaction.
3. As the reactants and products approach their equilibrium activity, the standard electrode potential provides a quantitative measure of the relative propensity to cause the half-reaction.
4. There is no change in the standard electrode potential regardless of the amount of reactants and products that are present in the balanced half-reaction.
5. When the electrode potential is positive, it indicates that the half-reaction that is being considered is thermodynamically beneficial in compared to the typical hydrogen electrode half-reaction.
6. It has been observed that the standard electrode potential for a half-reaction is observed to be dependent on temperature.

The electromotive force (EMF) of the cell can be determined by the utilization of either a calibrated potentiometer or a high resistance voltmeter.

5.12.1.1 Standard Hydrogen Electrode (SHE)

The hydrogen electrode serves as a standard reference for the determination of electrode potentials in many electrochemical studies. In theory, the electrode holds significant importance when employed in aqueous solutions (Figure 5.6).

At all temperatures, the potential of the reversible hydrogen electrode in a solution of hydrogen ions at unit activity is considered to be zero. The electrode is comprised of a platinum wire that is chemically unreactive, submerged in a solution containing hydrogen ions at a concentration of 1 M H^+ and saturated with hydrogen gas at a pressure of 1 atm H_2. The platinum specimen is fully submerged in an aqueous solution of hydrochloric acid that is free from

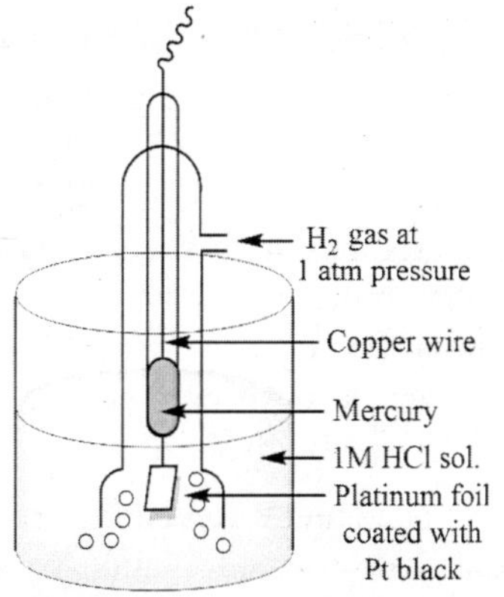

Figure 5.6 Standard hydrogen electrode (SHE).

arsenic. Additionally, a stream of hydrogen gas, which is devoid of both oxygen and carbon monoxide, is introduced to the surface of the platinum.

The displacement of air by hydrogen occurs gradually, leading to the attainment of a reversible potential. Unfortunately, this electrode exhibits several limitations. The maintenance of reversibility in hydrogen electrode is not possible in oxidizing fluids. In the event when current is extracted from the electrode, the electrode assumes the role of an anode due to the ionization of gas molecules.

$$H_2 \longrightarrow 2H^+ + 2e^- \quad E° = 0.00\,V$$

5.13 ELECTROCHEMICAL SERIES

The comparison of the potentials of various electrodes in relation to the standard hydrogen electrode (SHE) has resulted in the construction of a series of standard electrode potentials. The electrochemical series of elements, also known as the electromotive series or the activity series, is a series that is generated when the electrodes, which are made up of metals and non-metals individually, come into contact with the ions that are associated with them. In order to determine how these electrodes should be arranged, the values of their standard reduction potential or standard oxidation potential are taken into consideration.

According to international conventions, the standard potentials of electrodes are recorded during reduction half reactions, providing insights on the electrodes' propensities to function as cathodes with respect to the Standard Hydrogen Electrode (SHE). The series is a systematic arrangement of different metal electrodes based on their standard reduction potentials, or alternatively, their oxidation potentials in a decreasing order. Substances exhibiting positive standard reduction potentials (E^o) in half-cell reactions indeed function as cathodes relative to the standard hydrogen electrode (SHE), whereas substances with negative E^o values in reduction half-cell reactions assume the role of anodes with respect to the SHE.

5.13.1 Applications of Electrochemical Series

1. **Reactivity of Metals:** Electrodes exhibiting negative standard electrode potentials relative to hydrogen demonstrate a propensity for oxidation, whilst electrodes with positive values exhibit a tendency for reduction in aqueous solutions. The reactivity of a metal is directly proportional to its oxidation potential, as a higher oxidation potential indicates a greater ability to lose electrons and therefore a higher reactivity. Consequently, a metal possessing a higher oxidation potential has the ability to displace a metal with a lower oxidation potential from its corresponding salt solution. For instance, zinc (Zn) has the ability to displace copper (Cu) from its salt solution. In a manner consistent with the electrochemical series, metals positioned higher than hydrogen can effectively displace hydrogen when reacting with mineral acids.

2. **Displacement Reaction:** In order to make a prediction regarding the displacement of a metal from its salt solution by another metal. Zinc metal has the ability to displace copper from copper sulphate ($CuSO_4$). Metals with negative electrode potentials have the ability to displace hydrogen gas from acids. Metals such as iron (Fe), zinc (Zn), and aluminum (Al) have the ability to generate hydrogen gas (H_2) while undergoing a chemical reaction with acids.

3. Spontaneity or Feasibility of a Redox Reaction: The electromotive force (EMF) of the cell exhibits a positive value, indicating that the cell reaction occurs spontaneously. If the value of E_{cell} is negative, it indicates that the reaction is thermodynamically unfavorable and will not proceed spontaneously. As an illustration, let us examine the chemical reaction involving the elements copper (Cu) and zinc (Zn), represented as $Cu + Zn^{2+} \rightarrow Cu^{2+} + Zn$.

Considering the electrochemical cell reactions, it is observed that the standard reduction potential for the Zn^{2+}/Zn half-cell is –0.76 V, indicating the anode reaction. On the other hand, the standard reduction potential for the Cu^{2+}/Cu half-cell is 0.34 V.

In order for the aforementioned reaction to occur, it is necessary for copper (Cu) to undergo electron loss, while zinc (Zn) must undergo electron gain. This process is only viable if the oxidation potential of copper exceeds that of zinc. According to the electrochemical series, it has been observed that the oxidation potential of zinc (0.76V) is greater than that of copper (–0.34V). So the above reaction is not feasible

4. Calculation of EMF of a Cell: The electromotive force (EMF) of the cell can be easily determined by utilizing the equation $E_{cell} = E^{\circ}_{R} - E^{\circ}_{L}$.

5. Predicting the Anode and Cathode: In electrochemical systems, it is observed that the electrode with a larger reduction potential would function as the cathode, while the electrode with a lower electrode potential will function as the anode. For example, $E^{\circ}_{Zn^{2+}/Zn} = -0.76$ V; Anode; $E^{\circ}_{Cu^{2+}/Cu} = 0.34$ V; Anode.

6. Corrosion: The metal situated higher in the series exhibits an anodic or more active nature, rendering it more susceptible to corrosion. The metals situated at lower positions within the series are classified as noble metals, exhibiting reduced susceptibility to corrosion.

Half Reaction	E° V
$F_2(g) + 2e^- \longrightarrow 2F^-(aq)$	+2.87
$O_3(g) + 2H^+(aq) + 2e^- \longrightarrow O_2(g) + H_2O$	+2.07
$Co^{3+}(aq) + e^- \longrightarrow Co^{2+}(aq)$	+1.82
$H_2O_2(aq) + 2H^+(aq) + 2e^- \longrightarrow 2H_2O$	+1.77
$PbO_2(s) + 4H^+(aq) + SO_4^{2-}(aq) + 2e^- \longrightarrow PbSO_4(s) + 2H_2O$	+1.70
$Ce^{4+}(aq) + e^- \longrightarrow Ce^{3+}(aq)$	+1.61
$MnO_4^-(aq) + 8H^+(aq) + 5e^- \longrightarrow Mn^{2+}(aq) + 4H_2O$	+1.51
$Au^{3+}(aq) + 3e^- \longrightarrow Au(s)$	+1.50
$Cl_2(g) + 2e^- \longrightarrow 2Cl^-(aq)$	+1.36

Half Reaction	$E°$ V
$Cr_2O_7^{2-}(aq) + 14H^+(aq) + 6e^- \longrightarrow 2Cr^{3+}(aq) + 7H_2O$	+1.33
$MnO_2(s) + 4H^+(aq) + 2e^- \longrightarrow Mn^{2+}(aq) + 2H_2O$	+1.23
$O_2(g) + 4H^+(aq) + 4e^- \longrightarrow 2H_2O$	+1.23
$Br_2(l) + 2e^- \longrightarrow 2Br^-(aq)$	+1.07
$NO_3^-(aq + 4H^+(aq) + 3e^- \longrightarrow NO(g) + 7H_2O$	+0.96
$2Hg^{2+}(aq) + 2e^- \longrightarrow Hg_2^{2+}(aq)$	+0.92
$Hg_2^{2+}(aq) + 2e^- \longrightarrow 2Hg(l)$	+0.85
$Ag^+(aq) + e^- \longrightarrow Ag(s)$	+0.80
$Fe^{3+}(aq) + e^- \longrightarrow Fe^{2+}(aq)$	+0.77
$O_2(g) + 2H^+(aq) + 2e^- \longrightarrow H_2O_2(aq)$	+0.68
$MnO_4^-(aq) + 2H_2O + 3e^- \longrightarrow MnO_2(s) + 4OH^-(aq)$	+0.59
$I_2(s) + 2e^- \longrightarrow 2I^-(aq)$	+0.53
$O_2(g) + 2H_2O + 4e^- \longrightarrow 4OH^-(aq)$	+0.40
$Cu^{2+}(aq) + 2e^- \longrightarrow Cu(s)$	+0.34
$AgCl(s) + e^- \longrightarrow Ag(s) + Cl^-(aq)$	+0.22
$SO_4^{2-}(aq) + 4H^+(aq) + 2e^- \longrightarrow SO_2(g) + 2H_2O$	+0.20
$Cu^{2+}(aq) + e^- \longrightarrow Cu^+(aq)$	+0.15
$Sn^{4+}(aq) + 2e^- \longrightarrow Sn^{2+}(aq)$	+0.13
$2H^+(aq) + 2e^- \longrightarrow H_2(g)$	+0.00
$Pb^{2+}(aq) + 2e^- \longrightarrow Pb(s)$	–0.13
$Sn^{2+}(aq) + 2e^- \longrightarrow Sn(s)$	–0.14
$Ni^{2+}(aq) + 2e^- \longrightarrow Ni(s)$	–0.25

Half Reaction	$E°$ V
$Co^{2+}(aq) + 2e^- \longrightarrow Co(s)$	–0.28
$PbSO_4(s) + 2e^- \longrightarrow Pb(s) + So_4^{2-}(aq)$	–0.31
$Cd^{2+}(aq) + 2e^- \longrightarrow Cd(s)$	–0.40
$Fe^{2+}(aq) + 2e^- \longrightarrow Fe(s)$	–0.44
$Cr^{3+}(aq) + 3e^- \longrightarrow Cr(s)$	–0.74
$Zn^{2+}(aq) + 2e^- \longrightarrow Zn(s)$	–0.76
$2H_2O + 2e^- \longrightarrow H_2(g) + 2OH^-(aq)$	–0.83
$Mn^{2+}(aq) + 2e^- \longrightarrow Mn(s)$	–1.18
$Al^{3+}(aq) + 3e^- \longrightarrow Al(s)$	–1.66
$Be^{2+}(aq) + 2e^- \longrightarrow Be(s)$	–1.85
$Mg^{2+}(aq) + 2e^- \longrightarrow Mg(s)$	–2.37
$Na^+(aq) + e^- \longrightarrow Na(s)$	–2.71
$Ca^{2+}(aq) + 2e^- \longrightarrow Ca(s)$	–2.87
$Sr^{2+}(aq) + 2e^- \longrightarrow Sr(s)$	–2.89
$Ba^{2+}(aq) + 2e^- \longrightarrow Ba(s)$	–2.90
$K^+(aq) + 2e^- \longrightarrow K(s)$	–2.93
$Li^+(aq) + e^- \longrightarrow Li(s)$	–3.05

The tables offer a thorough compilation of the potentials of reduction half-reactions under standard conditions. These conditions include a concentration of 1 M, a temperature of 25 °C, and a pressure of 1 atm (1.01325×10^5 Pa). When referring to the oxidation half-reaction, the standard hydrogen electrode (SHE) is utilized as the reference mechanism. Standard reduction potentials are the entity names that are usually used to refer to the entities indicated above. When attempting to obtain conventional oxidation potentials, it is necessary to make use of the opposite sign. The tabular numbers that are presented can be utilized in the process of determining the standard cell potential for electrochemical cells. The numerical number that represents the difference between the reduction potential presented by the cathode and the reduction potential displayed by the anode is the standard cell potential. This value can be defined as the percentage of the difference between the two potentials.

$$E^\circ_{cell} = E^\circ_{cathode} - E^\circ_{anode}$$

5.14 CELL POTENTIAL OR EMF OF A CELL

The electromotive force (EMF) of the cell, also known as the cell potential or cell voltage, often refers to the difference in electrode potentials that are displayed by the two half-cells. Other names for the cell potential include cell voltage and cell potential. By utilizing the electrode potentials of the two half cells that comprise the cell, it is possible to ascertain the electromotive force (EMF) of a cell, which is also referred to as the cell potential. When the potential of the half-cell on the left-hand side is subtracted from the potential of the half-cell on the right-hand side, the electromotive force (EMF) of the cell is found to be positive. This can be determined by subtracting the positive potential from the negative potential.

1. When considering the oxidation potential of both the anode and cathode

$$E^\circ_{cell} = E^\circ_{left} - E^\circ_{right}$$

$$E^\circ_{cell} = E^\circ_{anode} - E^\circ_{cathode}$$

2. When considering the reduction potential of the cathode and anode

$$E^\circ_{cell} = E^\circ_{right} - E^\circ_{left}$$

$$E^\circ_{cell} = E^\circ_{cathode} - E^\circ_{anode}$$

The magnitude of the electrical energy produced by the spontaneous reaction is directly proportional to the cell potential. The cell potential, denoted as E, serves as a quantitative indicator of the ability of a cell reaction to facilitate the movement of electrons in a circuit by exerting both attractive and repulsive forces. The Nernst equation is employed in order to account for deviations from standard circumstances and calculate the cell potential (E_{cell}).

5.15 NERNST EQUATION

The impact of electrolytic concentration and temperature on the electromotive force (EMF) of a cell. The cell potential under any given conditions is contingent upon the potential observed under standard state conditions, as well as an additional term accounting for the potential observed under nonstandard state conditions. The electrode potential and electromotive force (EMF) of a cell are influenced by several factors, including the characteristics of the electrode, the temperature, and the activities (concentrations) of the ions present in the solution. Elements with stronger negative reduction potentials exhibit a greater propensity to function as reducing agents, hence undergoing oxidation themselves. Reduction reactions take place at the electrode with a greater potential, while oxidation reactions occur at the electrode with a lower potential. The volt (V) is the standard unit of potential, while the Coulomb (C) is the accepted unit of charge. These are related by: 1V = 1J/C.

The Faraday constant, denoted by the letter F, is the force that determines the charge of one mole of electrons ($F = 96{,}500\ C\ mol^{-1}$). By combining the standard cell potential and the Faraday constant, we are able to obtain an equation that represents the value of $\Delta G°$.

$$\Delta G° = -nF\, E°_{cell}$$

In this equation, the variable $\Delta G°$ represents the change in Gibbs free energy, n represents the number of moles of electrons, F represents the Faraday constant, and $E°_{cell}$ is the standard cell potential.

Walther Hermann Nernst (1864–1941) Nernst's initial investigations in the field of electrochemistry were motivated by Arrhenius' postulation of the dissociation theory, which explains the behavior of ions in solution. In 1889, the individual in question provided a comprehensive explanation of the theory pertaining to galvanic cells. This explanation was based on the assumption of an "electrolytic pressure of dissolution," which facilitates the movement of ions from electrodes into a solution. This process is counteracted by the osmotic pressure exerted by the dissolved ions. Furthermore, in the year 1889, he demonstrated the manner in which the attributes of the generated electrical current may be employed for the purpose of determining the alteration in free energy during the chemical reaction that gives rise to said current.

Walther Hermann Nernst

Through the use of thermodynamic analysis, one is able to make a determination regarding the connection that exists between the concentration of ions in solution and the electrode and cell potential. Between the Gibbs free energy and the equilibrium constant, there is a correlation that may be demonstrated.

$$\Delta G = \Delta G° + RT \ln Q$$

Where Q is the reaction quotient $Q = \frac{\text{Product}}{\text{Reactant}}$

But $\Delta G = -nF\, E°_{cell}$ and $\Delta G° = -nF\, E°_{cell}$

This means

$$-nFE = -nF\, E° + RT \ln Q$$

Dividing both sides by $-nF$, we get the Nernst equation:

$$E = E° - \frac{RT}{nF} \times \ln Q$$

using base-10 logarithms

$$E = E° - \frac{2.303RT}{nF} \times \log Q$$

All solutions are expressed in units of molarity, and the concentration of gases is expressed as the partial pressure in atmospheres. At room temperature (T = 298 K), the gas constant (R) is equal to 8.314 J/mol K, and Faraday's constant (F) is equal to 96,485 J/V-mol. In addition,

n represents the number of electrons that are involved in the reaction, and Q represents the reactant quotient.

$$\frac{2.303RT}{F} = 0.0591\,\text{V}$$

The final form of the Nernst Equation becomes

$$E = E^\circ - \frac{0.0591}{n} \times \log Q$$

This equation is called Nernst equation. It is clear that electrode potential is directly proportional to molar concentration of ions in the same solution and temperature.

Effect of temperature on EMF consider the Daniel cell reaction, $Zn + Cu^{2+} \longrightarrow Zn^{2+} + Cu$ Nernst equation is given by

$$E = E^\circ - \frac{0.0591}{n} \times \log \frac{Zn^{2+}}{Cu^{2+}}$$

Case 1: When $[Zn^{2+}] = [Cu^{2+}]$, the logarithmic term become zero, i.e. $E_{cell} = E^\circ_{cell}$ and EMF of the cell will be independent of temperature.

Case 2: When $[Zn^{2+}] > [Cu^{2+}]$, the logarithmic term remains negative, i.e. $E_{cell} < E^\circ_{cell}$ and EMF decreases with rise of temperature.

Case 3: When $[Zn^{2+}] > [Cu^{2+}]$, the logarithmic term become positive, i.e. $E_{cell} > E^\circ_{cell}$ and EMF increases with rise of temperature.

Consider the reversible half-reaction

$$aA + bB + \cdots + ne^- \longleftrightarrow cC + dD + \cdots$$

$$E = E^\circ - \frac{RT}{nF} \ln \frac{[C]^c [D]^d \cdots}{[A]^a [B]^b \cdots}$$

$$E = E^\circ - \frac{0.0591}{n} \log \frac{[C]^c [D]^d \cdots}{[A]^a [B]^b \cdots}$$

Half-cell potential is given by:

$$E = E^\circ - \frac{RT}{nF} \ln \frac{[\text{Red}]}{[\text{Ox}]}$$

E° is the thermodynamic driving force for the reaction, which occurs when the other reaction in the half cell is a typical hydrogen electrode reaction (where E° is equal to 0 V).

Please take note that each of these E° values is derived from the equation being written as a reduction (this is also referred to as the standard reduction potential).

5.15.1 Applications of Nernst Equation

1. To investigate the impact of electrolyte content on electrode potential.
2. Calculation of equilibrium constant of a reaction.

3. Determination of EMF.
4. Determination of unknown concentration.
5. Determination of a pH of a solution.
6. Used for finding out valency of an ion.

SOLVED PROBLEMS

PROBLEM 5.1 What is the difference between electrochemical cell and electrolytic cell?

Answer Differences between electrolytic and electrochemical cells

Electrolytic Cell	Electrochemical Cells
Electrolytic cells utilize electricity.	Galvanic cells serve as a means of storing electrical energy.
Electrical energy is transformed into chemical energy.	Conversion of chemical energy to electrical energy.
The anode is the positively charged plate, whereas the cathode is the negatively charged plate.	The anode is the negatively charged plate, while the cathode is the positively charged plate.
The cell receives electrons from the external power supply.	Electrons are extracted from the cell.
Not a spontaneous reaction. For example, in the process of electroplating, the change in Gibbs free energy (ΔG) is greater than zero.	Spontaneous reaction. Generate an electric current by causing electrons to go from the anode to the cathode through an external conductor. Example: Corrosion
Faraday's law of electrolysis regulates the magnitude of the chemical reaction taking place at the electrode.	The electromotive force (EMF) of the cell is determined by the concentration of the electrolyte and the chemical properties of the electrode, as described by the Nernst Equation.

PROBLEM 5.2 The resistance of 0.01N NaCl solution at 25°C is 200 ohm. Calculate the equivalent conductance and molar conductance of the solution (Given that: Cell constant = 1).

Solution

Conductance of the cell, $C = 1/R$

$= 1/200 = 0.005$ S

Specific conductance = conductance × cell constant

$= 0.005 \times 1$

$= 0.005 \text{ S cm}^{-1}$

Equivalent Conductance = Specific conductance × (1000/N)

$= 0.005 \times 1000/0.01$

$= 500 \text{ ohm}^{-1}\text{cm}^{2}\text{eq}^{-1}$

Molar Conductivity = Equivalent conductivity × n-factor

= 500 × 1

= 500 $ohm^{-1}mol^{-1}cm^2$

PROBLEM 5.3 A decinormal solution of NaCl has specific conductivity equal to 0.0092. If ionic conductance of Na^+ and Cl^- ions are 43.0 and 65.0 ohm^{-1} respectively. Calculate the degree of dissociation of NaCl solution.

Solution

Normality= 0.10 N

Equivalent conductance of NaCl

$$\lambda_m^C = \frac{\kappa \times 1000}{\text{Molarity}}$$

$$= 0.0092 \times 10000$$

$$= 92\ \text{ohm}^{-1}$$

$$\Lambda_m^{\infty} = \lambda_{Na^+}^{\infty} + \lambda_{Cl^-}^{\infty}$$

$$\alpha = \frac{\lambda_m^C}{\lambda_m^{\infty}} = \frac{92}{108} = 0.85$$

PROBLEM 5.4 Equivalent conductance of NaCl, HCl and C_2H_5COONa at infinite dilution are 126.45, 426.16 and 91 ohm^{-1} cm^2 respectively. Calculate the equivalent conductance of C_2H_5COOH.

Solution We know that,

$$\Lambda_m^{\infty}(C_2H_5COOH) = \Lambda_m^{\infty}(CH_3COONa) + \Lambda_m^{\infty}(HCl) - \Lambda_m^{\infty}(NaCl)$$

$$= 91 + 426.16 - 126.45$$

$$= 390.71\ \text{ohm}^{-1}\ \text{cm}^2$$

PROBLEM 5.5 The equivalent conductance of acetic acid at 25°C and at infinite dilution is 390.7 mho cm^2 Eq^{-1}. The equivalent conductance of a 5.9 × 10^{-3} M solution of acetic acid is 14.4 mho cm^2 Eq^{-1}. What's the degree of dissociation of acetic acid at this concentration?

Solution We know that,

$$\alpha = \frac{\Lambda_c}{\Lambda_{\infty}}$$

$$= 14.4/390.7$$

$$= 0.037 \text{ or } 3.7\%$$

PROBLEM 5.6 A 0.01 D solution of KCl has a specific conductance value of 0.00141 mho cm^{-1}. A cell filled with this solution has a resistance of 4.2156 ohms.

(a) What is the cell constant?

(b) The same cell filled with a solution of HCl has a resistance of 1.0326 ohms. What is specific conductivity of the HCl solution?

(A one demal (1 D) solution contains 1 mole of salt in 1 litre solution at 0°C)

Solution

(a)

$$\text{Cell constant} = \frac{\text{Specific conductance}}{\text{Conductance}}$$

$$= \frac{\text{Specific conductance}}{1/\text{Resistance}}$$

$$= \frac{0.00141}{1/4.2156} = 0.00594\,\text{cm}^{-1}$$

(b) Conductance of HCl solution $= \dfrac{1}{1.0326}$ mho

Since the same cell is used, we shall take the same value of cell constant in this case.

$$\text{Specific conductance} = \text{Cell constant} \times \text{Conductance}$$

$$= 0.00594 \times \frac{1}{1.0326}$$

$$= 0.00575\,\text{mho}\,\text{cm}^{-1}$$

PROBLEM 5.7 The equivalent conductance at infinite dilution are sodium benzoate = 82.4; hydrochloric acid = 426.2 and sodium chloride = 126.5 mho cm². Calculate Λ^{∞} for benzoic acid.

Solution From Kohlrausch's law, we get,

$$\Lambda^{\infty}_{\text{benzoic acid}} = \Lambda^{\infty}_{\text{sod. benzoate}} + \Lambda^{\infty}_{\text{hydrochloric acid}} - \Lambda^{\infty}_{\text{sodium chloride}}$$

$$= 82.6 + 426.2 - 126.5$$

$$= 382.1\,\text{mho}\,\text{cm}^{2}$$

PROBLEM 5.8 The standard potential of the Ag^{+}/Ag electrode is +0.80 V and the standard potential of the cell $Fe_{(s)}|Fe^{2+}{}_{(aq)}||Ag^{+}{}_{(aq)}|Ag_{(s)}$ is +1.24 V. What is the standard potential of the Fe^{2+}/Fe electrode?

Solution Half reactions

$$\text{Fe}(s) \longrightarrow \text{Fe}^{2+}(aq) + 2\text{e}^{-}$$

$$\text{Ag}^{+}(aq) + \text{e}^{-} \longrightarrow \text{Ag}(s)$$

Oxidation reaction—Anode

Reduction reaction—Cathode

$$E^{\circ}_{\text{Cell}} = E^{\circ}_{\text{Cathode}} - E^{\circ}_{\text{Anode}}$$

$$E^{\circ}_{\text{Anode}} = E^{\circ}_{\text{Cathode}} - E^{\circ}_{\text{Cell}}$$

$$E^{\circ}_{\text{Anode}} = 0.80\text{ V} \quad 1.24\text{ V}$$

$$E^{\circ}_{\text{Anode}} = -0.44\text{ V}$$

PROBLEM 5.9 The equilibrium constant for the reaction

$$Ni(s) + Hg_2Cl_2(s) \longrightarrow 2Hg(l) + 2Cl^{-1}(aq) + Ni^{2+}(aq)$$

is 1.8×10^{19} at 298K. What is the value of the standard cell potential $E°_{cell}$ for this reaction?

Solution We know that

$$\begin{aligned}\Delta G^\circ &= -RT \ \ln K \\ &= -(8.314 \text{ J}K^{-1}\text{mol}^{-1}) \ (298 \text{ K}) \ \ln(1.8 \times 10^{19}) \\ &= -109847.8 \text{ J mol}^{-1} \\ &= -1.098 \times 10^5 \text{ J mol}^{-1}\end{aligned}$$

By using Gibs free energy equation

$$\Delta G^\circ = -nF \, E^\circ_{cell}$$

$$\begin{aligned}E^\circ_{cell} &= \frac{-\Delta G^\circ}{nF} \\ &= -\frac{(-1.098 \times 10^5 \text{ J mol}^{-1})}{(2 \text{ mol}) \ (96500 \text{ C})} \\ &= 0.57 \text{ J/C} = 0.57 \text{ V}\end{aligned}$$

PROBLEM 5.10 Write the name of the cell which is generally used in hearing aids. Write the reactions taking place at the anode and the cathode of this cell.

Solution Mercury cells are used in hearing aids.

Reaction at anode:

$$Zn(s) + 2OH^{-1} \longrightarrow ZnO(s) + H_2O + 2e^{-1}$$

Reaction at cathode:

$$HgO + H_2O + 2e^- \longrightarrow Hg(l) + 2OH^{-1}$$

QUESTIONS

5.1 Find the valency of mercurous ions with the help of following cell:

Hg	0.001 N Mercurous nitrate in 0.1N HNO_3	0.01 N Mercurous nitrate in 0.1N HNO_3	Hg
	Anode	Cathode	

When the EMF measured at 18 °C is 0.029 Volt (F = 96,500 C mol^{-1}).

5.2 Consider the reaction,

$$2Ag^+(aq) + Cd(s) \longrightarrow 2Ag(s) + Cd^{2+}(aq)$$

Given that $E^{\circ}_{Ag^+/Ag} = 0.80\,V$ and $E^{\circ}_{Cd^{2+}/Cd} = -0.40\,V$. What is the standard potential E° for this reaction?

(***Answer:*** 1.20 volt)

5.3 To find the standard potential of M^{3+}/M electrode, the following cell is constituted:

$$Pt\,|M|\,M^{3+}(0.0018\ mol^{-1}L)\,|Ag^{+}(0.01\ mol^{-1}L)|\,Ag$$

The EMF of this cell is found to be 0.42 volt. Calculate the E°_{Anode}. Given that E°_{Anode} = 0.80 volt.

5.4 The standard electrode potential for Daniell cell is 1.1 V. Calculate the standard Gibbs energy for the cell reaction. (F = 96,500 C mol^{-1})

(**Answer:** 212,300 CV mol^{-1})

5.5 Calculate the degree of dissociation (α) of acetic acid if its molar conductivity (Λ_m) is 39.05 S cm^2 mol^{-1}. Given: $\lambda°(H^+)$ = 349.6 S cm^2 mol^{-1} and $\lambda°(CH_3COO^-)$ = 40.9 S cm^2 mol^{-1}.

5.6 The cell in which the following reaction occurs :

$$2Fe^{3+}(aq) + 2I^{-}(aq) \longrightarrow 2Fe^{2+}(aq) + I_2(s)$$

The above equation has E°_{cell} = 0.236V at 298K. Calculate the standard Gibbs energy and the equilibrium constant of the cell reaction.
(Antilog of 6.5 = 3.162 × 10^6; Antilog of 8.0 = 1 × 10^8; Antilog of 8.5 = 3.162 × 10^8)

(**Answer:** –45.548 kJ/mol^{-1})

5.7 Calculate the emf of the following cell at 298 K:

$$Fe(s)\,|\,Fe^{2+}(0.001\ M)\,||\,H^{+}(1M)\,|\,H_2(g)(1\ bar),\ Pt(s)$$

(Given E°_{cell} = +0.44V)

(**Answer:** 0.529 V)

5.8 The resistance of 0.01 M NaCl solution at 25°C is 200 Ω. The cell constant of the conductivity cell used is unity. Calculate the molar conductivity of the solution.

(**Answer:** 5 × 10^{-4} Sm^2 mol^{-1})

5.9 Conductivity of 2.5 × 10^{-4} M methanoic acid is 5.25 × 10^{-5} S cm^{-1}. Calculate its molar conductivity and degree of dissociation.
Given: $\lambda^0(H^+)$ = 349.5 Scm^2 mol^{-1} and $\lambda^0(HCOO^-)$ = 50.5 Scm^2 mol^{-1}.

(**Answer:** 400 Sm^2 mol^{-1}, 0.525)

5.10 The conductivity of 0.001 mol L^{-1} solution of CH_3COOH is 3.905 × 10^{-5} S cm^{-1}. Calculate its molar conductivity and degree of dissociation (α).
Given: λ^0 (H^+) = 349.6 S cm^2 mol^{-1} and λ^0 (CH_3COO^-) = 40.9 S cm^2 mol^{-1}

(**Answer:** 0.1)

5.11 Describe the construction and working of Galvanic cell. Calculate the EMF of the following cell and also write the cell reactions.

$$Zn\ |\ Zn^{2+}\ (0.001M)\ ||\ Ag^{+}\ (0.1M)\ |\ Ag$$

The standard potential of Ag/Ag^+ half-cell is +0.80 V and Zn/Zn^{2+} is –0.76V.

5.12 Derive Nernst Equation. The voltage of the cell $Pb/PbSO_4/Na_2SO_4/Hg$ is 0.9647 V at 25°C the temperature is 1.74×10^{-4} VK^{-1}. Calculate the values of ΔG, DS and ΔH.

5.13 Describe construction of a Galvanic cell. Write the electrode reaction and formula for its EMF.

5.14 A cell uses Zn^{2+}/Zn and Ag^{+}/Ag electrodes. Write the cell representation, half-cell reactions and net cell reaction. Calculate EMF of the cell. Given: $E^{\circ}_{Zn^{2+}/Zn} = -0.76$ V and $E^{\circ}_{Ag^{+}/Ag} = 0.8$ V.

5.15 Write equations for each half reaction and calculate E_{cell} for the following concentration $Zn/Zn^{2+}(1.0M) \| Zn^{2+}(0.15M)/Zn$ as the cell discharges. Does the concentration of two solutions becomes smaller or larger?

5.16 For a cell reaction $A(s) + 2B^{2+}(aq) \rightarrow A^{2+}(aq) + 2B(s)$ at 298 K, the equilibrium constant is 1.0×10^{4}. Calculate $E_{cell.}$

5.17 The standard EMF of the Daniel cell involving cell reaction $Zn(s) + Cu^{2+}(aq) \rightarrow Zn^{2+}(aq) + Cu(s)$ is 1.10 volts. Calculate the equilibrium constant of the cell reaction at 25°C.

5.18 A cell consists of two hydrogen electrodes. The negative electrode is in contact with a solution of 10 M hydrogen ions. The EMF of the cell is 0.118 V at 25°C. Calculate the concentration of hydrogen ion at positive electrode.

5.19 Derive Nernst equation. Calculate the mean ionic activity co-efficients of 0.1 mol/kg HCl, given the EMF of the cell H_2 (1 atm)/HCl(*aq*), AgCl(*s*)/Ag is 0.3524V and that standard electrode potential of Ag–AgCl is 0.2224V at 25°C.

CHAPTER 6

Acid-Base

6.1 INTRODUCTION

Arrhenius acids generate hydrogen ions when dissolved in water, while bases produce hydroxide ions. This definition presupposes that water functions as the solvent. Brønsted and Lowry built upon the Arrhenius explanations and established that acids are substances that donate protons, while bases are substances that accept protons. Additionally, they invented the notion of conjugate acid-base pairings.

A Brønsted–Lowry acid is defined as a substance that possesses a proton that can be easily removed. A Brønsted–Lowry base necessitates the presence of a set of nonbonding electrons. If a substance has the ability to act as either an acid or a base, it is considered amphiprotic. Examples of amphiprotic substances include HCO_3^-, HSO_4^-, H_2O. Water functions as a Brønsted–Lowry base by removing H^+ from the acid. Consequently, the acid produces a conjugate base and a hydronium ion.

$$HCl + H_2O \longrightarrow Cl^- + H_3O^+$$

In the case of a solvent capable of dissociating into a cation and an anion, the cation is considered the acid, while the anion is regarded as the base. An acid is any solute that raises the concentration of the cation, while bases are solutes that raise the concentration of the anion.

Water Ionization

Neutral water exhibits a propensity for ionization.

$$H_2O \rightleftharpoons H^+ + OH^-$$

The unbound proton combines with a water molecule to create the hydronium ion H_3O^+. The high ionic mobility is a result of the phenomenon known as proton leaping.

6.2 LEWIS ACIDS AND BASES

The Lewis definition provides additional elaboration to the existing definitions. A base is a substance that donates electron pairs, while an acid is a substance that accepts electron pairs. The two substances mix to create an adduct.

The addition of A and B results in the formation of an adduct A–B.

$$A + :B \longrightarrow A-B$$

Alternatively

$$\underset{\text{Acid}}{HA} + \underset{\text{Base}}{H_2O} \rightleftharpoons \underset{\text{Conjugated Acid}}{H_3O} + \underset{\text{Conjugated Base}}{A^-/OH^-}$$

If equilibrium is established, alterations in $[H^+]$ will modify the proportion of HA and A^-. The addition of extra H^+ ions will result in the consumption of A^- ions, leading to the formation of HA. The concentration of H_2O remains relatively constant, particularly in solutions with low acid content.

What is the molar concentration of water (H_2O)? The definition states that it is the number of moles per litre.

$$1 \text{ mole of } H_2O = 18 \text{ g} = 18 \text{ ml}$$

$$1\,\text{mol} \equiv 1\,\text{ml}$$

$$\frac{1000 \text{ g}}{18 \text{ g/mol}} = 55.5\,\text{M}$$

If there is enough amount of $[A^-]$, the surplus H^+ ions will be used and the concentration of $[H^+]$ will remain constant.

$$K_a = K[H_2O] = \frac{[H^+][A^-]}{[HA]}$$

Henceforth, we shall omit the letter "*a*" in "K_a". Weak acids have a dissociation constant (K) less than 1 ($K < 1$), while strong acids have a dissociation constant more than 1 ($K > 1$). A strong acid undergoes full dissociation and donates all of its protons to water molecules, resulting in the formation of hydronium ions (H_3O^+). The reaction between H^+ ions in aqueous solution and NH_3 in aqueous solution results in the formation of NH_4^+ ions.

$$H^+(aq) + :NH_3(aq) \longrightarrow NH_4^+(aq)$$

Additionally, it encompasses the interactions between metal ions or atoms and ligands, resulting in the formation of coordination compounds: The aqueous solution of Ag^+ reacts with two moles of NH_3 to form $Ag(NH_3)^{2+}$ in solution.

$$Ag^+(aq) + 2:NH_3(aq) \longrightarrow Ag(NH_3)^{2+}(aq)$$

Furthermore, substances that lack electrons, such as trivalent boron, are classified as Lewis acids.

$$B(CH_3)_3 + :NH_3(aq) \longrightarrow (CH_3)_3B \longleftarrow NH_3$$

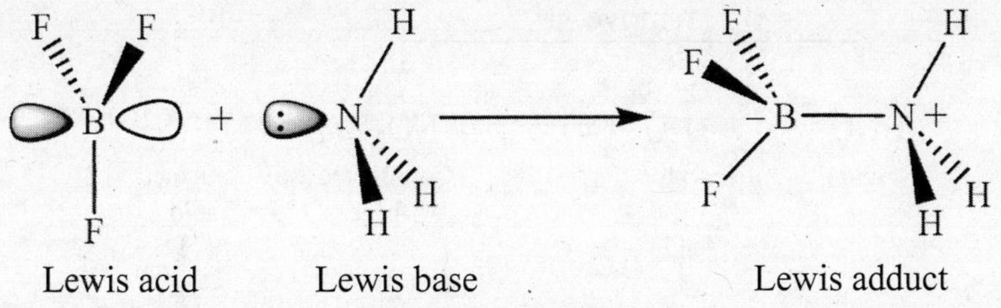

The highest occupied molecular orbital (HOMO) of the Lewis base interacts with the lowest unoccupied molecular orbital (LUMO) of the Lewis acid through their respective electron pairs. The molecular orbitals of the adduct have a lower energy level. If there is a decrease in energy, the adduct is considered stable. It can be easily defined by the following illustration:

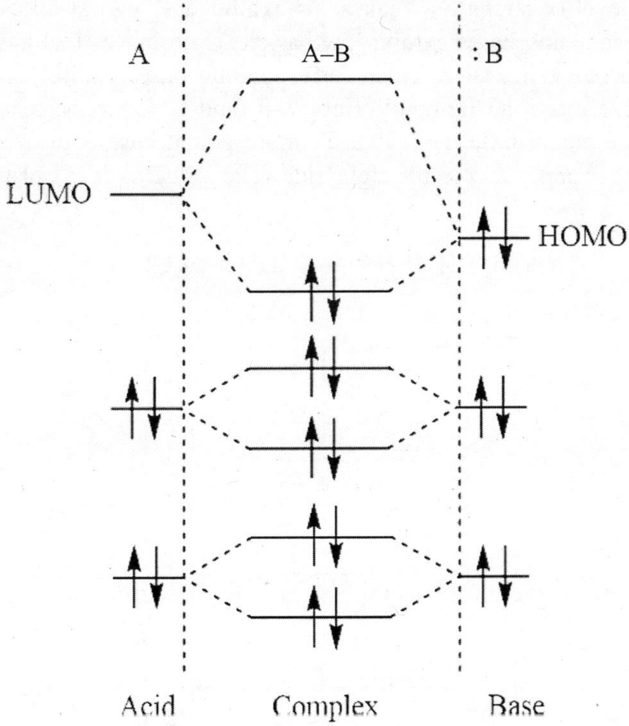

A Lewis base possesses an electron pair in its HOMO that has appropriate symmetry to engage with the LUMO of the Lewis acid. The proximity of the two orbitals in terms of energy directly correlates with the strength of the bond in the *adduct*.

6.3 CONJUGATE ACIDS AND BASES

The term "conjugate" is derived from the Latin word "*conjugare*" which means "to join together." The reaction between acids and bases always results in the formation of their respective conjugate bases and acids.

remove H^+

$$\underset{\text{Acid}}{HNO_2(aq)} + \underset{\text{Base}}{H_2O(l)} \rightleftharpoons \underset{\text{Conjugated base}}{NO_2^-(aq)} + \underset{\text{Conjugated acid}}{H_3O^+(aq)}$$

add H^+

6.4 STRENGTH OF ACID AND BASE

Powerful acids undergo total dissociation when mixed with water. Their corresponding bases exhibit a rather low level of strength. Weak acids exhibit only partial dissociation in aqueous solutions. Their corresponding bases exhibit low basicity. Substances that have minimal acidity do not undergo dissociation in water. Their corresponding bases exhibit exceptional strength. During every acid-base reaction, the equilibrium will tend to favor the reaction that transfers the proton to the more potent base. H_2O is a far more potent base compared to Cl^-, resulting in an equilibrium that strongly favors the right side. The value of K is not tested because it is much greater than 1 ($K \gg 1$).

$$HCl(aq) + H_2O(l) \longrightarrow H_3O^+(aq) + Cl^-(aq)$$

The strength of acid and bases are illustrated below:

	Acid	Base	
Strong	HCl	Cl^-	Negligible
Strong	H_2SO_4	HSO_4^-	Negligible
Strong	HNO_3	NO_3^-	Negligible
	H_3O^+	H_2O	
Weak	HSO_4^-	SO_4^{2-}	Weak
Weak	H_3PO_4	$H_2PO_4^-$	Weak
Weak	HF	F^-	Weak
Weak	$C_2H_3O_2$	$C_2H_3O_2^-$	Weak
Weak	H_2CO_3	HCO_3^-	Weak
Weak	H_2S	HS^-	Weak
Weak	H_2PO_4	HPO_4	Weak
Weak	NH_4^+	NH_3	Weak
Weak	HCO_3^-	CO_3^{2-}	Weak
Weak	HPO_4^-	PO_4^{2-}	Weak
	H_2O	OH^-	
Negligible	OH^-	O^{2-}	Strong
Negligible	H_2O	H^-	Strong
Negligible	CH_4	CH_3^-	Strong

6.5 HARD AND SOFT ACIDS AND BASES (HSAB)

The polarizability of an acid or base influences its reactivity. Hard acids and bases are characterized by their small size, compact structure, and lack of polarizability. Soft acids and bases have a greater size and a more spread out dispersion of electrons. Hard acids exhibit a tendency to selectively react with hard bases, while soft acids have a tendency to selectively react with soft bases, in addition to their inherent strength.

EXAMPLES: Solubility in water Silver halides

$$AgX(s) + H_2O(l) \rightleftharpoons Ag^+(aq) + X^-(aq)$$

Compound	Solubility product
AgF	205
AgCl	1.8×10^{-10}
AgBr	5.2×10^{-13}
AgI	8.3×10^{-17}

Solubility of Lithium Halides

$$LiBr > LiCl > LiI > LiF$$

Lithium bromide is greater than lithium chloride, which is greater than lithium iodide, which is greater than lithium fluoride. LiF exhibits a larger solvation (Δ_{solv}) compared to other salts, despite its lower solubility in water. This is a result of the robust interaction between a strong acid (Li^+) and a hard base (F^-).

EXAMPLE: Thiocyanate Bonding

Thiocyanate bonding refers to the formation of chemical bonds using thiocyanate ions.

The ligand SCN^- exhibits linkage isomerism by coordinating to metals through either the sulphur or the nitrogen atom. The Mercury (II) ion forms a link with the sulphur atom through a soft-soft contact, whereas the zinc ion forms a bond with the nitrogen atom.

EXAMPLE: K for ligand exchange reactions

Make a comparison between:

$$[MeHg(H_2O)]^+ + HCl \rightleftharpoons MeHgCl + H_3O^+ \quad K = 1.8 \times 10^{12}$$

$$[MeHg(H_2O)]^+ + HF \rightleftharpoons MeHgF + H_3O^+ \quad K = 4.5 \times 10^{-2}$$

Numerous attempts have been made to classify different metal ions and anions in order to forecast their reactivity, solubility, and other properties. R.G. Pearson (1963) classified acids and bases into two groups based on their hardness or softness, as determined by their K_f values. The bonding order for hard acids is as follows: $F^- > Cl^- > Br^- > I^-$. The bonding order for soft acids is as follows: $I^- > Br^- > Cl^- > F^-$.

Hard acids or bases are characterized by their compactness and the relatively strong attraction between their electrons and the nucleus. They exhibit low polarizability. F^- is a strong base, while metal ions like Li^+ are considered strong acids. Ions that are large and have a high degree of polarizability are classified as "soft." Iodide is classified as a soft base, while transition metals with low charge density, like Ag^+, are categorized as soft acids.

Hard acids have a tendency to form strong bonds with hard bases. Soft acids have a tendency to form chemical bonds with soft bases. Utilize the hard-soft acid/base technique to determine the solubility (either high or low) of silver fluoride, silver iodide, lithium fluoride, and lithium iodide. Classify each compound as a Lewis acid or Lewis base, and determine whether it is a hard or soft acid/base.

Charge Density—Hard Acids

Hard acids generally have a high level of charge density. They frequently consist of metal ions that possess a stronger positive charge and have a tiny ionic size. Their *d* orbitals are frequently inaccessible for participation in π bonding.

Charge Density—Soft Acids

Soft acids generally exhibit reduced charge density, which is characterized by lower ionic charge and larger ionic size. Their *d* orbitals are accessible for π bonding. Soft acids often refer to transition metals from the 2nd and 3rd rows of the periodic table, which have a positive charge of +1 or +2 and *d* orbitals that are either filled or almost filled.

Acids		
Hard Acids	**Borderline**	**Soft Acids**
H^+, Li^+, Na^+, K^+		
Be^{2+}, Mg^{2+}, Ca^{2+}		
BF_3, BCl_3, $B(OR)_3$	BBr_3, $B(CH_3)_3$	BH_3, Tl^+, $Tl(CH_3)_3$
Al^{3+}, $Al(CH_3)_3$, $AlCl_3$, AlH_3		
Cr^{3+}, Mn^{2+}, Fe^{3+}, Co^{3+}	Fe^{2+}, Co^{2+}, Ni^{2+}	Cu^+, Ag^+, Au^+
	Cu^{2+}, Zn^{2+}, Rh^{3+}	Cd^{2+}, $Hg_2{}^{2+}$
	Ir^{3+}, Ru^{3+}, Os^{2+}	Hg^{2+}, Pd^{2+}, Pt^{2+},
SO_3	SO_2	Pt^{4+}

Bases		
Hard Bases	**Borderline**	**Soft Bases**
F^-, Cl^-	Br^-	H^-, I^-
H_2O, OH^-, O^{2-}		H_2S, HS^-, S^{2-}
ROH, RO^-, R_2O, $CH_3CO_2{}^-$		RSH, RS^-, R_2S
NO_3^-, ClO_4^-	$\underline{N}O_2^-$, N_3^-, N_2	$\underline{S}CN^-$, $\underline{C}N^-$, RNC, CO
CO_3^{2-}, SO_4^{2-}, PO_4^{3-}	SO_3^{2-}	$S_2O_3{}^{2-}$
NH_3, RNH_2	$C_6H_5NH_2$, pyr	R_3P, C_6H_6

Effect of Linkage Site

$$SC\underline{N}^- \text{ vs. } NC\underline{S}^-$$

Nitrogen has a tendency to form coordination bonds with acids that are more electronegative, such as silicon, while sulphur has a preference for coordinating with acids that are less electronegative, such as platinum ions with a charge of +2.

Effect of Oxidation Number

The presence of Cu^{2+}/Cu^+ affects the acidity hardness, whereas the presence of SO_3/SO_2 also affects the acidity hardness. On the other hand, the presence of NO_3^-/NO_2^- affects the basicity hardness, and the presence of SO_4^{2-}/SO_3^{2-} also affects the basicity hardness.

Acid or Base Strength

It is crucial to understand that the concepts of hard/soft considerations are unrelated to the strength of acids or bases. An acid or a base can exhibit characteristics of hardness or softness, as well as possess varying degrees of strength, ranging from weak to strong. When comparing the reactivity of two bases towards the same acid in a competitive reaction, it is important to take into account both the relative strength of the bases and the hard/soft characteristics of each base and acid.

Consider the chemical reaction involving zinc oxide (ZnO) and lithium butyl (LiC_4H_9).

$$\underset{\text{soft-hard}}{ZnO} + \underset{\text{hard-soft}}{2LiC_4H_9} \rightleftharpoons \underset{\text{soft-soft}}{Zn(C_4H_9)_2} + \underset{\text{hard-hard}}{Li_2O}$$

The zinc ion has high Lewis acid properties, while the oxide ion demonstrates strong Lewis basic properties. Nevertheless, the reaction progresses towards the right ($K > 1$) due to the dominance of hard/soft factors over acid-base strength considerations.

The Characteristics of the Adduct

Adducts formed between hard acids and hard bases have a higher degree of ionic character in their bonding. These are typically more energetically preferred. Adducts formed between soft acids and soft bases have a higher degree of covalent character.

Additional Factors to Consider

Upon formation of the adduct, a typical occurrence involves a modification in the spatial arrangement surrounding the Lewis acid site.

$$BX_3 + :N(CH_3)_3 \longrightarrow X_3B \longleftarrow :N(CH_3)_3$$

The relative stability of the adduct is as follows: $BBr_3 > BCl_3 > BF_3$.

The order of this arrangement appears to be contrary to what one would anticipate based on the size or electronegativity of halogens. BF_3 exhibits a reactivity pattern that indicates the presence of π bonding to some extent. Steric factors can exert an influence. A notable instance is the adverse chemical response between $:N(C_6H_5)_3$ and BCl_3. The presence of the sizable phenyl groups has a destabilizing effect on the product by interacting with the chlorine atoms on boron.

6.5.1 Practical Uses of Hard/Soft Theory

The Qualitative Scheme is a chemical process that employs a sequence of processes to effectively isolate and detect many metal ions. This scheme primarily relies on the distinguishing characteristics of the metal ions, specifically their hard and soft qualities. The less malleable metals are separated as chlorides or sulphides, while the more rigid ions are produced as carbonates.

Qualitative Analysis Separation					
	Group 1	**Group 2**	**Group 3**	**Group 4**	**Group 5**
HSAB acids	**Soft**	**Borderline and soft**	**Borderline**	**Hard**	**Hard**
Reagent	HCl	H_2S (acidic)	H_2S (basic)	$(NH_4)_2CO_3$	Soluble
Precipitates	AgCl $PbCl_2$ Hg_2Cl_2	HgS CdS CuS SnS As_2S_3 Sb_2S_3 Bi_2S_3	MnS FeS CoS NiS ZnS $Al(OH)_3$ $Cr(OH)_3$	$CaCO_3$ $SrCO_3$ $BaCO_3$	Na^+ K^+ NH_4^+

Natural Evidence

The elements found in the earth's crust are categorized in geochemistry as either lithophiles or chalcophiles. The lithophile elements, such as Li^+, Mg^{2+}, Ti^{3+}, Al^{3+} and $Cr^{2+,\,3+}$, are commonly present in the form of silicates, where they are linked to oxygen atoms. These substances are highly reactive Lewis acids. The chalcophile elements are commonly seen in the form of sulphides or in chemical compounds linked to Se^{2-} or Te^{2-}. The ions included are: Cd^{2+}, Pb^{2+}, Sb^{3+}, and Bi^{3+}. These substances are classified as mild Lewis acids. Zinc ions, which have borderline properties, are commonly encountered in the form of sulphides.

6.6 PH

The pH is the logarithm of the hydronium ion concentration, expressed as the negative base-10 value. The "p" in pH indicates the application of a negative logarithm to the number of hydrogen ions.

$$pH = -\log [H_3O^+]$$

To obtain less precise measurements, Litmus paper might be utilized. The color of "Red" paper changes to blue when the pH level exceeds approximately 8, while the color of "Blue" paper changes to red when the pH level drops below approximately 5. These papers serve as indicators.

The equilibrium constant for the auto ionization of water, K_w, is equal to the product of the concentrations of hydronium ions (H_3O^+) and hydroxide ions (OH^-), which is 1.0×10^{-14} in pure water. $K_w = [H_3O^+]\,[OH^-] = 1.0 \times 10^{-14}$. Since the concentration of hydronium ions ($[H_3O^+]$) is equal to the concentration of hydroxide ions ($[OH^-]$) in pure water, the value of $[H_3O^+]$ can be calculated as $[H_3O^+] = [OH^-]$, $[H_3O^+] = (1.0 \times 10^{-14})^{1/2} = 1.0 \times 10^{-7}$. Thus, in the case of pure water, the pH value is calculated as $pH = -\log(1.0 \times 10^{-7}) = 7.00$. An acid contains a greater concentration of hydronium ions ($[H_3O^+]$) compared to pure water, resulting in a pH value below 7. A base contains a higher concentration of hydroxide ions ($[OH^-]$) than pure water, resulting in a lower concentration of hydronium ions ($[H_3O^+]$). As a result, the pH of a base is greater than 7. Additional instances include $pOH = -\log[OH^-]$ and $pK_w = -\log K_w$.

Relationships among $[H^+]$, $[OH^-]$, and pH at 25°C

Solution	$[H^+]$	$[OH^-]$	pH
Acidic	$> 1.0 \times 10^{-7}$	$< 1.0 \times 10^{-7}$	< 7.00
Neutral	$= 1.0 \times 10^{-7}$	$= 1.0 \times 10^{-7}$	$= 7.00$
Basic	$< 1.0 \times 10^{-7}$	$> 1.0 \times 10^{-7}$	> 7.00

The pH value is often represented as the negative logarithm of the concentration of hydrogen ions, $[H^+]$. It calculates the negative logarithm (base 10) of the molar concentrations of hydrogen ions (H^+) in a solution, which are actually hydronium ions (H_3O^+). The HCl solution with a pH of 2.0 has a hydronium ion concentration of 1×10^{-2} (1/100). When comparing a less concentrated solution of HCl with a pH of 5.0, it has a hydronium ion concentration of 1×10^{-5} (1/100,000). The concentration of hydrogen ions in pure water is 10^{-7}, resulting in a pH value of 7. Acids possess a significant concentration of hydrogen ions ($[H^+]$) and hence have a low pH value. Bases possess a reduced concentration of hydrogen ions ($[H^+]$) and hence exhibit an elevated pH level. Bases donate hydroxide (OH^-) ions upon dissociation.

These molecules attach to the H^+ ions that are generated during the dissociation of water. Therefore, the hydroxide (OH^-) ions in solution effectively absorb the hydrogen (H^+) ions, leading to a decrease in their concentration. The presence of NaOH with a pH of 12.0 results in a high concentration of –OH ions, causing the majority of H+ ions to combine with water molecules. This leads to a significant reduction in the concentration of free H^+ (and hydronium) ions, which becomes 1×10^{-12} (equivalent to 1 trillionth).

The following are the pH values of many common compounds.

Common Substances	pH	pOH	Common Substances	pH	pOH
Gastric juice	1	13	Human blood, tears	7	7
Lemon juice	2	12	Egg white, seawater		
Cola, vinegar	3	11	Baking soda	8	6
Wine			Borax	9	5
Tomatoes	4	10	Milk of magnesia	10	4
Banana			Lime water	11	3
Black Coffee	5	9	Household ammonia	12	2
Rain			Household bleach		
Saliva	6	8	NaOH, 0.1 M	13	1
Milk					

$$[H_3O^+][OH^-] = K_w = 1.0 \times 10^{-14}$$

We know that

$$-\log[H_3O^+] + -\log[OH^{-1}] = -\log K_w = 14.00$$

or, in other words,

$$pH + pOH = pK_w = 14.00$$

6.7 BUFFER SOLUTIONS

A buffer solution is a solution that maintains a constant pH value even when a little amount of acid or base is added. The solution with a pH less than 7 is prepared by combining a weak acid, ethanoic acid, with either sodium or potassium salt, namely sodium ethanoate. The Alkaline Buffer is created by combining ammonia, a weak base, with ammonium chloride.

Applications

Buffer solution is used in calibration of pH meters, regulation of biological systems such as blood and to control pH of shampoos. A weak acid is necessary for the establishment of an equilibrium, as it facilitates the removal and production of ions. The dissociation is minimal and the presence of ions is limited. When an excessive amount of a slightly soluble ionic compound is combined with water, a state of equilibrium is reached between the solid compound and the ions present in the saturated solution. This equilibrium is represented by a temperature-dependent constant.

6.8 SOLUBILITY PRODUCT

The solubility product constant, K_{sp}, refers to the equilibrium constant for particular operation.

$$K_{sp} = [Ca^{2+}][C_2O_4^{2-}]$$

K_{sp} is essentially a measure of the concentration of ions in a solution at the point of saturation. When the product of the ion concentrations equals the K_{sp} value, precipitation occurs.

The solubility product constant, K_{sp}, represents the equilibrium constant for the solubility equilibrium of an ionic substance that is very slightly soluble or almost insoluble. The constant in question pertains to the quantity of substance that is dissolved in a solution at its saturation point, rather than the total amount present in the flask. This constant is determined by multiplying the equilibrium concentrations of the ions in the compound. The solid's activity is 1.

The concentration of each ion is exponentiated by the number of occurrences of that ion in the compound's formula. Lead iodide, with the chemical formula PbI_2, is an example of a sparingly soluble salt. The state of balance or stability that it maintains is:

$$PbI_2(s) \xrightleftharpoons{H_2O} Pb^{2+}(aq) + 2I^-(aq)$$

The equation representing the solubility product constant is:

$$K_{sp} = [Pb^{2+}][I^-]^2$$

The solubility of the substance can be determined by calculating it using the known solubility product constant, K_{sp}. The water solubility of an ionic chemical refers to the quantity of the compound that dissolves in a given volume of saturated solution, usually measured in grams per litre (g/L). The term "molar solubility" refers to the solubility of a substance expressed in moles per litre (mol/L).

6.8.1 Factors that Affect Solubility

1. The solubility of a substance often increases as the temperature rises.
2. The presence of common ions tends to decrease the solubility of a substance.
3. The presence of salt has a minor impact on increasing solubility.
4. The solubility of ionic compounds, where the anions are conjugate bases of weak acids, is influenced by the pH of the solution.
5. Solubility is enhanced by the creation of complex ions.

SOLVED PROBLEMS

PROBLEM 6.1 Write an equation for the reaction of HSO_3^- with water. In both cases identify the conjugate acid-base pairs.

Solution The hydrogen sulphite ion (HSO_3^-) is amphiprotic in nature. The equation can be written as

$$HSO_3^-(aq) + H_2O(l) \rightleftharpoons SO_3^{2-}(aq) + H_3O^+(aq)$$

The conjugate pairs in this equation are HSO_3^- (acid) and SO_3^{2-} (conjugate base); and H_2O (base) and H_3O^+ (conjugate acid).

$$HSO_3^-(aq) + H_2O(l) \rightleftharpoons H_2SO_3(aq) + OH^-(aq)$$

The conjugate pairs in this equation are H_2O (acid) and OH^- (conjugate base), and HSO_3^- (base) and H_2SO_3 (conjugate acid).

PROBLEM 6.2 The reaction of the oxide ion (O^{2-}) with water when lithium oxide (Li_2O) is dissolved in it causes the solution to become basic. Write down the reaction that takes place and list the conjugate acid-base combinations.

Solution

$$O^{2-}(aq) + H_2O(l) \rightleftharpoons OH^-(aq) + OH^-(aq)$$

OH^- is the conjugate acid of the base O^{2-}. OH^- is also the conjugate base of the acid H_2O.

PROBLEM 6.3 Determine the solubility of calcium fluoride in water (measured in gm/L) using the K_{sp} (3.4×10^{-11}).

Solution Let s be the molar solubility of CaF_2.

$$\underset{\text{Equilibrium}}{CaF_2(s)} \underset{}{\overset{H_2O}{\rightleftharpoons}} \underset{s}{Ca^{2+}(aq)} + \underset{2s}{2F^-(aq)}$$

After substitution into the equilibrium-constant equation

$$K_{sp} = [Ca^{2+}][F^-]^2$$

$$3.4 \times 10^{-11} = (s)(2s)^2 = (s)(4s^2) = 4s^3$$

$$3.4 \times 10^{-11} = 4s^3$$

$$s^3 = \frac{3.4 \times 10^{-11}}{4}$$

$$s = \sqrt[3]{\frac{3.4 \times 10^{-11}}{4}} = 2.0 \times 10^{-4}\,M$$

$$= [Ca^{2+}] = [CaF_2]$$

$$\text{Solubility} = \frac{2.0 \times 10^{-4}\,\text{mol } CaF_2}{L} \times \frac{78.1g\ CaF_2}{1\ \text{mol } CaF_2}$$

$$= 1.6 \times 10^{-2}\,g\ CaF_2/L = 0.016g\ CaF_2/L$$

PROBLEM 6.4 Determine the pH of a buffer with [HA] and [A⁻] of 0.1 mol dm^{-3} and 0.1 mol dm^{-3}, respectively. The weak acid HA has a K_a of 2 × 10^{-4} mol dm^{-3}.

Solution

$$K_a = \frac{[H^+(aq)][A^-(aq)]}{[HA(aq)]}$$

Rearranging

$$[H^+(aq)] = \frac{[HA(aq)] \times K_a}{[A^-(aq)]}$$

Given that [A⁻] = 0.1 mol dm^{-3}, [HA] = 0.1 mol dm^{-3}, If the K_a of the weak acid HA is 2 × 10^{-4} mol dm^{-3}.

$$[H^+(aq)] = \frac{0.1 \times 2 \times 10^{-4}}{0.1} = 2 \times 10^{-4}\ \text{mol dm}^{-3}$$

$$pH = -\log_{10}[H^+(aq)] = 3.699$$

PROBLEM 6.5 Determine the pH of the mixture that results from mixing 500 cm^3 of weak acid HX (0.1 mol dm^{-3}) with 500 cm^3 of its salt NaX (0.2 mol dm^{-3}). K_a = 4 × 10^{-5} mol dm^{-3}.

Solution

$$K_a = \frac{[H^+(aq)][X^-(aq)]}{[HX(aq)]}$$

Rearranging

$$[H^+(aq)] = \frac{[HX(aq)] \times K_a}{[X^-(aq)]}$$

After mixing the solutions, the volume is now 1 dm^3, which means that $[X^-] = 0.10$ mol dm^{-3} and [HX] = 0.05 mol dm^{-3}.
Substituting

$$[H^+(aq)] = \frac{0.05 \times 4 \times 10^{-5}}{0.1} = 2 \times 10^{-5} \text{mol dm}^{-3}$$

$$pH = -\log_{10}[H^+(aq)] = 4.699$$

PROBLEM 6.6 The pH of a freshly squeezed apple juice sample is 3.76. Compute $[H^+]$.

Solution We must compute $[H^+]$ using pH. To calculate, we shall utilise the equation pH = $-\log [H^+]$. pH = $-\log [H^+]$ = 3.76. Thus, $\log [H^+] = -3.76$

To find $[H^+]$, we need to determine the antilog of –3.76.

$$[H^+] = \text{antilog}[-3.76] = 10^{-3.76} = 1.7 \times 10^{-4} \text{ M}$$

PROBLEM 6.7 What is the pH of an $HClO_4$ solution at 0.040 M?

Solution Strong acid $HClO_4$ is entirely ionised, resulting in $[H^+] = [ClO_4^-] = 0.040$ M. The solution's pH is determined by

$$pH = -\log [0.040] = 1.40$$

PROBLEM 6.8 The pH of a 0.10 M solution of formic acid, HCOOH, at 25 °C is 2.38. Calculate K_a for formic acid at this temperature.

Solution: The equilibrium concentrations of all three components are required in order to compute K_a. The pH allows us to find $[H_3O^+]$, which is the same as $[HCOO^-]$.

We know that

$$K_a = \frac{[H_3O^+][COO^-]}{[HCOOH]}$$

$$pH = -\log [H_3O^+]$$

$$2.38 = -\log [H_3O^+]$$

$$-2.38 = \log [H_3O^+]$$

$$10^{-2.38} = 10^{\log [H_3O^+]} = [H_3O^+]$$

$$4.2 \times 10^{-3} = [H_3O^+] = [HCOO^-]$$

	[HCOOH], M	$[H_3O^+]$, M	$[HCOO^-]$, M
Initially	0.10	0	0
Change	-4.2×10^{-3}	$+4.2 \times 10^{-3}$	$+4.2 \times 10^{-3}$
At Equilibrium	$0.10 - 4.2 \times 10^{-3} = 0.0958 = 0.10$	4.2×10^{-3}	4.2×10^{-3}

$$K_a = \frac{[4.2 \times 10^{-3}][4.2 \times 10^{-3}]}{[0.10]}$$

$$= [1.8 \times 10^{-4}]$$

QUESTIONS

6.1 Calculate the pH of a solution containing 4.0 g of NaOH in 500ml water.

6.2 AgI^{2-} is stable but AgF^{2-} does not exist, why?

6.3 Explain why:

(i) Phenol is more acidic than alcohols

(ii) Compared to acetic acid, chloroacetic acid is a stronger acid.

(iii) Halo alkanes are more reactive than haloarencs

(iv) Aniline is less basic than ammonia

6.4 Write the conjugate acid formula for each of the following:
HSO_3^-, F^-, PO_4^{3-}, CO.

(**Answers:** H_2SO_3, HF, HPO_4^{2-}, HCO^+)

6.5 For each of the following acids, what is its conjugate base: $HClO_4$, H_2S, PH_4^+, HCO_3^- ?
(**Answers:** ClO_4^-, HS^-, PH_3, and CO_3^{2-}).

6.6 For each of the bases below, what is the conjugate acid: CN^-, SO_4^{2-}, H_2O, HCO_3^-?
(**Answers:** HCN, HSO_4^-, H_3O^+, and H_2CO_3).

6.7 (a) A lemon juice sample has $[H^+] = 3.8 \times 10^{-4}$ M. What is the pH level?

(b) The $[H^+]$ of a widely used window-cleaning solution is 5.3×10^{-9} M. What is the pH level?

(**Answer:** (a) 3.42, (b) 8.28)

6.8 The pH of a solution created by dissolving an antacid pill is 9.18. Compute $[H^+]$.
(**Answer:** $[H^+] = 6.6 \times 10^{-10}$ M)

6.9 The pH of an aqueous solution of HNO_3 is 2.34. What is the acid's concentration?
(**Answer:** 0.0046 M)

6.10 In a solution containing (a) KOH with a pH of 11.89, and (b) $Ca(OH)_2$ with a pH of 11.68, what is the concentration?

(**Answers:** (a) 7.8×10^{-3} M, (b) 2.4×10^{-13} M)

6.11 The chemical structure of niacin, one of the B vitamins, is as follows:

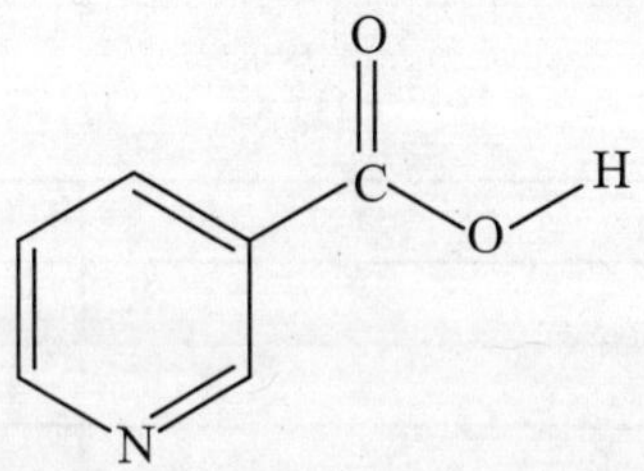

The pH of a niacin 0.020 M solution is 3.26. (a) What proportion of the acid in this solution is ionised? (b) What is niacin's acid-dissociation constant (K_a)?
(**Answers:** (a) 2.7%, (b) 1.5×10^{-5})

6.12 Explain Hard-Soft Acid-Base (HSAB) theory with the help of suitable examples.

CHAPTER 7

Ellingham Diagrams

7.1 INTRODUCTION

The Gibbs free energy (ΔG) of a process serves as a quantitative indicator of the thermodynamic potential that propels a reaction to take place. A negative number for the change in Gibbs free energy (ΔG) signifies the ability of a reaction to occur spontaneously in the absence of external influences. Conversely, a positive value for ΔG shows that the reaction cannot continue spontaneously. The equation representing Gibbs free energy is given by:

$$\Delta G = \Delta H - T\Delta S,$$

where ΔH denotes the enthalpy, T represents the absolute temperature, and ΔS signifies the entropy.

Enthalpy (ΔH) is a quantification of the energy released during a reaction, commonly referred to as the "heat of reaction." In the event of a negative value, the reaction releases energy, whilst a positive value indicates that the reaction necessitates energy.

Entropy (ΔS) is a quantitative measure that characterizes the alteration in the potential for disorder between the products and the reactants. As an illustration, when a solid substance, characterized by its ordered state, undergoes a reaction with a liquid substance, which possesses a relatively lower degree of order, resulting in the formation of a gaseous state, which exhibits a significantly higher level of disorder, it is typically accompanied by a substantial increase in the entropy of the system.

7.2 CONSTRUCTION OF AN ELLINGHAM DIAGRAM

The Ellingham diagram is a graphical representation that illustrates the relationship between the Gibbs free energy change (ΔG) and temperature (T). The constancy of ΔH and ΔS with temperature, except during phase transitions, allows for the construction of a free energy vs temperature graph consisting of a sequence of linear segments. In this graph, the slope represents

ΔS and the y-intercept represents ΔH. The gradient of the line undergoes alteration upon the occurrence of melting or vaporization of any of the constituent components.

The majority of metal oxides exhibit a negative free energy of formation. Consequently, the figure is constructed with a ΔG value of zero at the apex, and all ΔG values depicted are negative numerical values. The graphic indicates the temperatures at which either the metal or oxide undergo melting or vaporization.

The provided Ellingham diagram pertains to the reaction of metals with oxygen to produce oxides. It is worth noting that analogous diagrams may be constructed for metals reacting with other elements such as sulfur or chlorine. However, the oxide version of the diagram is typically encountered more frequently. The oxygen partial pressure is assumed to be 1 atmosphere, and all reactions are standardized to consume one mole of O_2. The upward slope of the majority of the lines can be attributed to the presence of both the metal and the oxide in condensed phases, such as solid or liquid states. The aforementioned reactions involve the conversion of a gaseous substance and a condensed phase substance into another condensed phase substance, resulting in a decrease in entropy. One noteworthy anomaly in this context is to the process of solid carbon oxidation. The queue for the reaction

$$C + O_2 \longrightarrow CO_2$$

When a solid reacts with one mole of gas to yield one mole of gas, the resulting change in entropy is minimal, resulting in a nearly horizontal line. In regards to the chemical reaction

$$2C + O_2 \longrightarrow 2CO_2$$

In this experimental setup, a solid substance has a chemical reaction with a gaseous substance, resulting in the production of two moles of gaseous products. Consequently, a significant augmentation in entropy occurs, leading to a pronounced downward slope in the corresponding line. Analogous patterns may be observed in the characteristics of lead and lithium, wherein the oxides of both elements exhibit boiling points that are marginally lower than those of the respective metals.

The Ellingham diagram (Figure 7.1) finds application in three primary areas:

1. Assess the comparative level of difficulty in the reduction process of a specific metallic oxide to its corresponding metal.
2. Ascertain the equilibrium partial pressure of oxygen when in contact with a metal oxide at a designated temperature.
3. Calculate the proportion of carbon monoxide to carbon dioxide required to effectively reduce the oxide to its metallic form at a given temperature.

7.3 EASE OF REDUCTION

The placement of the line corresponding to a certain reaction on the Ellingham diagram indicates the thermodynamic stability of the oxide in relation to temperature. Metals positioned higher in the picture exhibit reactions that are considered more "noble" in nature, such as gold and platinum. The oxides of these metals are characterized by their instability and susceptibility to reduction. As we descend along the diagram, the reactivity of the metals increases gradually, while the reducibility of their respective oxides diminishes.

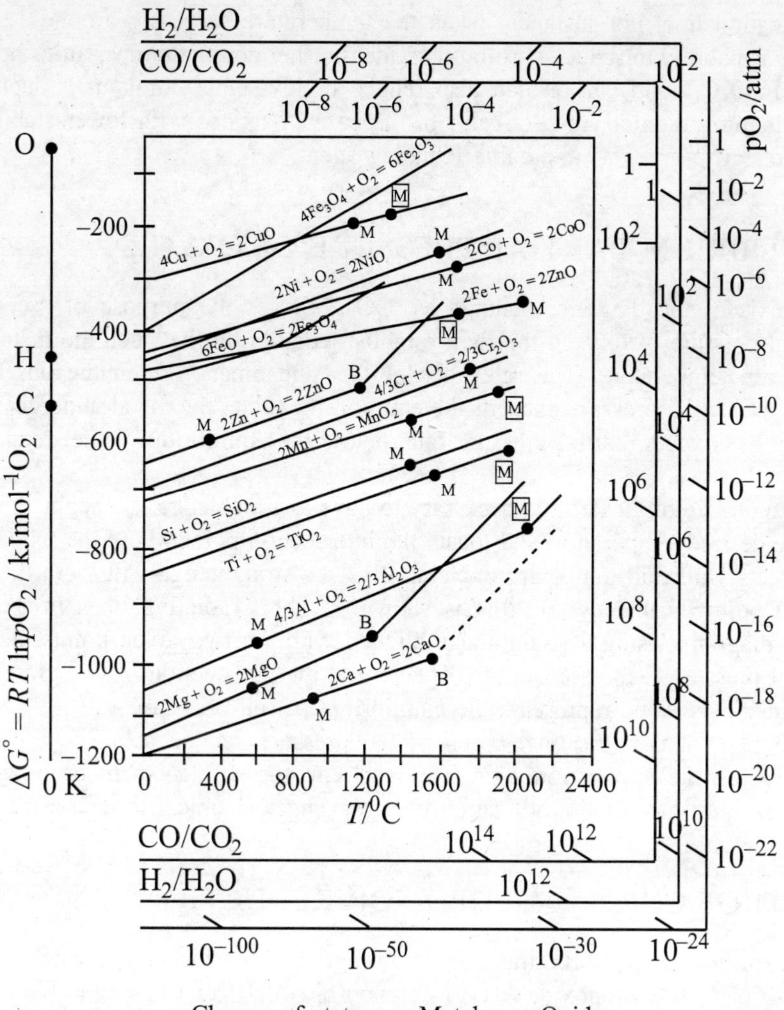

Change of state	Metal	Oxide
Melting point	M	[M]
Boiling point	B	[B]

Figurere 7.1 Ellingham diagram.

A specific metal has the ability to undergo reduction reactions with the oxides of all other metals that are positioned higher than it on the diagram. As an illustration, it can be observed that the position of the line representing magnesium $2Mg + O_2 \rightarrow 2MgO$ sits beneath the line representing titanium $Ti + O_2 \rightarrow TiO_2$ in the electrochemical series. Consequently, magnesium possesses the ability to facilitate the reduction of titanium oxide, leading to the formation of metallic titanium.

Due to its negative slope $2C + O_2 \rightarrow 2CO$, the line intersects with the lines representing many other metals. The exceptional use of carbon as a reducing agent arises from its ability to convert metal oxides to their corresponding metals once the carbon oxidation line falls below the metal oxidation line. For instance, when the temperature surpasses around 1225°C, solid carbon has the capability to reduce chromium oxide. Furthermore, at temperatures beyond about 1620°C and 1650°C, solid carbon can also reduce highly-stable compounds such as silicon dioxide and titanium dioxide, respectively. In the case of oxides with lower stability, carbon monoxide frequently serves as a suitable reducing agent.

7.4 EQUILIBRIUM PARTIAL PRESSURE OF OXYGEN

The vertical axis of the diagram, denoted as "pO_2," serves the purpose of ascertaining the specific partial pressure of oxygen that will establish equilibrium between the metal and metal oxide at a given temperature. The relevance of this phenomenon is in the observation that when the oxygen partial pressure exceeds the equilibrium value, the metal undergoes oxidation, whereas when the oxygen partial pressure falls below the equilibrium value, reduction of the oxide occurs.

In order to utilize this scale, it is necessary to have a straightedge. To begin, it is necessary to identify the desired temperature and locate the intersection point where the oxidation line of interest intersects with said temperature. Next, align the straightedge with both the designated point and the point labeled as "0" that is shown by short radiating lines in the upper left corner of the diagram. Using a straightedge aligned with the two given points, determine the oxygen partial pressure (expressed in atmospheres) at the intersection of the straightedge with the "pO_2" scale. This value represents the equilibrium partial pressure.

The attainment of equilibrium oxygen partial pressure can be achieved through several methods, including the application of a hard vacuum, the displacement of oxygen through purging with an inert gas, or the utilization of a scavenger chemical to devour the oxygen.

7.5 RATIO OF CO/CO_2 NEEDED FOR REDUCTION

In the context of employing carbon as a reducing agent, it is necessary to establish a specific minimum ratio of carbon monoxide (CO) to carbon dioxide (CO_2) that can effectively reduce a particular oxide. The more challenging it is to decrease the oxide, the higher the requirement for CO in the gaseous mixture. In order to ascertain the CO/CO_2 ratio for the reduction of a metal oxide at a specific temperature, the procedure employed for determining the equilibrium pressure of oxygen can be utilized. However, it is necessary to align the straightedge with the designated point labeled as "C" situated at the center of the left side of the diagram. Subsequently, the CO/CO_2 ratio can be obtained by referring to the scale marked as "CO/CO_2".

An often observed chemical process that signifies oxidation can be represented by the following generic reaction:

$$2xM(s) + O_2(g) \longrightarrow 2M_xO(s)$$

The observed response indicates a decrease in the gaseous reactant quantity from the left to the right, as a solid metal oxide product is created on the right side. Therefore, it may be concluded that there is a decrease in molecular randomness from left to right. Therefore, the change in entropy (ΔS) exhibits a negative value, causing the change in Gibbs free energy (ΔG) to move towards the upper side, even in the presence of an increasing temperature (T).

Therefore, in the majority of the aforementioned reactions involving the synthesis of $M_xO(s)$, the curve exhibits a positive trend.

With the exception of phase transition processes, each graphic exhibits a linear relationship. The occurrence of phase transition is denoted by a positive elevation in the gradient of temperature. A dramatic change in the curve within the Zn, ZnO plot serves as an indication of the melting process.

The stability of metal oxide (M_xO) is observed when the value of ΔG is negative, as indicated by a specific point on the curve. Prior to this threshold, the metal oxide exhibits instability and undergoes spontaneous decomposition.

The determination of the feasibility of reducing the oxide of the upper line by the element represented by the lower line relies on the disparity between the two $\Delta_r G^0$ values subsequent to the point of crossing in the Ellingham diagram.

7.6 FEATURES OF ELLINGHAM DIAGRAM

The Ellingham diagram pertaining to oxides has several significant characteristics:

1. The graph depicting the production of a metal oxide exhibits a linear relationship characterized by a positive slope.
2. There is an abrupt alteration in the inclines observed in certain metal oxides, such as MgO, ZnO, and HgO.
3. In the Ellingham diagram, it can be observed that certain metal oxides, such as Ag_2O and HgO, are positioned in the upper region of the graph.
4. The graph depicting the generation of carbon monoxide (CO) exhibits a linear relationship characterized by a negative slope. The aforementioned line connects the lines of numerous metal oxides.
5. The graph depicting the generation of carbon dioxide (CO_2) exhibits a linear relationship that is nearly parallel to the temperature axis.

7.7 SIGNIFICANCE OF ELLINGHAM DIAGRAM

1. The observation of a positive slope in the stability of metal oxides indicates a reduction in their stability as temperature increases.
2. The decline in the stabilities of the entities can be attributed to an elevation in the $\Delta G°$ value.
3. The abrupt transition depicted in the graph indicates a phase change, namely the conversion from a solid state to a liquid state or from a liquid state to a vapor state.
4. The negative slope observed in the CO compound indicates that its stability increases as the temperature rises, which contrasts with the behavior observed in metal oxides.
5. The metals' relative propensity to undergo oxidation is arranged in the following order:

   According to the given sequence, the relative order of the elements in terms of their reactivity is as follows: magnesium (Mg) exhibits higher reactivity than aluminum (Al), chromium (Cr), iron (Fe), mercury (Hg), and silver (Ag), respectively. The observed trend can be attributed to the progressive elevation in the magnitude of the negative free energy shift associated with the creation of oxide, as follows: $MgO > Al_2O_3 > Cr_2O_3 > FeO > Ag_2O$.

It is important to note that the standard Gibbs free energy ($\Delta G°$) values of HgO and Ag_2O exhibit positive values, even when considering low temperatures. The graphical representation illustrating the creation of silver and mercury oxides is seen in the upper region of the Ellingham diagram. This suggests that Ag_2O and HgO exhibit instability and can undergo decomposition when subjected to moderate temperatures, specifically through heating at 600 K and 700 K, respectively.

7.8 LIMITATIONS OF ELLINGHAM DIAGRAM

The Ellingham diagram is constructed based on fundamental thermodynamic principles. The primary function of this indicator is to determine the feasibility of a reaction, whereas it lacks the ability to forecast the reaction's kinetics, such as the rate of reduction or the duration required for the process to take place.

The understanding of $\Delta G°$ relies on the equilibrium constant (K) as expressed by the equation $\Delta G° = -RT \ln K$. Therefore, it is postulated that the reactants and products have reached a state of equilibrium. Nevertheless, it is important to note that this assertion does not hold true in all cases, as there are instances where the reactant or product involved in a chemical reaction may exist in a solid state.

QUESTIONS

7.1 Explain, using the Ellingham diagram, how a mixture of Cl_2/O_2 gas may be used to separate Zn from Fe in a galvanised scrap.

7.2 Using Ellingham diagram, predict which metal can readily reduce the Cu_2O to Cu at 400 °C? **(Answer:** Al)

7.3 Give the Significance of the Ellingham diagram.

7.4 Write two limitations of the Ellingham diagram.

7.5 Write down the application of Ellingham diagram.

CHAPTER 8

Thermochemistry

8.1 INTRODUCTION

The scientific study of heat transfer in chemical reactions is known as thermochemistry. It investigates the correlations between chemical reactions and alterations in energy. Heat refers to the process of transferring thermal energy from one body to another when they have different temperatures. Temperature is a quantitative assessment of the amount of thermal energy present.

8.2 EXOTHERMIC PROCESS

Any process that releases heat and transfers thermal energy from the system to the environment is called an exothermic process. When the pressure of a system remains constant, the release of heat causes a drop in its enthalpy (Figure 8.1). Therefore, the change in enthalpy is negative, indicated by $\Delta H < 0$. For example, the reaction between 2 moles of hydrogen gas (H_2) and 1 mole of oxygen gas (O_2) produces 2 moles of liquid water (H_2O) and releases energy. The gaseous form of H_2O undergoes a phase change to its liquid form, releasing energy in the process. The value of ΔH for this reaction is negative.

$$2H_2(g) + O_2(g) \longrightarrow 2H_2O(l) + \text{Energy}$$

$$H_2O(g) \longrightarrow H_2O(l) + \text{Energy}$$

Another example, the chemical equation for the reaction between carbon and oxygen to form carbon dioxide is as follows:

$$C + O_2 \longrightarrow CO_2$$

The enthalpy change (ΔH) for this reaction is –94050 calories.

Chemical heat packs are another typical example, where iron filings are oxidised by an exothermic reaction. Because our hands absorb the heat emitted by the process, they get heated.

8.3 ENDOTHERMIC PROCESS

An endothermic process is defined as a process that requires the input of heat from the surroundings, resulting in a decrease in temperature of the surroundings (Figure 8.1). In this scenario, the heat being absorbed causes a rise in the enthalpy of the system under constant pressure. As a result, the change in enthalpy is positive, indicated by $\Delta H > 0$. For example,

$$\text{Energy} + 2HgO(s) \longrightarrow 2Hg(l) + O_2(g)$$

$$\text{Energy} + H_2O(s) \longrightarrow H_2O(l)$$

The reaction can be represented as the conversion of energy and water in solid form to liquid water. The value of ΔH for this reaction is positive.

Another example, an endothermic reaction occurs when nitrogen (N_2) and oxygen (O_2) combine to form two molecules of nitric oxide (2NO), with an enthalpy change (ΔH) of +42000 calories.

$$N_2 + O_2 \longrightarrow 2NO$$

Other common example, chemical cold packs consist of NH_4NO_3, which undergoes dissolution in water through an endothermic reaction. The reason our hands become cold is due to the transfer of heat from our hands to the surrounding environment.

Common Representation of Exo and Endothermic Reactions

Assume that in a reaction, A undergoes a reaction with B to produce C, resulting in the release of Q units of heat (exothermic). Thus, the equation can be expressed as follows:

$$A + B \longrightarrow C + \Delta H$$

Similarly, if the reaction is endothermic, it can be symbolized as $A + B \rightarrow C - \Delta H$.

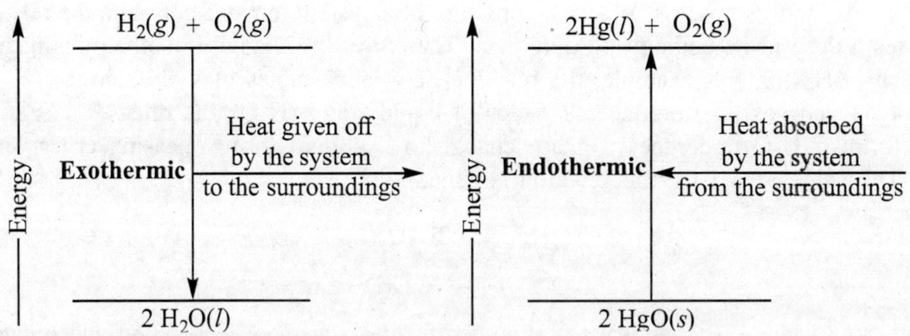

Figure 8.1 Diagram illustrating Exothermic and Endothermic Processes.

When a chemical reaction occurs under constant pressure, the heat change is represented by the symbol q_p. Conversely, when the reaction takes place under constant volume, the heat change is signified by q_v. However, the magnitudes of q_p and q_v may not be equivalent for the identical chemical reaction.

For example,

$$C_7H_8 + 9O_2 \rightleftharpoons 7CO_2 + 4H_2O$$

The values for q_p and q_v are –930 kcal and –928.8 kcal, respectively.

The q_p represents the change in enthalpy (ΔH) and q_v represents the change in internal energy (ΔU), it is customary to describe the heat changes on the right side of the chemical equation using ΔH and ΔU.

8.4 SPECIFIC HEAT

The amount of heat (q) required to raise the temperature of one gram of a substance by one degree Celsius is known as the specific heat (s) of that substance. The amount of heat (q) required to raise the temperature of a certain amount (m) of a substance by one degree Celsius is known as the substance's heat capacity (C). Table 8.1 displays the specific heats of several commonly found substances. Mathematically, heat capacity is represented as,

$$C = m \times s$$

Mathematically, heat (q) absorbed or released can be represented as,

$$q = m \times s \times \Delta t$$

so,

$$q = C \times \Delta t$$

where, $\Delta t = t_{\text{final}} - t_{\text{initial}}$

Table 8.1 The specific heats of several commonly found substances

Substances	Specific Heat (J/g. °C)	Substances	Specific Heat (J/g. °C)
Al	0.900	Fe	0.444
Au	0.129	Hg	0.139
C (graphite)	0.720	H_2O	4.184
C (diamond)	0.502	C_2H_5OH (ethanol)	2.46
Cu	0.385		

8.5 STANDARD ENTHALPY OF FORMATION AND REACTION

8.5.1 Heat of Reaction

The heat of reaction refers to the amount of heat released or absorbed during a chemical reaction. The heat of reaction is the quantity of heat released or absorbed at a specific temperature when the reactants are transformed into products, as indicated by the balanced chemical equation. By applying the first law of thermodynamics, one may express the relationship between q, the heat of reaction, and W, the work done by the reactants during the reaction. The equation is written as $q = \Delta U - W$. ΔU is the disparity in energy between the internal energy of the products and the internal energy of the reactants, as seen below.

$$\Delta U = \Sigma U \text{ (products)} - \Sigma U \text{ (reactants)}$$

The change in internal energy (ΔU) is equal to the sum of the internal energies of the products (ΣU) and the sum of the internal energy of the reactants.

If a chemical reaction takes place in a bomb calorimeter at a constant volume, the work done (W) is zero. In this scenario, the heat transferred (q) is equal to the change in internal energy (ΔU).

$$q_v = \Delta U$$

Hence, the heat of reaction at constant volume, q_v, represents the alteration in the system's internal energy that occurs during a chemical reaction. Nevertheless, when a chemical reaction takes place under constant pressure conditions in an open calorimeter, the work done (W) is equal to $-P\Delta V$. In this scenario, the heat transferred at constant volume (q_v) can be expressed as the change in internal energy (ΔU) minus the negative work done ($-P\Delta V$), which simplifies to $\Delta U + P\Delta V$, which is equivalent to the change in enthalpy (ΔH). So,

$$q_v = \Delta U - (-P\Delta V) = \Delta U + P\Delta V = \Delta H$$

The enthalpy of reaction is the quantity of heat released or absorbed when a chemical reaction takes place under constant pressure and at a specific temperature. Let's examine a generic chemical reaction in which a moles of A react with b moles of B to yield c moles of C and d moles of D. This reaction may be expressed as:

$$a\text{A} + b\text{B} \longrightarrow c\text{C} + d\text{D}$$

The change in enthalpy of the system during the conversion of reactants into products is determined by subtracting the total enthalpy of the reactants from the total enthalpy of the products, while keeping the temperature and pressure constant. The term used to refer to this amount is known as the heat of reaction. The change in enthalpy (ΔH) is equal to the sum of the enthalpies of the products (ΣH) minus the sum of the enthalpy of the reactants,

$$\Delta H = \Sigma H \text{ (Product)} - \Sigma H \text{ (reactants)}$$

$$\Delta H = (c\text{H}_\text{C} - d\text{H}_\text{D}) - (a\text{H}_\text{A} - b\text{H}_\text{B})$$

where, H_A, H_B, H_C and H_D are enthalpies of A, B, C and D.

The change in volume ΔV is negligible for reactions involving solids and liquids. Consequently, since the change in volume ($P\Delta V$) is negligible, we can disregard it and conclude that A is equivalent to ΔU. Nevertheless, when the reactants and products are in gaseous state, it is imperative to consider the changes in volume and provide explicit details regarding the procedure of the reaction. Typically, reactions that are predominantly composed of gases are carried out under conditions of constant pressure, allowing for the measurement of the change in enthalpy (ΔH).

Conditions under which the heat added at constant pressure (q_p) is equal to the heat added at constant volume (q_v) or the change in enthalpy (ΔH) is equal to the change in internal energy (ΔU) are as follows ($q_p = q_v$ or $\Delta H = \Delta U$):

1. During a reaction occurring in a sealed container, the volume remains unchanged, so that $\Delta V = 0$.
2. In chemical reactions where only solids, liquids, or solutions are present and there are no gaseous reactants or products, the changes in volume are insignificant.
3. In a reaction, where the number of moles of gaseous reactants and products are the same, shown by $\Delta n = 0$. As an illustration, the chemical equation can be written as

follows: Hydrogen gas (H_2) reacts with chlorine gas (Cl_2) to produce two molecules of hydrogen chloride gas (2HCl).

$$H_2(g) + Cl_2(g) \longrightarrow 2HCl(g)$$

8.5.2 Standard Enthalpy of Formation

The standard enthalpy of formation ($\Delta H°$) is the amount of heat released or absorbed when one mole of a compound is produced from its constituent elements under standard conditions of 1 atm pressure. The standard enthalpy of production for any element in its most stable state is precisely zero. For example, the standard enthalpy change of formation ($\Delta H°$) for O_2 is 0 kJ/mol, for C in graphite form is 0 kJ/mol, for O_3 is 142 kJ/mol, and for C in diamond form is 1.90 kJ/mol. The standard enthalpy of reaction ($\Delta H°$) refers to the enthalpy change that occurs during a reaction conducted at a pressure of 1 atmosphere.

The chemical equation represents the reaction between substances A and B, resulting in the formation of substances C and D. The coefficients *a*, *b*, *c*, and *d* represent the stoichiometric ratios between the reactants and products.

$$aA + bB \longrightarrow cC + dD$$

The change in enthalpy for the reaction is equal to the sum of the enthalpy of formation of compound C multiplied by the stoichiometric coefficient *c*, and the enthalpy of formation of compound D multiplied by the stoichiometric coefficient *d*. The change in enthalpy of a reaction (ΔH°_{rxn}) can be calculated using the formula:

$$\Delta H^\circ_{rxn} = [c\,\Delta H^\circ_f(C) + d\,\Delta H^\circ_f(D)] - [a\,\Delta H^\circ_f(A) + b\,\Delta H^\circ_f(B)]$$

$$\Delta H^\circ_{rxn} = \Sigma n\,\Delta H^\circ_f(\text{products}) - \Sigma m\,\Delta H^\circ_f(\text{reactants})$$

Where ΔH°_f represents the standard enthalpy of formation for a substance.

8.6 THE LAWS OF THERMOCHEMISTRY

8.6.1 Law of Lavoisier and Laplace

The first law of thermochemistry was established based on the empirical findings of A.L. Lavoisier and P.S. Laplace in 1780. The law can be expressed as follows: "The heat change associated with a chemical process in one direction is equal in magnitude but opposite in sign to the heat change associated with the same reaction in the opposite direction."

Example,

$$H_2(g) + \frac{1}{2}O_2(g) \longrightarrow H_2O(l) \qquad \Delta H = -68300\,\text{cal}$$

$$H_2O(l) \longrightarrow H_2(g) + \frac{1}{2}O_2(g) \qquad \Delta H = +68300\,\text{cal}$$

Consider two reactants, A and B, which undergo a transformation into C and D, releasing heat (q). Subsequently, C and D react to revert back to the initial state of A and B, absorbing heat (q'). As per the law, the initial heat content (H) or energy content (U) of A and B will eventually revert to the same magnitude. Consequently, the net change in heat content or energy content would be null.

$$q + q' = 0 \quad \text{or} \quad q = q'$$

8.6.2 Hess's Law of Constant Heat Summation

Hess's law of constant heat summation, sometimes known as Hess's law, is a principle in thermodynamics that states that the total enthalpy change of a chemical reaction is independent of the pathway taken to achieve that reaction.

G.H. Hess (1840) empirically established the second law of thermochemistry, which has significant practical implications and is sometimes referred to as the law of constant heat summation. The law can be expressed as follows: "The total heat change (q_p or q_v) for a specific chemical process remains constant, regardless of whether the process takes place in a single stage or multiple stages." The change in enthalpy during the conversion of reactants to products remains constant, regardless of whether the reaction occurs in a single step or through a series of steps.

Enthalpy is a thermodynamic property that depends only on the current state of a system and is a state function. The means of transportation are inconsequential; what truly matters is the initial and final points of departure and arrival.

Example: The thermochemical reactions

$$C + \frac{1}{2}O_2 \longrightarrow CO \qquad q'_p = -26.42\,\text{kcal}$$

$$CO + \frac{1}{2}O_2 \longrightarrow CO_2 \qquad q''_p = -67.63\,\text{kcal}$$

$$C + O_2 \longrightarrow CO_2 \qquad q_p = q'_p + q''_p = -94.05\,\text{kcal}$$

8.7 BORN HABER CYCLE

The Born-Haber cycle is a concept in chemistry that is used to calculate the lattice energy of an ionic compound. In 1919, Born and Haber devised a simplified and cyclic approach to connect the lattice energies of ionic crystals to other thermodynamic parameters, using Hess's formula as a basis. As an illustration, the energy changes that occur when metallic sodium and chlorine gas combine to generate sodium chloride can be symbolized by ΔH_f.

$$Na(s) + \frac{1}{2}Cl_2(g) \longrightarrow NaCl$$

The total process is considered to take place in following intermediary steps:

Intermediate Steps:

1. Conversion of metallic sodium into gaseous sodium atoms.

2. The energy needed to convert one mole of sodium from a solid to a gaseous state is referred to as the enthalpy of sublimation, denoted as ΔH_s. This stage is a highly energy-intensive process.

$$Na(s) + \Delta H \longrightarrow Na(g)$$

3. Chlorine molecule dissociates into chlorine atom. The energy per mole of chlorine required for dissociation is known as the "enthalpy of dissociation" and is symbolized by the symbol ΔH_d.

$$Cl_2(g) + \Delta H_d \longrightarrow 2Cl(g)$$

or $$\frac{1}{2}Cl_2(g) + \frac{1}{2}\Delta H_d \longrightarrow Cl(g)$$

4. Conversion of a gaseous sodium atom into a gaseous cation. The energy needed for this process is referred to as Ionization Energy (IE).

$$Na(g) + IE \longrightarrow Na^+(g) + e^-$$

5. Conversion of a gaseous chlorine atom into a gaseous anion. This stage entails the liberation of energy known as Electron Affinity (EA).

$$Cl(g) + e^- \longrightarrow Cl^-(g) + EA$$

The process of combining gaseous ions with opposite charges to create a solid crystal. This process entails the liberation of energy known as lattice energy (U).

$$Na^+(g) + Cl^-(g) \longrightarrow NaCl(s) + U$$

The energy transformations at each phase are as follows:

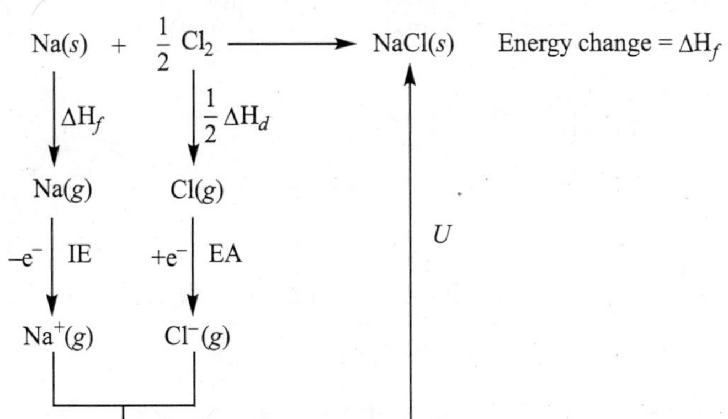

The total energy changes occurring during different phases are equivalent to ΔH_f, which represents the heat of production of NaCl(s) in accordance with Hess's law.

$$\Delta H_f = \Delta H_s + \frac{1}{2}\Delta H_d + IE + EA + U$$

The different values for sodium chloride (NaCl) are as follows:

The heat of sublimation of sodium, denoted as ΔH_s, is equal to 108.5 kJ mole^{-1}. The dissociation energy of chlorine, represented by ΔH_d, is equal to 243.0 kJ mole^{-1}. The ionization energy of sodium, represented by IE, is equal to 495.8 kJ mole^{-1}. The electron affinity (EA) of chlorine is –349 kJ mole^{-1}, while the lattice energy (U) of NaCl is –769.8 kJ mole^{-1}.

Replacing these values

$$\Delta H_f = \Delta H_s + \frac{1}{2}\Delta H_d + \text{IE} + \text{EA} + \text{U}$$

$$= 108.5 + \frac{1}{2}(243.0) + 495.8 + (-349) + (-769.8)$$

$$= -393 \text{ kJ mole}^{-1}$$

The Born-Haber cycle for the combustion of propane. The balanced chemical equation for the reaction is

$$C_3H_8(g) + 5O_2(g) \longrightarrow 3CO_2(g) + 4H_2O(l)$$

The change in enthalpy (ΔH) for the reaction may be calculated using the equation

$$\Delta H = [3(-393.5\text{ kJ}) + 4(-285.8\text{ kJ})] - [1(-103.85\text{ kJ}) + 5(0\text{ kJ})]$$

The energy transformations at each phase are as follows:

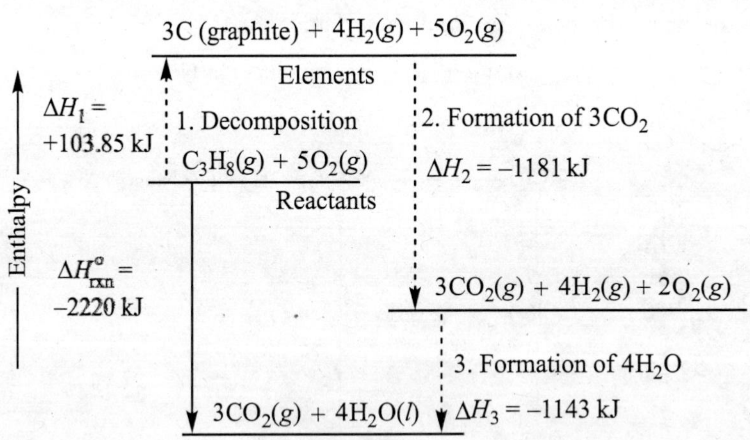

The Born-Haber cycle for the combustion of methane. The energy transformations at each phase are as follows:

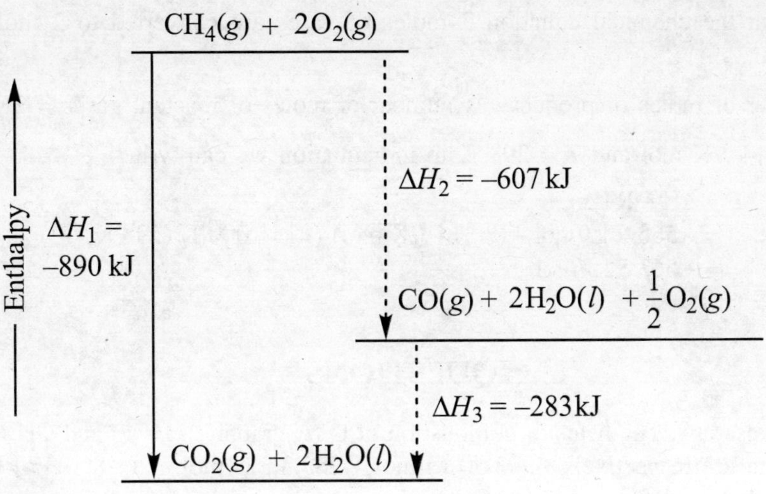

The Born-Haber cycle for the reaction of argon with chlorine is given below. The energy transformations at each phase are as follows:

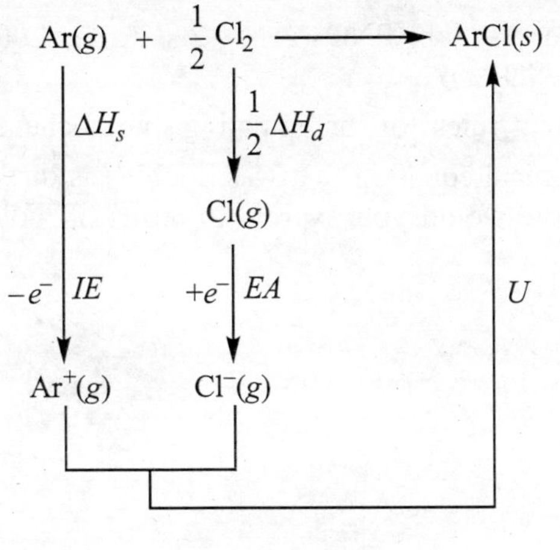

SOLVED PROBLEMS

EXAMPLE 8.1 A 466-gram water sample is heated to 74.60°C from 8.50°C. Determine how much heat the water has absorbed (in kilojoules).

Solution: Using Equation, we write

$$q = m \times s \times \Delta t$$
$$= (466\ g)(4.184\ \text{J/g °C})(74.60\ \text{°C} - 8.50\ \text{°C})$$
$$= 1.29 \times 10^5\ \text{J} \times (1\ \text{kJ}/1000\ \text{J})$$
$$= 129\ \text{kJ}$$

EXAMPLE 8.2 Determine the internal energy change that results from the conversion of two moles of CO to two moles of CO_2 at 1 atm and 25°C.

$$2CO(g) + O_2(g) \longrightarrow 2CO_2(g) \quad \Delta H = -566.0\ \text{kJ/mol}$$

Solution From the chemical equation 3 moles of gases are converted to 2 moles of gases so that

$$\Delta n = \text{number of moles of product gas number of moles of reactant gases} = 2 - 3 = -1$$

Using $R = 8.314$ J/K mol and $T = 298$ K in the equation we can write

$$\begin{aligned}\Delta E &= \Delta H - RT\Delta n \\ &= -566.0 \text{ kJ/mol} - (8.314 \text{ J/K mol}) (1 \text{ kJ/1000 J}) (298 \text{ K})(-1) \\ &= -563.5 \text{ kJ/mol}\end{aligned}$$

QUESTIONS

8.1 Define Hess law. The heats of combustion of C (*s*, graphite), $H_2(g)$ and $CH_4(g)$ at 298 K and 1 atm are respectively −393.50 kJ/mol, −285.83 kJ/mol and −890.36 kJ/mol. What is the enthalpy of formation for CH_4?

8.2 What happens to the value of ΔH of a thermochemical reaction if the reaction is reversed?

8.3 ΔH = 114 kJ for the reaction: $2NO_2(g) \rightarrow 2NO(g) + O_2(g)$. What is the ΔH for $NO(g) + \frac{1}{2}O_2(g) \rightarrow NO_2(g)$?

8.4 What are two important rules for manipulating thermochemical equations?

8.5 When the thermochemical equation that was doubled is dissociating moles back into its initial elements how would you express in terms of moles of product?

UNIT 3

Water Chemistry and Corrosion

CHAPTER 9

Water Chemistry

9.1 INTRODUCTION

H_2O is the chemical formula for water, which is present on Earth in the form of ice, water vapor, and liquid. In addition to being a universal solvent, water is a fundamental component in the advancement of numerous industries and a crucial substance for engineering applications. The Swiss chemist Werner Stumm (1924–1999) is commonly regarded as the father of water chemistry. His contributions to the field's advancement are both extensive and profound. His co-authorship (1970) of the important textbook Aquatic Chemistry with James Morgan, his first Ph.D. student, propelled him to the forefront of his discipline.

Werner Stumm

Evapotranspiration, which includes both evaporation and transpiration, is a recurring cycle in which water continuously moves from the land surface to the atmosphere of the Earth. Evaporation refers to the process by which water from liquid or sediment sources enters the atmosphere. The process by which water moves within a plant and is subsequently lost as vapor through stomata in the leaves is known as transpiration.

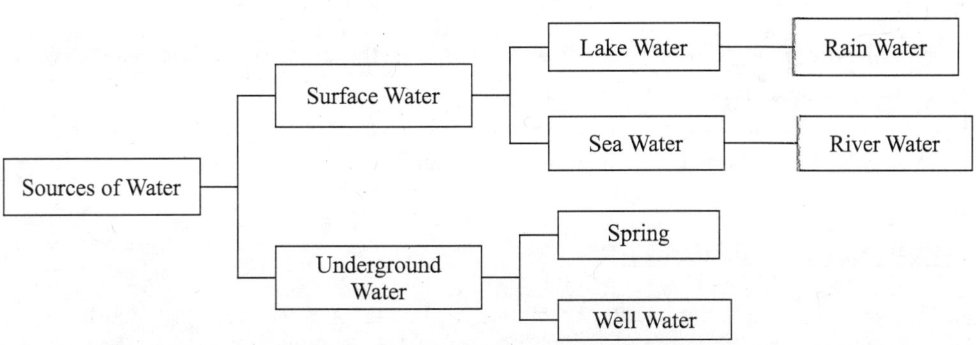

Figure 9.1 Sources of water in Earth.

Water possesses profound spiritual significance across various cultures, being linked to concepts such as mortality, spiritual purification, and initiation. Land covers only 30% of the Earth's surface, while water covers approximately 70%. It causes one to question whether the more appropriate nomenclature for Planet Earth would be Planet Water. Approximately fifty percent of inland water (including lakes, rivers, and groundwater) is saline. Therefore, freshwater comprises a negligible proportion (0.8%) of the overall water resources on Earth.

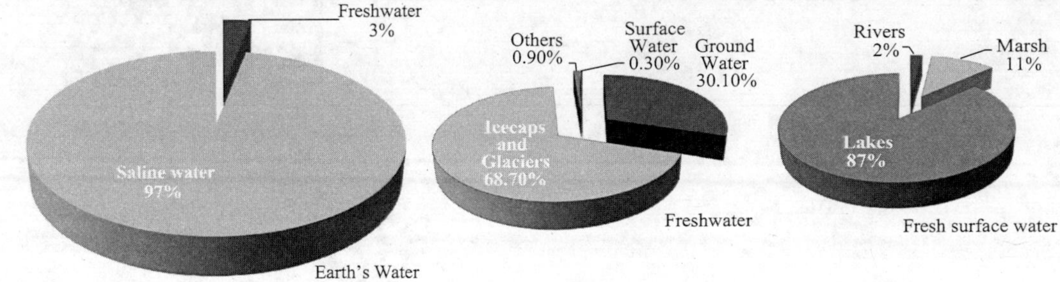

Figure 9.2 Distribution of Earth Water

9.2 Types of Impurities

Water contaminants can be attributed to a variety of human activities, dissolved inorganic salts (e.g., sulphates and chlorides of calcium, magnesium, and iron), gases (e.g., carbon dioxide, nitrogen, oxygen, and hydrogen), and organic salts. Mud, clay, as well as plant and animal matter, are examples of suspended impurities. Colloidal impurities, such as alumina and silica, as well as organic residues, result from their minute size. Soluble substances, excluding gases, are resistant to removal through conventional filtration systems. Parasites, bacteria, germs, pathogens, microbes, and viruses were all examples of bacterial impurities. Additionally, it comprises beneficial bacteria and algae that decompose debris, including cholera-causing bacteria.

9.2.1 Effects of Impurities

Water impurity induces numerous alterations in physicochemical properties. Among them are the subsequent:

1. Dissolved organic substances induce a change in the color of surface water.
2. New compound formation alters the odor and flavor of water.
3. Impurities that are suspended in water contribute to increased turbidity.
4. Waterborne microorganisms increase pathogenicity.

9.3 HARDNESS OF WATER

The attributes of water that inhibit the formation of leather are referred to as hardness. This is the result of salts such as bicarbonates, sulphates, calcium chlorides, magnesium chlorides, and other metal chlorides. In water, two distinct forms of hardness are detected.

Units of Hardness

1. Parts per million (ppm): The unit weight of the solute per million weight units of solution is denoted as parts per million (ppm). Additionally, the unit is denoted in milligrams per liter (mg/L).

$$1 \text{ ppm} = 1 \text{ mg/L}$$

2. Degree Clark (°Cl): This parameter denotes the quantity of grains per gallon of water hardness. The hardness metric that is still employed in the water utility industry is grains per gallon.

$$1 \text{ gain} = 64.8 \text{ mg of } CaCO_3 = 1/7000 \text{ lb}$$
$$1 \text{ ppm} = 0.07 \text{ °Cl}$$

$$\text{A grain per gallon} = \frac{64.8 \text{ mg } CaCO_3/\text{gal}}{3.78 \text{ L/gal}} = 17.1 \text{mg/L as } CaCO_3.$$

3. Degree French (°Fr): Degree French denotes the quantity of hardness units that induce an ion reaction in 10^5 parts of water.

$$1 \text{ ppm} = 0.1 \text{ °Fr}$$

It is important to note that the unit hardness, expressed in parts per million (ppm), $CaCO_3$, is determined by the formula weight of 100 (40 + 12 + 3 × 16), which simplifies the calculation of an equivalent weight of 50. For example, Ca+Mg (expressed in mg/L as $CaCO_3$) ≈ total hardness (in mg/L as $CaCO_3$). The following formula is utilized to convert between units of mg/L for Ca^{2+} and Mg^{2+} and units of mg/L for $CaCO_3$.

$$Ca^{2+}(\text{mg/L as } CaCO_3) = \frac{\text{mg/L as Ca}}{20 \text{ mg/meq of Ca}} \quad 5\,0 \text{ mg/meq of } CaCO_3$$

$$Mg^{2+}(\text{mg/L as } CaCO_3) = \frac{\text{mg/L as Mg}}{12.15 \text{ mg/meq of Mg}} \quad 50 \text{ mg/meq of } CaCO_3$$

9.3.1 Temporary Hardness (TH)

This phenomenon is alternatively referred to as carbonate hardness. The act of boiling the water alone will result in its removal. Temporary or carbonate hardness is caused by the presence of bicarbonate of calcium, magnesium, or both in water. The process of boiling water leads to the conversion of Ca^{2+} and Mg^{2+} ions, which are coupled with HCO_3^-, and CO_3^{2-} ions, into insoluble carbonates. This phenomenon will result in the precipitation and subsequent formation of sediment at the bottom.

$$Ca(HCO_3)_2 \xrightarrow{\Delta} CaCO_3 \downarrow + CO_2 \uparrow + H_2O$$
$$Mg(HCO_3)_2 \xrightarrow{\Delta} Mg(OH)_2 \downarrow + 2CO_2 \uparrow$$

9.3.2 Permanent Hardness (PH)

The term "permanent hardness" or "non-carbonate hardness" refers to water that contains chlorides or sulphates of calcium, magnesium, or both. The substance does not undergo removal with heating of the water. The cost of removing non-carbonate hardness is significantly higher compared to carbonate hardness. In the case of hard water, the presence of permanent Ca^{2+} and Mg^{2+} ions is observed in association with other ions such as Cl^-, NO_3^-, and SO_4^{2-}. The categorisation of water in terms of hardness is tabulated in Table 9.1.

$$CaCl_2 \longrightarrow Ca^{2+} + 2Cl^-$$

$$MgSO_4 \longrightarrow Mg^{2+} + SO_4^{2-}$$

Note: permanent hardness can be obtained by subtracting the value of temporary hardness from the total hardness. Permanent Hardness = Total Hardness – Temporary Hardness

Table 9.1 Hardness Table

Water Nature	Hardness, mg/L	Sodium added, mg/L
Soft	0–49	0–22.5
Medium Hard	50–149	23–68.5
Hard	150–249	69–115
Very Hard	250–229	115–138
Extremely Hard	>300	>138

9.3.3 Disadvantages of Hard Water

Water serves as a fundamental element in various routine household tasks, including but not limited to cleansing, bathing, hydration, and culinary purposes. The presence of hard water during washing and bathing might hinder the creation of lather due to the inhibitory effects of the hardness-producing components, resulting in the formation of white scum. The subsequent reactions will justify it.

$$C_{17}H_{35}COONa + H_2O \longrightarrow C_{17}H_{35}COOH + NaOH$$

$$C_{17}H_{35}COOH + C_{17}H_{35}COONa \longrightarrow \text{Lather}$$

$$2C_{17}H_{35}COONa + \underset{\text{(in water)}}{CaSO_4} \longrightarrow \underset{\text{(White Scum)}}{(C_{17}H_{35}COO)_2\,Ca \downarrow} + Na_2SO_4$$

Water is utilized in several industrial applications, such as boiler feed, paper mills, the sugar business, the dyeing industry, and laundries. It is imperative that water be devoid of scale and sludges in order to prevent the degradation of boiler capacity. The presence of iron and lime in water can have detrimental effects on the soap resin, leading to its degradation. The absence of sulphates and alkaline carbonates is necessary due to the deliquescent nature of the substance. The presence of iron and hardness in water should be avoided.

9.4 ESTIMATION OF HARDNESS

9.4.1 Determination of Total Hardness by EDTA Method

The determination of calcium and magnesium content in water will be conducted using the process of titration, using Ethylene Diamine Tetra Acetic Acid (EDTA) with a predetermined concentration. A measure or metric that is used to assess or evaluate a certain phenomenon or situation. The addition of Erichrome Black-T at a pH of 10 is performed in the water sample. The wine red color observed in water can be attributed to the creation of an unstable combination containing calcium (Ca^{2+}) and magnesium (Mg^{2+}) ions. The solution is subsequently subjected to titration with an EDTA solution, resulting in a transition of the wine red color of the solution to a steel blue or violet color, which serves as an indication of the endpoint.

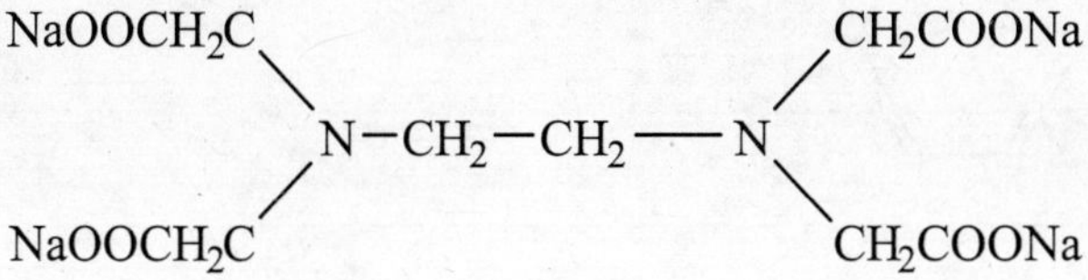

EDTA

$$\left.\begin{matrix} Ca^{2+} \\ Mg^{2+} \end{matrix}\right\} + EBT \longrightarrow \left[\begin{matrix} Ca^{2+} \\ Mg^{2+} \end{matrix} EBT\right] \text{unstable complex (wine red)}$$

$$\left[\begin{matrix} Ca^{2+} \\ Mg^{2+} \end{matrix} EBT\right] \text{unstable complex (wine red)} \xrightarrow{\text{EDTA}} \left[\begin{matrix} Ca^{2+} \\ Mg^{2+} \end{matrix} EDTA\right] \text{Complex} + \text{EBT (blue color)}$$

The hexa-dentate ligand EDTA exhibits the ability to form a complex by trapping a metal ion M, specifically calcium (Ca^{2+}) or magnesium (Mg^{2+}), at its central coordination site. These compounds consist of a Central Metal atom that is encompassed by Ligands. The core metal atom functions as a Lewis acid, exhibiting an affinity for accepting electron pairs, while the ligands act as Lewis bases, offering electron pairs for donation.

Titration reaction: $HY^3\ (aq) + Ca^{2+}(aq) \longrightarrow CaY^2\ (aq) + H^+(aq)$ (also for Mg^{2+})

End point reaction: $HY^3\ (aq) + MgIn\ (aq) \longrightarrow MgY^2\ (aq) + HIn^2\ (aq)$

wine red   sky blue

Formation of octahedral complex due to interaction of EDTA and metal atom.

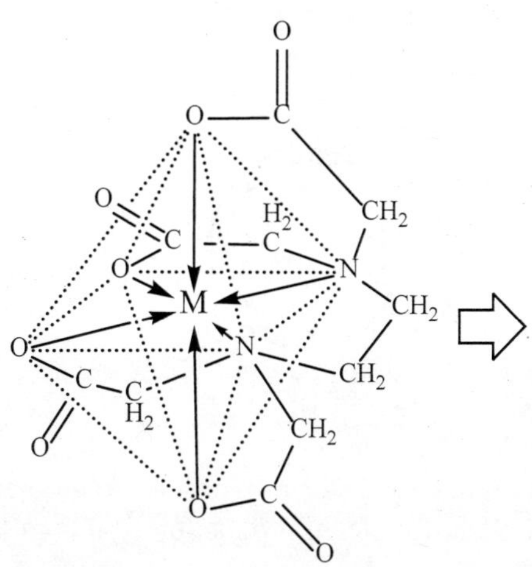

Formation of Octahedral complex

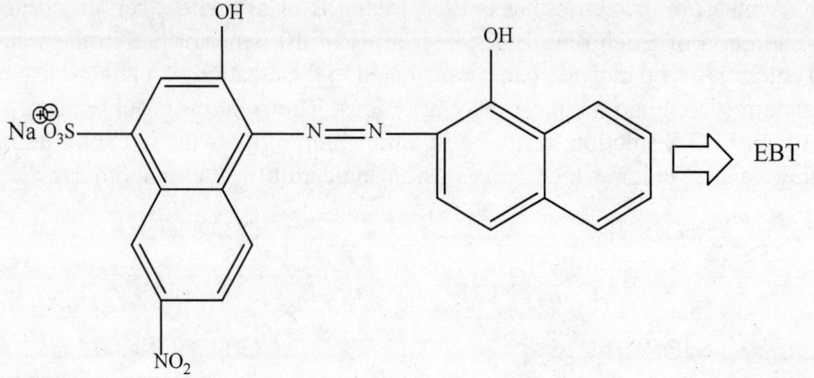

The indicator used in this study is a chemically complex organic compound known as sodium-1 (1-hydroxy-2-naphthylazo)-6-nitro-2-naphthol-4-sulphonate, which is usually referred to as Erichrome Black-T.

Method:

1. Pour 50 ml of water into a conical flask.
2. To initiate the reaction, it is necessary to introduce a buffer solution together with a little quantity of EBT-T.
3. The process of titration against EDTA is to be conducted.
4. The red color undergoes a transformation to a shade like steel blue or violet.

9.5 SCALE AND SLUDGE

Boilers are employed inside industrial sectors for the purpose of steam generation. The process of water evaporation involves the continuous transformation of liquid water into gaseous steam. During this procedure, the formation of dissolved salt is also observed. As concentration levels rise, the substances stick to the inner surface of the boiler or are expelled from the water as precipitate. The boiler experiences difficulties, specifically the occurrence of scale and sludge buildup.

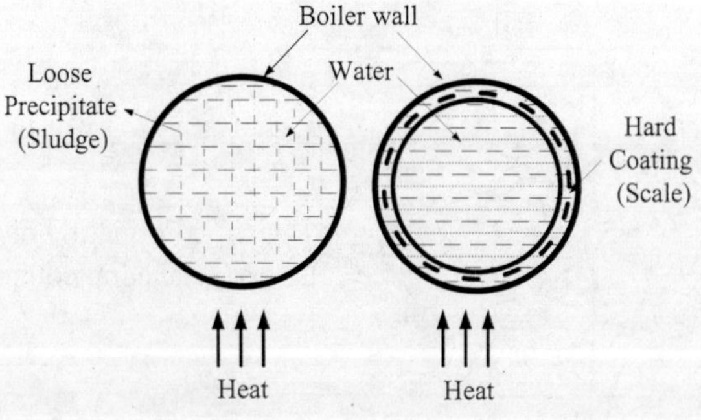

Scale: The concept of scale refers to the relative size or proportion of an object, phenomenon, or the constant evaporation of water in boilers for steam production results in a steady increase in the concentration of salts present in the water. Once the concentration of salts in the solution reaches a saturation point, the salts precipitate and adhere to the inner surface of the boiler.

Sludge: The phenomenon of sludge formation occurs when salts undergo precipitation in a loose and slimy manner during the process of boiling within a boiler. The introduction of magnesium chloride results in the formation of a rigid deposit and undergoes decomposition, yielding hydroxide and hydrochloric acids, hence causing a reduction in pH levels.

$$MgCl_2 + 2H_2O \longrightarrow 2HCl + Mg(OH)_2$$

The deposition of magnesium sulphate results in the formation of a rigid layer on the surfaces subjected to heat. The deposition of calcium sulphate results in the formation of a rigid layer on the inner surfaces of the boiler. The decomposition of calcium bicarbonate occurs at a relatively low temperature, resulting in the liberation of carbon dioxide. The heating surface experiences the formation of calcium carbonate deposits in the form of a soft scale.

$$Ca(HCO_3)_2 \longrightarrow CaCO_3 + CO_2 + H_2O$$

9.5.1 Disadvantages of Scale Formation

The drawbacks associated with scale formation is its negative impact on industrial processes are discussed below.

1. The presence of scale creates a thermal barrier that separates heat from water, resulting in an increased amount of heat needed to bring the water to its boiling point. A certain proportion of fuel is wasted.
2. The presence of additional material in a boiler might lead to a decrease in its efficiency and performance due to the obstruction of pipes.
3. The safety of the boiler is compromised due to a decrease in its structural integrity caused by overheating, resulting in distortion of the metal framework, commonly referred to as bagging.
4. The excessive build-up of scale has the potential to pose a risk of explosion.
5. The addition of scales to a system results in a decrease in pH, leading to an increase in acidity. Additionally, the presence of ferric ions in combination with oxygen in water results in the formation of $Fe(OH)_3$, which is responsible for the occurrence of rust in a boiler.

9.5.2 Removal of Scales

1. The removal of loose scale adhered to the interior walls of boilers can be effectively achieved by the use of wire brushes for scraping purposes.
2. The removal of the substance can be achieved through the application of thermal shocks to the boiler.
3. The removal of scales can be accomplished through the process of dissolution, wherein the scales are dissolved using appropriate chemicals. For instance, calcium carbonates ($CaCO_3$) can be effectively dissolved by employing a solution of 5–10% hydrochloric

acid (HCl). The internal treatment of water in a boiler involves the direct addition of chemicals to the water.

4. Complexing agents can be employed to eliminate scales that generate elements. The addition of complexing agents facilitates the production of soluble complexes, such as calcium sulphate ($CaSO_4$), through the use of EDTA.
5. Another technique that might be employed is the blow down procedure. The process involves the substitution of a water-filled measuring instrument with freshwater.

9.5.3 Disadvantages of Sludge Formation

1. The issue of fuel wastage is significant due to the low heat conductivity of sludge, resulting in a considerable amount of heat being wasted.
2. The efficiency of a boiler falls in direct proportion to the amount of sludge accumulated on its inner walls, resulting in a reduction in steam generation.
3. The excessive accumulation of sludge has a detrimental impact on the safety of the boiler. It precipitates and obstructs the flow of water.
4. On occasion, the pipes may become obstructed, leading to an elevated risk of explosion.
5. The reduction in pH levels during the creation of sludge has the potential to induce corrosion in boilers. Boiler corrosion refers to the process of material degradation caused by chemical or electrochemical reactions. The existence of dissolved gases, such as oxygen (O_2) and carbon dioxide (CO_2), is responsible for this phenomenon. Additionally, this phenomenon might be attributed to the generation of acids during the process of hydrolysis, as well as the existence of unbound acid molecules.

9.5.4 Prevention of Sludge Formation

One method for mitigating sludge formation involves the extraction of a sample of sludge containing water, followed by the introduction of fresh water in a process commonly referred to as blow down process.

The prevention of this issue can be achieved through the utilization of soft water. The implementation of pre-boiler water treatment measures can effectively mitigate the formation of sludge by facilitating the elimination of calcium (Ca), magnesium (Mg), and silica (SiO_2) from the water prior to its introduction into the boiler system.

9.6 CAUSTIC EMBRITTLEMENT

Caustic embrittlement, also known as caustic cracking, is a phenomenon characterized by the embrittlement of materials due to exposure to caustic substances. The presence of corrosive compounds, such as sodium hydroxide (NaOH), leads to the development of brittleness in the boiler material. The phenomenon being referred to is known as caustic embrittlement. The process of water softening, known as the lime soda process, involves the use of sodium carbonate (Na_2CO_3) to eliminate calcium (Ca) and magnesium (Mg) salts from the water. As a result, sodium carbonate is left behind in the water. The compound sodium carbonate (Na_2CO_3)

undergoes a process known as hydrolysis, resulting in the formation of sodium hydroxide (NaOH), commonly referred to as caustic soda.

$$Na_2CO_3 + H_2O \longrightarrow 2NaOH + CO_2$$

As the concentration of sodium hydroxide (NaOH) grows, water infiltrates into small fissures, resulting in the brittleness of the surrounding area. The concentration of NaOH grows as evaporation progresses, leading to a reaction with the iron in the boiler and resulting in embrittlement.

Prevention: The prevention of caustic soda generation can be achieved with the inclusion of sodium sulphate (Na_2SO_4) or sodium phosphate (Na_2PO_4), as well as tannin and lignin, which effectively obstruct hairline fractures. The formation of $CaSO_4$ is prevented by avoiding an excess of Na_2SO_4.

9.7 PRIMING AND FOAMING

The phenomena of priming and foaming are two important processes that occur in boilers.

Priming

Priming refers to the phenomenon wherein the continual boiling of water in a boiler can result in the entrainment of minute water droplets that are carried over to the engine. The phenomenon of having different levels of water content inside steam, resulting in the development of wet steam in the form of foam or mist, is referred to as priming, specifically characterized by the presence of droplets in the steam. The accumulation of salt crystals resulting from this phenomenon contributes to the reduction in energy efficiency of the boiler.

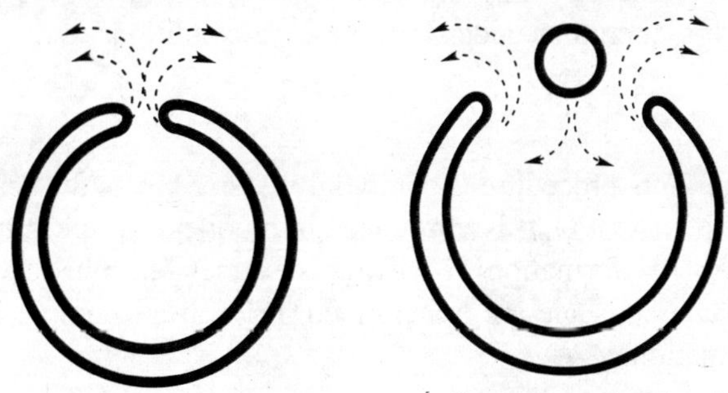

Normal Bubble　　　Carry over droplet

Priming occurs as a result of the presence of suspended contaminants and dissolved salts. The likelihood of priming is heightened by the production of high steam velocity and the rapid boiling of water. The occurrence of high water levels and a flawed boiler design are additional factors contributing to the occurrence of leaks due to priming. The prevention of boiler-related issues can be achieved by the use of effective boiler design practices and the avoidance of abrupt temperature fluctuations. It is imperative to maintain a low water level during the process of continual heating. The implementation of routine maintenance and the proper installation of a mechanical steam purifier serve to mitigate the occurrence of priming development.

Foaming

The phenomenon of foaming, characterized by the formation of small bubbles and froth on the surface of water, occurs as a result of the elevated concentration of solid particles within the water. The formation of this phenomenon is attributed to the disparity in solute concentration and the presence of suspended substances such as oil and grease. Foaming occurs as a result of the presence of chemicals that reduce surface tension. The prevention of foaming can be achieved through the incorporation of antifoaming substances such as castor oil, as well as the elimination of oily particles via the utilization of silicic acid and sodium aluminate.

9.8 TREATMENT METHODS OF WATER

9.8.1 Internal Treatment Method

1. Phosphate Conditioning: A limited quantity of phosphate ions is introduced to induce the precipitation of calcium ions. The selection of phosphate is contingent upon the pH conditions prevailing within the boiler system.

$$3CaCl_2 + 2Na_3PO_4 \longrightarrow Ca_3(PO_4)_2 \downarrow + 6NaCl$$

2. Colloidal Conditioning: The mitigation of scale formation in water boilers can be achieved through the application of various substances such as kerosene, tannin, and starch. The particles responsible for scale formation are effectively coated and can be readily eliminated by the blow down process.

3. Carbonate Conditioning: In this experimental procedure, sodium carbonate (Na_2CO_3) is introduced into a water boiler. Calcium salts are precipitated in the form of calcium carbonate ($CaCO_3$). The removal is achieved using the blow down procedure. This particular substance finds application in boilers operating under low pressure conditions.

$$CaSO_4 + Na_2CO_3 \longrightarrow CaCO_3 \downarrow + Na_2SO_4$$

4. Calgon Conditioning: This procedure is commonly referred to as the "calcium gone" phase. Sodium hexametaphosphate ($(Na_2PO_3)_6$) is introduced into boilers, where it undergoes a reaction with calcium, resulting in the formation of calcium hexametaphosphate ($(Ca_2PO_3)_6$). Salts that have the propensity to generate scale are transformed into soluble complexes, hence inhibiting the process of scale formation.

$$Na_2[Na_4(PO_3)_6] \rightleftharpoons 2Na^+ + [Na_4(PO_3)_6]^{2-}$$

$$2CaSO_4 + [Na_3(PO_3)_6]^{2-} \longrightarrow [Ca_2(PO_3)_6] + 2Na_2SO_4$$

5. Sodium Aluminate Treatment: The application of sodium aluminate as a treatment method of water can be understand as follows.

$$NaAlO_2 + 2H_2O \longrightarrow NaOH + Al(OH)_3 \downarrow$$

$$2NaOH + MgCl_2 \longrightarrow Mg(OH)_2 \downarrow + 2NaCl$$

6. Radioactive Conditioning: Radioactive tablets are introduced to emit radiation energy, hence impeding the production of scales.

7. Electrical Conditioning: When a mercury bulb is placed in a boiler, it creates an electrical discharge. The prevention of scale formation is observed.

9.8.2 External Treatment Method

Various techniques are employed for the purpose of water softening, which involve the elimination of chemicals responsible for water hardness.

1. **Zeolite Process:** The zeolite process refers to a chemical process that utilizes zeolites as catalysts for various industrial applications.
2. **Ion Exchange Process:** The ion exchange process is a method used in several industries to separate and purify ions in a solution.
3. **Mixed Bed Deionization:** Mixed bed deionization is a process used for the removal of ions from water.

9.8.2.1 Zeolite or Permutit Process

The term "zeolite" originates from the combination of two Greek words, namely "*zein*" and "*lithos*" which respectively translate to "to boil" and "stone." The term was first utilized in 1756 by Granstedt, a Swedish geologist, to designate a specific category of naturally occurring minerals. Upon being subjected to heat, the substances underwent expansion and released the water molecules that were previously bound to them. The zeolites under consideration are classified as hydrated double silicates, which are composed of an alkali or alkali earth oxide, alumina, silica, and water. Zeolites are a class of hydrated sodium alumino silicate minerals. The chemical formula provided consists of sodium oxide (Na_2O), aluminum oxide (Al_2O_3), silicon dioxide (SiO_2), and water (H_2O), denoted by X and Y. The variable X represents the range of 2 to 10, while the variable Y represents the range of 2 to 6.

$$Na_2O \cdot Al_2O_3 \cdot XSiO_2 \cdot YH_2O \text{ where } X = 2\text{–}10,\ Y = 2\text{–}6$$

Natural Zeolite

Non-porous natural zeolites are obtained through a series of processes involving the washing, heating, and treatment with sodium hydroxide of green sand. The organism possesses the ability to engage in the exchange of sodium ions, with several examples being as follows:

1. Natrolite $Na_2O.Al_2O_3.4SiO_2.2H_2O$
2. Laumontite $CaO.Al_2O_3.4SiO_2.4H_2O$
3. Harmotome $BaO.K_2O.Al_2O_3.5SiO_2.5H_2O$

Johann Nepomuk von Fuchs

Synthetic Zeolite

The initial production of synthetic zeolite can be attributed to the efforts of German chemist Fuchs in 1818. Fuchs successfully synthesized zeolite by combining solutions containing sodium aluminate and sodium silicate.

The substance is derived from a solution containing sodium silicate and aluminum hydroxide. The material has a greater exchange capacity and possesses a porous structure.

Process

The process involves the utilization of a steel tank that is equipped with a substantial coating of Zeolite material. Zeolite effectively retains calcium and magnesium ions, which are the primary contributors to water hardness, during the process of water filtration. The process involves the exchange of calcium and magnesium ions with an equivalent quantity of zeolite, resulting in the formation of sodium zeolite.

The chemical reactions involved in the process of water softening.

$$\left.\begin{matrix}Ca\\Mg\end{matrix}\right](HCO_3)_2 + Na_2Ze \longrightarrow \left.\begin{matrix}Ca\\Mg\end{matrix}\right]Ze + 2NaHCO_3$$

$$\left.\begin{matrix}Ca\\Mg\end{matrix}\right]SO_4 + Na_2Ze \longrightarrow \left.\begin{matrix}Ca\\Mg\end{matrix}\right]Ze + Na_2SO_4$$

$$\left.\begin{matrix}Ca\\Mg\end{matrix}\right](NO_3)_2 + Na_2Ze \longrightarrow \left.\begin{matrix}Ca\\Mg\end{matrix}\right]Ze + 2NaNO_3$$

$$\left.\begin{matrix}Ca\\Mg\end{matrix}\right]Cl_2 + Na_2Ze \longrightarrow \left.\begin{matrix}Ca\\Mg\end{matrix}\right]Ze + 2NaCl$$

The exchange of sodium (Na) ions persists until the depletion of available sodium ions.

1. Regeneration: Following an extended duration of the zeolite process, when all the binding sites of zeolite undergo substitution with calcium and magnesium, the zeolite (indicated by Ze in chemical reactions) becomes fully depleted. Currently, the regeneration of zeolite is required. Sodium chloride (NaCl) is introduced into the process tank for the purpose of regeneration. The subsequent reaction is implicated in the process of regeneration.

$$\left.\begin{matrix}Ca\\Mg\end{matrix}\right]Ze + 2NaCl \longrightarrow Na_2Ze + \left.\begin{matrix}Ca\\Mg\end{matrix}\right]Cl_2$$

The addition of calcium chloride ($CaCl_2$) and magnesium chloride ($MgCl_2$) resulted in drainage, while sodium zeolite (Na_2Ze) exhibited potential for reusability.

2. Advantages of zeolite process

1. The approach described herein effectively eliminates water hardness.
2. The aforementioned procedure has the capability to be executed autonomously, devoid of any external intervention, with the sole influence of gravity force.
3. In this particular procedure, the involvement of specialists is not necessary and the operation is straightforward to execute.
4. There is an absence of sludge formation and the production of soft water occurs in a relatively short period of time.
5. A limited amount of space is necessary for the installation of a tank.

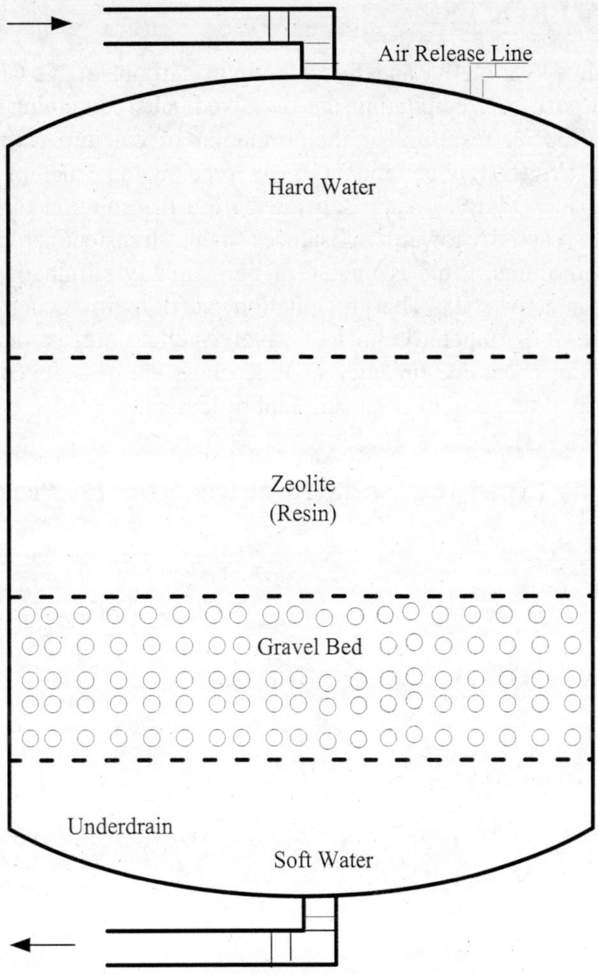

Figure 9.3 Softening of hard water by zoelite.

3. Disadvantages of zeolite process: One of the drawbacks associated with the zeolite process is its inherent limitations.

1. Only calcium and magnesium ions have the capability of being eliminated.
2. There is a significant concentration of sodium ions observed in the treated water.
3. The treatment does not address the corrosive effects caused by other acids.
4. The treatment of water containing Fe^{2+} and Mn^{2+} ions is challenging due to the inability to regenerate Fe and Mn Zeolite.
5. In order to avoid obstruction in Zeolite beds, it is imperative that water is devoid of any suspended contaminants.
6. Treated water often exhibits a higher concentration of dissolved solids.

9.9 LIME-SODA PROCESS

The combination of lime ($Ca(OH)_2$) and Soda, Sodium carbonate (Na_2CO_3), is employed as the reagent for the purpose of precipitating the dissolved salts containing calcium ions (Ca^{2+}) and magnesium ions (Mg^{2+}), resulting in the formation of calcium carbonate ($CaCO_3$) and magnesium hydroxide ($Mg(OH)_2$) precipitates. The precipitated calcium carbonate ($CaCO_3$) and magnesium hydroxide ($Mg(OH)_2$) are separated from the solution through the process of filtration. Lime exhibits reactivity towards substances such as transient hardness, carbon dioxide, acids, bicarbonates, and alums. Lime is unable of permanently eliminating calcium hardness, a task that is better suited for soda. The precipitation reactions involving lime and soda have a notably sluggish rate. It is imperative to exclusively include precise quantities of lime and soda. The presence of an excessive quantity of lime and soda in a boiler system can lead to the occurrence of boiler issues, such as caustic embrittlement.

9.9.1 Calculation of Lime and Soda Required for the Process

Amount of lime required for softening

$$= \frac{74}{100}\left[\begin{array}{l}\text{TH of } Ca^{2+} + 2\times\text{TH of } Mg^{2+} + \text{PH of } Mg^{2+} + CO_2 + \frac{1}{2}HCl + H_2SO_4 + \frac{1}{2}NaHCO_3 \\ + \frac{1}{2}KHCO_3 + FeSO_4 + \{3\times Al_2(SO_4)_3\} - \frac{1}{2}NaAlO_2\end{array}\right]$$

Amount of soda required for softening:

$$= \frac{106}{100}\left[\begin{array}{l}\text{PH of } Ca^{2+} + 2\times\text{PH of } Mg^{2+} + \frac{1}{2}\ HCl + H_2SO_4 - \frac{1}{2}\ NaHCO_3 - \frac{1}{2}\ KHCO_3 \\ + FeSO_4 + \{3\times Al_2(SO_4)_3\}\end{array}\right]$$

All the values are in terms of chemical equivalents of $CaCO_3$. Where TH = Temporary Hardness, PH = Permanent Hardness.

The aforementioned formulas are derived from the chemical reactions that occur in the process, as described subsequently.

9.9.2 Treatment with Lime

1. Removal of temporary hardness of Ca^{2+} and Mg^{2+}

The stoichiometric ratio for the elimination of temporary hardness of Ca^{2+} is one mole of lime per mole of Ca^{2+}.

$$Ca(HCO_3)_2 + Ca(OH)_2 \longrightarrow 2CaCO_3 + 2H_2O$$

The elimination of one mole of temporary hardness of Mg^{2+} necessitates the utilization of two moles of lime.

$$Mg(HCO_3)_2 + 2Ca(OH)_2 \longrightarrow Mg(OH)_2 + 2CaCO_3 + 2H_2O$$

2. Removal of Permanent hardness of Mg^{2+}

The stoichiometric ratio for the elimination of persistent Mg^{2+} hardness is one mole of lime per mole of hardness.

$$MgCl_2 + Ca(OH)_2 \longrightarrow Mg(OH)_2 + CaCl_2$$

$$MgSO_4 + Ca(OH)_2 \longrightarrow Mg(OH)_2 + CaSO_4$$

$$Mg(NO_3)_2 + Ca(OH)_2 \longrightarrow Mg(OH)_2 + Ca(NO_3)_2$$

$$Mg(CO_3) + Ca(OH)_2 \longrightarrow Mg(OH)_2 + CaCO_3$$

3. Removal of CO_2

The stoichiometric ratio for the elimination of one mole of CO_2 hardness necessitates the use of one mole of lime.

$$CO_2 + Ca(OH)_2 \longrightarrow CaCO_3 + H_2O$$

4. Removal of acids

The stoichiometric ratio for the neutralization reaction between lime and hydrochloric acid is 1:0.5, indicating that one mole of lime is needed to neutralize half a mole of HCl.

$$2HCl + Ca(OH)_2 \longrightarrow CaCl_2 + 2H_2O$$

The stoichiometric ratio between lime and H_2SO_4 is 1:1, meaning that one mole of lime is necessary to neutralize one mole of H_2SO_4.

$$H_2SO_4 + Ca(OH)_2 \longrightarrow CaSO_4 + 2H_2O$$

5. Removal of bicarbonates of Na^+ and K^+

One mole of bicarbonates of Na^+ and K^+ each can be removed by utilizing half a mole of lime.

$$2NaHCO_3 + Ca(OH)_2 \longrightarrow CaCO_3 + Na_2CO_3 + 2H_2O$$

$$2KHCO_3 + Ca(OH)_2 \longrightarrow 2CaCO_3 + K_2CO_3 + 2H_2O$$

6. Removal of alums

The stoichiometric ratio between lime and H_2SO_4 is 1:1, meaning that one mole of lime is necessary to neutralize one mole of H_2SO_4.

$$FeSO_4 + Ca(OH)_2 \longrightarrow CaSO_4 + Fe(OH)_2$$

The stoichiometric ratio for the reaction between lime and $Al_2(SO_4)_3$ is 3 moles of lime per 1 mole of $Al_2(SO_4)_3$.

$$Al_2(SO_4)_3 + 3Ca(OH)_2 \longrightarrow 3CaSO_4 + 2Al(OH)_3$$

The compound NaOH, which is generated from the reaction of $NaAlO_2$, undergoes a chemical reaction with salts that contribute to water hardness, resulting in the formation of lime. Therefore, it is necessary to deduct the hardness of the substance from the lime requirement.

$$NaAlO_2 + 2H_2O \longrightarrow Al(OH)_3 + NaOH$$
$$2NaOH + CaCl_2 \longrightarrow Ca(OH)_2 + 2NaCl$$

In the process of treating lime to eliminate the presence of Mg^{2+} ions, the introduction of acids and alums inadvertently results in the generation of permanent hardness caused by Ca^{2+} ions in water. Therefore, it is necessary to implement both lime and soda treatment methods.

9.9.3 Treatment with Soda

The chemical reaction involved in the elimination of the permanent hardness caused by the presence of Ca^{2+} ions is as follows.

$$CaCl_2 + Na_2CO_3 \longrightarrow CaCO_3 + 2NaCl$$
$$CaSO_4 + Na_2CO_3 \longrightarrow CaCO_3 + Na_2SO_4$$
$$Ca(NO_3)_2 + Na_2CO_3 \longrightarrow CaCO_3 + 2NaNO_3$$

9.10 ION EXCHANGE PROCESS

The ion exchange process, also known as ion exchange chromatography, is a widely used technique in various scientific and industrial applications. It involves the exchange of ions between a solid phase, known as the ion exchange resin. This procedure is commonly referred to as demineralization or deionization, as it effectively eliminates mineral ions from water. The aforementioned technique involves the utilization of ion exchange resins that have been specifically manufactured for this purpose. Resin possesses the ability to effectively bind minerals and selectively remove salts from water.

Advantage:

One advantage is that it allows for increased efficiency and productivity.

1. One of the notable characteristics of this product is its ability to generate soft water.
2. The capacity to effectively address water with excessive presence of acidity or alkalinity.
3. The potential for regenerating ion resins exists.
4. The expense of maintenance is minimal.

Disadvantage:

1. One potential drawback is that the expense associated with equipment is substantial.
2. Water with a high level of turbidity poses challenges in terms of treatment feasibility.
3. The treatment of such water necessitates the use of costly chemical agents.

9.11 MIXED BED DEIONIZATION

The equipment is comprised of a singular column that contains a combination of cation and anion exchangers. When water flows through a bed, it undergoes many interactions with both exchangers. The water that is produced as a result has a reduced quantity of dissolved salt.

Regeneration

Regeneration of the system is achieved through the process of backwashing. This involves the use of dilute sodium hydroxide (NaOH) to regenerate the anions, and dilute to regenerate the cations. Sulphuric acid (H_2SO_4) is a highly corrosive and strong.

Enhanced convenience and heightened effectiveness. There are several drawbacks associated with this phenomenon. The expense of regeneration is higher. The cost of equipment is significantly elevated.

Desalination

Desalination refers to the process of removing salt and other impurities from seawater. Desalination is a procedure that involves the extraction of a certain quantity of salt and other minerals from brackish water, which typically contains approximately 3.5% salts. The process of separating salts from water involves the initial step of evaporation, followed by the subsequent step of condensation.

SOLVED PROBLEMS

PROBLEM 9.1 A sample of water is found to contains following dissolving salts in milligrams per litre: $Mg(HCO_3)_2$ = 73, $CaCl_2$ = 111, $Ca(HCO_3)_2$ = 81, $MgSO_4$ = 40 and $MgCl_2$ = 95. Calculate temporary and permanent hardness and total hardness.

Solution We can write down given values in tabulated form as follows:

Name of the hardness causing salts	Amount of the hardness causing salts(mg/Lit)	Molecular weight of hardness causing salts	Amounts equivalent to $CaCO_3$ (mg/Lit)
$Mg(HCO_3)_2$	73	146	73 × 100/146 = 50
$CaCl_2$	111	111	111 × 100/111 = 100
$Ca(HCO_3)_2$	81	162	81 × 100/162 = 50
$MgSO_4$	40	120	40 × 100/120 = 33.3
$MgCl_2$	95	95	95 × 100/95 = 100

Temporary hardness = $Mg(HCO_3)_2$ + $Ca(HCO_3)_2$ = 50 + 50 = 100 mgs/Lit. Permanent hardness = $CaCl_2$ + $MgSO_4$ + $MgCl_2$ = 100 + 33.3 + 100 = 233.3 mgs/Lit.

Total hardness = Temporary hardness + Permanent hardness = 100 + 233.3 = 333.3 mgs/Lit.

PROBLEM 9.2 A sample of water is found to contains following dissolving salts in milligrams per litre $Mg(HCO_3)_2$ = 16.8, $MgCl_2$ = 12.0, $MgSO_4$ = 29.6 and NaCl = 5.0. Calculate temporary and permanent hardness of water.

Solution We can write down given values in tabulated form as follows:

Name of the hardness causing salts	Amount of the hardness causing salts(mg/Lit)	Molecular weight of hardness causing salts	Amounts equivalent to $CaCO_3$ (mg/Lit)
$Mg(HCO_3)_2$	16.8	146	16.8 × 100/146 = 11.50
$MgCl_2$	12	95	12.0 × 100/95 = 12.63
$MgSO_4$	29.6	120	29.6 × 100/120 = 24.66
NaCl	5	NaCl does not contribute any hardness to water hence it is ignored.	

Temporary hardness = $Mg(HCO_3)_2$ = 11.50 mgs/Lit.

Permanent hardness = $MgCl_2 + MgSO_4$ = 12.63 + 24.66 = 37.29mgs/Lit.

PROBLEM 9.3 A sample of water is found to contains following analytical data in milligrams per litre: $Mg(HCO_3)_2$ = 14.6, $MgCl_2$ = 9.5, $MgSO_4$ = 6.0 and $Ca(HCO_3)_2$ = 16.2. Calculate temporary and permanent hardness of water in parts per million, Degree Clarke's and Degree French

Solution We can write down given values in tabulated form as follows:

Name of the hardness causing salts	Amount of the hardness causing salts (mg/Lit)	Molecular weight of hardness causing salts	Amounts equivalent to $CaCO_3$ (mg/Lit)
$Mg(HCO_3)_2$	14.6	146	14.6 × 100/146 = 10
$MgCl_2$	9.5	95	9.5 × 100/95 = 10
$MgSO_4$	6	120	6.0 × 100/120 = 5
$Ca(HCO_3)_2$	16.2	162	16.2 × 100/162 =10

Temporary hardness [$Mg(HCO_3)_2 + Ca(HCO_3)_2$] = 10 + 10 = 20 mg/Lit = 20 ppm = 20 × 0.07°Cl = 1.4°Cl = 20 × 0.1°Fr = 2°Fr

Permanent hardness [$MgCl_2 + MgSO_4$] = 10 + 5 = 15 mg/Lit = 15 ppm = 15 × 0.07°Cl = 1.05°Cl = 15 × 0.1°Fr = 1.5°Fr

PROBLEM 9.4 Calculate the temporary and permanent hardness of water sample containing $Mg(HCO_3)_2$= 7.3 mg/L, $Ca(HCO_3)_2$ = 16.2 mg/L, $MgCl_2$ = 9.5 mg/L, $CaSO_4$ = 13.6 mg/L.

Answer: TH = 15 and PH =20

PROBLEM 9.5 Calculate the temporary and permanent hardness of water sample containing $Mg(HCO_3)_2$ = 7.3 mg/L, $Ca(HCO_3)_2$ = 16.2 mg/L, $MgCl_2$= 9.5 mg/L, $CaSO_4$ = 13.6 mg/L.

Answer: TH = 150 and Total H = 350

PROBLEM 9.6 50 ml of a sample water consumed 15 ml of 0.01 EDTA before boiling and 5 ml of the same EDTA after boiling. Calculate the degree of hardness, permanent hardness and temporary hardness.

Solution

50 ml of water sample = 15 ml of 0.01M EDTA

= (15 × 100)/50 ml of 0.01EDTA = 300 ml of 0.01M EDTA

= 2 × 300 ml of 0.01 N EDTA (Molarity of EDTA = 2 × Normality of EDTA)

= 600 ml or 0.6 L of 0.01 eq. of $CaCO_3$.

= 0.6 × 0.01 × 50 g$CaCO_3$ eq.

Hence total hardness = 0.30 g or 300 mg of $CaCO_3$ eq = 300 mg/L or ppm.

Now 50 ml of boiled water = 5 ml of 0.01M EDTA

∴ 1000 ml of boiled water = (5 × 1000)/50 ml of 0.01M EDTA

= 100 mL of 0.01M EDTA

= 200 mL or 0.2 L of 0.01N EDTA

= 0.2 × 0.01 × 50g of $CaCO_3$ eq

= 0.1 g or 100 mg of $CaCO_3$ eq

Hence permanent hardness = 100 mg/L or ppm

∴ Temporary hardness = 300 –100 = 200 ppm.

PROBLEM 9.7 0.5 g of $CaCO_3$ was dissolved in HCl and the solution made up to 500 ml with distilled water. 50 ml of the solution required 48 ml of EDTA solution for titration. 50 ml of hard water sample required 15 ml of EDTA and after boiling and filtering required 10 ml of EDTA solution. Calculate the hardness.

Solution

500 ml of SHW = 0.5 g or 500 mg $CaCO_3$ eq

∴ 1 ml SHW = 1 mg $CaCO_3$ eq

Now 48 ml of EDTA solution = 50/48 mg $CaCO_3$ eq

∴ 1ml of EDTA solution = 50/48 mg $CaCO_3$ eq

Calculation of the total hardness of water:

50 ml hard water = 15 ml EDTA = 15 × 50/48 mg of $CaCO_3$ eq

∴ 1000 ml of hard water = (15.625 × 1000)/50 = 312.5 mg $CaCO_3$ eq

Hence total hardness = 312.5 mg/L or 312.5 ppm.

Numerical Problems based on Lime-Soda Process

PROBLEM 9.8 Calculate the amount of lime and soda required for the treatment of 10,000 L of water containing the following salts dissolved per litre. $CaCO_3$ = 50 mg, $CaCl_2$ = 11.1 mg, $MgSO_4$ = 12 mg, $NaHCO_3$ = 7.25 mg, silica = 10 mg.

Solution Conversion of the hardness in terms of $CaCO_3$ equivalents:

Name of the hardness causing salts	Amount of the hardness causing salts (mg/Lit)	Molecular weight of hardness causing salts	Amounts equivalent to $CaCO_3$ (mg/Lit)
$CaCO_3$	50	100	50 × 100/100 = 50
$CaCl_2$	11.1	111	11.1 × 100/111 = 10
$MgSO_4$	12	120	12 × 100/120 = 10
$NaHCO_3$	7.25	84	7.25 × 100/84 = 8.63

Amount of lime required $= 74/100\ [CaCO_3 + \frac{1}{2} NaHCO_3 + MgSO_4]$
$= 74/100\ [50 + 4.3 + 10] = = 47.58$ mg/L

Amount of lime required for 10,000 L of water = 47.58 × 10,000 = 4,75,800 mg/10,000 L
= 0.4758 kg/10,000L.

Amount of Soda required $= 106\ [CaCl_2 + MgSO_4 - \frac{1}{2} NaHCO_3]/100$
= 106 [10 + 10 – 4.3] /100 = 16.64 mg/L

Soda required for 10,000 L of water = 16.64 × 10,000 = 166400 mg/10,000 L
= 0.1664 kg/10,000 L

PROBLEM 9.9 Explain with chemical equations and calculate the amount of lime and soda required for softening of 1,00,000 L of water containing the following: HCl = 7.3 mg/L, $Al_2(SO_4)_3$ = 34.2 mg/L, $MgCl_2$ = 9.5 mg/L, NaCl = 29.25 mg/L. Purity of lime is 90% and that of soda is 98%. 10% of chemicals are to be used in excess in order to complete the reaction quickly.

Solution: Conversion of the hardness in terms of $CaCO_3$ equivalents:

Name of the hardness causing salts	Amount of the hardness causing salts (mg/Lit)	Molecular weight of hardness causing salts	Amounts equivalent to $CaCO_3$ (mg/Lit)
HCl	7.3	73	7.3 × 100/73 = 10
$Al_2(SO_4)_3$	34.2	114	34.2 × 100/114 = 30
$MgCl_2$	9.5	95	9.5 × 100/95 = 10

Lime requirement $= 74/100\,[HCl + 3Al_2(SO_4)_3 + MgCl_2]$ × volume of water × 100/%purity

= 74/100 (10 + 90 + 10) × 1,00,000 × 100/90 = 9.04 kg

∴ Lime required (using 10% excess) = 9.04 × 110/100 = 9.944 kg

Soda requirement $=106/100\,[HCl + Al_2(SO_4)_3 + MgCl_2]$ × volume of water × 100/% purity

= 106/100 (10 + 90 + 10) × 1,00,000 × 100/98 = 11.8 kg

∴ Soda required (using 10% excess) = 11.8 × 110/100 = 12.98 kg.

QUESTIONS

9.1 Calculate temporary hardness and permanent hardness of a sample of water containing: $Mg(HCO_3)_2$ = 7.3 mg/l; $Ca(HCO_3)_2$ = 16.2 mg/l; $MgCl_2$ = 9.5 mg/l; $CaSO_4$ = 13.6 mg/l.

9.2 10000 litres of hard water was softened by passing it through zeolite softener. The exhausted zeolite softener bed was regenerated by passing it through 58 litres of NaCl solution containing 100 g/l of brine solution. Calculate the hardness of water in term of $CaCO_3$ equivalent.

9.3 A water sample is found to contain 40.5 mg/L $Ca(HCO_3)_2$; 14.6 mg/L $Mg(HCO_3)_2$; 22.2 mg/L $CaCl_2$; 24 mg/L $MgSO_4$ and 18mg/L NaCl. Calculate the temporary and permanent hardness of the water sample.

9.4 Explain the zeolite process of water softening. What are the advantages and limitations of this process? Calculate the amount of lime and soda required for the treatment of 10,000 litres of water whose analysis is as follows (in mg/L): $Mg(HCO_3)_2$ = 73; $CaSO_4$ = 102; $MgCl_2$ = 95; $MgSO_4$ = 24; $Ca(HCO_3)_2$ = 121.5; NaCl = 55.

9.5 Write the different chemical reactions taking place in soda-lime process. A sample of water on analysis give following result: Analysis of Raw water: Ca^{2+} = 80 mg/l; Mg^{2+} = 24 mg/l, CO_2 = 33 mg/l, HCO_3^- = 132 mg/l, H^+ = 10 mg/l & NaCl = 4.3 ppm. Analysis of treated water: CO_3^{2-} = 12 mg/l and OH^- = 34 mg/l. Calculate the quantity of lime (74% pure) and soda (92% pure) needed for softening 50000 L of water.

9.6 With the help of neat sketch, explain continuous cold Lime soda process for softening of hard water. Also write the chemical reactions involved. Calculate the quantity of lime (74% pure) and soda (90% pure) for softening 50,000 liters of water containing the following salts: $Mg(HCO_3)_2$ = 50 mg/L, $MgCl_2$ = 6 ppm, $Ca(HCO_3)_2$ = 81 mg/L, CO_2 = 44 ppm, Na_2SO_4 = 10°F, HCl = 73 mg/L, $Al_2(SO_4)_3$ = 57 mg/L.

9.7 Explain the basic principle of lime-soda process. Calculate the amount of lime and soda required for softening 30000 liters of water, using 20 ppm of sodium aluminate as coagulant. Impurities in water are as follows: Ca^{2+} = 160 ppm, Mg^{2+} = 96 ppm, dissolved CO_2 = 34 ppm and HCO_3^- = 403 ppm.

9.8 100 ml of water sample has hardness equivalent to 12.5 ml of 0.08 N $MgSO_4$ solutions. Calculate the hardness of this water sample.

9.9 Write the different chemical reactions taking place in soda-lime process. A sample of water on analysis give following result: Analysis of Raw water: Ca^{2+} = 80 mg/l; Mg^{2+} = 24 mg/l, CO_2 = 33 mg/l, HCO_3^- = 132 mg/l, H^+ = 10 mg/l & NaCl = 4.3 ppm. Analysis of treated water: CO_3^{2-} = 12 mg/l and OH^- = 34 mg/l. Calculate the quantity of lime (74% pure) and soda (92% pure) needed for softening 50000 L of water.

9.10 Calculate the hardness of water sample containing impurity of $Ca(HCO_3)_2$ = 81 mg/l. Give your answer in °F also.

9.11 Explain reverse osmosis with its advantages. A water sample on analysis gives the following data: Ca^{2+} = 20 ppm, Mg^{2+} = 25 ppm, CO_2 = 30 ppm, HCO_3^- = 150 ppm, K^+ = 10 ppm. Calculate the lime (87% pure) and soda (91% pure) required to soften 10^6 liter of sample water.

9.12 What is hardness of water? What do you mean by term permutit? Describe Zeolite or Permutit process for softening of hard water.

9.13 A 100 mL sample of water required 13.5 mL of 0.02M EDTA solution for titration using Eriochrome black-T as indicator. Another 100 mL of water from the same source was boiled and precipitate removed by filtration. The filtrate required 6.0 mL of 0.02M EDTA solution. Calculate the total hardness, permanent hardness and temporary hardness of water sample.

9.14 Calculate the amount of lime and soda required to softened 15000 litres of water, which analysed as follows: temporary hardness = 20 ppm; permanent hardness = 15 ppm; permanent Mg hardness = 10 ppm.

9.15 An exhausted zeolite softener was regenerated by passing 150 liter of NaCl solution having strength of 1.5 gm/L NaCl. Find the total volume of water that can be softened by this zeolite softener. If the hardness of water is 600 ppm.

9.16 Calculate the amount of lime (84% pure) and Soda (92% pure) required for treatment of 20,000 liters of water, whose analysis is as follows $Ca(HCO_3)_2$ = 40.5 ppm; $Mg(HCO_3)_2$ = 36.5 ppm; $MgSO_4$ = 30.00 ppm; $CaSO_4$ = 34.0 ppm; $CaCl_2$ = 27.75 ppm and NaCl = 10.00 ppm. Also calculate the temporary and permanent hardness of water.

9.17 Calculate the amount of lime and soda required for softening 90,000 liters of water containing the following salts per liter : $Ca(HCO_3)_2$ = 162 mg, $CaSO_4$ = 136 mg and NaCl = 56.1 mg. Purity of lime is 92 % and soda is 99%.

9.18 Calculate the amount of lime (84% pure) and soda (92% pure) required for treatment of 20,000 litres of water whose analysis is as follows:
$Ca(HCO_3)_2$ = 40.5 ppm; $Mg(HCO_3)_2$ = 36.5 ppm; $MgSO_4$ = 30 ppm; $CaSO_4$ = 34 ppm; $CaCl_2$ = 27.75 ppm; NaCl = 10 ppm. Also calculate temporary and permanent hardness of water sample. [Given atomic masses of H = 1, Na = 23, Ca = 40, Mg = 24, O = 16, C = 12, S = 32, Cl = 35.5]

9.19 Calculate the amount of lime (91% pure) and soda (97.2% pure) required for softening one million litres of water containing: H^+ (free acidity) = 1.5 ppm; HCO_3^- = 396.5 ppm; Mg^{2+} = 42 ppm; Ca^{2+} = 90 ppm; $FeSO_4.7H_2O$ = 14 ppm. (Given atomic masses of H = 1, Ca = 40, Mg = 24, O = 16, C = 12, S = 32, Fe = 56)

CHAPTER 10

Corrosion

10.1 INTRODUCTION

Corrosion is the slow degradation or erosion of a metal's surface caused by an undesired chemical or electrochemical reaction between the metal and its surroundings. Corrosion of metals refers to the natural deterioration of metals due to their chemical, electrochemical, or biological interactions with the surrounding environment. Alternatively, corrosion refers to the process of decay or destruction that occurs when a solid metallic material is chemically or electrochemically attacked by its surrounding environment, resulting in the loss of material from its surface. Therefore, it is precisely the opposite of the process of extracting metals from ores.

Corrosion is a detrimental phenomenon. Corrosion refers to the process by which metals and alloys are gradually destroyed. The presence of corrosion poses a significant constraint on advancements across various domains. The expense associated with replacing materials and equipment damaged by corrosion is boundless. Metals and alloys serve as fabrication or construction materials in engineering applications. Insufficient maintenance of metals or alloy structures leads to gradual deterioration caused by air gases, moisture, and other substances. An instance of this is the development of a reddish coating or hydrated ferric oxide (Fe_3O_4) on the surface of iron, sometimes referred to as "iron rusting". When iron is exposed to the elements, it undergoes a chemical reaction that results in the formation of a reddish layer of Fe_3O_4 scale and powder. The formation of a green coating composed of basic carbonate [$CuCO_3$ + $Cu(OH)_2$] occurs on the surface of copper when it is exposed to wet air that contains CO_2.

10.2 DRAWBACKS OF CORROSION

Corrosion is a gradual phenomenon that exclusively affects the surface of metals, however it results in significant losses like dismantling of machinery, equipment, construction materials, and various metals items and constructions. Consequently, the financial losses resulting from corrosion are substantial, with an estimated global annual cost of around 2 to 2.5 billion dollars.

10.3 CORROSION CONSEQUENCES

The consequences of corrosion are outlined below:

1. Corrosion has significant economic and societal repercussions.
2. The creation of corrosion product on the machinery causes a decrease in machine efficiency, ultimately resulting in the shutdown of the plant.
3. The potential contamination or loss of products as a result of corrosion.
4. There is a reduction in the rate of production due to decreased efficiency and the time-consuming process of replacing corroded equipment or machinery.
5. The deteriorated equipment needs to be substituted.
6. It is necessary to do preventive maintenance, such as applying metallic coating or organic coating. Rise in expenses related to maintenance and manufacturing.
7. Corrosion results in the release of hazardous substances.
8. Health risks can arise from pollution caused by corrosion products or the release of chemicals from corroded equipment.

10.4 METALLIC CORROSION

Metals are obtained by extracting them from their metallic compounds, sometimes known as ores. Metals naturally occur in the form of carbonates, sulphides, and sulphates. Ore refers to the chemically coupled states of metal that exhibit low energy and are therefore considered thermodynamically stable. Metallurgy necessitates a substantial quantity of energy. The metal obtained by extraction possesses greater energy and is therefore in a thermodynamically unstable state. Metals experience corrosion by chemically or electrochemically interacting with their environment, resulting in the formation of compounds on the surface.

During the extraction process, ores are transformed into their metallic states through the application of energy using various methods. Metals in their pure metallic state are inherently unstable due to their excited state. Consequently, after the metals are obtained from their ores, they undergo a reverse reaction and transform into thermodynamically stable metallic complexes, characterized by a lower energy state. Consequently, when metals are utilized in different configurations, they come into contact with the environment, leading to the initiation of corrosion on the metal surface (resulting in the transformation into a more stable compound). This is the fundamental cause of metallic corrosion.

$$\text{Metal} \xrightleftharpoons[\text{Metallurgy-Reduction}]{\text{Corrosion-Oxidation}} \text{Metallic Compound} + \text{Energy}$$

Thermodynamically, corroded metal is more stable than pure metal. However, corrosion leads to the loss of desirable qualities such as malleability, ductility, hardness, luster, and electrical conductivity.

$$\text{Ore of Metal} \xrightarrow[+\text{ Energy}]{\text{Metallurgical Process}} \text{Pure Metal} \xrightarrow{\text{Environment}} \text{Corroded Metal}$$

Ore of Metal	Pure Metal	Corroded Metal
Thermodynamically stable state	Thermodynamically unstable state	More stable than pure metal

Metallic elements exist in two distinct states in the natural world:

1. **Native State:** Metals naturally occur in the Earth's crust and can be found in their elemental form. The term "native state" refers to a state in which a substance exists in its free or uncombined form. These metals exhibit inert properties. These metals are characterized by their noble nature and exceptional resistance to corrosion. Examples include gold (Au), platinum (Pt), silver (Ag), and so on.
2. **Combined State:** With the exception of noble metals, all other metals exhibit a high level of reactivity and readily react with their surroundings to produce stable compounds known as ores and minerals. This is the amalgamation of metallic states. Examples of chemical compounds are Fe_2O_3, ZnO, PbS, $CaCO_3$, and so on.

10.5 THEORIES OF CORROSION

The phenomenon of corrosion can be elucidated by the application of the following theoretical frameworks:

1. Acid theory
2. Dry or direct chemical corrosion theory
3. Electrochemical or wet corrosion theory.

10.5.1 Acid Theory

This theory is particularly applicable to rusting of iron in the atmosphere. According to this theory, rusting of iron is due to the continued action of oxygen, carbon dioxide and moisture, converting the metal into a soluble ferrous bicarbonate which is further oxidised to the basic ferric carbonate and finally converted to hydrated ferric oxide.

This idea is specifically relevant to the corrosion of iron in the atmosphere. According to this idea, the process of rusting in iron occurs as a result of the ongoing interaction between oxygen, carbon dioxide, and moisture. This interaction causes the metal to transform into a soluble compound known as ferrous bicarbonate, which is subsequently oxidized to form the basic ferric carbonate. Finally, the ferric carbonate is converted into hydrated ferric oxide. Chemical reactions involve in this theory are as follows:

$$Fe + O + 2CO_2 + H_2O \longrightarrow Fe(HCO_3)_2$$

$$2Fe(HCO_3)_2 + H_2O + O \longrightarrow 2Fe(OH)CO_3 + 2CO_2 + 2H_2O$$

$$2Fe(OH)CO_3 + 2H_2O \longrightarrow 2Fe(OH)_3 + 2CO_2$$

This theory is substantiated by the subsequent facts:

1. The analysis of rust indicates the existence of ferrous and ferric carbonates, as well as hydrated ferric oxide.
2. Inhibition of corrosion when lime or NaOH is introduced to the water in which iron is submerged.

10.5.2 Dry or Chemical Corrosion

Dry or Chemical corrosion refers to the process of deterioration caused by the reaction of a material with its environment, specifically in the absence of moisture or by chemical reactions. This form of corrosion mostly arises from the direct chemical interaction between the environment and the metal surface. It involves the direct assault of atmospheric gases like O_2, halogens, H_2S, SO_2, N_2, or anhydrous inorganic liquids on the metal surface in contact with them. For instance, silver materials are subject to chemical corrosion when exposed to atmospheric H_2S gas, while iron metal is susceptible to chemical corrosion when exposed to HCl gas.

There are three types of dry corrosion and chemical corrosion:

1. Oxygen-induced corrosion, also known as oxidation corrosion.
2. Corrosion caused by gases such as hydrogen (H_2), sulphur dioxide (SO_2), carbon dioxide (CO_2), hydrogen sulphide (H_2S), and fluorine (F_2), among others.
3. Corrosion caused by the interaction of liquid metal.

10.5.2.1 Corrosion by Oxygen (or) Oxidation Corrosion

This phenomenon occurs as a result of the direct influence of oxygen at either low or high temperatures, typically in the absence of moisture. At elevated temperatures, oxygen reacts with and oxidizes all metals, with the exception of noble metals such as silver, gold, and platinum. At typical temperatures, most metals experience a modest level of corrosion. Nevertheless, alkali metals such as Li, Na, K, Rb, and alkaline earth metals like Be, Ca, Sr, are highly susceptible to fast oxidation and are readily attacked.

The reactions involved in oxidative corrosion are:

$$2M \xrightarrow{\text{loss of electrons}} \underset{\text{Metal ions}}{2M^{n+}} + 2ne^-$$

$$\frac{n}{2}O_2 + 2ne^- \xrightarrow{\text{Gain of electrons}} \underset{\text{Oxide ions}}{nO^{2-}}$$

The comprehensive reaction can be expressed as:

$$2M + \frac{n}{2}O_2 \longrightarrow 2M^{n+} + nO^{2-} \rightarrow \text{Metal Oxide}$$

Mechanism of Oxidation Corrosion

Initially, oxidation takes place at the metal's surface, resulting in the formation of a layer of metal oxide. This layer acts as a barrier, impeding further oxidation. Hence, in order for oxidation to persist, the metal must either diffuse outward through the scale towards the surface or the oxygen must diffuse inward through the scale towards the underlying metal. Both transfers take place, however, the metal's outward diffusion is often far faster than the inward diffusion of oxygen. Due to the significant size difference between the metal ion and the oxide ion, the metal ion exhibits considerably more mobility.

Nature of the Oxide Formed

It has a significant impact on the subsequent oxidation corrosion process.

1. Stable Layer: A stable layer possesses a fine-grained structure and has the ability to strongly attach to the surface of the parent metal. This layer will be impermeable and acts as a protective covering, effectively safeguarding the metal surface. As a result, more oxidative corrosion is inhibited. For example, elements such as Aluminium (Al), Tin (Sn), Lead (Pb), Copper (Cu), and others, create persistent oxide coatings on their surfaces, which effectively inhibit further oxidation.

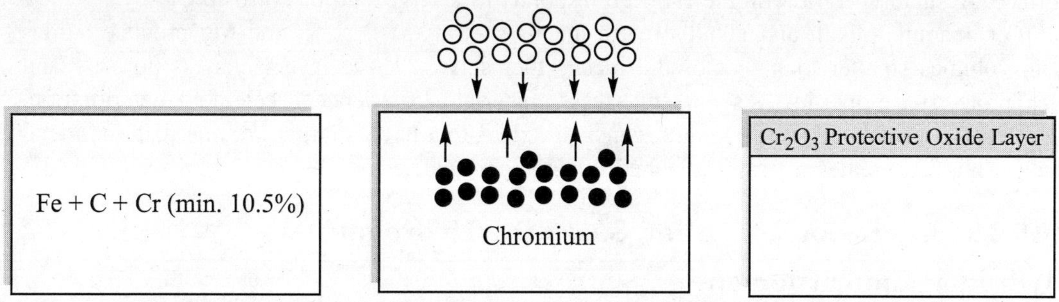

2. Unstable Layer: The layer of oxide that is created breaks down into its constituent elements, namely metal and oxygen. Therefore, oxidation corrosion cannot occur in such instances. For example, elements such as silver (Ag), gold (Au), and platinum (Pt) are resistant to oxidative corrosion.

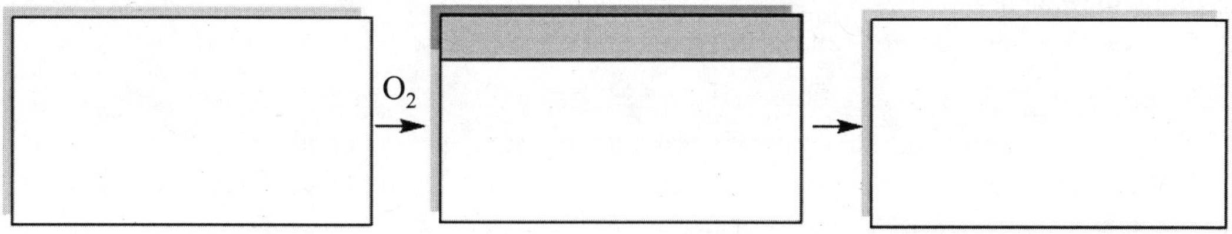

3. Volatile Layer: The oxide layer is inherently volatile and rapidly evaporates upon formation. Thus, the underlying metal surface is left exposed, making it vulnerable to further corrosion. This results in quick and ongoing corrosion, which leads to an excessive amount of corrosion. For example, the element molybdenum generates a volatile layer of MoO_3.

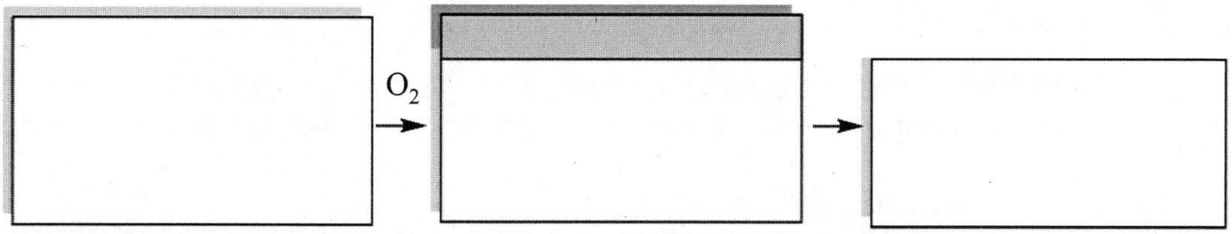

4. Porous Layer: Characterized by the presence of pores and fissures. In this scenario, ambient oxygen can reach the metal's underlying surface through the holes or cracks in the layer, resulting in ongoing corrosion until the entire metal is transformed into its oxide form. For example, when iron is exposed to hydrogen sulphide (H_2S) at high temperatures, it creates a porous coating of iron sulphide (FeS).

Pilling–Bedworth Rule

If the volume of the oxide layer is equal to or greater than the volume of the metal it is generated from, it will serve as a protective barrier that is impermeable. Conversely, if the volume of the oxide layer is smaller than the volume of the metal, the oxide layer is porous and therefore not protective. Due of its inability to impede the oxygen's access to the underlying fresh metal surface. A small specific volume ratio corresponds to a large rate of corrosion.

For example, alkali and alkaline earth metals such as Li, Na, K, and Mg produce oxides with volumes smaller than that of the metals themselves. These layers possess porosity and lack protective properties. As a result, these materials experience accelerated deterioration. Conversely, metals such as aluminum generate oxides that have a larger volume than the metal itself. Thus, Al creates a firmly attached impermeable barrier.

10.5.2.2 Corrosion by Other Gases (by Hydrogen)

Hydrogen Embrittlement

Hydrogen embrittlement refers to the reduction in ductility of a material caused by the presence of hydrogen.

Mechanism: This form of corrosion happens when a metal is exposed to an atmosphere containing hydrogen. Iron releases atomic hydrogen when reacting with hydrogen sulphide in the following manner:

The reaction between Fe^{2+} and H_2S produces FeS and $2H^+$.

$$Fe^{2+} + H_2S \longrightarrow FeS + 2H^+$$

Hydrogen permeates the metal matrix in its atomic state and accumulates within the voids within the metal. Moreover, the process of atomic hydrogen diffusing leads to their combination and the subsequent formation of hydrogen gas.

The reaction equation is:

$$H^+ + H^+ \longrightarrow H_2\uparrow$$

The accumulation of these gasses in the empty spaces results in extremely high pressure, leading to the formation of cracks or blisters on the metal surface.

Decarburisation

Carbon imparts adequate strength to steel. However, when steel is subjected to a hydrogen-rich atmosphere at elevated temperatures, it undergoes a process in which atomic hydrogen is generated.

The reaction is represented by the equation:

$$H_2 \longrightarrow 2H^+$$

The carbon in the steel undergoes a reaction with atomic hydrogen, resulting in the formation of methane gas. The chemical equation represents the reaction where one molecule of carbon (C) reacts with four hydrogen ions (H^+) to produce methane (CH_4).

$$C + 4H^+ \longrightarrow CH_4$$

Therefore, the carbon content in steel reduces. Decarburization refers to the reduction of carbon content in steel. The accumulation of methane gas within the empty spaces of steel leads to the generation of significant pressure, resulting in the formation of cracks. Consequently, steel experiences a decrease in its structural integrity.

10.5.2.3 Liquid Metal Corrosion

Liquid metal corrosion refers to the process of deterioration or degradation caused by the interaction between a liquid metal and another material. This phenomenon occurs as a result of the chemical interaction between molten metal and solid metal or alloy at elevated temperatures. Corrosion of this nature can manifest in nuclear power equipment. The corrosion process encompasses either: (i) the dissolving of a solid metal by a liquid metal or (ii) the infiltration of the liquid metal into the solid metal. Both of these corrosion mechanisms result in the degradation of the structural integrity of the metal.

10.5.3 Electrochemical or Wet Corrosion Theory

It occurs in environments with high moisture or wetness due to the creation of galvanic cells that are short-circuited. Electrochemical or aqueous corrosion is more prevalent than non-aqueous corrosion. Electrochemical corrosion occurs when

1. There are distinct 'anodic' and 'cathodic' regions that allow current to flow through the conducting medium.
2. Oxidation (corrosion) happening in the positively charged sites, resulting in the production of metallic ions.
3. Cathodic areas give rise to non-metallic ions such as OH^- or O^{2-}.
4. Metallic and non-metallic ions diffuse towards each other in a conducting media, leading to the development of corrosion products between the anodic and cathodic areas.
5. The oxidation reaction occurs in the anodic area, resulting in the release of free electrons. As a result, the anodic metal is either dissolved or transformed into a mixed state, such as an oxide. Therefore, corrosion consistently takes place at the anodic regions.

The metal M undergoes oxidation, resulting in the formation of M^{n+} metal ions and the release of ne^- electrons. These M^{n+} metal ions can dissolve in solution and subsequently form compounds, such as oxides.

$$\text{M (metal)} \longrightarrow M^{n+} + ne^-$$

$$M^{n+} \text{ (metal ion)} \rightarrow \text{Dissolves in solution} \rightarrow \text{forms compounds such as oxide}$$

10.5.3.1 Corrosion at Cathodic Area

At the cathodic region, a reduction reaction occurs, resulting in the gain of electrons. Typically, cathode reactions do not have an impact on the cathode itself, as most metals cannot undergo further reduction. At the cathodic region, the dissolved components in the conducting medium undergo electron acceptance, resulting in the formation of ions such as OH^- and O^{2-}.

The cathodic reaction involves the consumption of electrons through:

1. The process of hydrogen evolution.
2. The absorption of oxygen is influenced by the characteristics of the corrosive environment.

Hydrogen Evolution Type

Metals positioned higher than hydrogen in the electrochemical series exhibit a propensity to dissolve in acidic solutions, accompanied by the release of hydrogen gas. It takes place in an acidic atmosphere. Let's examine the case of iron.

At the anode: $Fe \longrightarrow Fe^{2+} + 2e^-$

The reaction results in the production of two electrons. These electrons move through the metal, traveling from the anode to the cathode.

At the cathode, hydrogen ions (H^+) from the acidic solution are converted into hydrogen gas.

The reaction at the cathode is: $2H^+ + 2e^- \longrightarrow H_2\uparrow$

The general reaction can be expressed as follows: $Fe + 2H^+ \longrightarrow Fe^{2+} + H_2$

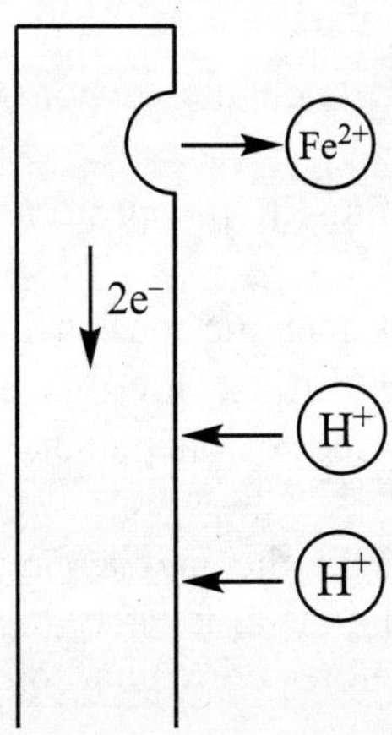

Oxygen Absorption Type

An instance of this sort of corrosion is the oxidation of iron in a neutral aqueous solution of electrolytes, such as a solution of NaCl, in the presence of ambient oxygen. Iron surfaces typically exhibit a thin layer of iron oxide. However, the formation of cracks in the iron oxide film leads to the creation of anodic patches on the surface, while the unaffected metal sections function as cathodes.

At the anode, the metal undergoes dissolution, resulting in the formation of ferrous ions and the release of electrons.

$$Fe \longrightarrow Fe^{2+} + 2e^-$$

The element Fe undergoes oxidation, resulting in the formation of Fe^{2+} ions and the release of two electrons.

At the cathode, the freed electrons are captured by the dissolved oxygen.

$$\frac{1}{2}O_2 + H_2O + 2e^- \longrightarrow 2OH^-$$

The reaction results in the formation of hydroxide ions (OH^-) when oxygen, water, and electrons are combined. The Fe^{2+} ions and OH^- ions move through a substance by diffusion. When these ions come into contact with one other, they react to form ferrous hydroxide, which is then precipitated.

$$Fe^{2+} + 2OH^- \longrightarrow Fe(OH)_2$$

The reaction involves the combination of one Fe^{2+} ion with two OH^- ions. In the presence of sufficient oxygen, the process of oxidizing ferrous hydroxide to ferric hydroxide occurs quickly. The chemical equation is as follows:

$$4Fe(OH)_2 + O_2 + 2H_2O \longrightarrow 4Fe(OH)_3 \text{ (Yellow rust } Fe_2O_3{\cdot}H_2O)$$

The chemical formula for the corrosion product is $Fe_2O_3.H_2O$. In cases when there is a restricted supply of oxygen, the corrosion result may be black anhydrous magnetite, Fe_3O_4.

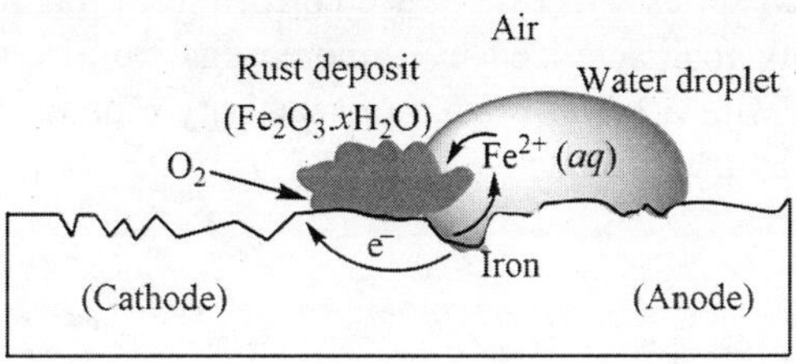

For example

The corrosion of iron occurs in a neutral aqueous solution of electrolyte when oxygen is present, or in an acidic environment where hydrogen is released. The Fe^{2+} ion is generated at the anode, whereas the OH^- ions are produced at the cathode. The smaller Fe^{2+} ions exhibit a higher rate of diffusion compared to the larger OH^- ions. As a result, corrosion takes place at the anode, while the rust formed as a corrosion product is deposited near the cathode.

The electrons liberated at the anode are conducted towards the cathode and play a role in numerous cathodic processes, such as the generation of hydrogen gas through the process of hydrogen evolution. Following reactions occur in the absence of oxygen:

1. In an acidic environment: $H^+ + 2e^- \longrightarrow H_2$
2. In a neutral or alkaline environment: $2H_2O + 2e^- \longrightarrow H_2 + 2OH^-$

Oxygen absorption occurs under the following conditions:

1. In the presence of dissolved oxygen in an acidic medium: $4H^+ + O_2 + 4e^- \longrightarrow 2H_2O$.
2. In the presence of dissolved oxygen in a neutral or weakly alkaline media: $2H_2O + O_2 + 4e \longrightarrow 4OH^-$.

Electroplating: During electroplating, metal ions are attracted to the cathode where they gain electrons and are subsequently deposited onto the cathode's surface.

10.5.3.2 Types of Electrochemical Corrosion

Electrochemical corrosion can be categorized into two distinct types:

1. Galvanic corrosion, also known as bimetallic corrosion, occurs when two different metals come into contact and undergo a chemical reaction resulting in the deterioration of one or both metals.
2. Corrosion caused by differential aeration or concentration cells.

Galvanic Corrosion

It occurs when multiple metals are collectively exposed to a corrosive atmosphere. The metal positioned higher in the electrochemical series, with a stronger negative electrode potential, will function as the anode and experience corrosion. For instance, combinations of Zn and Cu, Fe and Cu, and Zn and Ag. In the Zn-Cu galvanic cell, Zn functions as the anode, undergoing oxidation and corrosion, while Cu serves as the cathode and is shielded from these processes.

When two metals with different properties (such as zinc and copper) are connected electrically and exposed to a substance that conducts electricity (electrolyte), the metal that is higher in the electrochemical series will experience corrosion. In this procedure, the metal with a higher degree of reactivity (characterized by a more negative electrode potential) serves as the anode, while the metal with a lower degree of reactivity (characterized by a less negative electrode potential) serves as the cathode.

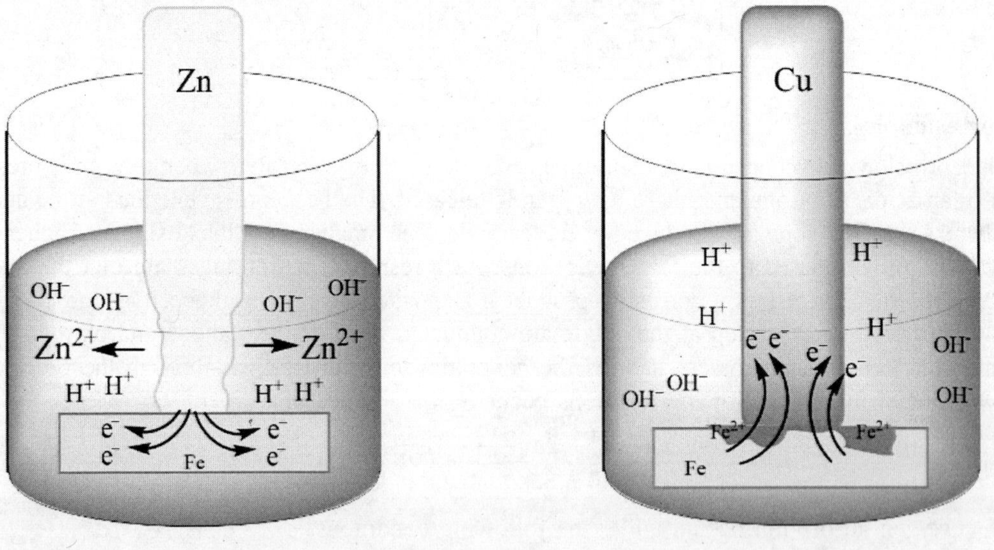

In the above illustration, zinc, which is positioned higher in the electrochemical series, serves as the anode and undergoes corrosion and dissolution. Conversely, copper, which is positioned lower in the electrochemical series, functions as the cathode.

Mechanism

In an acidic solution, corrosion takes place through the process of hydrogen evolution, whereas in a neutral or slightly alkaline solution, corrosion happens by oxygen absorption. The flow of electron-current occurs from the anode, which is made of zinc, to the cathode, which is made of copper.

$$Zn \rightarrow Zn^{2+} + 2e^- \text{ (Oxidation)}$$

Zinc (Zn) is oxidized to form Zinc ions (Zn^{2+}) and release two electrons ($2e^-$). Hence, it is apparent that corrosion takes place specifically at the anode metal, while the cathodic component remains shielded from the detrimental effects. For example: (i) Steel screws used in brass maritime hardware. (ii) Lead-antimony solder applied around copper wire. (iii) A steel propeller shaft inserted into a bronze bearing. (iv) Steel pipe connected to copper plumbing.

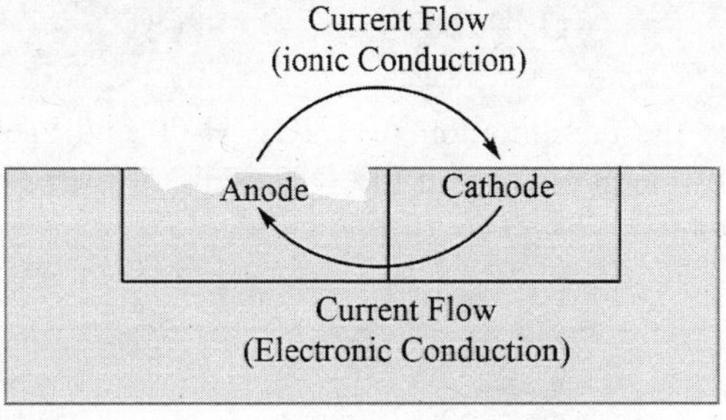

Galvanic corrosion can be reduced by:

1. Avoiding the formation of galvanic couples.
2. Introducing an insulating substance to separate the two metals.

Concentration Cell Corrosion or Differential Aeration Corrosion

Cell corrosion, also known as differential aeration corrosion, refers to the process of corrosion that occurs when there is a difference in the oxygen concentration between different areas of a metal surface. The cause of this phenomenon is the electrochemical corrosion of the metal surface, which occurs when it is exposed to an electrolyte with different concentrations or levels of aeration. It happens when a metallic surface is partially submerged in an electrolyte and partially exposed to air.

This phenomenon arises when one portion of metal is subjected to a dissimilar air composition compared to another portion. This results in a disparity in electrical potential between locations with varying levels of aeration. Experimental findings have revealed that regions with low oxygen levels exhibit anodic behavior.

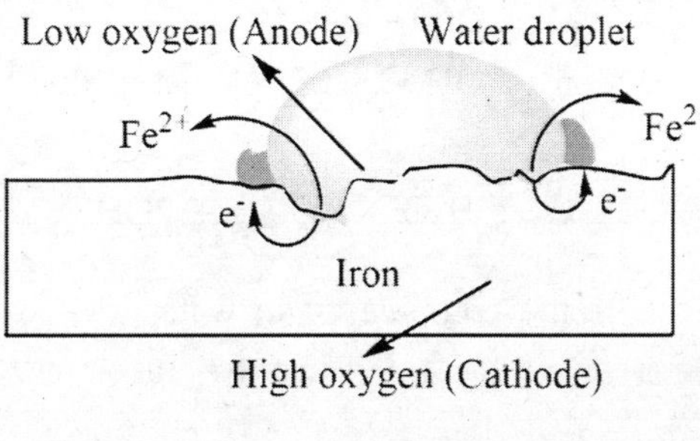

Examples: (i) Water line corrosion (ii) The metallic component partially submerged in the ground.

The metallic component with low oxygen content becomes an anode and experiences oxidation.

$$Zn \longrightarrow Zn^{2+} + 2e^-$$

The oxidation of zinc results in the formation of zinc ions (Zn^{2+}) and the release of two electrons ($2e^-$). The oxygen-rich metallic portion then acts as the cathode. At the cathode, oxygen (O_2) undergoes reduction by accepting electrons to generate hydroxide (OH^-) ions.

$$\frac{1}{2}O_2 + H_2O + 2e^- \longrightarrow 2OH^-$$

The reaction involves the combination of $1/2O_2$, H_2O, and 2 electrons to produce $2OH^-$ ions. Additionally, the Zn^{2+} ions react with the OH^- ions to form Zn $(OH)_2$. Hence, corrosion takes place at the anode.

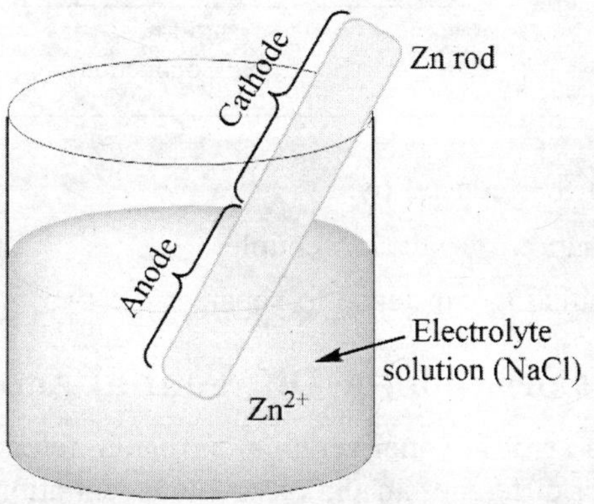

Explanation

When a metal is partially submerged in a conducting solution, the part of the metal that is above the solution becomes more exposed to air and undergoes cathodic polarization. The metal component submerged in the solution experiences reduced aeration, leading to an anodic state and subsequent corrosion.

At the anode, corrosion takes place under conditions of reduced aeration. The reaction involves the conversion of a metal (M) into its corresponding cation (M^{2+}) by losing two electrons ($2e^-$).

$$\text{M (metal)} \longrightarrow M^{2+} + 2e^-$$

At the cathode, the production of OH^- ions is increased due to aeration. The reaction is as follows:

$$\frac{1}{2}O_2 + H_2O + 2e^- \longrightarrow 2OH^-$$

One-half of an oxygen molecule combines with water and two electrons to produce two hydroxide ions.

Types of concentration cell corrosion are as follows:

1. Pitting or localized corrosion
2. Crevice corrosion
3. Stress corrosion
4. Pipeline corrosion
5. Corrosion on wire fence

1. Pitting Corrosion: Pitting corrosion is a specific type of corrosion that occurs when certain areas of a metal experience rapid attack, leading to the production of cavities. These cavities are surrounded by comparatively unaffected metal, resulting in the formation of pinholes, pits, and cavities in the metal. Pitting often occurs when the protective coating on the metal deteriorates or fractures at certain locations. This leads to the creation of minor anodic regions and major cathodic regions.

Pitting can also occur due to the presence of external contaminants, such as sand, dust, and scale that are embedded on the surface of metals. Due to the variation in oxygen levels in contact with the metal, the smaller portion (located beneath the impurity) becomes the anodic region, while the larger surrounding portions become the cathodic regions.The corrosion begins rapidly due to the presence of impurities. Once a minor cavity is created, the rate of corrosion will escalate.

Pitting corrosion can be understood by examining the presence of a little amount of water or brine solution (a solution of sodium chloride in water) on a metal surface, particularly iron.

The region encompassed by the salt solution droplet has reduced oxygen levels and functions as the anode. Corrosion occurs in this region, where the exposed part serves as the cathode because to the elevated oxygen levels.

Research has revealed that the corrosion rate is higher when the cathode area is larger and the anode area is smaller. Therefore, the presence of a larger amount of material surrounding the small anodic area leads to the creation of a hole or pit.

At the anode, the iron (Fe) undergoes oxidation and is converted into ferrous ions (Fe^{2+}), while simultaneously releasing electrons.

$$Fe \longrightarrow Fe^{2+} + 2e^-$$

The chemical equation represents the process of iron (Fe) losing two electrons ($2e^-$) to form a positively charged iron ion (Fe^{2+}).

At the cathode, oxygen undergoes a conversion to hydroxide ions through the following reaction:

$$\frac{1}{2}O_2 + H_2O + 2e^- \longrightarrow 2OH^-$$

The overall reaction can be represented as: $Fe + 2OH^- \longrightarrow Fe(OH)_2$.

The above mechanisms can be verified by employing ferroxyl indicator, which is a blend including phenolphthalein and potassium ferricyanide. Due to the formation of OH^- ions at the cathode, the presence of phenolphthalein indicator causes the area to exhibit a pink color. Iron undergoes oxidation at the anode, resulting in the formation of Fe^{2+} ions. These ions then react with ferricyanide, leading to the manifestation of a blue color.

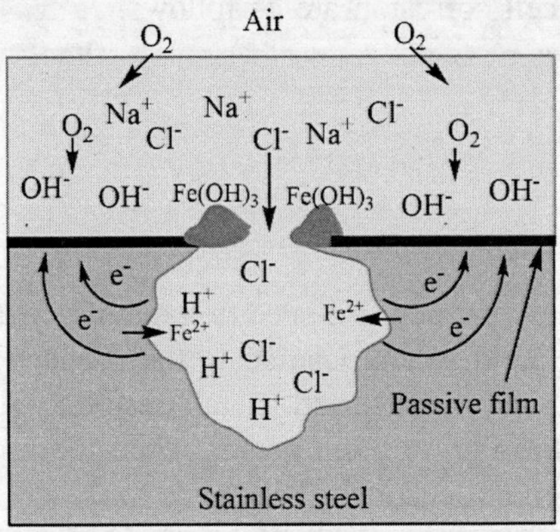

2. Crevice Corrosion: Crevice corrosion refers to a type of corrosion that occurs in confined spaces or crevices, such as gaps or cracks, where there is limited access to oxygen and stagnant conditions. When a narrow opening, known as a crevice, exists between metallic and non-metallic materials and is exposed to a liquid, the crevice acts as the anodic zone and experiences corrosion. Therefore, the amount of oxygen reaching the crevice is reduced. The uncovered region possesses a significant amount of oxygen supply and functions as the cathode. Bolts, nuts, rivets, and joints exemplify this form of corrosion.

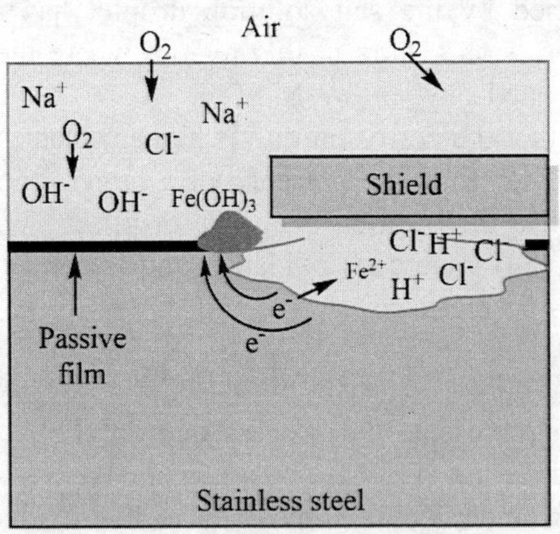

3. Stress Corrosion: Stress corrosion occurs when the combination of static tensile strains and a corrosive environment leads to cracking in metal. These stresses might arise from residual cold working and quenching, welding, thermal treatment, or applied loads during service. Under such circumstances, the metal experiencing stress becomes more anodic, leading to an accelerated corrosion rate.

Stress corrosion requires the existence of both tensile stress and a certain corrosive environment. The corrosive agents exhibit a high degree of specificity and selectivity, such as the use of caustic alkalis and strong nitrate solutions to corrode mild steel.

1. Detection of ammonia in brass.
2. Solution of acid chloride for use on stainless steel.

This type of corrosion is a sort of localized electrochemical corrosion that occurs along narrow routes, resulting in the formation of anodic patches relative to the more cathodic portions on the metal surface. The existence of stress leads to the formation of strains, which in turn create specific areas with elevated electrode potential. These substances undergo significant chemical reactivity, making them susceptible to corrosion even in mildly corrosive surroundings. This leads to the development of cracks, which continue to expand and spread within a structure until it ultimately fails, although in other cases, the crack growth may cease after reaching a certain distance.

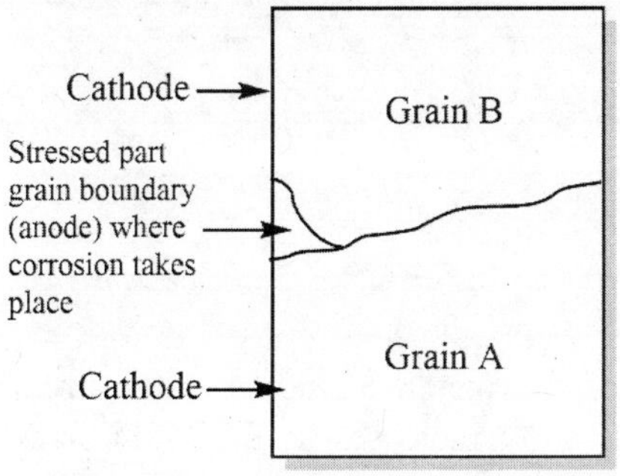

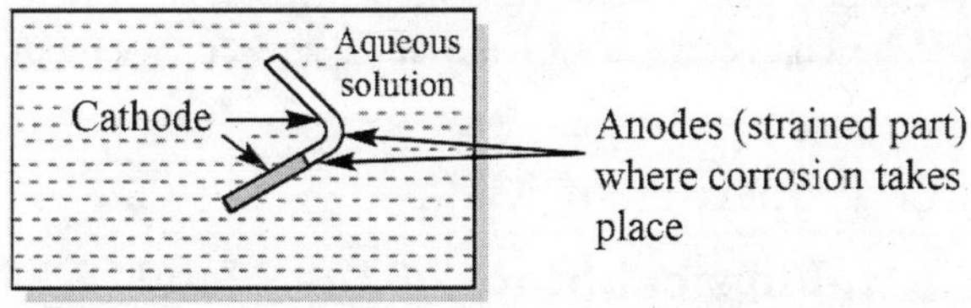

4. Water-line or Pipeline Corrosion: In a steel tank with stagnant water, the concentration of Oxygen is higher above the water surface compared to below it. This creates an oxygen concentration cell. The metal located immediately above the water level in this cell is cathodic with respect to the metal situated below the water level. Corrosion takes place in the anodic region, specifically in the metal located directly beneath the water surface. Pipelines or cables that are buried and transition from clay soil, which has less aeration, to sand soil, which has more aeration, may experience corrosion as a result of differential aeration.

5. Corrosion on Wire Fence: A wire fence is characterized by places where the wires intersect (anodic) that have lower levels of aeration compared to the rest of the fence (cathodic). Therefore, corrosion occurs at the point where the wires intersect. Other instances are corrosion that takes place beneath metal washers and a lead pipeline that passes through clay to cinders (ash). The cathodic region is entirely immune to corrosion.

Passivity

Passivity refers to the absence of activity in situations when a metal would normally be anticipated to react rapidly. Some metals exhibit passivity towards specific corrosive chemicals. For instance, iron exhibits passivity towards concentration nitric acid. Aluminium does not react with concentrated substances and Nitric acid (HNO_3) in the absence of chlorides and other similar compounds. Metals that are prone to corrosion can be become passive by combining them with one or more metals that are already passive or resistant to corrosion. Iron can be made passive by combining it with transition metals like chromium, nickel, and molybdenum.

The amount of inert metal to be introduced varies depending on the surrounding conditions. For instance, when the iron-chromium alloy contains a chromium content ranging from 12 to 20%, it exhibits passivity towards neutral salt solutions. However, increasing the proportion of chromium in the alloy does not render it passive to hydrochloric acid.

10.6 FACTORS INFLUENCING CORROSION

The rate and amount of corrosion are contingent upon the subsequent factors. Two elements determine the rate of corrosion. Therefore, it is crucial to understand these parameters and the mechanism by which they influence the corrosion rate, as the rate of corrosion varies in different atmospheric conditions.

1. Properties of Metal
2. The characteristics of the corrosive environment

10.6.1 Properties of Metal

10.6.1.1 Physical State of the Metal

The corrosion rate is affected by the physical properties of the metal, including grain size, crystal orientation, and stress. The solubility of a metal or alloy increases as the grain size decreases, resulting in increasing corrosion susceptibility. Furthermore, regions experiencing strain, even within a metal that is not alloyed, have a tendency to become anodic, leading to the occurrence of corrosion in these regions.

10.6.1.2 Purity of Metal

The rate of corrosion increases as the % purity of a metal decreases. Impurities present in a metal result in heterogeneity and give rise to minuscule electrochemical cells at the exposed areas, leading to corrosion of the anodic regions. The metal that is 100% pure will not experience any form of corrosion. An observation is made on the rise in the rate of corrosion of aluminium in hydrochloric acid as the percentage of impurities increases.

10.6.1.3 Overvoltage

The corrosion rate in a corrosive environment is inversely related to the overvoltage of a metal. For instance, when zinc metal is immersed in 1 M sulphuric acid, the overvoltage of hydrogen is measured to be 0.7 V, and the corrosion rate is observed to be low. By introducing a little quantity of copper sulphate into a solution of dilute sulphuric acid, the hydrogen overvoltage is effectively diminished to a value of 0.33 V. Consequently, this leads to an escalated rate of corrosion for zinc metal.

10.6.1.4 Nature of Surface Film

When exposed to air, nearly all metals develop a small layer of metal oxide on their surface, typically measuring a few angstroms in thickness. The specific volume ratio refers to the ratio between the volumes of the metal oxide and the metal. The oxidation corrosion rate decreases as the specific volume ratio increases.

The specific volume ratios of nickel (Ni), chromium (Cr), and tungsten (W) are 1.6, 2.0, and 3.6, respectively. Therefore, the rate of oxidation of tungsten is minimal, even when exposed to high temperatures.

10.6.1.5 Relative Areas of the Anodic and Cathodic Parts

The corrosion of the anodic component is exactly proportional to the ratio of areas between the cathodic part and the anodic part when two different metals or alloys are in contact. Corrosion occurs at a faster and more intense rate when the anodic area is small, such as when a small steel pipe is placed in a large copper tank. This is due to the significantly higher current density at a smaller anodic area, which can only be sustained by undergoing corrosion more vigorously.

10.6.1.6 Position in Galvanic Series

The higher a metal is positioned in the galvanic series, the larger its oxidation potential and, consequently, its inclination to become anodic. This leads to a higher rate of corrosion.

10.6.1.7 Solubility of Corrosion Products

In electrochemical corrosion, if the corrosion product is soluble in the corroding medium, the corrosion rate increases. As an illustration, when lead (Pb) is present in a medium containing sulphuric acid (H_2SO_4), it reacts to generate lead sulphate ($PbSO_4$), which is not soluble in the corrosive media. As a result, the corrosion process occurs at a slower pace.

10.6.1.8 Volatility of Corrosion Products

If the corrosion product is volatile, metal will undergo rapid and ongoing corrosion. This occurs because once corrosion product is generated, it evaporates, thereby exposing the underlying metal surface to more damage.

10.6.2 Nature of the Corroding Environment

10.6.2.1 Temperature

The corrosion rate has a direct correlation with temperature, meaning that an increase in temperature leads to an increase in the rate of corrosion. The reason for this is because the rate of ion diffusion escalates with an increase in temperature.

10.6.2.2 Humidity of Air

Corrosion will occur at a higher rate in environments with high relative humidity. The presence of moisture functions as a solvent for gases such as oxygen, carbon dioxide, and sulphur dioxide in the atmosphere, resulting in the formation of an electrolyte that is necessary for the establishment of a corrosion cell.

10.6.2.3 Presence of Impurities in Atmosphere

Metals have greater corrosion in proximity to industrial and coastal regions. The enhanced conductivity of the liquid layer in contact with the metal surface is a result of corrosive gases such as H_2S, SO_2, CO_2, and fumes of H_2SO_4 and HCl in industrial regions, as well as NaCl from sea water. This leads to an elevated corrosion rate.

10.6.2.4 Presence of Suspended Particles in Atmosphere

The presence of suspended particles in the atmosphere can contribute to atmospheric corrosion.

1. If the suspended particles possess chemical reactivity, such as NaCl or Ammonium sulphate, they have the ability to absorb moisture and function as potent electrolytes, leading to increased corrosion.
2. In the case when the suspended particles are chemically inert (such as charcoal), they have the ability to absorb both sulphur gasses and moisture, which gradually increases the pace of corrosion.

10.6.2.5 The Impact of pH

The corrosion of metals susceptible to acid attack can be mitigated by raising the pH of the corrosive environment. Acidic media, with a pH below 7, are typically more corrosive compared to alkaline and neutral media. Amphoteric metals such as aluminum, zinc, and lead dissolve in alkaline liquids by forming complex ions. The rate of corrosion of iron in water devoid of oxygen is gradual, until the pH drops below 5. The corrosion rate is significantly elevated in the presence of oxygen.

Corrosion of metals that are easily affected by acid can be mitigated by raising the pH of the corrosive environment. For example, zinc, which corrodes quickly even in mildly acidic solutions like carbonic acid, experiences minimal corrosion at a pH of 11.

10.6.2.6 Nature of ions present

10.6.2.7 Conductance of the corroding medium

10.6.2.8 Formation of oxygen concentration cell

10.6.2.9 Flow velocity of process stream

10.6.2.10 Polarization of electrodes

10.7 CORROSION CONTROL (PROTECTION AGAINST CORROSION)

Given the significant harm caused by the corrosion process and the substantial losses it entails, it is imperative to minimize or regulate the corrosion of metals. Corrosion can only be entirely halted under optimal circumstances. However, achieving optimal conditions is unattainable. Nevertheless, it is only feasible to significantly reduce corrosion.

Given the wide range of corrosion types and the varying conditions in which corrosion takes place, a variety of strategies are employed to manage and prevent corrosion. Corrosion, which occurs when metal or alloy reacts with the environment, necessitates corrosion control methods that focus on altering either the metal or the environment.

10.7.1 Choice of Metals and Alloys

1. One option is to utilize noble metals such as gold and platinum. They exhibit the highest level of resistance to corrosion. Due to their inherent value, they are unsuitable for common or ordinary use.

2. An alternative option is to utilize the most refined metal available. However, producing a metal with a high level of chemical purity is often unattainable in numerous instances. Therefore, even a little quantity of impurities results in corrosion.
3. Therefore, the subsequent option is to utilize corrosion resistant alloys. Various corrosion-resistant alloys have been formulated to cater to specific applications and environmental conditions.

   For example,

   (a) Stainless steel that contains chromium forms a highly cohesive oxide coating that shields the steel from additional corrosion.

   (b) Cupro-nickel alloys, consisting of 70% copper and 30% nickel, are currently employed in the manufacturing of condenser tubes and bubble trays utilized in fractionating columns within oil refineries.

   (c) Nimonic alloys, which are composed of nickel, chromium, and molybdenum, exhibit exceptional resistance to hot gases and are commonly employed in gas turbines under high stress conditions.

10.7.2 Proper Designing

Proper geometrical design plays a vital role in the control of corrosion of equipment and structures. The general guidelines of the design of materials and components to control corrosion are the following:

1. Always employ a straightforward and uncomplicated style and structure.
2. The design should refrain from including complex shapes with several angles, edges, corners, and so on.
3. Prevent the interaction of metals with different properties, as this might result in galvanic corrosion. To address this issue, insulation might be employed.
4. For the contact between two metals of different types, it is important to maximize the size of the anodic area while minimizing the size of the cathodic area.
5. To the greatest extent feasible, it is advisable to prevent the presence of fissures, which are gaps or cracks, between adjoining components of a building.
6. It is recommended to substitute bolts and rivets with appropriate welding.
7. Substitute metal washers with rubber or plastic washers, as they are non-absorbent of water.
8. Corrosion in pipes can be mitigated by employing bends with a smooth surface.
9. Annealing, a form of heat treatment, reduces stress corrosion.
10. An optimal water storage container design allows for effortless drainage and cleaning. This design prevents the collection of dirt and other substances.

Using Pure Metal

The presence of impurities in a metal leads to a heterogeneous condition, which in turn increases the rate of corrosion. Enhancing the purity of a metal can enhance its resistance to corrosion.

The corrosion resistance of a purified metal is also influenced by the characteristics of the corrosive environment.

Using Metal Alloys

By incorporating suitable components into metal alloys, the corrosion resistance of most metals can be enhanced. For iron, chromium (Cr) is the most appropriate alloying metal. Surgical equipment use steel with a chromium content of 13%.

10.7.3 Cathodic Protection

The principle is to induce the metal to exhibit cathodic behavior in order to prevent corrosion. Cathodic protection refers to the process of minimizing or preventing corrosion by designating the metallic structure as the cathode in an electrolytic cell. Corrosion is prevented due to the absence of an anodic region on the metal. There exist two approaches for implementing cathodic protection on metallic constructions.

(a) Sacrificial anodic protection (galvanic protection)

(b) Impressed current cathodic protection

10.7.3.1 Sacrificial Anodic Protection Method

In this technique, the metallic structure that needs protection is connected to a more reactive metal (anodic metal) to make it the cathode. Consequently, corrosion will solely target the active metal. The parent structure is therefore safeguarded. The metal used in this process is referred to as a sacrificial anode. The more reactive metal undergoes electron loss and experiences gradual corrosion, thus providing protection for the parent cathodic metal. For example, the galvanization process involves the protection of iron by coating it with zinc.

A new sacrificial anode block is used to replace the corroded one. The metals typically used as sacrificial anodes are magnesium, zinc, aluminium, and their alloys. Magnesium possesses the greatest unfavourable potential and may deliver the maximum current output, making it extensively employed in high resistivity electrolytes such as soil. This technology can be applied to various structures such as underground wires and water tanks.

1. Applications

1. Protection against soil corrosion for buried pipelines and underground wires.
2. Providing protection against corrosion in marine environments for cables, ship hulls, piers, and other structures.
3. Introducing magnesium sheets into household water boilers as a preventive measure against rust formation.
4. Calcium metal is used to reduce engine corrosion.

2. Advantages

1. Minimal expenses for installation and operation.

2. Ability to safeguard intricate formations.
3. Applicable to a broad spectrum of highly corrosive substances.

3. Limitations

1. A significant amount of initial current is necessary.
2. Unprotected components are unable to be safeguarded.
3. Restricted capacity for driving, so unsuitable for transporting sizable goods.

10.7.3.2 Impressed Current Cathodic Protection Method

This approach involves the application of an impressed current in the opposite direction to neutralize the corrosion current and change the corroding metal from anode to cathode. Typically, the impressed current is obtained from a direct current source, such as a battery or rectifier connected to an AC line. This current is then applied to an insoluble and inert anode, which can be made of materials like graphite, scrap iron, stainless steel, platinum, or high silica iron.

An enough direct current is administered to an inactive anode, which is concealed in the soil (or submerged in the corroding substance) and linked to the metallic structure that requires safeguarding. The anode often consists of a backfill material, such as coke breeze or gypsum, which serves to enhance electrical conductivity with the surrounding soil. Impressed current cathodic protection has been utilized on many structures such as open water box coolers, water tanks, subsurface oil or water pipes, condensers, transmission line towers, marine piers, and laid up ships. This protective approach is especially advantageous for expansive constructions throughout extended periods of operation.

10.7.4 Protective Coatings

To prevent corrosion of metals, it is essential to apply protective coatings to the surface. These coatings function as a tangible barrier that separates the metal surface being coated from the surrounding environment. They provide an attractive appearance and confer unique characteristics like as durability, resistance to rusting, and insulation against heat.

Categorization: Protective coatings can be roughly divided into two distinct categories. They exist:

1. Coatings made of inorganic materials.
2. Coatings made of organic materials.

Inorganic coatings can be further categorized into two distinct categories. They exist:

Coating with a Metallic Layer

1. Hot dipping refers to the process of galvanizing and tinning.
2. Metal cladding refers to the process of covering a surface with a layer of metal.
3. The processes of Cementation-Sherardizing, Chromating, and Calorizing.
4. Electroplating is the process of depositing a layer of metal onto a surface using an electric current.

Non-metallic Coating

1. Surface coating refers to the application of chromate, phosphate, or oxide coatings through chemical conversion processes.
2. Anodizing
3. Enamel, vitreous, or porcelain coating.

Organic coverings comprise Coatings, such as paints, varnishes, lacquers, and enamels.

10.7.5 Modifying the Environment

The corrosion rate is also influenced by the corrosive environment. The corrosive nature of the environment can be mitigated by:

1. **Deaeration:** Deaeration refers to the process of removing dissolved oxygen from a substance by adjusting its temperature and using mechanical agitation.
2. **Deactivation:** This process entails the introduction of compounds that have the ability to react with oxygen in a liquid solution. For example,

$$2Na_2SO_3 + O_2 \longrightarrow 2Na_2SO_4$$

$$N_2H_4 + O_2 \longrightarrow N_2 + 2H_2O$$

3. **Dehumidification:** Dehumidification involves the process of decreasing the amount of moisture present in the air. For example, substances such as alumina or silica gel have the ability to absorb moisture. These substances are commonly used in air-conditioning shops.

QUESTIONS

10.1 Explain sacrificial coating method for the prevention of corrosion.

10.2 Discuss cathodic protection methods for the prevention of corrosion.

10.3 Explain the type of corrosion which is brought about by direct action of oxygen at low (or) high temperatures, usually in the absence of moisture?

10.4 What are the types of Corrosion? Explain briefly the factors affecting corrosion.

10.5 Discuss the mechanism of electrochemical theory of corrosion by absorption of oxygen. What effect will increased oxygen supply have on such corrosion? How can corrosion be minimized by proper design?

10.6 Discuss the differences between anodic and cathodic metallic coatings. Explain the processes of Galvanizing and Electroplating? What will happen if an iron ship travelling in the sea is attached through an insulated metallic wire to a small sheet of magnesium?

10.7 Comment on the use of Al in place of Zn for cathodic protection of iron from rusting.

10.8 Define the term corrosion. Describe the mechanism of electrochemical corrosion with the help of hydrogen evolution and oxygen absorption reactions. How it can be prevented using sacrificial anodic protection?

10.9 Why Iron Nail present on the door undergoes corrosion?

10.10 Outline the mechanism involved in electrochemical theory of corrosion. How corrosion is prevented by anodic and cathodic inhibitors?

10.11 Describe the mechanism of electrochemical or wet corrosion with help of reactions? Explain the cathodic protection method of prevention of corrosion.

10.12 Comment on the use of aluminium in place of Zinc for Cathodic protection of iron from rusting.

10.13 What is the effect of the following factors on the rate of corrsion:

(i) Temperature and humidity

(ii) pH

(iii) Presence of suspended particles in atmosphere?

10.14 Distinguish between dry and wet corrosion.

10.15 What are corrosion inhibitors? Explain how corrosion is controlled by using anodic and cathodic inhibitors.

10.16 Discuss the factors which effect carrion. Also explain the methods for prevention of corrosion.

10.17 What is meant by corrosion? Explain the electrochemical theory of corrosion.

10.18 State two conditions for wet corrosion to take place.

Unit 4(A)

Chemistry of Engineering Materials

Macromolecule and Glass

CHAPTER 11

Polymers

11.1 INTRODUCTION

Polymers are large-scale macromolecules created by the linkage of repeating structural components. The term 'Polymers' is derived from the combination of two Greek words: '*poly*' meaning many, and '*mer*' meaning unit or part. Polymers are macromolecules, which are characterized by their huge size and high molecular mass. Polymerization refers to the repetitive covalent joining of monomers, resulting in the formation of larger molecules. The process by which polymers are formed from their separate monomers is known as polymerization. The characteristic chain structure of many repeat units in polystyrene is chemically connected as follows. Different representations are equivalent where x is the number of repeat units.

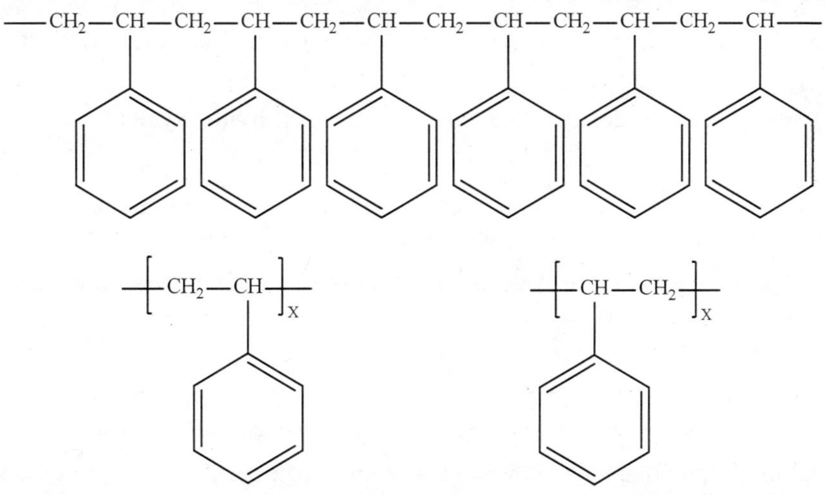

Polystyrene

11.2 CLASSIFICATION OF POLYMERS

The polymers are classified into the following manner for easy understanding:

1. Based on Source
2. Based on Nature of Repeat Units
3. Based on Structural Characteristics
4. Based on Stereochemistry
5. Based on Functional Group
6. Based on Mode of Polymerization
7. Based on Physical Properties

11.2.1 Classification Based on Source

Polymers can be classified into three categories, as indicated by the sources:

1. **Natural Polymers:** Natural polymers are organic compounds that occur naturally in plants and animals. Instances include proteins, cellulose, starch, some resins, and rubber.
2. **Semi-synthetic Polymers:** Cellulose derivatives as cellulose acetate (rayon) and cellulose nitrate are typical examples of this particular subcategory.
3. **Synthetic Polymers:** Synthetic polymers, such as polythene (plastic), nylon 6, 6 (synthetic fibers), and Buna-S (synthetic rubber), are artificial polymers that are widely utilized in both everyday life and many industries.

11.2.2 Classification Based on Nature of Repeat Units

The behavior of polymers is mostly determined by the repeating unit of monomers.

1. Homopolymer: Homopolymers are polymers composed of identical repeat units of monomers. For instance, all identical monomers are denoted by the symbol 'A'.

-A-A-A-A-A-A-A-A-A-A-A-

2. Copolymer: Copolymers are polymers composed of distinct repeat units of monomers. These are classified into the following categories:

(a) **Alternating Copolymer**: An alternating copolymer refers to a polymer in which different monomers are organized in an alternating manner. For example, the pattern is repeated in the sequence A-B-A-B-A-B-A-B-A-B.

(b) **Block Copolymer:** A polymer in which one group or cluster of monomers is followed by another group of monomers, e.g., AAAAAAAA-BBBBBBBBB.

(c) **Random Copolymer:** A polymer in which different monomers are organized in a random manner, e.g. AA-B-A-B-B-A-A-A-B-B-A.

(d) **Graft Copolymer:** A graft copolymer refers to a polymer in which a distinct monomeric chain is organized in a grafting way.

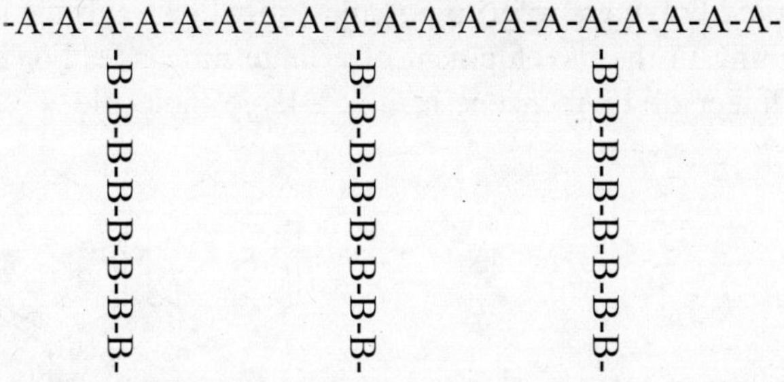

11.2.3 Classification Based on Structural Characteristics

This category is intricately linked to the qualities of materials.

1. Linear: Linear polymers are characterized by a continuous and straight chain structure. For example, linear addition of cyclohexene with itself.

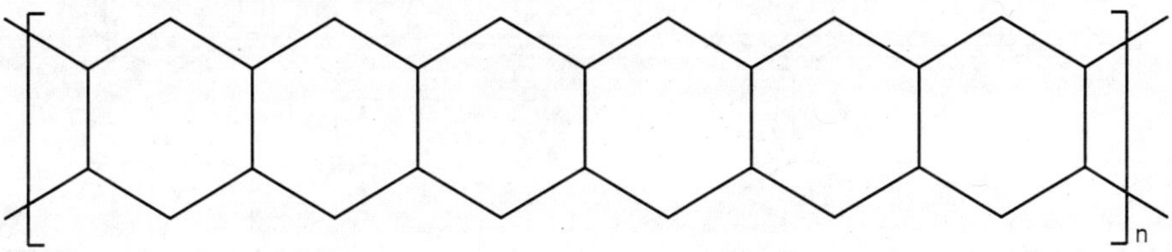

2. Branched: Branched polymers are characterized by the presence of sporadic branches on a longer chain. For example, branched addition of styrene.

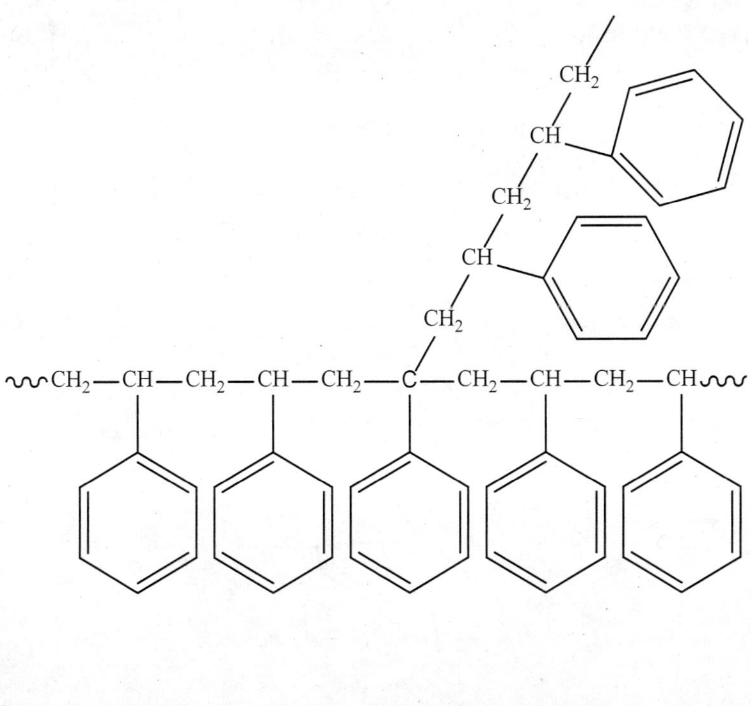

3. Networked: Networked polymers refer to polymers in which multiple interconnected linear chains are present, resulting in the development of a large molecule. For example, addition of polystyrene with cross-linker divinylbenzene results a large molecule.

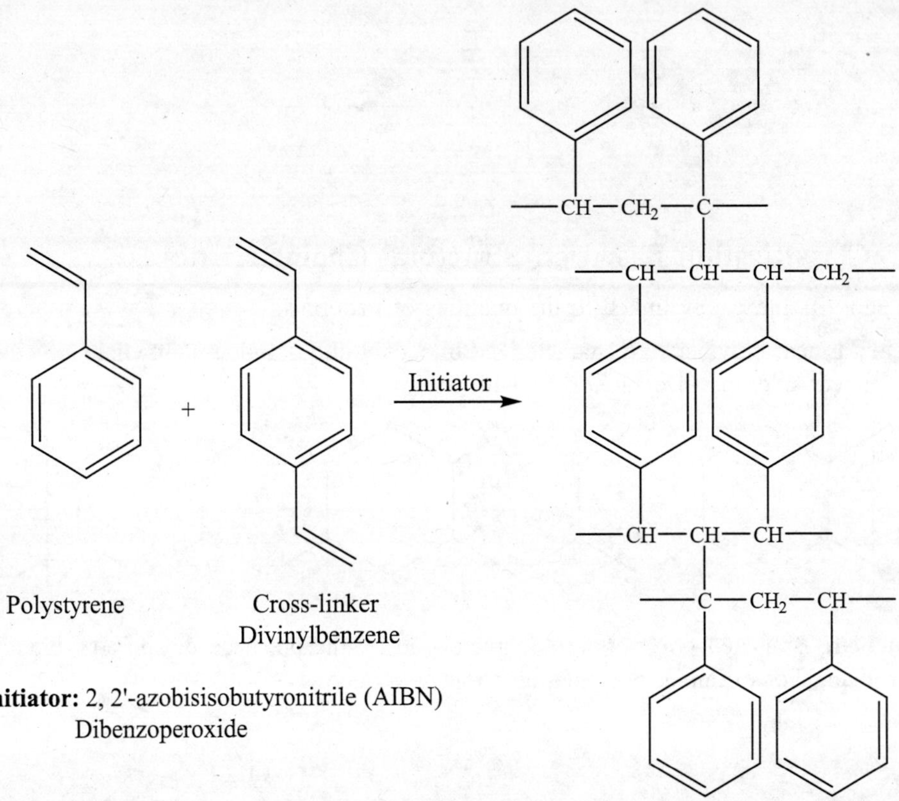

Initiator: 2, 2'-azobisisobutyronitrile (AIBN)
Dibenzoperoxide

4. Dendrimers: Dendrimers are a type of highly branched macromolecules. It looks like as follows:

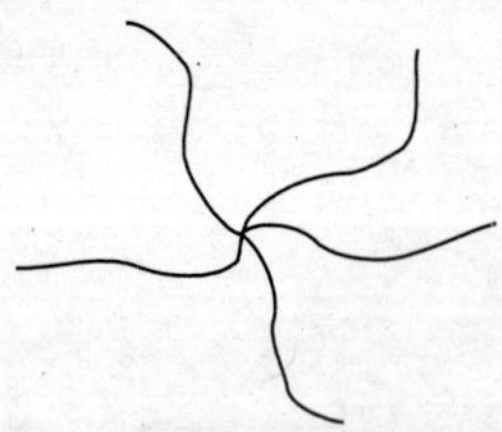

5. Combed Polymers: Polymer strands that have been carefully arranged and aligned. It looks like as follows:

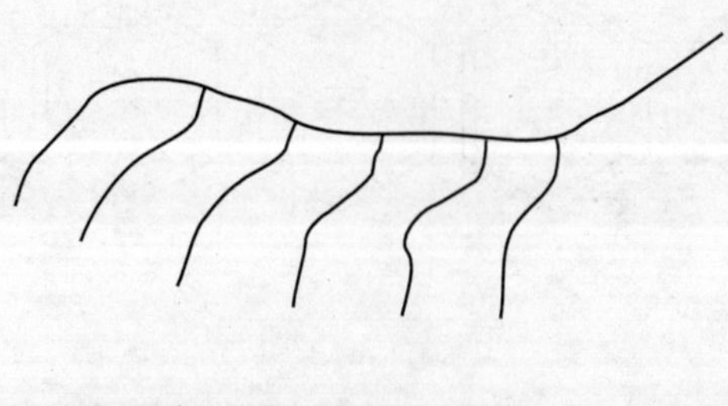

6. Star Polymers: Star polymer looks like as follows:

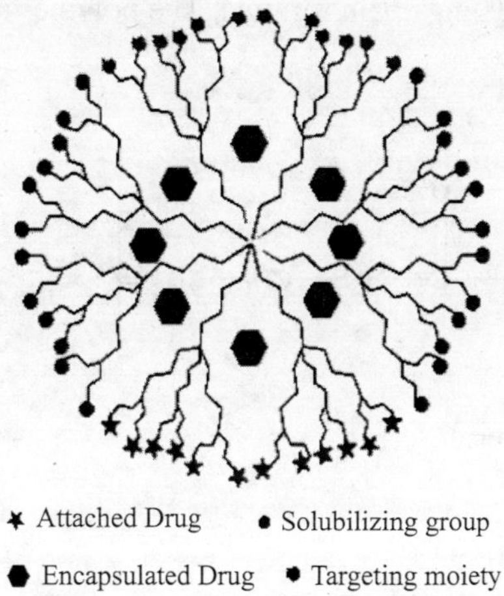

11.2.4 Classification Based on Stereochemistry of Linkages

The tacticity of polymers refers to the spatial arrangement of the connected alkyl groups in the polymer's structure. There are three types of alkyl groups based on their sites of attachment.

1. Isotactic: It refers to a polymer in which the branched alkyl group is connected to the same side or symmetrically to the polymer backbone.

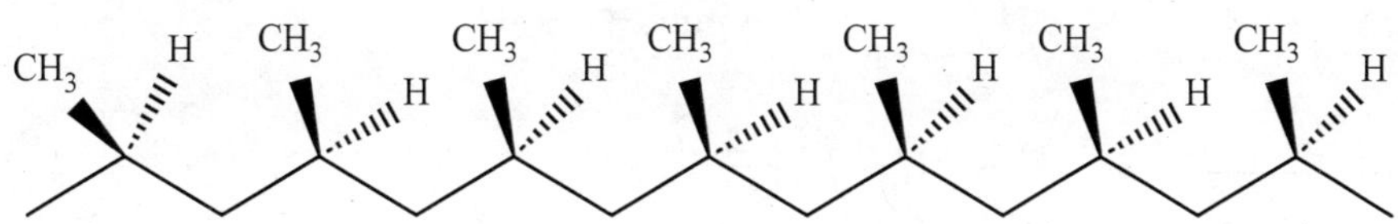

2. Syndiotactic: Syndiotactic refers to a molecular structure where alkyl groups are positioned on alternating sides of the backbone.

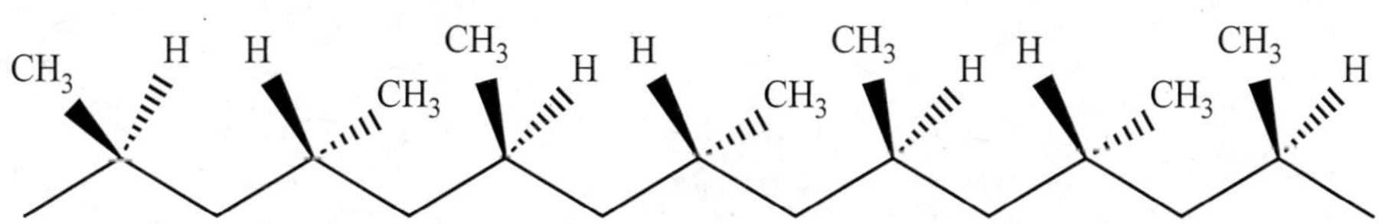

3. Atactic: Atactic polymers are characterized by the random arrangement of all attached alkyl groups, which is the most prevalent configuration in polymers.

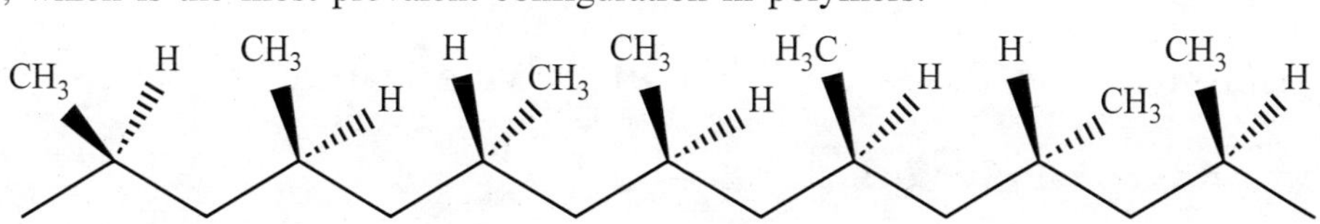

11.2.5 Classification Based on Functional Group

Polymers are denoted by the repeating chain formula. The polymer is characterized by functional groups present in its repeat units.

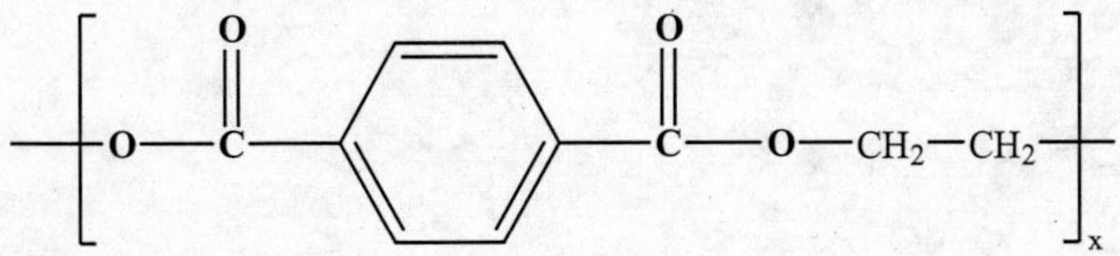

Polyester

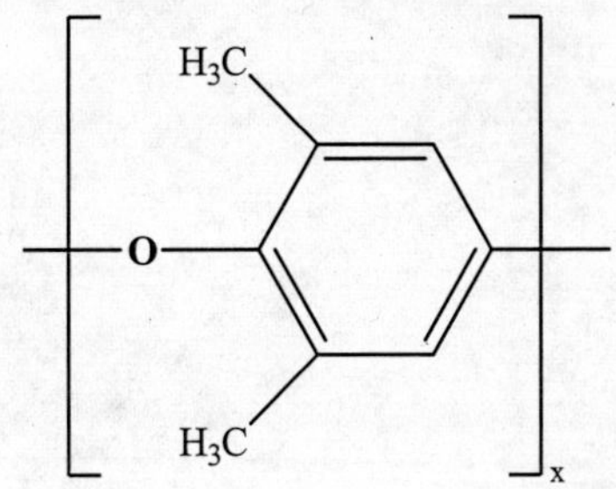

Polyether

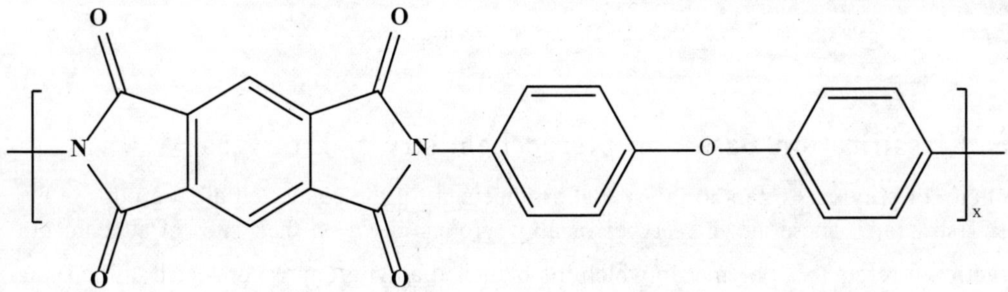

Polyimide

$$CH_2{=}CH_2 \longrightarrow \left[CH_2 - CH_2 \right]_x$$

Polyethylene

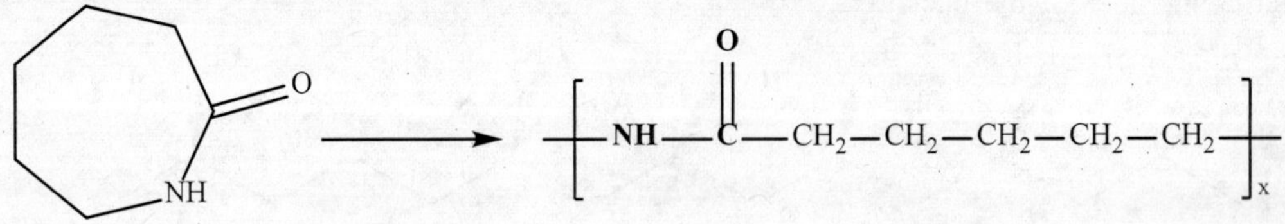

Caprolactam polyamide

11.2.6 Classification Based on Mode of Polymerization

11.2.6.1 Addition Polymers

Addition polymers are created through the repetitive addition of monomer molecules that contain double or triple bonds. Homopolymers, such as polythene, are addition polymers that result from the polymerization of a single kind of monomer.

$$n\,CH_2{=}CH_2 \longrightarrow \left[CH_2{-}CH_2 \right]_n$$

Ethene Polythene

Copolymers refer to the polymers produced during addition polymerization using two distinct monomers, such as Buna-S and Buna-N.

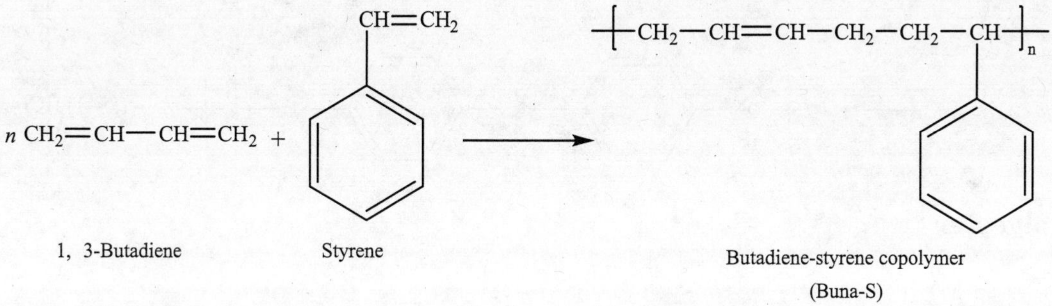

11.2.6.2 Condensation Polymers

These polymerization reactions involve the removal of tiny molecules such as water, alcohol, hydrogen chloride, etc. Some examples are terylene (dacron), nylon 6, 6, and nylon 6.

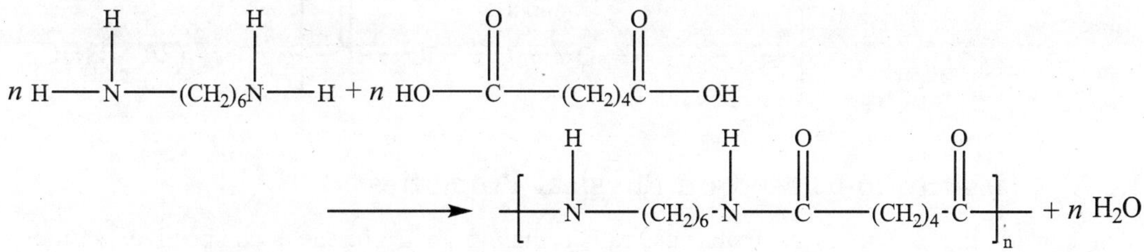

Natural polymers: cellulose, starch, wool and silk

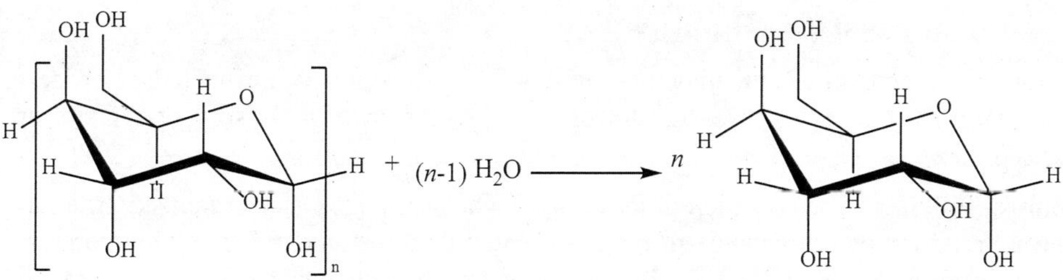

Cellulose Glucose

Polyamide

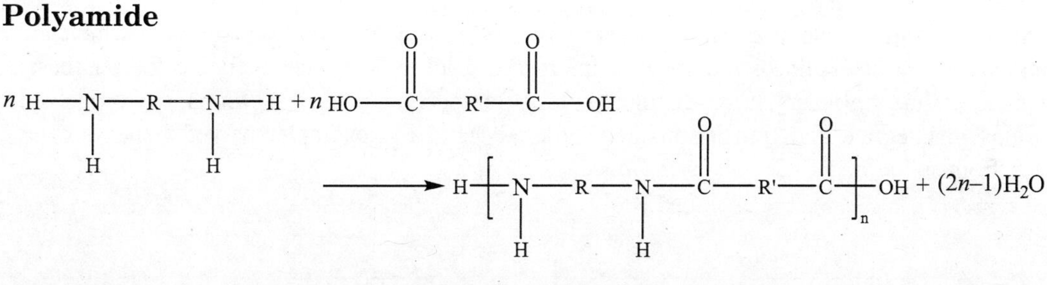

Polycarbonates

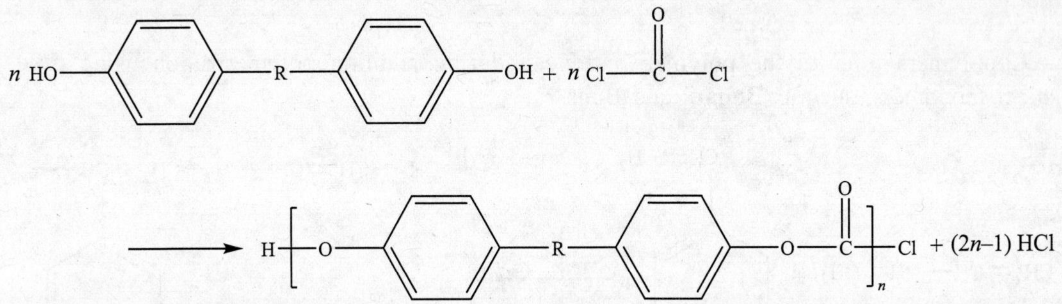

Polyester

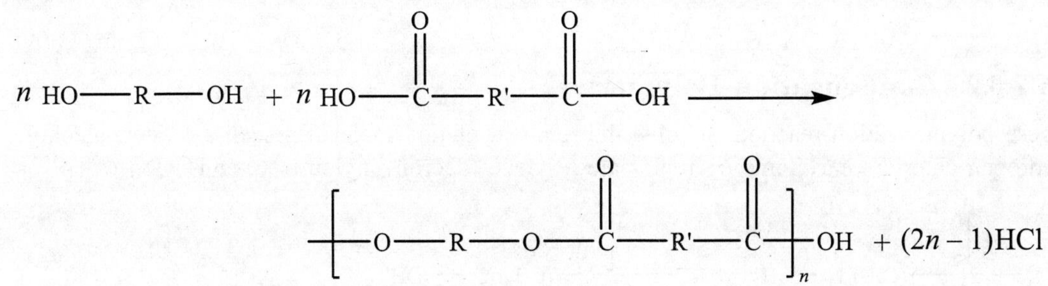

11.2.7 Classification Based on Physical Properties

When considering polymers as materials, rather than from a chemical perspective, various classifications are commonly employed based on temperature.

11.2.7.1 Thermoplastics (Plastics)

Linear or branched architecture can undergo melting and reforming when heated, a property referred to as thermoplasticity. These substances can dissolve in organic solvents. There are two distinct types.

1. Amorphous: Polymers are considered amorphous when they lack any organized structure. They exhibit a broad temperature range throughout which they become softer. At temperatures higher than this, polymers exist in a liquid state. For example, polystyrene and poly(methyl methacrylate).

2. Semi-crystalline: Semi-crystalline materials are made up of small crystallites, which are domains with a crystalline structure at a microscopic level. They are available for purchase. Semi-crystalline polymers have distinct melting temperatures (T_m) at which the organized portions disintegrate and transit into a disordered state. For example, polyethylene, textiles made from nylon and PET polyester fibers.

11.2.7.2 Thermosets (resins)

Thermosets are polymers that are highly cross-linked and have significant rigidity. They undergo degradation when exposed to heat. Not capable of being dissolved in solvents. Typically produced through molding and subsequently formed through machining processes such as grinding and drilling. For example, items such as tires and rubber bands.

Physical Property

Characteristic of a thermosets that may be observed or measured without changing its chemical composition.

11.2.7.3 Elastomers (rubbers)

Elastomers are materials that can be stretched and quickly return to their original size due to their moderate cross-linking. Elastomers are elaborated later in the chapter.

11.2.7.4 Thermoplastic Elastomers

Thermoplastic elastomers are materials that exhibit both thermoplastic and elastomeric properties. Thermoplastic elastomers are elastic polymers that have the ability to be melted, such as the soles of tennis shoes.

11.2.7.5 Dendrimers

Dendrimers are polymers with many branches that are arranged in a regular and consecutive manner.

11.3 POLYMERIZATION PROCESS

The fundamental reaction process for the synthesis of polymers is fully contingent upon the participating species. The process of polymerization can be classified into three distinct groups.

1. Radical polymerization
2. Cationic polymerization
3. Anionic polymerization

11.3.1 Radical Polymerization

During the process of polymerization, the generation of the chemically active species is facilitated by a substance known as an initiator. Radical polymerization is a type of polymerization when the reactive species involved is a radical. The primary distinction between chain polymerization and step polymerization is in the instantaneous formation of high molecular weight polymer. After the reactive center is formed, the monomers quickly join together to form a big structure. The high molecular weight polymer is not achieved until the completion of the polymerization process. In order for a molecule to undergo chain polymerization, it often has to possess double bonds, such as C=C or C=O. Primarily, the reaction must have a thermodynamically negative change in Gibbs free energy (ΔG) and be kinetically viable.

Mechanism (Steps in) of Radical Polymerization

Step I: Initiation

The initial stage of radical polymerization involves the generation of unpaired electrons, known as free radicals. This occurs as a result of the homolytic dissociation of the starting species, which produces a pair of free radicals.

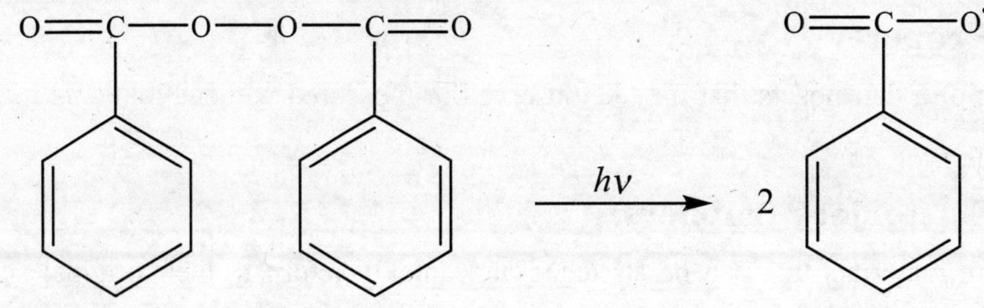

The subsequent step of initiation entails the incorporation of this radical into the monomer molecule (M) in order to commence the chain reaction.

$$\underset{\text{Free radical}}{2R^{\bullet}} + \underset{\text{monomer}}{M} \longrightarrow M^{*}$$

The M* is the radical that initiates or serves as the primary radical.

Step II: Propagation

The initial free radical initiates an attack on the monomer, resulting in the formation of the propagating radical. This process can be represented by the following reactions:

$$R\cdot + M \longrightarrow \cdot M_1$$
$$\cdot M_1 + M \longrightarrow \cdot M_2$$
$$\cdot M_2 + M \longrightarrow \cdot M_3$$
$$\cdot M_3 + M \longrightarrow \cdot M_4 \text{ and so on}$$

The propagation mechanism involves the interaction between the radical M* and its corresponding monomer M.

$$\dot{M} + CH_2{=}\underset{\substack{|\\ Y}}{CH} \longrightarrow M{-}CH_2{-}\underset{\substack{|\\ Y}}{\dot{C}H}$$

By continuously adding additional monomers in this manner, a polymer chain will ultimately be formed where the substituents are positioned on alternating atoms.

Step III: Termination

The most prevalent terminations occur as a result of the collision between two expanding chains.

(i) **Collision between two growing chains:** The collision of a developing chain with an initiator radical when the latter is relatively abundant.

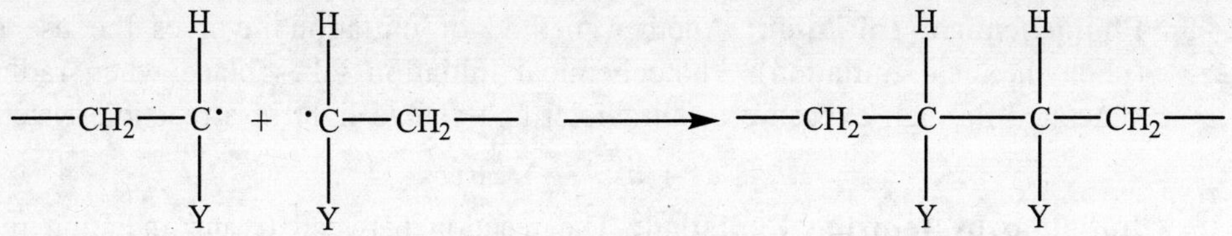

(ii) The collision of a developing chain with an initiator radical when the latter is relatively abundant.

(iii) Impact between an expanding chain and contaminants. In certain cases, the polymer's chain length can be regulated by intentionally introducing inhibitors.

Less frequently, through disproportionation, an individual hydrogen atom from one radical is transferred to another, resulting in the formation of two polymers: one saturated and the other unsaturated.

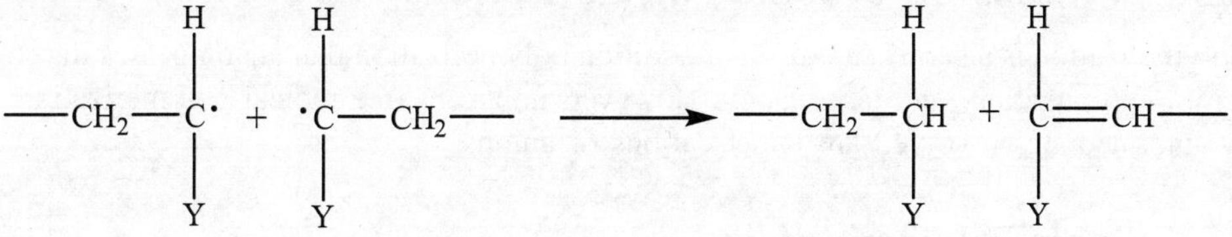

Radical Polymerization Initiation Methods

1. **Thermal Initiation:** Thermal initiation refers to the process of starting a chemical reaction through the use of heat. The generation of an initiating radical is achieved by subjecting the reaction media to heat. The following are examples of common initiators:

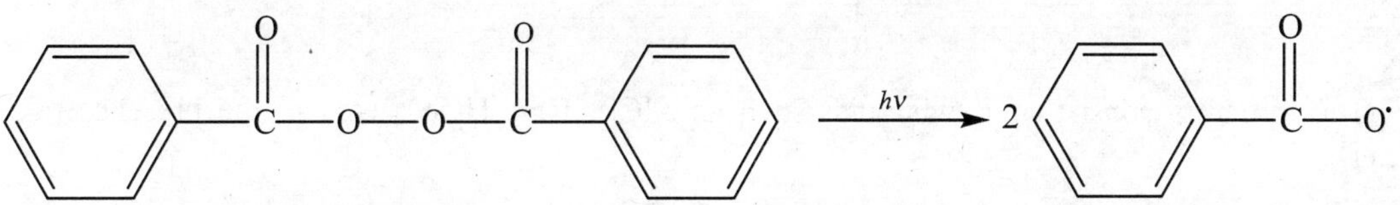

dibenzoylperoxide

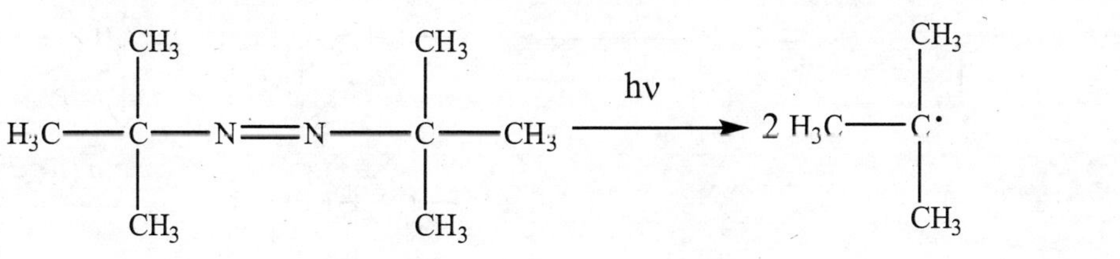

2, 2'-Azobisisobutyronitrile (AIBN)

2. **Redox Initiation:** These compounds are often produced through the reaction of peroxides with Fe^{2+} ions. The reactions can be represented as follows:

$$H_2O_2 + Fe^{2+} \longrightarrow HO^- + HO\cdot + Fe^{3+}$$

$$R\text{—}OO\text{—}R + Fe^{2+} \longrightarrow RO^- + RO\cdot + Fe^{3+}$$

$$R\text{—}OOH + Fe^{2+} \longrightarrow OH^- + RO\cdot + Fe^{3+}$$

3. **Photochemical Initiation:** Another method of initiation involves the use of light (photochemical initiation). Photochemical initiation takes place when radicals are generated through exposure to ultraviolet and visible light in a reaction system.

$$I + h\nu \longrightarrow 2R\cdot$$

4. **Initiation by Ionizing Radiation:** The reaction between C and radiation results in the formation of C+· and the release of one electron (e^-). The energy of this reaction ranges from 10 kiloelectron volts (keV) to 100 millielectron volts (meV).

$$C + \text{radiation} \longrightarrow C^{+\cdot} + e^- \;(E = 10 \text{ keV} - 100 \text{ meV})$$

The reaction C+· decomposes into A· and B^+

$$C^{+\cdot} \longrightarrow A\cdot + B^+$$

Mechanism (Steps in) of Ionic Polymerization

Ionic polymerization is a significant form of addition polymerization that encompasses the stages of initiation, propagation, and termination. However, unlike in free radical polymerization, the reactive intermediates involved are either cations or anions.

11.3.2 Cationic Polymerization

The complete process of cationic polymerisation is processed by following steps:

Step I: Initiation

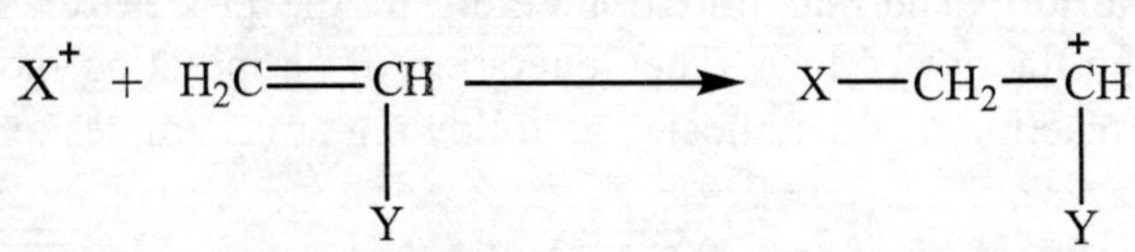

The initiators consist of Lewis acids such as $AlCl_3$, BF_3, H_2SO_4 etc., in the presence of water.

Step II: Propagation

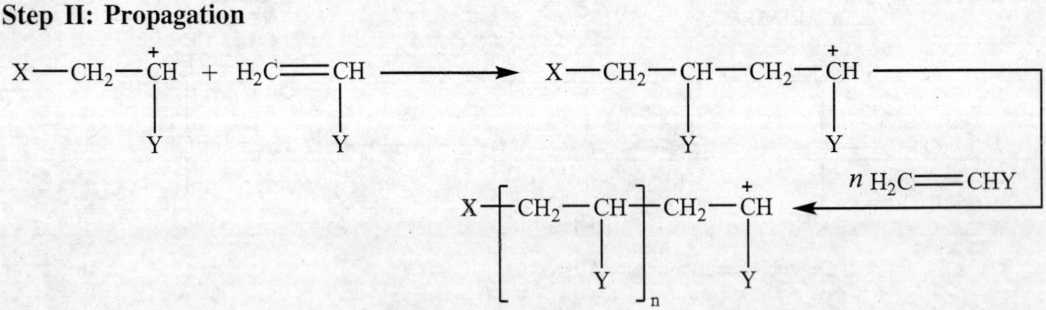

Step III: Termination

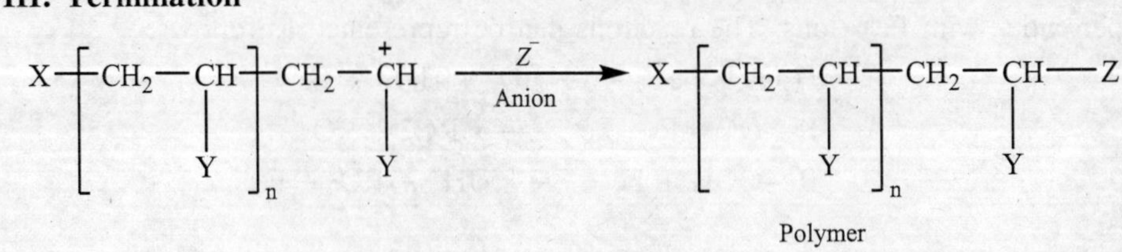

When the medium has a high dielectric strength, it results in a higher rate of polymerization and a greater degree of polymerization. This is because there is improved separation of charges between ions and counter ions.

11.3.3 Anionic Polymerization Mechanism

The complete process of anionic polymerisation is processed by following steps:

Step I: Initiation

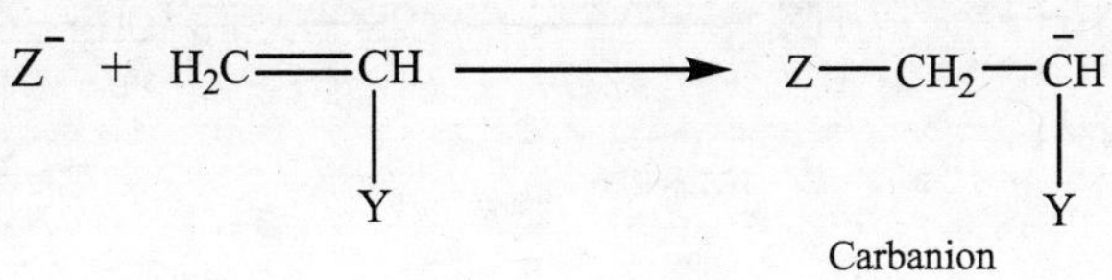

Step II: Propagation

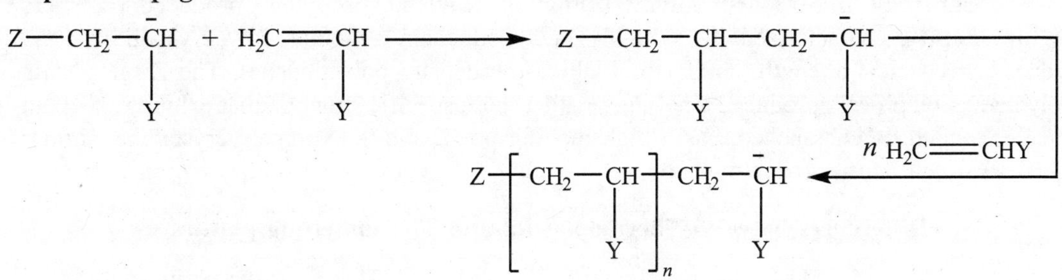

Step III: Termination

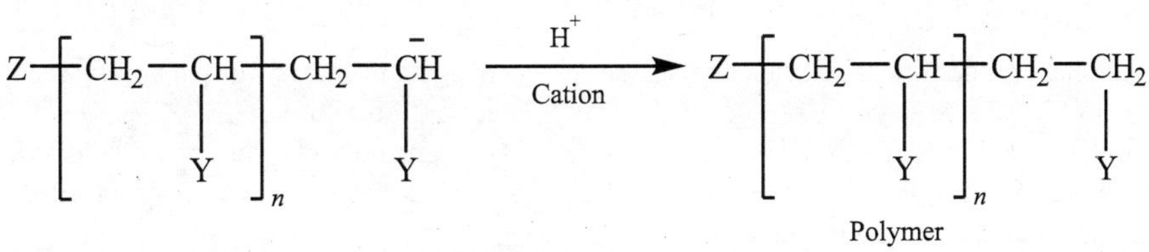

The polymerization of olefins using metallic sodium and sodium ethoxide occurs through this method.

11.3.4 Ziegler-Natta Polymerisation or Coordination Polymerization

Zeigler and Natta noted that the presence of a combination of transition metal halides (such as $TiCl_4$, $ZnBr_3$, and halides of V, Zr, Cr, Mo, etc.) along with orgnometallic compounds (such as triethyl/trimethyl aluminium) results in stereospecific polymerization of olefins.

Mechanism of Ziegler-Natta Polymerization

Step I: Initiation

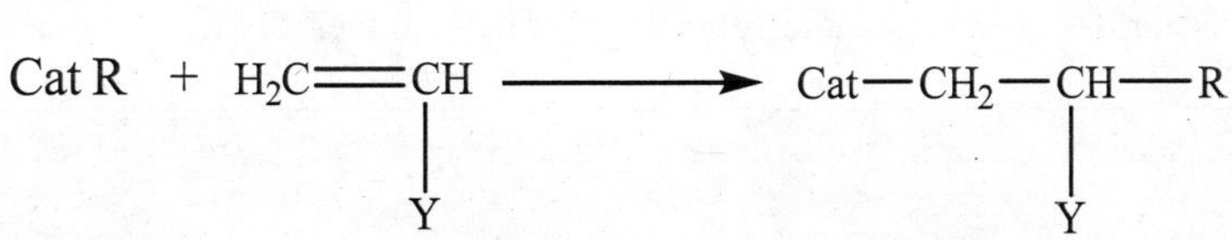

Step II: Propagation

$$\text{Cat}-CH_2-\underset{\displaystyle Y}{\underset{|}{CH}}-R + n\,H_2C=\underset{\displaystyle Y}{\underset{|}{CH}} \longrightarrow \text{Cat}-CH_2-\underset{\displaystyle Y}{\underset{|}{CH}}\left[CH_2-\underset{\displaystyle Y}{\underset{|}{CH}}\right]_n R$$

Step III: Termination

$$-CH_2-\underset{\displaystyle Y}{\underset{|}{CH}}\left[CH_2-\underset{\displaystyle Y}{\underset{|}{CH}}\right]_n R + HX \longrightarrow \text{Cat}-X + \underbrace{CH_3-\underset{\displaystyle Y}{\underset{|}{CH}}\left[CH_2-\underset{\displaystyle Y}{\underset{|}{CH}}\right]_n R}_{\text{Polymer}}$$

Through prior free radical polymerization of ethylene, propylene produced polymers of low density, characterized by a highly branched chain structure, low crystallinity, and reduced mechanical strength, similar to LDPE (low density polyethylene). The Ziegler-Natta polymerization process produced a polymer with a stereospecific structure that exhibits excellent tensile strength, is unbranched, has a high melting point, and is extremely crystalline, similar to HDPE (high density polyethylene).

Differences between Thermoplastics and Thermosetting plastics

S. No.	Thermoplastics	Thermosetting plastics
1	Formed either by addition or by condensation polymerization reactions.	Formed by condensation polymerization reactions
2	They have either linear or branched structures.	They have three dimensional cross-linked network structures.
3	Adjacent polymer chains are held together by either Vander Waals forces, or by dipole-dipole forces or by hydrogen bonds.	Adjacent polymer chains are held together by strong covalent bonds called crossed-links.
4	They soften on heating and stiffen on cooling	They do not soften on heating.
5	Low molecular weight thermoplastics are soluble in their suitable solvents.	They are insoluble in any solvent.
6	They can be remolded, re-shaped and re-used.	They can be reclaimed from waste i.e., they can be recycled.
7	There is no change in chemical composition and structure during molding process.	They undergo chemical changes such as further polymerisation and cross-linking during moulding process.
8	They are soft and flexible	They are hard, rigid and infusible.
9	They can be reclaimed from waste i.e., they can be recycled.	They cannot be reclaimed from waste. They cannot be recycled.
10	They undergo reversible changes, on the application of heat.	They undergo irreversible changes on the application of heat.
11	They swell or dissolve in organic solvents.	They neither dissolve nor swell in organic solvents.

S. No.	Thermoplastics	Thermosetting plastics
12	They are tough materials.	They are brittle materials
13	The molded articles are to be cooled to room temperature before taking out from the moulds to avoid deformation.	The molded articles can be taken out of the molds even when they are still hot without any deformations.
14	Curing can be done by cooling.	Curing can be done by applying heat and pressure.
15	Examples: Polyethylene (PE), Bakelite.	Examples: Phenol-formaldehyde resin (PF), urea-formaldehyde resin.

11.4 PREPARATION, PROPERTIES AND USES OF THERMOPLASTICS

11.4.1 Polyethylene (PE)—HDPE, LDPE

The desired outcome can be achieved through the process of polymerizing ethylene at high pressure (1500 atm) and at a temperature range of 150 to 250 degrees Celsius. This process requires the presence of small amounts (0.01%) of oxygen or peroxides as catalysts. Polyethylene, a waxy substance, is produced by polymerizing ethylene and is subsequently discharged out the perforated bottom of the container.

There are two varieties of polyethylenes that can be obtained:

1. LDPE (Low Density Polyethylene) and
2. HDPE (High Density Polyethylene).

HDPE is manufactured by low-pressure techniques utilizing ionic catalysts, while LDPE is manufactured through high-pressure techniques employing free radical initiators. The density of LDPE ranges from 0.91 to 0.925 g/cm3. The density of HDPE ranges from 0.941 to 0.965 g/cm3. High-density polyethylene (HDPE) is a linear polymer that has superior resistance to chemicals.

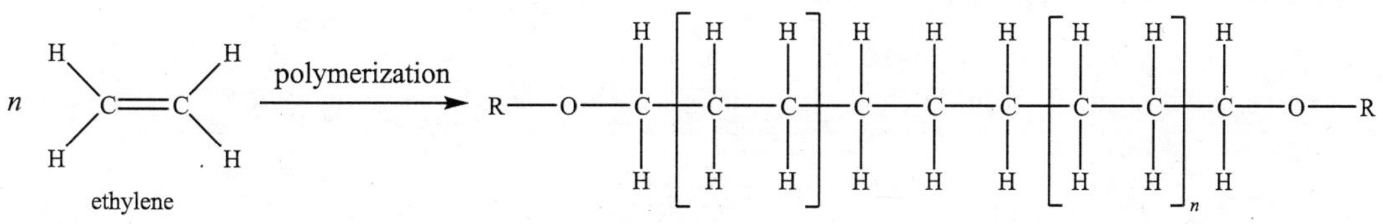

Properties

1. The substance is a solid that is hard, waxy, and white in color. It is also translucent and does not have any polarity.
2. It readily forms crystals.
3. It has excellent chemical resistance to acids, alkalies, and salt solutions, as well as possessing strong insulating characteristics.
4. It is vulnerable to organic solvents, especially kerosene.
5. It exhibits permeability to numerous organic solvents.

Uses

1. These materials are beneficial for manufacturing insulator components, bottle caps, pliable bottles, pipes, and other related products.

2. LDPE is utilized in the production of film and sheeting. LDPE pipes are utilized for agricultural, irrigation, and domestic water line connections.
3. HDPE is utilized in the production of toys and various household items.

Packaging films

11.4.2 Polypropylene

The process involves the polymerization of propylene using a Zeigler-Natta Catalyst.

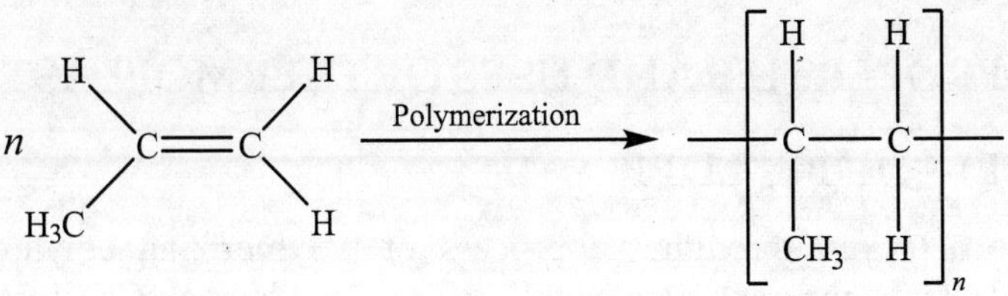

Propylene Polypropylene

Properties: Polypropylene is a polymer that is both isotactic and extremely crystalline. It surpasses polyethylene in terms of chemical resistance, hardness, strength, and moisture-proofing capabilities. It has the lowest weight compared to other high polymers.

Uses: Polypropylene fibers possess exceptional strength, making them ideal for manufacturing ropes, carpets, blankets, bags, and garments. Additionally, it is utilized for the packaging of pharmaceutical and cosmetic products, as well as for the production of sterilizable hospital equipment, washing machine parts, pipes, TV and radio components, molded baggage, and vibrant furniture.

11.4.3 Polyvinyl Chloride (PVC)

Vinyl chloride is typically synthesized through the reaction of acetylene with hydrogen chloride in the presence of a metal chloride catalyst, under conditions of 1 to 1.5 atmospheres of pressure and temperatures ranging from 60°C to 80°C.

$$HC \equiv CH + HCl \xrightarrow{HgCl_2} CH_2 = CH - Cl$$

Polyvinyl Chloride (PVC) is synthesized through the thermal polymerization of a water emulsion containing vinyl chloride, with the addition of a minute quantity of benzoyl peroxide or hydrogen peroxide, under a pressurized autoclave.

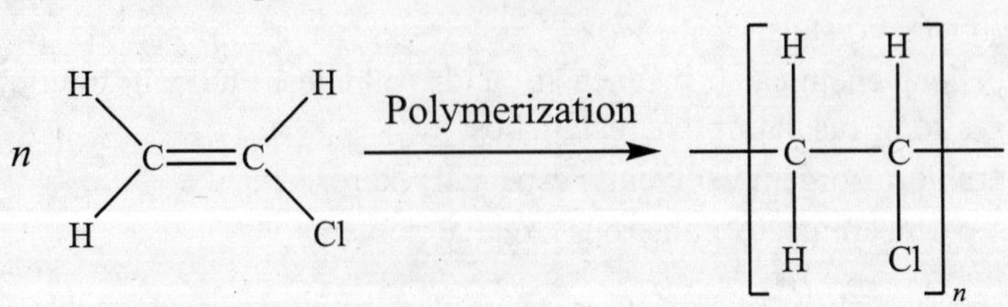

Vinyl chloride Polyvinyl chloride (PVC)

Properties

1. It manifests as a transparent, inflexible, durable, and resistant substance.
2. It possesses a high density and a low softening point.
3. It exhibits resistance to light, ambient oxygen, inorganic acids and alkalies, oil, petrol, etc. However, it is soluble in heated chlorinated hydrocarbons.
4. It is the most often utilized synthetic polymer.
5. It is cost-effective, waterproof, and has the ability to extinguish itself.

Uses

1. It is utilized within the construction industry for applications like as pipelines and turbine pumps.
2. Flexible PVC is commonly used for wire and cable coatings, low pressure flexible tubing and hose, footwear, gaskets, seals, sheet coatings, and ATM cards.
3. Its primary applications include wire insulation, leather cloth, packaging, and toys.
4. This substance is employed in the production of film, sheet, and floor covering.
5. PVC pipes are utilized for transporting corrosive chemicals within petrochemical plants.

Flexible PVC Applications

PVC pipe sections

Water pipes

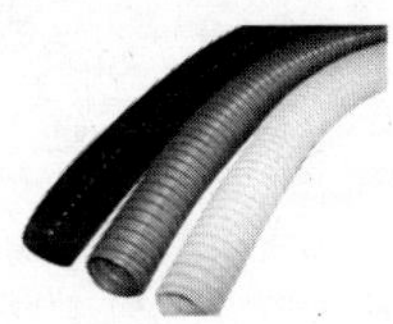

Pneumatic tubing

11.4.4 Polymethyl Methacrylate (PMMA)

It is alternatively referred to as Lucite or Plexiglass. The substance is derived through the process of polymerization of methyl methacrylate in the presence of either acetyl peroxide or hydrogen peroxide. It is a compound made of acrylic polymers.

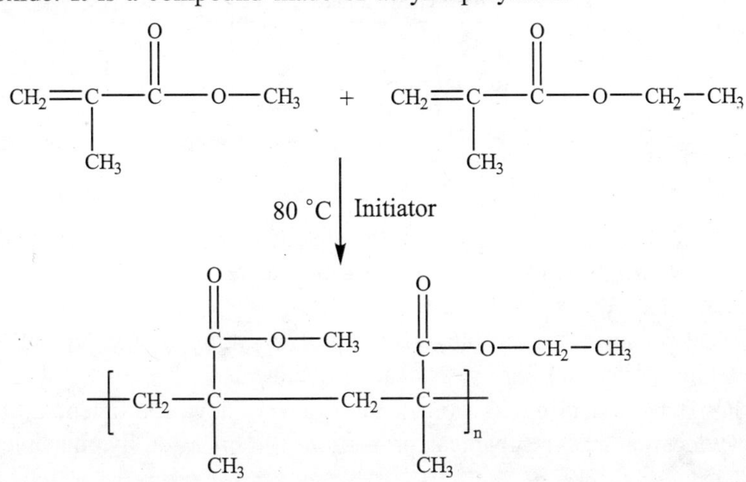

Properties

1. It is a durable and rather inflexible substance with a high temperature at which it becomes soft, around 130–140°C. However, it takes on a rubbery consistency when exposed to temperatures above 65°C.
2. It has a moderate level of strength.
3. It possesses favorable optical characteristics.
4. It has excellent durability and chemical resistance.
5. It is clear (opaque) and possesses excellent electrical resistance.

Uses: This material finds application in various fields like as lenses, illuminated signs, displays, window glazing, vehicle lenses, windshields, lighting fixtures, and furniture.

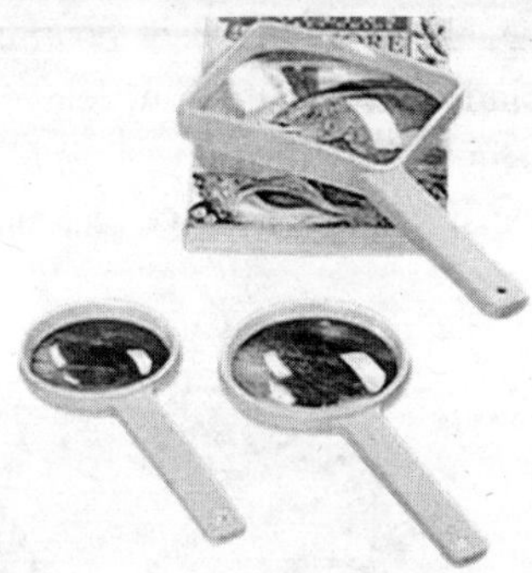

Magnifying glass

11.4.5 Polytetrafluoroethylene (TEFLON)

Teflon is produced through the polymerization of tetrafluoroethylene in a water emulsion, using benzoyl peroxide as a catalyst, while applying pressure.

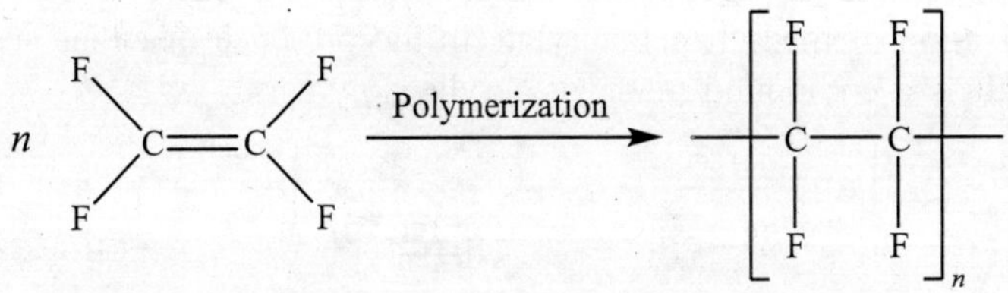

Tetrafluoroethylene　　Teflon

Properties

1. The strong attraction forces between the distinct chains of polytetrafluoroethylene are a result of the presence of highly electronegative fluorine atoms and the regular arrangement of its molecule.
2. The material possesses strong attractive forces, resulting in exceptional toughness, a high softening point, and high resistance to chemicals.
3. It also has a high density, a waxy texture, a very low coefficient of friction, and excellent electrical and mechanical properties. It can be easily machined, punched, and drilled.

4. However, the material has the drawback of being insoluble and unable to exist in a genuine molten state.
5. At around 350°C, it undergoes sintering, resulting in the formation of a very viscous and opaque mass. This mass can be shaped into certain forms by subjecting it to high pressures.

Uses

1. It serves as an insulating material in motors, transformers, cables, wires, fittings, etc. It is also used for manufacturing gaskets, packaging, pump components, tank linings, chemical-carrying pipes, tubing, and tanks.
2. This substance is utilized to apply a protective layer on glass fibers, asbestos fibers, and cloths. It is also used in applications such as non-lubricating bearings and non-sticking stop-cocks.

Teflon coated cookware

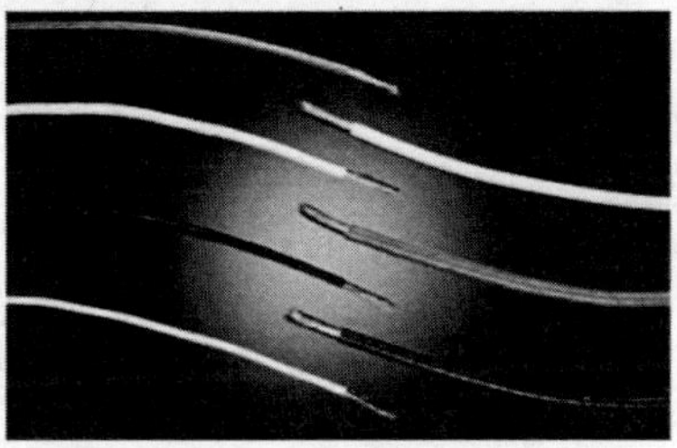

Teflon insulated wires

11.4.6 Polyamides

Polyamides are synthetic polymers that include repeating amide groups and have a structure like that of polypeptides.

Nylon: Nylon is a synthetic polymer that is commonly used in the production of various materials. Nylons refer to synthetic polyamides used in the formation of fibers.

11.4.6.1 Nylon 6

Preparation: Nylon 6 can be synthesized through the process of self-condensation of amino caproic acid. Nylon-6 is synthesized through the process of ring opening polymerization of caprolactam. Caprolactam undergoes ring-opening polymerization when subjected to heating at approximately 533 K in a nitrogen environment for a duration of around 4-5 hours.

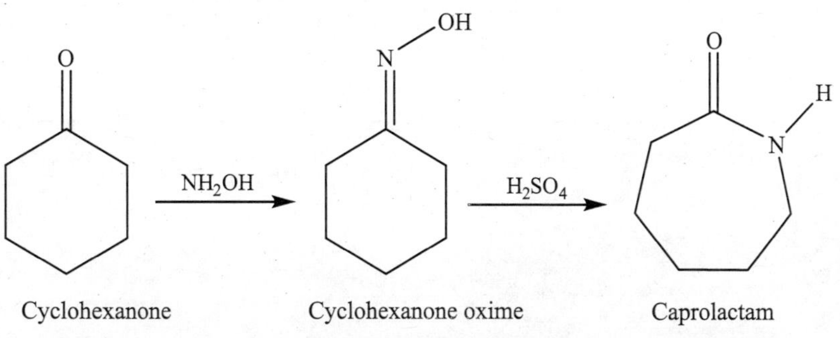

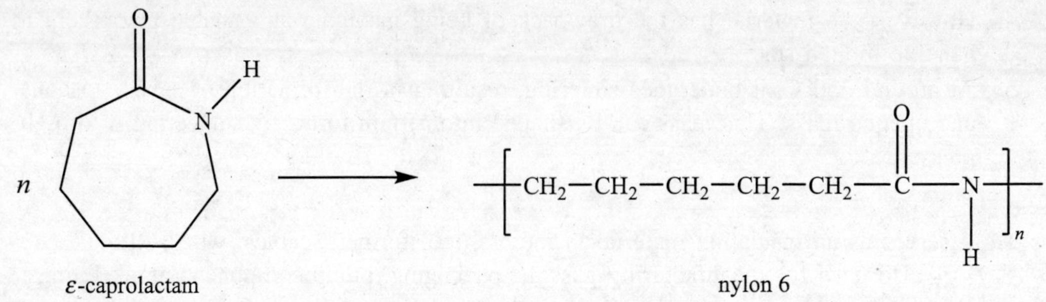

11.4.6.2 Nylon-6, 6

Nylon-6, 6 is synthesized through the polymerization of Hexamethylene diamine with Adipic acid in a 1:1 ratio. The commercial production of adipic acid involves the oxidation of cyclohexane, which is obtained from either benzene or petroleum.

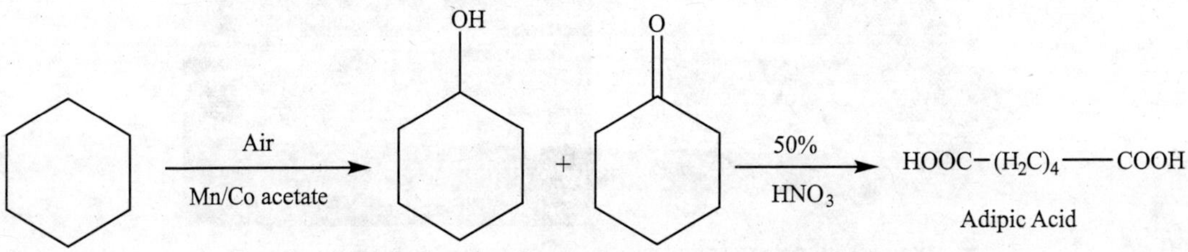

Hexamethylene is synthesized through the ultimate hydrogenation of adiponitrile.

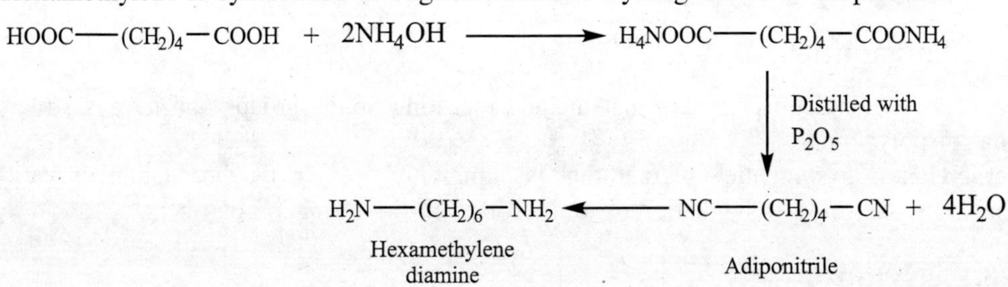

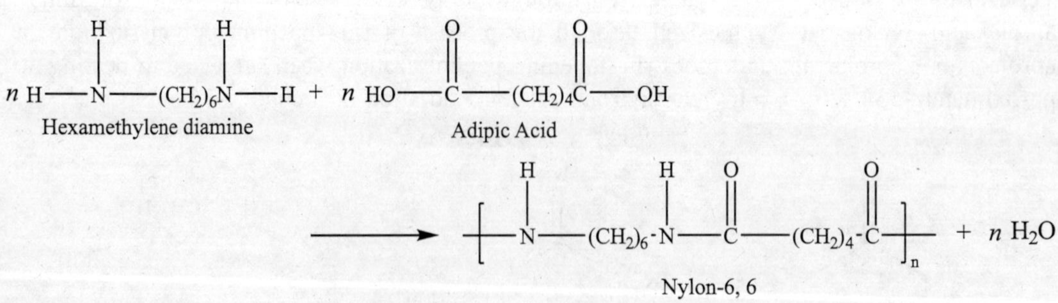

11.4.6.3 Nylon 6, 10

Nylon 6, 10 is synthesized through the reaction of 1, 6-diaminohexane and adipoyl chloride in the presence of NaOH, which serves to eliminate the produced HCl.

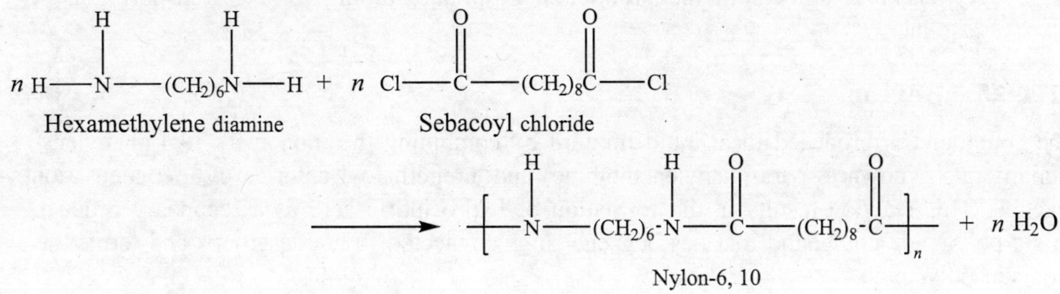

11.4.6.4 Nylon-11

Nylon-11 is synthesized through the self-condensation of ω-aminoundecanoic acid.

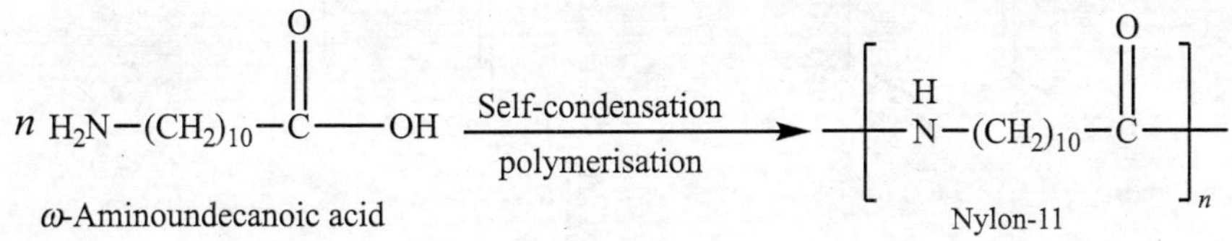

Properties

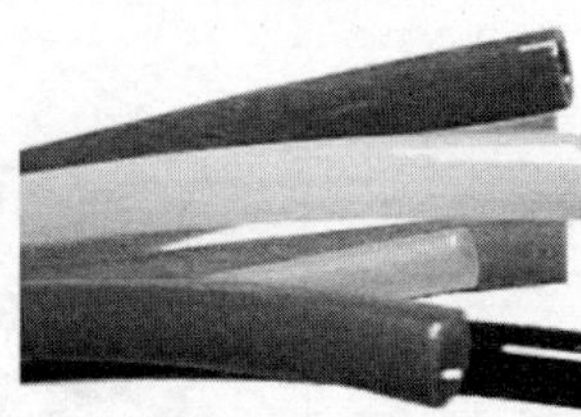
Nylon tubing

1. They exhibit favorable mechanical characteristics.
2. These materials provide both abrasion and chemical resistance.
3. They possess the ability to lubricate themselves.
4. All nylons have hygroscopic qualities, which can lead to a decrease in mechanical properties and an increase in part dimensions.
5. Oxidizing chemicals such as hydrogen peroxide and potassium permanganate have the ability to break down the fibers.

Uses

1. Nylon 6, 6 is predominantly utilized in the production of fibers that are commonly employed in the manufacturing of socks, undergarments, carpets, and similar products.
2. Nylon 6, 6 is commonly utilized in mechanical engineering for various applications such as gears, bearings, bushes, and cans.
3. Nylon 6, 6 is utilized in the jacketing of electrical equipment to offer a durable and resistant outer covering that safeguards the primary electrical insulation from abrasion.
4. Nylon 6 is primarily utilized in the production of tire cords.
5. Nylon 6, 10 is appropriate for producing monofilaments utilized in bristles, brushes, and similar applications.

6. Reinforced with glass fibers Nylon polymers are utilized in the enclosures and casings of household items, the radiator components of automobiles, and as coil formers for relays.
7. Nylon 11 is utilized in the production of pliable tubing for the transportation of gasoline.

11.4.6.5 Kevlar

The compound is produced in a liquid mixture by combining the monomers 1, 4-phenylene-diamine (also known as para-phenylenediamine) and terephthaloyl chloride by a condensation reaction. This reaction results in the formation of hydrochloric acid as a secondary product. It is a polyamide compound that has benzene rings connected to amide groups and termed as aromatic polyamide.

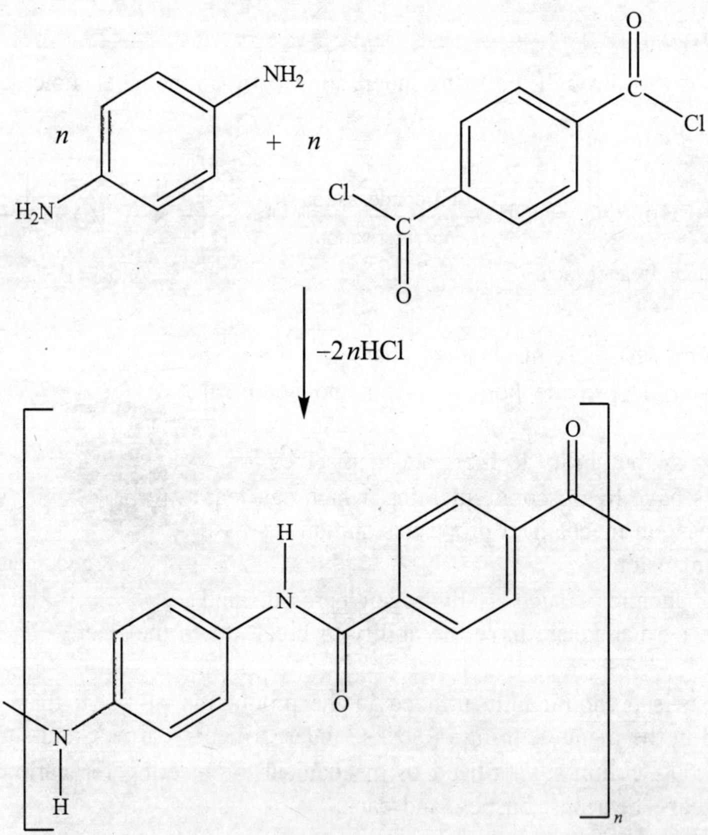

Properties

1. It exhibits exceptional thermal stability and remarkable flexibility.
2. It possesses extraordinary strength, being 10 times stronger than aluminum and 5 times stronger than steel.

Uses

1. This is employed in the aerospace and aircraft sectors.
2. This is very useful in the production of ropes, cables, protective clothing, bulletproof vests, helmets, car parts, and other related items.

Bulletproof vest Combat helmet

11.4.7 Polyurethane

They are generated through the chemical reaction between an isocyanate and a polyol that contains, on average, two or more hydroxyl groups. This reaction takes place either in the presence of a catalyst or by being activated with ultraviolet radiation.

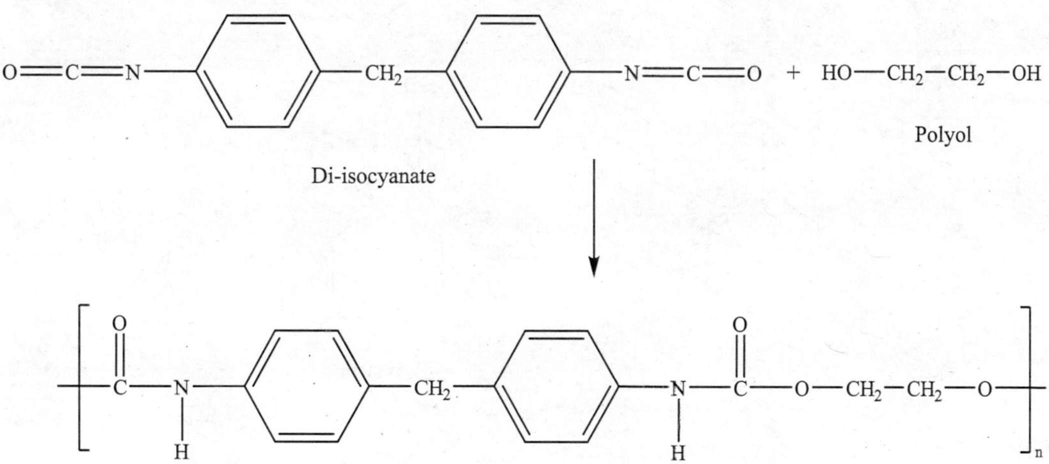

Polyurethane

Properties

1. It has the ability to be expanded like polystyrene, but unlike thermocoal, it possesses a soft and spongy texture, sometimes referred to as 'U'-foam.
2. It has a low heat conductivity, low viscosity, and excellent abrasion resistance.
3. The presence of aromatic groups, long alkyl chains, branching, cross-linking, and the degree of secondary bonding forces significantly affect mechanical qualities such as elongation break, rip strength, and tensile strength.

Uses

1. It serves as an insulating material in freezers. Because of its porous texture, it is utilized in the production of pillows and mattresses.
2. A layer of polyurethane is placed onto the leather, enhancing its aesthetic and overall quality. Consequently, the bags created from this material become visually appealing, while the shoes offer a comfortable sensation.
3. Additionally, it is employed in the production of chairs and other similar items.

11.4.8 Polystyrene

It is synthesized by the process of polymerization of styrene and is the most extensively utilized kind of plastic.

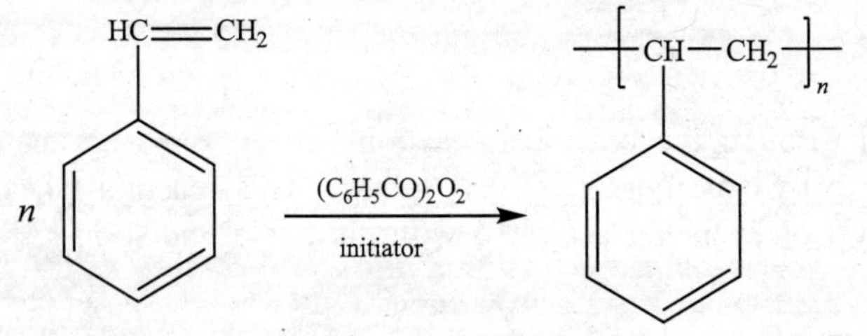

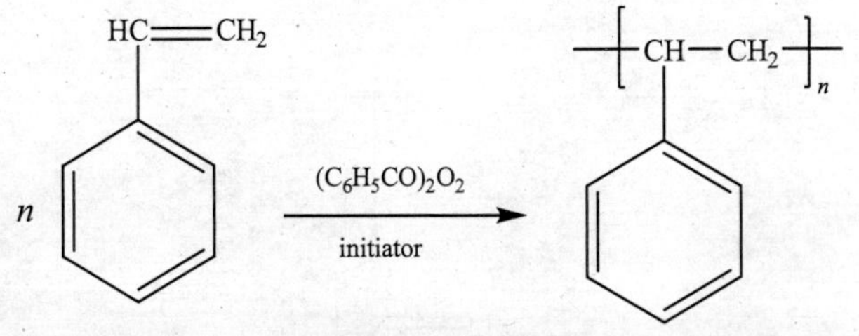

Styrene Polystrene

Properties

1. It is a rigid, highly clear, and durable resin.
2. At ambient temperature, the thermoplastic polymer exists in a solid state, but it transitions into a flowing state when exposed to temperatures over 100°C. It reverts to a hard state upon cooling.
3. Polystyrene exhibits low solubility in water.

Uses

1. In the medical field, it is utilized for the sterilization of test tubes, diagnostic components, and other medical devices.

2. This material is utilized in the production of automotive components such as knobs, instrument panels, and sound insulation foam.
3. Polystyrene foodservice packaging effectively preserves the freshness of food for an extended duration and is more cost-effective compared to other options.
4. It is employed in the packaging of consumer items, such as DVD cases and egg cartons, to safeguard against deterioration or harm.
5. It offers thermal insulation and is employed in refrigerators, freezers, and similar applications.
6. This material is utilized in housing for all IT equipment, including televisions, computers, and so on.

11.5 EPOXY RESIN

Epoxy resins are derived by the polycondensation of epichlorohydrin with polyphenols or polyalcohols.

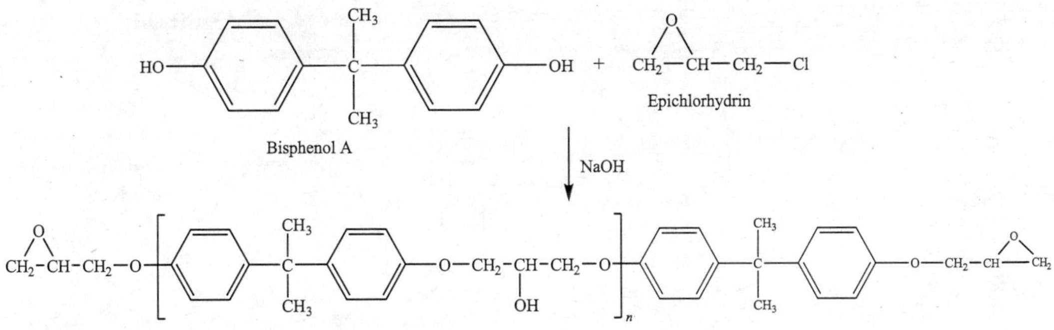

Properties and Uses

1. Enhanced adhesive characteristics (the capacity to adhere to the reinforcing or core material)
2. Adaptability
3. Chemical inertness refers to the lack of reactivity of a substance with other chemicals.
4. High temperaturc stability
5. Exceptional mechanical characteristics, notably in terms of strength and stiffness.
6. Enhanced durability against wear and micro cracking
7. Mitigated deterioration caused by water infiltration (reduction of characteristics resulting from water penetration)
8. Enhanced resistance to osmosis (deterioration of the surface caused by water permeability)

11.6 PHENOLIC RESIN

Phenolic resin is a type of synthetic polymer that is derived from phenol and formaldehyde. Phenol-formaldehyde resin is alternatively referred to as phenolic resin. The synthesis of synthetic resins involves the reaction between phenol and formaldehyde in the presence of an acid-alkaline catalyst in an aqueous solution. Under the influence of an acid catalyst, the first reaction results in the production of *o*- and *p*-hydroxyl methyl phenol, which forms a linear polymer.

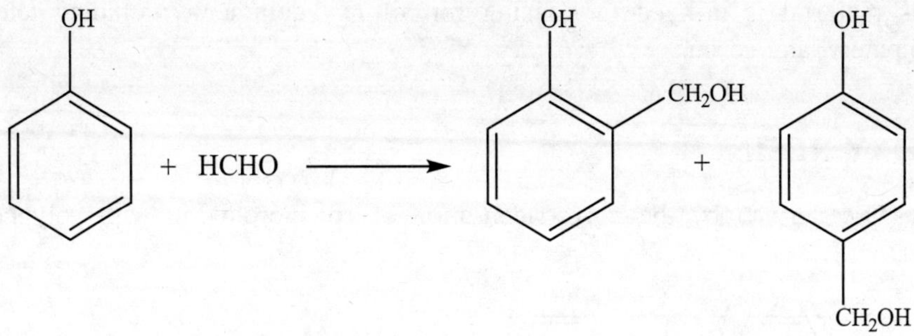

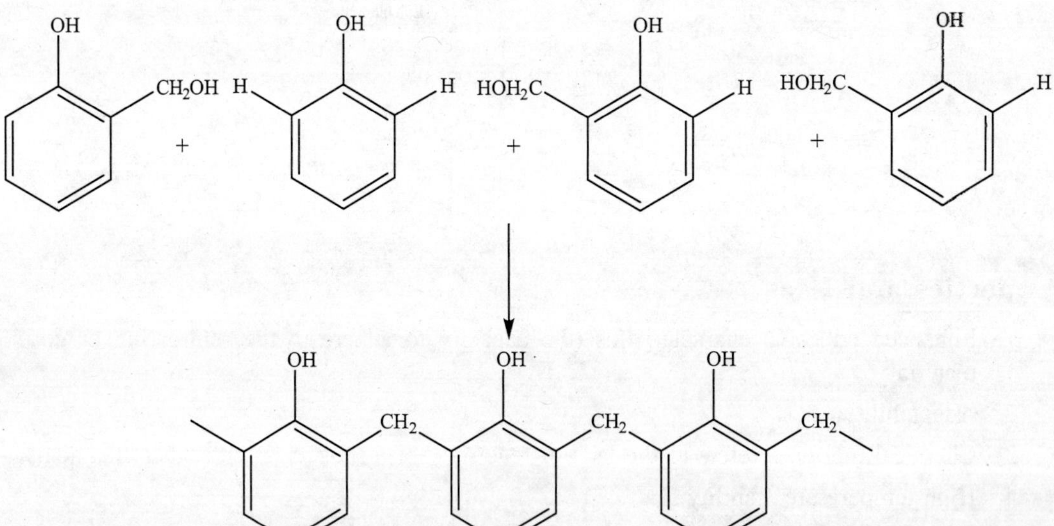

Novolac

In the second stage, a minute quantity of hexamethylene tetramine is introduced. Under the influence of high temperature, the polymer chain elongates and a three-dimensional resin is generated, resulting in the formation of crosslinked structures. The soluble novolac undergoes a chemical transformation to become the insoluble Bakelite.

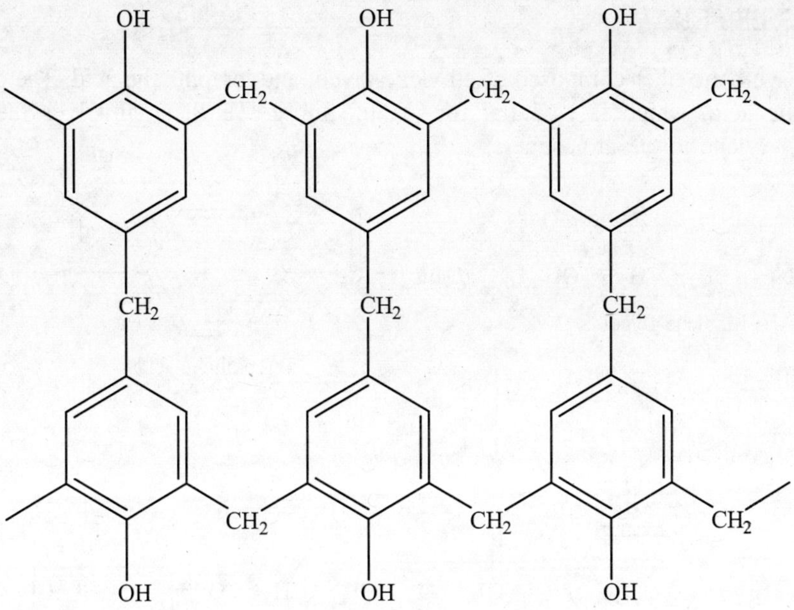

Bakelite

Properties

1. These phenolic resins are rigid, hard, and water resistant.
2. They are resistant to non-oxidising acids, organic solvents but are susceptible to alkalies.
3. Solubilities and melting point of the resins gradually change with rise of molecular weight.
4. These resins possess electric insulating properties.

Uses

1. It has extensive use as a replacement for metal in situations where great tensile strength is not required.
2. It has the ability to replace glass as an inert substance.
3. It is suitable for fabricating insulator components such as switches, plugs, and heater handles.
4. It can be shaped to form cabinets for televisions, radios, and telephone components.
5. It serves as an adhesive and is also utilized in paints and varnishes. Additionally, it functions as a cation exchanger resin for water softening and finds application in the paper industry for propeller shafts.

11.7 POLYESTER

Terylene is a synthetic fiber composed of ethylene glycol and terephthalic acid. The production of Terephthalic acid, which is essential for manufacturing Terylene, involves the catalytic oxidation of *p*-xylene in the atmosphere.

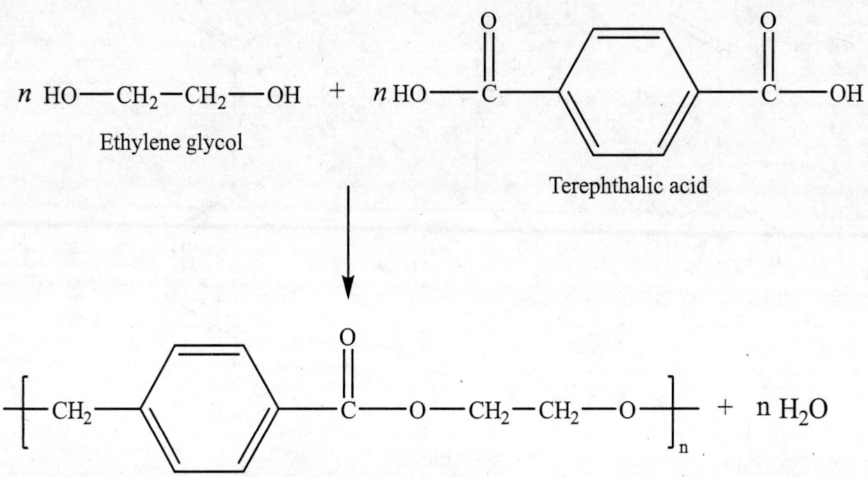

Polyethylene terephthalate

Properties

1. This substance is colorless and stiff.
2. This substance has high resistance to mineral and organic acids, but its resistance to alkalis is comparatively lower.
3. This substance has hydrophobic properties. The high melting point of this substance can be attributed to the presence of an aromatic ring.

Uses

1. Its primary application lies in the production of synthetic fibers.
2. It has the ability to be combined with wool and cotton to enhance its functionality and resistance to wrinkles.
3. Polyethylene terephthalate film is also utilized in electrical insulation.

11.8 SILICONE RESINS

Silicone resins consist of a silicone-oxygen structure with organic radicals bonded to silicon atoms. Consequently, their organization is as follows:

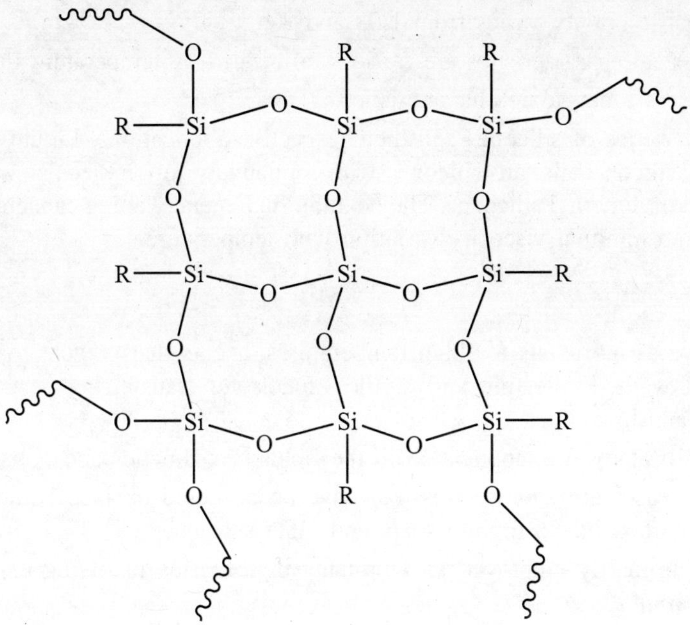

Where R= alkyl or phenyl radical.

Preparation

The process involves the reaction of silicon with either alkyl halide or silicon halide using a Grignard reagent. The resulting product is subjected to fractional distillation in order to obtain various organo-silicon chlorides. These chlorides are further polymerized using meticulously regulated hydrolysis.

1. Dimethyl-silicon dichloride is a compound that has two functional groups and is capable of producing polymers with extremely long chains.
2. Trimethyl silicon chloride is classified as 'monofunctional' and hence acts as a 'chain-stopper'. Consequently, this is employed in specific ratios to restrict the length of the chain.
3. Monomethyl silicon chlorides are considered 'trifunctional' as they undergo cross-linking to form the final polymer. The fraction utilized in polymerization directly influences the extent of cross-linking achievable.

Properties

1. Silicones can exist in various physical states, such as liquids, viscous liquids, semi-solids, rubber-like substances, or solids, depending on the relative amounts of different alkyl silicon halides utilized in their production.
2. Due to the presence of silicon-oxygen bonds, these materials have exceptional stability at elevated temperatures, excellent resistance to water, and great resistance to oxidation. However, their chemical resistance is often inferior to that of other types of plastics.

3. Their specific gravity varies from 1.03 and 2.1.
4. Their physical characteristics are far less influenced by temperature fluctuations.
5. They possess a non-toxic characteristic.
6. Various varieties of silicones and their respective applications: Liquid silicones, often known as silicone oils, are silicones with a relatively low molecular weight, typically composed of dimethyl silicones. They exhibit high metal wetting capability, low surface tension, and minimal viscosity variation with temperature.

Uses

1. They serve as lubricants for high temperatures, agents that prevent foaming, coatings that repel water for leather and textiles, media for transferring heat, and fluids for damping and hydraulics.
2. Additionally, they find application in the fields of cosmetics and polishes.
3. Silicone greases are silicone oils that have been altered by incorporating fillers such as silica, carbon black, lithium soap, and other substances.
4. They are primarily employed as lubricants in scenarios involving extreme high and low temperatures.

11.9 RUBBERS (ELASTOMERS) POLYMER

Elastomers are a type of high polymers that have the ability to undergo significant elongation (ranging from 500 to 1000%) when subjected to stress. However, they are able to fully return to their original size once the stress is released. Elastomers possess the characteristic of elasticity. This phenomenon occurs as a result of the helical configuration of elastomers.

Characteristics of Elastomers

Elastomers possess a structure that is both linear and strongly coiled. The intermolecular cohesive forces between the chains are insignificant due to the hydrocarbon nature of the molecules. They possess a significant level of flexibility. They have the ability to retract quickly.

Rubbers, sometimes referred to as Elastomers, are a type of high polymers that has elastic characteristics above 300%. Natural rubbers are composed of latex, a dispersion of isoprene. Throughout the treatment process, the isoprene molecules undergo polymerization, resulting in the formation of elongated and coiled chains of *cis*-polyisoprene. Natural rubber is derived from the latex of many botanical species.

Latex

Hevea brasiliensis, a plant, produces a milky white fluid that is released when the plant's stem is severed. The latex is diluted by adding water. Next, a solution of either acetic or formic acid is introduced into the mixture, with a ratio of 1 kilogram of acid per 200 kilograms of latex, in order to create the coagulum. This is processed to get either crepe rubber or smoked rubber.

Hevea brasiliensis (Rubber Tree)

Natural Rubber

The primary source of natural rubber is *Hevea brasiliensis*, a species of rubber trees cultivated in tropical nations. The rubber obtained from this species is commonly referred to as "Hevea" rubber. Rubber is kept as latex, a milky white fluid, in the cambium layer beneath the bark of the rubber tree.

Isoprene (2-methyl-1, 3-butadiene) forms a high polymer known as natural rubber. The substance is a hydrocarbon polymer with the chemical formula $(C_5H_8)_n$. Polyisoprene is present in two distinct geometric isomers, namely the *cis* and *trans* forms. Natural rubber possesses a soft texture and exhibits a *cis*-configuration, whereas 'Gutta percha' or 'Ballata' displays a *trans* configuration.

The raw rubber powder undergoes catalytic reaction with H_2 gas, resulting in the decolorization of bromine water, the formation of ozonide with ozone, and reaction with HCl. Each C_5H_8 unit in the rubber molecule undergoes a reaction with one molecule of Br_2, O_3, and HCl, resulting in the formation of a single double bond. Isoprene molecules undergo head and tail linkage during the process of linkage. Rubber is susceptible to the corrosive effects of powerful alkaline substances and oxidizing agents. Nitric acid (HNO_3) has a corrosive effect on rubber.

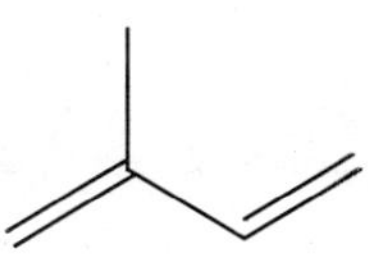

Isoprene

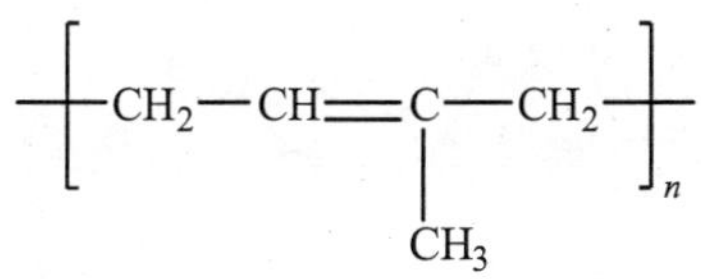

Isoprene

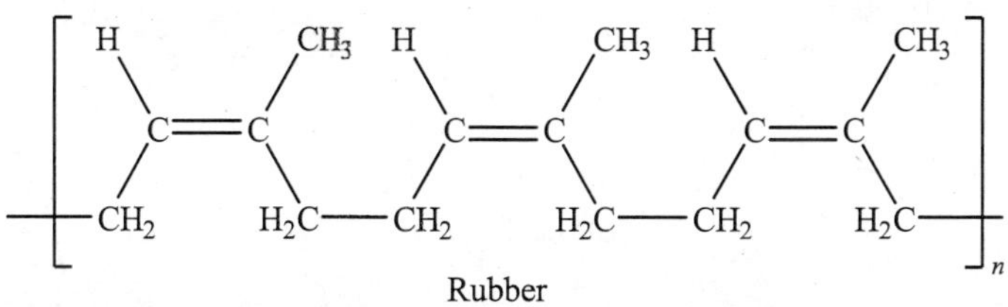

Rubber

Deficiencies of Natural Rubber

Raw natural rubber lacks practical utility. The presence of numerous intrinsic weaknesses is the cause.

1. It exhibits high hardness and brittleness at low temperatures, while displaying softness and stickiness at high temperatures.
2. The substance possesses a smooth texture and can dissolve easily in various hydrocarbon solvents as well as carbon disulphide (CS_2).
3. It exhibits a great capacity for water absorption.
4. It is a viscous substance, which poses challenges in terms of handling. It is easily susceptible to assault by both acids and alkalies.
5. It has a poor level of tensile strength.
6. It has little resistance to abrasion.
7. It is susceptible to oxidative destruction.

11.10 VULCANIZATION

Vulcanization refers to the process of treating rubber or similar materials with sulphur or other chemicals to improve its strength, elasticity, and durability.

The process of vulcanization was discovered by Charles Goodyear in 1839. Vulcanization is a thermal process in which raw rubber is heated at temperatures ranging from 100 to 140 degrees Celsius, along with the addition of sulphur, for a duration of 1 to 4 hours. Sulphur chemically reacts with the double bonds of various rubber molecules, resulting in the formation of cross-links between the chains.

The extent of vulcanization is contingent upon the quantity of sulphur employed. The tire business utilizes sulphur at a concentration of 3–5%. When the sulphur content is raised to 30%, it results in the production of a durable and inflexible rubber known as "ebonite".

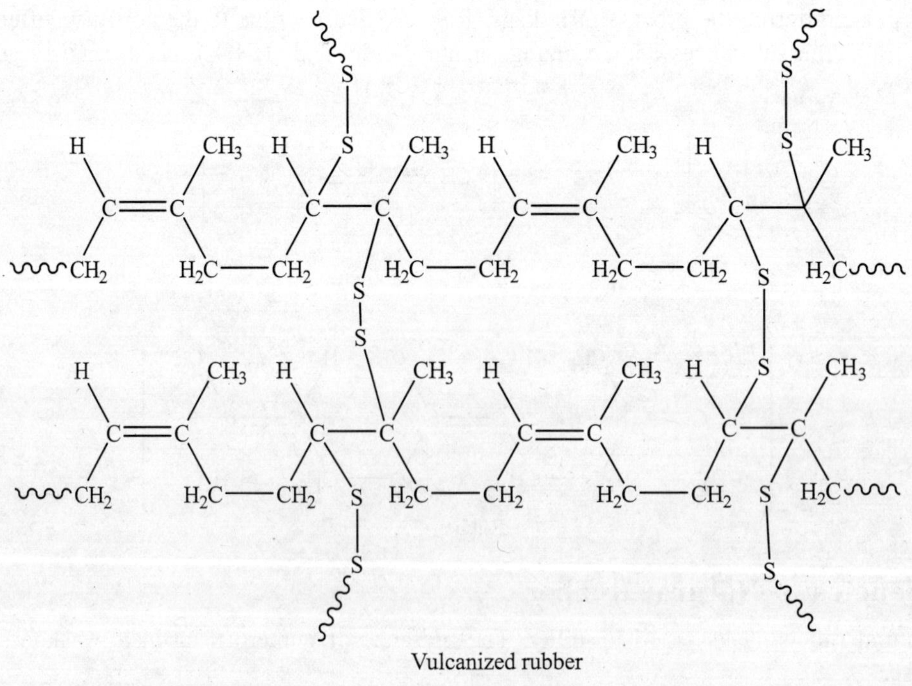

Vulcanized rubber

Unvulcanized rubber products exhibit low tensile strength and possess a texture akin to that of chewing gum. Currently, both natural and synthetic rubber undergo the process of vulcanization. The vulcanization process is predominantly carried out using sulphur in natural rubber, ethylene-propylene-diene (EPDM), butyl rubbers, and nitrile rubber, accounting for 90% of all cases.

Advantages of Vulcanization

1. Vulcanized rubber exhibits excellent tensile strength. It possesses a tensile strength that is tenfold higher than that of raw rubber.
2. It has exceptional resilience, meaning that objects manufactured from it regain their former shape once the deforming load is removed.
3. It has superior resistance to moisture, oxidation, and abrasion.
4. It has significantly greater durability than untreated rubber.
5. The useful temperature range of this material is wider (–40 to 100°C) compared to the useful temperature range of raw rubber (10–60°C).
6. Ebonite is a superior electrical insulator.
7. This material exhibits resistance to organic solvents such as petrol, benzene, and CCl_4. Additionally, it is resistant to fats and oils, but it may undergo swelling in their presence.
8. It exhibits minimal adhesiveness.
9. It has a modest level of flexibility. The property of the material is contingent upon the degree of vulcanization. Ebonite exhibits minimal flexibility.
10. Manipulating vulcanized rubber to achieve the desired forms is a straightforward process.

Rubber Compounding

The compounding of plastics refers to the mechanical mixing of different additives with polymers, without any chemical reaction, in order to enhance specific attributes of the final product. It has the following attributes.

1. The additives are integrated into the polymer to create a uniform blend.
2. Additives serve the purpose of altering the characteristics of the polymer.
3. To incorporate novel characteristics into the manufactured items.
4. To facilitate the manipulation of polymers during the manufacturing process.
5. To enhance the aesthetic appeal and provide vibrant hues to the polymer items. The products should be affordable and adequately fulfill their intended purposes.
6. The materials must exhibit stability during both processing and service circumstances. They must be devoid of any taste or odor.
7. The finished product, after processing, must be non-toxic for food grade application.
8. They should not have a negative impact on the characteristics of the polymer.
9. The main additives utilized in compounding include: (a) Fillers, (b) Plasticizers, (c) Lubricants, (d) Accelerators, (e) Coloring agents, and (f) Vulcanizing agents.

11.11 BUNA-S OR STYRENE RUBBER

Buna-S rubber is a highly significant form of synthetic rubber that is created through the copolymerization of butadiene (constituting about 75% by weight) and styrene (constituting 25% by weight).

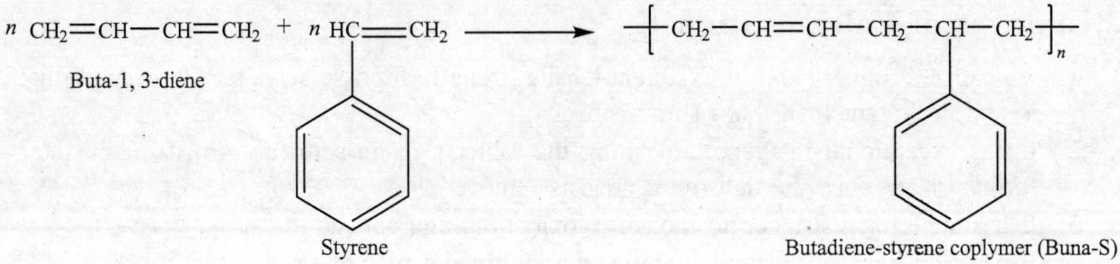

Properties

1. Styrene rubber has similar processing characteristics and quality of finished goods as natural rubber.
2. It exhibits exceptional resistance to abrasion, can withstand heavy loads, and has excellent resilience.
3. Nevertheless, it undergoes rapid oxidation, particularly in the presence of even small amounts of ozone found in the atmosphere.
4. Furthermore, it expands when exposed to oils and solvents.
5. It can undergo vulcanization using either sulphur or sulphur monochloride (S_2Cl_2), similar to natural rubber.
6. Nevertheless, it necessitates a reduced amount of sulphur, yet a greater quantity of accelerators for the process of vulcanization.

Uses

1. Primarily employed in the production of automobile tires.
2. These elastomers are also utilized in many applications such as floor tiles, shoe soles, gaskets, footwear components, wire and cable insulations, carpet backing, adhesives, and tank linings.

11.12 NITRILE RUBBER OR GR-A OR BUNA-N OR NBR

Preparation: The emulsion system is used to copolymerize butadiene and acrylonitrile, resulting in its preparation.

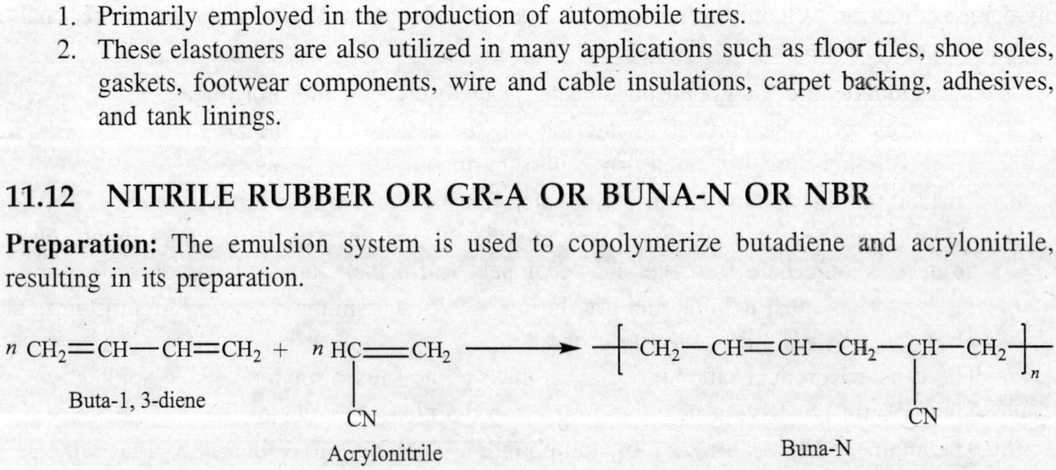

Properties

1. Nitrile rubber has lower resistance to alkalis compared to natural rubber due to the presence of a cyano group.
2. It demonstrates exceptional resistance to oils, chemicals, and aging caused by sunlight. Increasing the acrylonitrate proportion in nitrile rubber enhances its resistance to acids, salts, oils, solvents, and other substances. However, the ability to withstand low temperatures is compromised.
3. Compared to natural rubber, vulcanized nitrile rubber exhibits greater heat resistance and can withstand high temperatures.
4. It has excellent resistance to abrasion, even when subjected to immersion in gasoline or lubricants.

Uses

1. It is used for manufacturing conveyor belt, tank lining and gasket production.
2. Rollers for printing applications, foams with resistance to oil.
3. Automotive components and components for high-altitude aircraft flexible tubing and bonding agents.

11.13 BUTYL RUBBER

Preparation

Butyl rubber is synthesized using the cationic copolymerization of isobutene and a tiny quantity (1–5%) of isoprene, started by aluminum chloride.

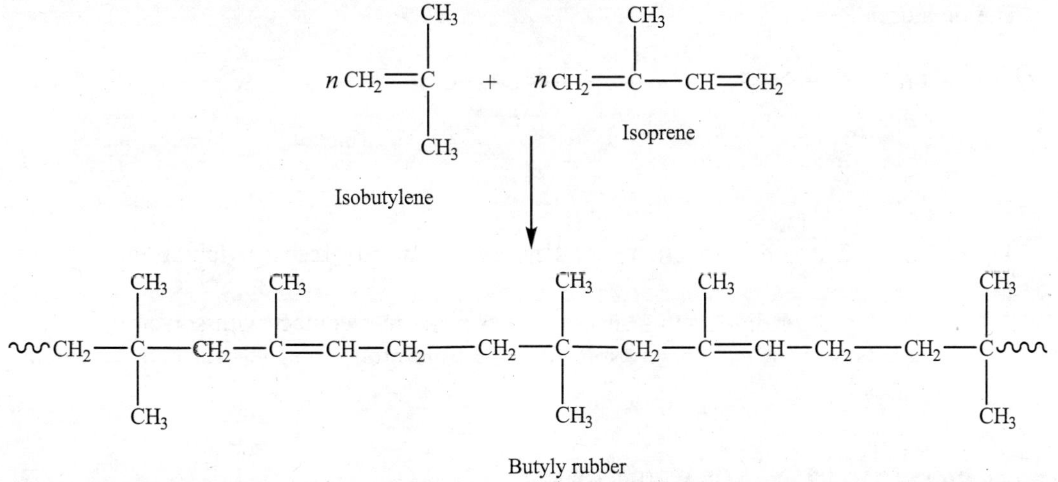

Properties

1. Under typical circumstances, it exists in an amorphous state, but it undergoes crystallization when subjected to stretching.

2. It exhibits high resistance to oxidation as a result of its low level of unsaturation. Because of its extremely low unsaturation, the material is capable of undergoing vulcanization, however it has limited potential for hardening.
3. It can undergo degradation due to heat or light, resulting in the formation of sticky substances with low molecular weight. Therefore, it is necessary to stabilize it.
4. Due to its non-polarity, it exhibits excellent electrical insulating qualities.
5. When compared to natural rubber, it has exceptional resistance to the passage of air and other gases.
6. It retains its flexibility even at temperatures as low as –50°C due to its inability to crystallize upon cooling.
7. Due to its hydrocarbon composition, butyl rubber can dissolve in hydrocarbon solvents such as benzene. However, it exhibits strong resistance to polar solvents like alcohol and acetone, as well as to aging agents such as HCl, HF, HNO_3, and H_2SO_4.

Uses

1. It is utilized for the insulation of high voltage lines and cables, as well as the inner tubes of automobile tires.
2. Conveyor belts designed for the transportation of food and several other goods.
3. Tank lining hoses.

11.14 THIOKOL

This substance is commonly known as polysulphide rubber (or Gr-P). The synthesis of this compound involves the condensation polymerization of sodium polysulphide (Na_2S_x) with ethylene dichloride.

$$n\ Cl{-}CH_2CH_2{-}Cl + Na_2S_4 \xrightarrow{Mg(OH)_2} \left[CH_2{-}CH_2{-}S{-}S{-}S{-}S \right]_n + 2NaCl$$

Thiokol

Uses

1. It is utilized in the production of oil hoses, chemically resistant tubing, and engine gaskets.
2. It is employed for diaphragms and seals that come into contact with solvents, printing rolls, containers for transferring solvents, and solid propellant fuels for rockets, among other uses.

11.15 POLYURETHANE RUBBERS

Polyurethane, also known as isocyanate rubber, is created through the chemical reaction between polyalcohol and di-isocyanates.

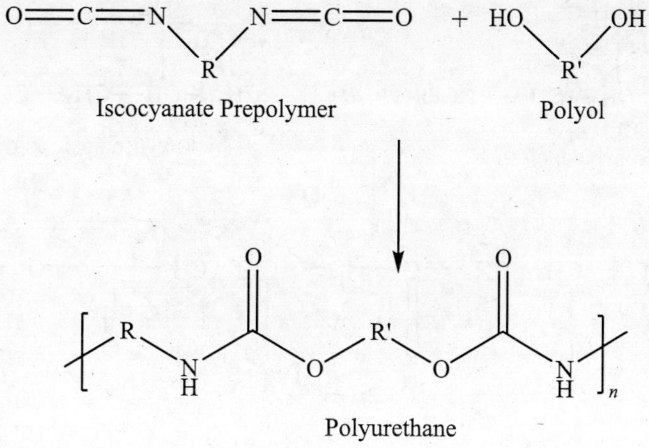

Properties

1. Polyurethanes have exceptional resistance to oxidation due to their saturated nature.
2. Additionally, they exhibit strong resistance to numerous organic solvents, but are susceptible to damage from acids and alkalis, particularly when they are concentrated and heated.
3. The polyurethane foams has the characteristics of being lightweight, durable, and capable of withstanding high temperatures, abrasion, chemical exposure, and weathering.

Uses

It is primarily used for surface coatings and the production of foams and spandex fibers.

11.16 BIODEGRADABLE AND NANOPARTICLE POLYMER

Biodegradable Polymer

These polymers are capable of undergoing decomposition in the presence or absence of oxygen, due to the activity of microorganisms or enzymes. This polymer undergoes degradation and experiences a loss of its original structural integrity. The components involved in its development include starch, cellulose, and polyesters. Aliphatic polyesters are the predominant polymers in this category.

11.16.1 Poly β-hydroxybutyrate–co-β-hydroxyvalerate (PHBV)

It is a type of aliphatic polyester that can naturally decompose. The compound is formed via the combination of 3-hydroxy butanoic acid and 3-hydroxy pentanoic acid, with the monomers being bonded together through an ester linkage.

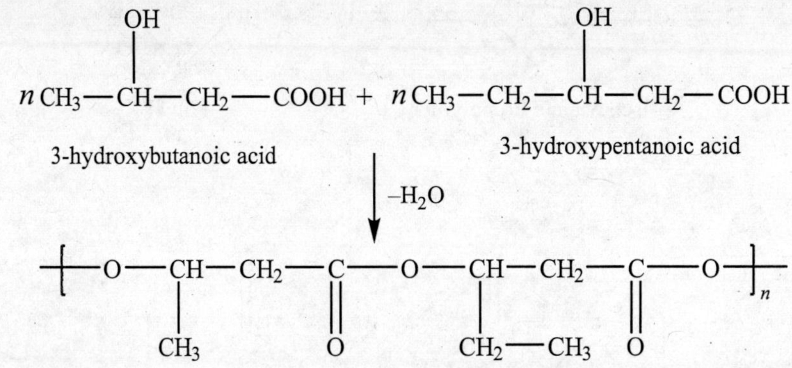

PHBV

Properties and Uses

It undergoes decomposition to produce carbon dioxide and water. It possesses a brittle characteristic and can be utilized in pharmaceutical production and bottle making.

11.16.2 Nylon 2–nylon 6

It is a polyamide copolymerisation of glycine ($H_2N–CH_2–COOH$) and amino caproic acid ($H_2N–(CH_2)_5–COOH$).

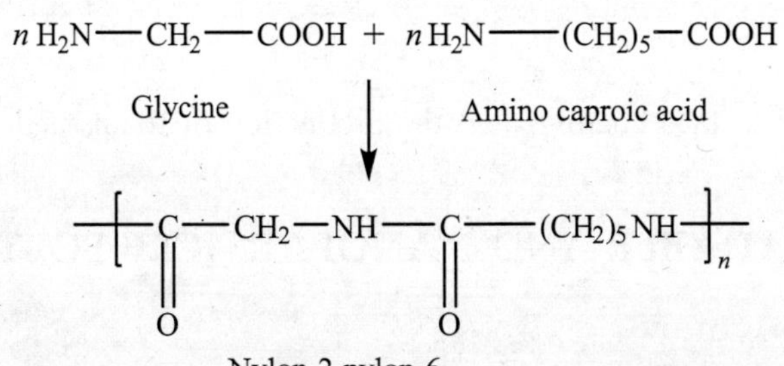

Nylon-2-nylon-6

These are utilized as orthopedic devices, implants, sutures, and drug release matrices.

11.16.3 Polyhydroxybutyrate (PHB)

The compound is created through the process of hydroxybutyric acid (3-hydroxy butanoic acid) molecules condensing together. It is a thermoplastic that can be melted and processed, and it is semi-crystalline. This thermoplastic is made from microorganisms through the fermentation of renewable carbohydrate feedstocks. PHB is a fully biodegradable and biocompatible plastic that serves as an appealing environmentally-friendly substitute for fossil-based thermoplastics like PE and PP.

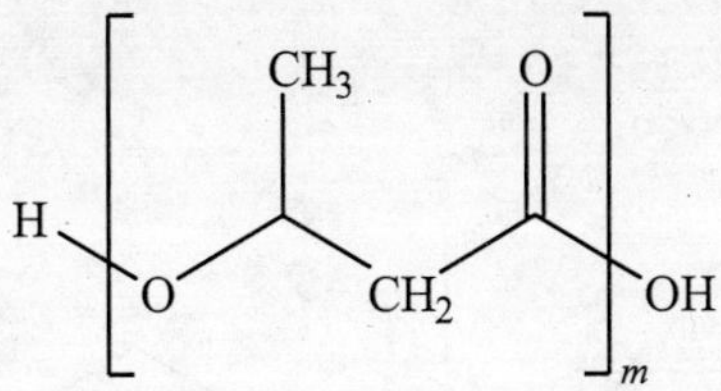

Properties and Uses

1. This substance is not soluble in water and is quite resistant to breaking down through hydrolysis. This characteristic sets PHB apart from the majority of other biodegradable polymers currently on the market, as they tend to either dissolve in water or are sensitive to moisture.
2. High oxygen permeability.
3. Excellent ultraviolet resistance, but limited resistance to acids and bases.
4. This substance can dissolve in chloroform and other chlorinated hydrocarbons.
5. This material is biocompatible, making it acceptable for use in medical applications.
6. Polypropylene sinks in water, while promoting its anaerobic biodegradation in sediments.
7. Nonhazardous.

11.16.4 Polyglycolide or Poly(glycolic acid) (PGA)

It is a type of polymer that can be broken down by natural processes and is made from a simple, straight chain polyester. Glycolic acid can be used as a starting material for the synthesis of this compound using either polycondensation or ring-opening polymerization.

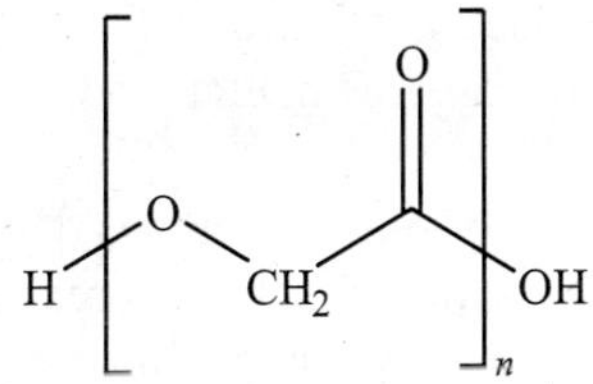

11.16.5 Polycaprolactone (PCL)

Polycaprolactone (PCL) is a biodegradable polyester that has a relatively low melting point of approximately 60°C and a glass transition temperature of around –60°C. Poly(ε-caprolactone) (PCL) is synthesized through the ring-opening polymerization of ε-caprolactone, utilizing a catalyst such as stannous octoate. A comprehensive evaluation has been conducted on several catalysts used for the ring opening polymerization of caprolactone.

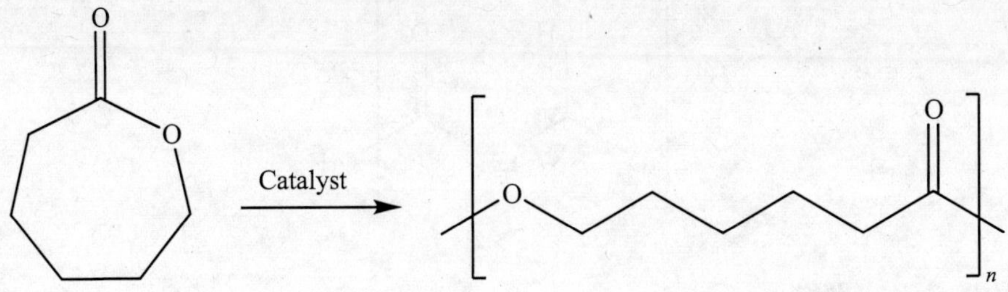

Polycaprolactone is primarily utilized in the manufacturing of specialized polyurethanes. Polycaprolactones confer excellent resistance to water, oil, solvent, and chlorine to the resulting polyurethane.

Uses of Biodegradable Polymers

1. Approximately 86% of plastic packaging is utilized for a single purpose before being disposed of, resulting in a continuous flow of waste that remains in rivers and landfills, emitting pollutants and causing harm to wildlife.
2. Conventional polymers, such as polyethylene and polypropylene, possess inherent durability and can endure for numerous years following disposal. They are suitable for products that necessitate a prolonged duration of use, such as plastic tables and chairs. However, they are unsuitable for applications where plastics are utilized for brief durations and subsequently discarded, such as packaging items.
3. In addition, plastics are frequently contaminated by food and other organic compounds, rendering the physical recycling of these materials unfeasible and generally unfavorable. Biodegradable polymers (BPs) can be discarded in specific bioactive settings where they will degrade due to the enzymatic activities of microorganisms such as bacteria, algae, and fungi.
4. The polymer chains can also be degraded through non-enzymatic mechanisms, such as chemical hydrolysis. Biofuels are frequently produced by the conversion of atmospheric carbon dioxide by plants. Biodegradation transforms them into carbon dioxide (CO_2), methane (CH_4), water, biomass, organic matter, and other natural compounds. Biological activities naturally recycle BPs.
5. Biodegradable polymers consist of polymer chains that undergo hydrolytic or enzymatic cleavage, leading to the formation of soluble breakdown products. Biodegradability is highly sought after in biomedical applications, as it guarantees the polymer's breakdown and removal from the body, eliminating the need for retrieval.
6. Biodegradable polymers are utilized in various fields—Methods for controlled and sustained release of drugs; Tissue engineering scaffolds are structures used to support the growth and development of new tissues Provisional prosthesis.

11.17 CONDUCTIVE POLYMERS OR INTRINSICALLY CONDUCTING POLYMERS (ICPS)

These substances are organic polymers with the ability to conduct electricity. These compounds can exhibit either metallic conductivity or semiconductor behavior. Conductive polymers offer a significant benefit in terms of their capacity to be processed, mostly through dispersion. Conductive polymers typically lack thermoplastic properties, meaning they cannot be easily shaped through heating and molding. However, similar to insulating polymers, they belong to the category of organic materials. While they possess excellent electrical conductivity, these polymers do not exhibit comparable mechanical qualities to other polymers that are already on the market. The electrical characteristics can be precisely adjusted through the utilization of organic synthesis technologies and sophisticated dispersion techniques.

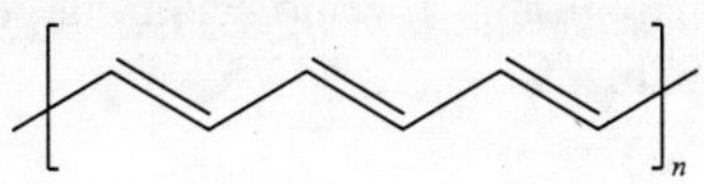

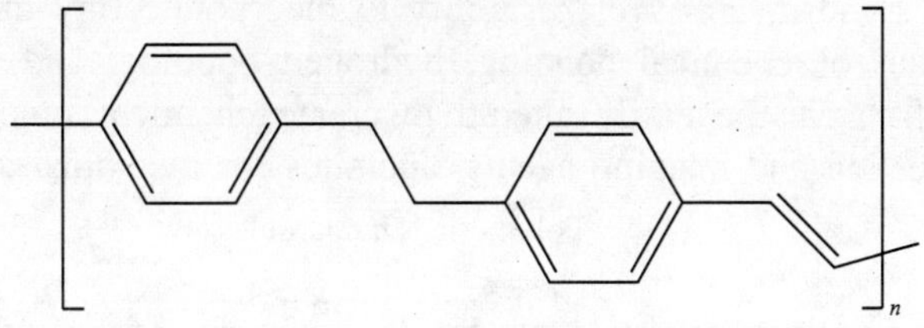

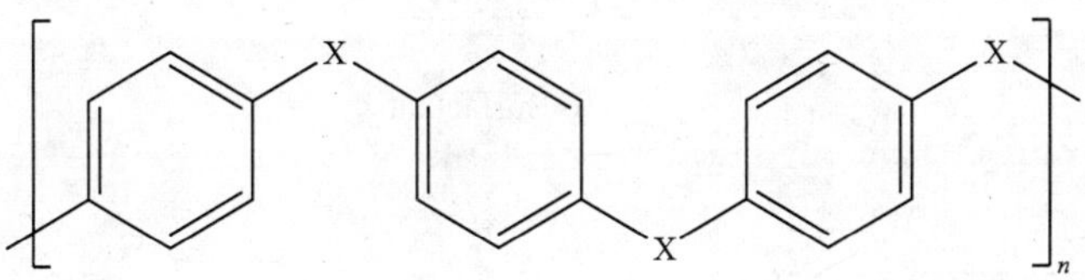

X = NH/N, S

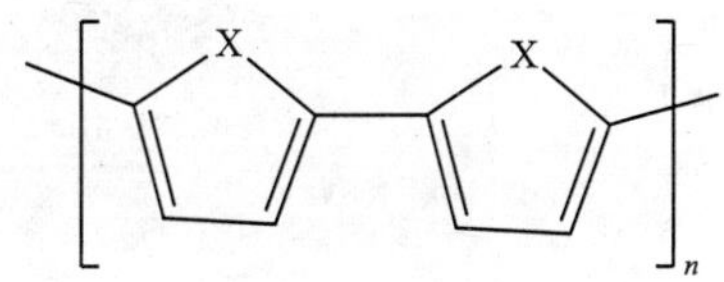

X = NH, S

Polypyrrole, with X representing NH, and polythiophene, with X representing S, are examples of conductive polymers. Similarly, polyaniline, with X representing NH, and polyphenylene sulphide, with X representing S, are also conductive polymers.

11.18 NANOPARTICLES

Nanoparticles can be designed with unique compositions, dimensions, configurations, and surface chemistries to facilitate innovative methodologies in various biological domains. The distinctive characteristics of nanoparticles and their interactions in biological environments also provide innovative and comprehensive methods for investigating fundamental biological inquiries.

Therapeutic carriers made from biocompatible and biodegradable polymers, known as polymeric nanoparticles (NPs), have undergone thorough investigation. Polymeric nanoparticles are created using block copolymers with varying degrees of hydrophobicity. The copolymers exhibit self-assembly, forming a micelle structure with a core-shell arrangement when placed in water. Polymeric nanoparticles have been developed to encapsulate both hydrophilic and hydrophobic tiny medicinal molecules, as well as proteins and nucleic acid macromolecules. The NP design enables precise and gradual delivery of medication at specific locations. Polymeric nanoparticles typically enhance the safety and effectiveness of the medications they transport. Investigating the functionalization of polymeric nanoparticles (NPs) with targeted ligands for enhanced drug delivery has been a significant research focus. This is because polymeric NPs possess the distinct advantage of being customizable before particle formation.

Dendrimers are a distinct category of polymeric nanoparticles. Dendrimers are highly-branched macromolecules composed of synthetic or natural building blocks such as amino acids, sugars, and nucleotides. The tree possesses a central nucleus, internal tiers of branches, and an outside covering. The diverse amalgamation of these constituents can produce dendrimers with precisely determined dimensions, morphology, and extent of branching. Dendrimers possess a distinctive structure, which enables their utilization as both sensors and carriers for drugs and genes.

Dendrimers have the ability to incorporate tiny molecules within the central cavities by means of chemical bonding, hydrogen bonding, and hydrophobic interactions. The outside surface can be easily altered to create chemical functional groups for molecular targeting, detection and imaging agents, and sites for medicinal attachment.

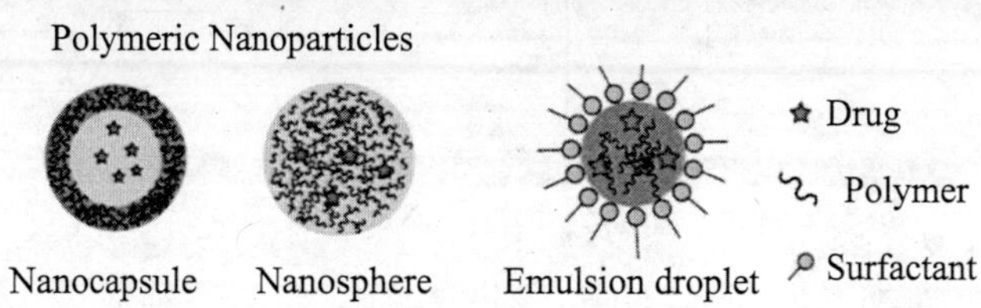

QUESTIONS

11.1 Distinguish between homopolymer and copolymer with one example each.

11.2 Arrange the following polymers in the increasing order of molecular forces
(a) Polyester (b) Polythene (c) Bakelite (d) Rubber

11.3 Explain the free radical mechanism of addition polymerisation.

11.4 Give the structure and name of the monomers of the following addition polymers. Give one use of each polymer.
(a) Teflon (b) Buna-S (c) Natural rubber (d) Neoprene
(e) Buna-N (f) polyvinyl chloride

11.5 What is meant by vulcanization of rubber?

11.6 With the help of suitable equations explain the formation of following polymers:
(a) Nylon 6 (b) Nylon 6,6 (c) Terelene (d) Glyptal
(e) Bakelite (f) Melamine formaldehyde polymer

11.7 What are biodegradable polymers? Give one example for it. Explain its method of preparation.

11.8 Define functionality of the monomer for the formation of polymer. Also explain why methanol do not polymerize.

11.9 Give the preparation and properties of Nylon 6, 6, PTFE, PET and Buna-N.

11.10 Differentiate between:
(a) Homopolymer and co-polymer
(b) Thermoplastic and Thermosetting polymers

11.11 Explain the reaction of Ziegler-Natta Catalyst? Explain the types of Tacticity and stereochemistry of linkages for the following structure:

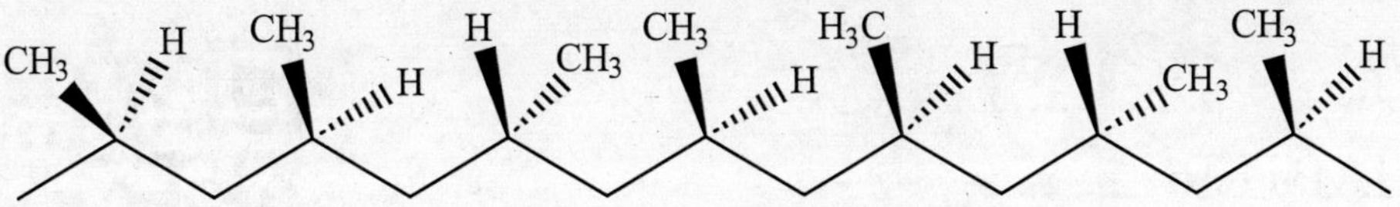

11.12 What are conducting polymers? How they are classified?

11.13 Distinguish between the following:
(a) Addition and condensation polymerization
(b) Anionic and cationic polymerization
(c) Thermoplastic and thermosetting resins
(d) LDPE and HDPE

11.14 What are the limitations of raw rubber? Explain the process of vulcanization.

11.15 Write the preparation, properties and uses of the following polymers:
(a) PMMA (b) Bakelite (c) Nylon 6, 6 (d) PAN

11.16 Discuss the preparation and uses of PTFE.

11.17 Differentiate between Chain Growth and Step Growth polymerization.

11.18 Outline the process of vulcanization of rubber. Describe the preparation, important properties and uses of Butyl rubber or Kevlar.

11.19 Giving examples write a brief note on conducting polymers.

11.20 Give the structure of vulcanized rubber.

11.21 Give two differences between addition and condensation polymers.

11.22 Write down the preparation (structure of monomers and polymers) and uses of:
(i) Nylon-6, 6 (ii) Dacron (iii) BUNA-N
(iv) Neoprene Rubber (v) Nylon-6

11.23 What are polymer composites? Discuss the classification and applications of polymer composites.

11.24 What is Biodegradable polymers? Discuss their importance.

11.25 Define functionality. What is the minimum functionality required for a compound to act as monomer?

CHAPTER 12

Glass

12.1 INTRODUCTION

Glass is a solid material that lacks a specific melting temperature and has a high viscosity, which inhibits crystallization. Upon the application of heat, glass undergoes a gradual softening process and eventually transforms into a liquid state. Glass is any substance that has solidified from a liquid condition without undergoing crystallization. The glass–liquid transition, also known as the glass transition, refers to the slow and reversible transformation that occurs in amorphous materials (or in amorphous portions within semicrystalline materials) when the temperature rises. This transition involves a change from a rigid and somewhat fragile "glassy" state to a more fluid or elastic one. A solid lacking a definite crystalline structure and undergoing a transition from a rigid to a more flexible state, known as a glass transition, is referred to as a glass. Vitrification refers to the process of achieving the reverse transition from a viscous liquid to the glass state by supercooling. The glass-transition temperature, T_g, is a parameter that defines the temperature range at which a material undergoes the glass transition. If a crystalline condition of the material exists, the temperature is always below its melting point, T_m.

Glass is an amorphous substance with a high level of hardness, brittleness, and transparency or translucency. It is formed by supercooling a liquid to a state of infinite viscosity. The substance is acquired from the fusion of a blend of various metallic silicates, often composed of sodium (Na), potassium (K), calcium (Ca), and lead (Pb). It lacks a distinct melting temperature, precise formula, or crystalline structure. It can be expressed as chemical formula $xR_2O.yMO.6SiO_2$ represents a compound where R represents an atom of a monovalent alkali metal such as Na or K, M represents an atom of a bivalent metal such as Ca, Pb, or Zn, and x and y are whole numbers. Thus, approximate composition of ordinary glass (called soda-lime glass) is $Na_2O.CaO.6SiO_2$. In some glasses, SiO_2 may be replaced by Al_2O_3. B_2O_3. P_2O_5, etc.

Edouard Benedictus

Edouard Benedictus, a polymath from France, is credited with the discovery of safety glass in 1903. He assembled "Triplex" by sandwiching a layer of Collodion between two panes of glass.

12.2 GENERAL PROPERTIES OF GLASS

Glass is a substance without a crystalline structure and lacks a specific temperature at which it transitions from a solid to a liquid state. It has the ability to absorb, reflect, or transmit light, and can achieve a high level of polish. It possesses excellent electrical insulation properties and is susceptible to the effects of alkalis. It remains unaffected by air, water, acids, and most chemical reagents, with the exception of hydrofluoric acid (HF), which transforms its silica into SiF_4. It has the ability to be moulded into complex forms, including products with intricate shapes. The material exhibits brittleness and undergoes softening with heating. Its lightweight nature can be attributed to its uniform internal structure, resembling that of liquids. Due to the absence of a crystal structure, there is no possibility for slippage to occur between planes. Therefore, glass exhibits a significant resistance to compression.

Window panes, dinnerware, and optoelectronics are just a few examples of its many functional, technical, and aesthetic applications. "Silicate glasses" derived from the chemical compound silica, also known as quartz or silicon dioxide, the main component of sand, are the most common and long-standing forms of man-made glass. Its main components are sand: 45.4 kg, soda ash (dense): 16 kg, salt cake: 4.5 kg, powdered coal: 0.2 kg, lime: 6.8 kg, cullet: 22.7 kg, and other materials: 0.5–1.0 kg.

12.3 COMPOSITION AND MAIN INGREDIENTS OF GLASS

1. Soda-lime glass, which is a type of silica-based glass, is commonly used for glazing and containers. It is made up of about 75% silicon dioxide (SiO_2), sodium oxide (Na_2O) derived from sodium carbonate (Na_2CO_3), calcium oxide (CaO) also known as lime from $CaCO_3$, and a few other small additives.
2. Silicon dioxide (SiO_2) is a prevalent elemental component of glass.
3. Sodium oxide (Na_2O) is often derived from Na_2CO_3, commonly known as "soda", which reduces the glass-transition temperature. The soda renders the glass water-soluble, a characteristic that is typically undesired. Transparent glass has a low melting point and a smooth texture.
4. Lime, also known as calcium oxide (CaO), is typically derived from limestone, which is composed of $CaCO_3$.
5. Magnesium oxide (MgO) and aluminium oxide (Al_2O_3) are included to enhance the chemical durability. The soda-lime glass, which is the end product, consists of approximately 70 to 74% silica by weight. Approximately 90% of glass produced is composed of soda-lime glasses.
6. Potassium is commonly referred to as potash, with its chemical formula being K_2CO_3. (Durable glass).
7. Calcium are limestone, chalk and lime.
8. Lead are litharge, and red lead. (Flint glass)
9. Silica are quartz, white sand, and ignited flint.
10. Zinc is zinc oxide. (Heat and shock-proof glass)
11. Borate are borax, and boric acid (Heat and shock-proof glass)

12. Cullet or pieces of broken glass to increase the fusibility
13. Colours: Yellow-ferric salt; Green–ferrous and chromium salts; Blue–cobalt salts; Purple-manganese dioxide; Red–nickel salts or Cu_2O; Lemon-yellow–CdS; Fluorescent greenish-yellow–uranium oxide; Opaque milky–white-cryolite (Na_3AlF_6) or calcium phosphate.

12.4 MANUFACTURE OF GLASS

The manufacturing process (displayed in Fig. 12.1) can be categorized into four primary stages:

1. Melting, 2. Shaping or forming, 3. Annealing, and 4. Finishing

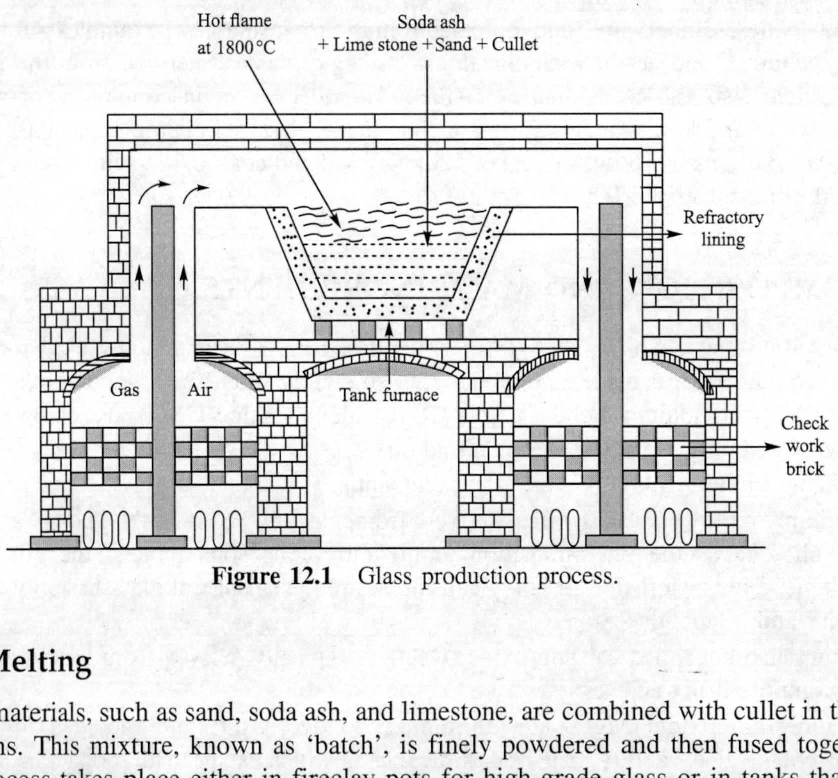

Figure 12.1 Glass production process.

12.4.1 Melting

The raw materials, such as sand, soda ash, and limestone, are combined with cullet in the correct proportions. This mixture, known as 'batch', is finely powdered and then fused together. The fusion process takes place either in fireclay pots for high-grade glass or in tanks that are part of an open-hearth furnace. The heating is achieved by burning a mixture of producer gas and air over the charge. Under high prevailing temperature conditions (about 1,800°C), the charge undergoes melting and fusion. The furnace is 38 meters in length, 9 meters in width, and 1.5 meters in height. It has a capacity of 1350 metric tons. The glass is created by liquefying in the center of the furnace, where the flames move back and forth from one side to the other. The melting process of common soda-glass consists of the following sequence of reactions:

$$CaCO_3 + SiO_2 \longrightarrow CaSiO_3 + CO_2\uparrow$$

$$Na_2CO_3 + SIO_2 \longrightarrow Na_2SiO_3 + CO_2\uparrow$$

Once all the carbon dioxide has been released from the molten mass, decolorizers like MnO_2 or nitre are introduced to eliminate any ferrous compounds and carbon, if they are present. At this point, if a tinted glass is desired, the coloring salts are introduced. The heating process is continued until the molten mass is completely devoid of bubbles and glass balls, and then it is cooled to around 800°C.

1. The refined glass product is extracted from the opposite end of the tank.
2. The durability of the tank and the longevity of the glass are contingent upon the construction glass's quality.
3. The furnace is often coated with refractory bricks.
4. The furnace is heated using either electrothermal or gas methods, namely as a regenerative furnace.
5. The system functions in two cycles, each with two sets of checker work chambers.
6. The combustion gases, having released a portion of their thermal energy while traversing the furnace housing the liquefied glass, proceed downwards through a series of chambers filled with exposed brickwork or checker work.
7. A significant amount of sensible heat in the gases is extracted, resulting in the checker work attaining temperatures ranging from 1500°C near the furnace to 650°C on the exit side.
8. At the same time, air is prepared by passing it through the other regenerative chamber that has already been heated, and it is then combined with the burned fuel gas.
9. The flame produced is of a higher temperature due to the preheating of the air, which would not have been achievable otherwise.
10. The air-fuel combination, or cycle, is periodically reversed and introduced into the furnace from the opposite side through the preheated checker work at consistent intervals of 20–30 minutes.
11. This regenerative mechanism allows for significant heat conservation, resulting in the attainment of a greater temperature.
12. The temperature of the furnace should be increased by certain increments each day until it reaches and maintains a temperature of at least 1200°C.
13. The majority of heat is dissipated through radiation from the furnace, whereas a significantly less portion is utilized in the melting process.
14. In order to mitigate the effects of glass melting, water cooling pipes are commonly installed in the walls of the furnace.

12.4.2 Forming and Shaping

The glass is currently formed by either mechanized or manual moulding techniques. In machine moulding, the glass mission design should prioritize the completion of the object within a short time frame. Within this very brief duration, the glass undergoes a transition from a viscous liquid state to a transparent solid one. Therefore, the design of a machine that satisfies the requirements of heat flow, metal stability, and bearing clearance is highly intricate. Now, thanks to technological advancements, the glass engineer is capable of successfully completing the mentioned responsibilities.

12.4.3 Annealing

Annealing is a process that involves heating a material and then slowly cooling it down to reduce its hardness and increase its ductility. Annealing is the method by which it is achieved. Annealing comprises two distinct operations:

1. Maintaining the glass at a temperature above a specific critical point for a sufficient duration to decrease internal tension through the process of plastic deformation, until it reaches a predetermined maximum level. It is crucial to minimize strains throughout the glass production process, whether using machine or manual moulding techniques.
2. Gradually cooling the bulk to room temperature at a rate that ensures the strain remains below the maximum limit. Subsequently, glass products are permitted to undergo a gradual cooling process to reach the ambient temperature by traversing various chambers with progressively decreasing temperatures. Due to its poor heat conductivity, glass cools down rapidly when allowed to cool quickly. As a result, the outer layer of glass cools down before the interior, causing the interior to become strained. Due to this uneven expansion, the articles are prone to fracturing into fragments. Increasing the duration of the annealing process directly correlates with an improvement in the overall quality of the glass.

12.4.4 Finishing

Following the annealing process, all glass articles undergo several physical operations and finishing processes, including cleaning, grinding, polishing, cutting, and sand-blasting, enamelling, grinding, and glazing.

12.5 TYPES OF GLASSES

Commercial glasses can be categorized as follows:

1. **Soda-lime or soft glass:** Soda-lime or soft glass refers to a type of glass that has soda and lime as its main components. The primary constituents include silica (sand), calcium carbonate, and soda ash. Their composition can be approximated as $Na_2O.CaO.6SiO_2$. They have a low cost, are resistant to devitrification, and exhibit a moderate level of water resistance. Due to their high melting point, they possess a low resistance to heat and can be easily manipulated through hot-working processes. However, these glasses are susceptible to assault by ordinary chemicals such as acids.

   Applications: They are extensively utilized in various industries like as window manufacturing, electric lighting, glassware production, construction materials, and affordable tableware. These materials are preferred when there is little need for great resistance to high temperatures or chemical stability.

2. **Potash-lime or hard glass:** This glass variant is derived from silica (sand), calcium carbonate, and potassium carbonate. Their composition can be approximated as $K_2O.CaO.6SiO_2$. These substances have a high melting point, are resistant to fusion, and are less reactive to acids, alkalis, and other solvents compared to regular glasses.

Applications: These glasses, which are more expensive than soda-lime glasses, are utilized in chemical apparatus and combustion tubes specifically designed for heating operations.

3. **Lead glass or flint glass:** This glass variant is produced by employing lead oxide, rather than calcium oxide, to fuse with silica. Up to 80% of PbO is included in dense optical glasses. Furthermore, K_2O is employed as a substitute for sodium oxide. The approximate composition of the substance is $K_2O.PbO.6SiO_2$. Lead glass is significantly costlier to produce than regular lime-soda glass, however considerably more malleable and manageable. Lead glass exhibits a lower softening temperature compared to soda-glass, as well as a greater refractive index. It possesses exceptional electrical characteristics. It has a high level of brightness, a shiny appearance, and a specific gravity ranging from 3 to 3.3.

   Applications: Lead glasses are extensively utilized for the production of premium tableware, optical components (such as lenses), neon sign tubes, cathode-ray tubes, electrical insulators, and artistic artifacts, owing to their exceptional luster. High lead-content glasses are utilized for manufacturing exceptionally dense optical glasses, which are employed in the production of windows and shields. These windows and shields serve the purpose of safeguarding personnel in the medical and atomic energy sectors from X-rays and gamma-rays, respectively.

4. **Borosilicate glass or Pyrex glass or jena glass:** This variety of glass is the prevailing type among the durable glasses found in the business sector. These glasses mostly consist of silica and boron, with a little presence of alumina and alkali oxides. A standard formula for such a glass would be as follows:

   $$80.5\%\ SiO_2,\ 13\%\ B_2O_3,\ 3\%\ Al_2O_3,\ 3\%\ K_2O,\ 0.5\%\ Na_2O$$

   Replacing the alkali (Na_2O) and basic alkaline-earth oxides (CaO) in soda-lime glasses with boron and aluminium oxides produces a glass with a low thermal coefficient of expansion and strong chemical resistance. Borosilicate glasses possess exceptionally elevated softening temperatures and have outstanding resistivity, rendering them highly resistant to shocks.

   Applications: They are widely employed in several industrial sectors for tasks such as transporting corrosive liquids through pipelines, providing transparent gauges for measuring liquids, manufacturing high-quality laboratory equipment, producing kitchenware, constructing chemical plants, fabricating television tubes, and creating electrical insulators.

5. **Aluminosilicate glass:** An ordinary examination of such a glass would proceed as follows:

   $$55\%\ SiO_2,\ 23\%\ Al_2O_3,\ 7\%\ B_2O_3,\ 9\%\ MgO,\ 5\%\ CaO,\ 1\%\ Na_2O + K_2O$$

   These glasses have remarkably high softening temperatures.

   Applications: High-pressure mercury discharge tubes, chemical combustion tubes, and certain home equipment.

6. **Vitreosil or 99.5% silica glass:** The production of this glass variant involves subjecting pure sand (SiO_2) to intense heat until it reaches its melting point, which is around

1,750°C. Due to the lack of fluxing agents, eliminating all air-bubbles is exceedingly challenging. Furthermore, due to the elevated viscosity of this glass at its operational temperature, the process of shaping becomes somewhat challenging. The end result is transparent. The approximate softening temperature of the material is 1,650 °C. The thermal expansion of this material is exceptionally low, specifically measuring approximately 5.5×10^{-7} $cm^{-1}°C^{-1}$. When vitreosil glass is heated above its melting point for extended duration, it undergoes a transformation and becomes transparent, referred to as "clear silica glass". This glass possesses significant light transmission qualities, allowing no less than 93% of light to travel through a 1 meter thick piece of this material.

Applications: This material finds primary usage in chemical plants, chemical laboratory equipment, and as electrical insulating materials in electrical heaters, furnaces, and similar applications.

7. **Safety glass:** This glass variant is manufactured by sandwiching two or three flat glass sheets together and inserting alternating layers of thin vinyl plastic in between. The entire assembly is then exposed to gentle pressure. Subsequently, the glass layers and plastic layers are subjected to heat, causing them to amalgamate and form a cohesive sandwich structure. Upon cooling, the glass undergoes a significant increase in toughness. When the glass shatters, it does not fragment because the inside plastic layer restrains the fractured glass particles.

   Applications: This material finds extensive usage in the automobile and aerospace sectors, particularly for manufacturing windshields and other related components.

8. **Optical or crookes glasses:** This glass is composed of phosphorus and lead silicate, together with a small amount of cerium oxide. The latter has the ability to absorb ultraviolet radiation, which can be harmful to the eyes. Furthermore, they are provided with comprehensive uniformity. This is ensured by a meticulous production procedure and by carefully heating the molten substance for an extended duration. Optical glasses typically exhibit low melting temperatures and possess a somewhat delicate nature. Their chemical resistance and durability are much inferior to that of regular glass.

   Applications: They are utilized in the production of lenses.

9. **Polycrystalline glass or pyroceram:** Polycrystalline glass or pyroceram are materials with a crystalline structure composed of several small crystals. This variety of glass is manufactured by incorporating one or more nucleating chemicals into a specialized or standard glass mixture, which is subsequently molded into desired shapes. Subsequently, the material undergoes a regulated heat-treatment process. The nucleating chemicals stimulate the generation of many sub-microscopic crystallites, which serve as nuclei for further crystal development. Crystalline glass lacks ductility, but possesses significantly higher impact resistance compared to regular glass. It demonstrates exceptional strength and significant hardness, and can be molded into various objects using any molding techniques.

10. **Toughened glass:** This glass is manufactured by immersing heated items in an oil bath, resulting in rapid cooling. By performing this action, the outer layers of the objects

undergo a reduction in size and become compressed, while the inner layers experience a state of stretching or tension. This type of glass exhibits more elasticity and is able to endure mechanical and thermal shocks. When a glass of this nature shatters, it does not propel through the air, but instead disintegrates into a thin substance like powder.

Applications: It is employed in the fabrication of window shields for high-speed vehicles such as cars, trucks, and airplanes, as well as for furnace window shields, automatic sliding doors, and huge display cases.

11. **Insulating glass:** This glass is a transparent structure formed by sandwiching two or more glass plates with a gap of 6 to 13 mm, which is filled with dehydrated air and then sealed tightly around the edges. This offers excellent thermal insulation. Therefore, if this type of glass is utilized to divide flats, it effectively blocks the transfer of heat, resulting in the apartments maintaining a cool temperature during the summer and a warm temperature during the winter.

12. **Wired glass:** Glass that has been reinforced with wire mesh or other materials to increase its strength and resistance to breakage. This glass variant is created by incorporating a wire mesh into the core of the glass sheet during the casting process. An inherent benefit of this glass is its ability to fracture without disintegrating into fragments. Furthermore, it has a higher level of fire resistance compared to standard glass.

    Applications: Utilized in the construction of fire-resistant doors, windows, skylights, roofs, and other similar structures.

13. **Laminated glass:** This glass variant is manufactured through the process of pressing or gluing many sheets or plates of glass together, with the inclusion of one or more alternating layers of bonding materials such as plastic resin, asphalt, or synthetic rubber. Laminated glass possesses two key properties. Firstly, it is shatter-proof, meaning that its fragments remain intact and do not disperse when broken abruptly. Secondly, it is shock-proof, enabling it to endure rapid temperature fluctuations and even some amount of pressure without fracturing.

    A bullet-resistant laminated glass is produced through the process of compressing multiple layers of glass with vinyl resins in alternating layers. Typically, the thickness of this type of glass ranges from 12.7 mm to 76.5 mm. Specialized applications require even more substantial variants.

    Applications: It is an optimal material for manufacturing safety glass in various transportation vehicles such as aircraft, automobiles, helicopters, submarines, and more. Automotive windshields and windows, among other uses, commonly incorporate bullet-resistant laminated glass.

14. **Glasswool:** Glasswool is a fibrous material made up of thin threads or filaments of glass that are mixed. The glass utilized for this objective must be entirely devoid of alkali. Glass filaments are produced by extruding molten glass through small orifices, typically ranging in diameter from 0.0005 to 0.007 mm, in a continuous manner. The glass filaments are draped over a swiftly rotating drum to transform the material into a wool-like texture.

Characteristics

1. It exhibits non-combustible and fire-resistant properties.
2. The electrical conductivity of the material is minimal.
3. It is entirely heat-resistant, meaning that heat does not impact its size or volume.
4. It is non-absorbent.
5. It exhibits resistance to the action of the majority of chemicals.
6. The density of the substance is rather low, measuring approximately 65 kg/m^3.
7. The thermal conductivity of this material is rather low, measuring approximately 0.033–0.034 $kcal/m^2/C/hr$.
8. The tensile strength of this material is approximately eight times greater than that of steel.

Applications

1. It is utilized for thermal insulation reasons, such as in household and industrial appliances like ovens, insulation of metal pipelines, motors, vacuum cleaners, and the walls and roofs of buildings.
2. Glasswool, because to its chemical resistance, is utilized for filtering corrosive liquids such as acid and acidic solutions.
3. It is employed for the purpose of electrical insulation.
4. Additionally, it is employed for the purpose of sound insulation.
5. It is also utilized in air filters as materials for filtering dust.
6. Additionally, it is employed in the production of fiberglass by the amalgamation with polymeric resins.

QUESTIONS

12.1 What are the raw materials used in the manufacture of glass?

12.2 Write short note on safety glass.

12.3 Discuss the manufacture of glass by pot furnace process.

12.4 Write an account of soda-lime glass and safety glass.

12.5 Give the composition, properties and uses of flint glass.

12.6 Discuss the manufacture of ordinary glass with the help of flow-sheet-diagram. Give the reactions involved.

12.7 Give the composition, properties and uses of borosilicate glass.

12.8 What are the functions of various ingredients used in the manufacture of glass?

12.9 Classify various types of glass and write short notes on: (i) crookes glass, (ii) pyrex glass, (ii) flint glass, (iv) glasswool.

12.10 Discuss the following in brief (any three): (i) Laminated glass, (ii) borosilicate glass, (iii) glasswool.

12.11 Give classification of glass with their respective uses. How is glass manufactured?

12.12 Describe the chemistry involved in formation of glass.

12.13 Write short note on types of glass.

12.14 What is glass? What are these raw materials used in the manufacture of glass

12.15 Describe the manufacturing process of borosilicate glass. Discuss the importance of silica glass.

Unit 4(B)

Chemistry of Engineering Materials

Fuel and Lubricants

CHAPTER 13

Fuels

13.1 INTRODUCTION

Chemical fuel refers to a carbon-based substance that, when burned correctly in the presence of air, produces a significant amount of heat that can be efficiently utilized for both household and industrial applications. Examples of fuel sources include wood, charcoal, coal, kerosene, petrol, diesel, producer gas, water gas, natural gas, and more.

13.2 FUEL COMBUSTION

During the process of combustion, the carbon (C) and hydrogen (H) in the fuel react with the oxygen in the air to produce carbon dioxide (CO_2) and water (H_2O) correspondingly. Chemical fuels lose heat through burning, as the heat content of the resulting products (such as CO_2 and H_2O) is lower than that of the initial reactants (such as carbon and hydrogen in the fuel).

Combustion is the process of exothermic oxidation of a fuel, either by air or oxygen that occurs at a rapid rate and generates a high temperature, typically accompanied by the presence of a flame.

Since the majority of fuels consist of carbon or a combination of carbon and hydrogen, the process of burning entails the carbon being oxidized into carbon dioxide and the hydrogen being oxidized into water. If sulphur is present, it undergoes oxidation to generate sulphur dioxide, while the mineral substance transforms into ash. During burning, complex fuels such as coal undergo thermal disintegration, resulting in the formation of simpler compounds. These products are further oxidized to carbon dioxide, water, and other substances. As an illustration, the process of burning Coke produces carbon dioxide.

$$\text{Coal} \longrightarrow \text{Coke} + \text{Coal gas}$$

$$\text{C(coke)} + O_2 \longrightarrow CO_2$$

13.3 CATEGORIZATION OF FUELS

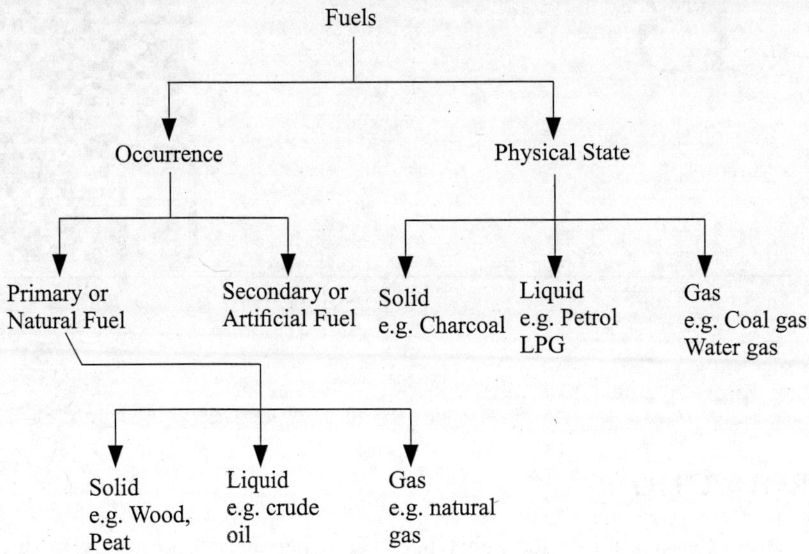

Fuels are categorized based on their occurrence and physical state.

Primary or Natural Fuel: These fuels are inherently present in nature, including coal, crude oil, and natural gas. Primary fuel, derived from ancient organic matter, are commonly known as fossil fuels.

Secondary or Artificial Fuels are produced by the treatment process of naturally existing fuels, such as coke, gasoline, coal gas, and others.

13.4 CHARACTERISTICS OF A GOOD FUEL

1. High energy content, i.e., calorific value.
2. The ignition temperature is moderate.
3. Minimal moisture level.
4. Insufficient presence of non-combustible material.
5. When using solid fuel, it is important for the ash content to be minimal and for the size of the fuel to be consistent.
6. Easily accessible in large quantities at a low price.
7. Combustion by-products should be non-toxic.
8. Combustion should exhibit high controllability.
9. The storage and transport should prioritize safety, convenience, and cost-effectiveness.

13.5 BIOGAS

Biogas is an environmentally friendly and cost-effective energy source, comprising primarily of 55–70% methane, which is highly combustible. The composition consists of methane (CH_4), carbon dioxide (CO_2), hydrogen (H_2), and hydrogen sulphide (H_2S). Biogas commonly denotes a gas generated by the biological decomposition of organic substances in the absence of oxygen. Biogas is generated through the anaerobic digestion or fermentation of organic waste, such as cattle dung, by bacteria. This process is frequently referred to as a gobar gas plant.

13.5.1 History of Biogas

The history of biogas refers to the chronological development and evolution of the production and utilization of biogas as an energy source. Biogas is a long-standing type of sustainable energy. According to Marco Polo, the utilization of this technology can be traced back to ancient Chinese literature, approximately 2000–3000 years ago. The concept of anaerobic digestion was initially elucidated by Benjamin Franklin in 1764. Sir Humphrey Davy ascertained in 1808 that methane is a byproduct of the decomposition process. The inaugural digestive facility was constructed in 1859 in Bombay, India specifically to serve a leper colony.

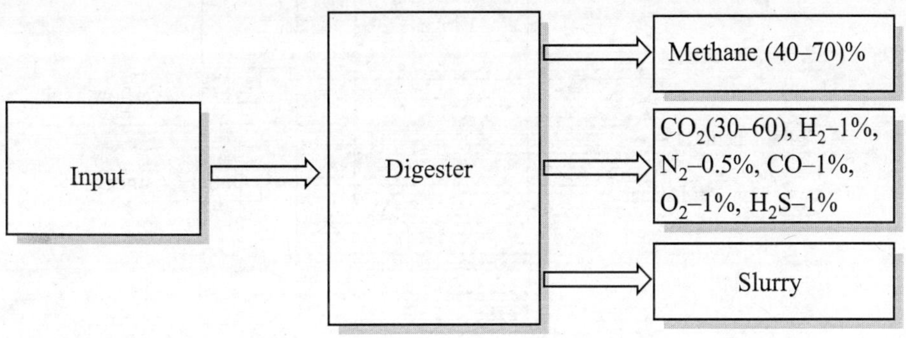

13.5.2 Biogas Plants (Digester Type)

Biogas is a versatile fuel that can be utilized for illumination, heating, or propulsion. The advantages of biogas include improved sanitation and cleanliness, as well as the provision of fuel and energy, which in turn reduces indoor air pollution. Additionally, biogas helps to mitigate deforestation, cut down on CO_2 emissions, and serves as a source of fertilizer. Biogas plants are utilized to supply fuel for cooking and organic fertilizer to rural homes, while also alleviating the burden on rural women, decreasing reliance on trees, and enhancing social advantages. The following types of bio-plants are commonly employed in our country.

1. Fixed-dome 2. Floating-drum 3. Balloon 4. Horizontal 5. Earth-pit 6. Ferro cement

13.5.2.1 Fixed-dome Plant

A fixed-dome plant is a type of structure used in waste management systems. The biogas plant is built underground using a framework made of bricks and cement, which provides protection against physical damage, saves space, and shields it from the impact of external temperatures. The plant features a fixed-dome roof that is constructed using brick materials, eliminating the presence of any moving parts or steel components that are susceptible to rust. This design ensures a prolonged lifespan of the plant, lasting for 20 years or more. The system comprises an elevated mixing tank, which is followed by an underground mixing tank that connects to a sloping intake chamber (intake tank). The inflow chamber extends downward into the tank.

Which is known as digester, a large tank with a domed ceiling. The digester's ceiling is equipped with an exit that includes a valve for the delivery of produced biogas. The digester is accessed through a lower opening that leads to an outflow chamber. The outflow chamber is connected to a tiny overflow tank at the top (Figure 13.1). Biomass raw materials, such as animal dung, poultry wastes, plant wastes (such as husk, grass, and weeds), human excreta, industrial wastes, and residential wastes (including vegetable peels and waste food ingredients), are combined with water.

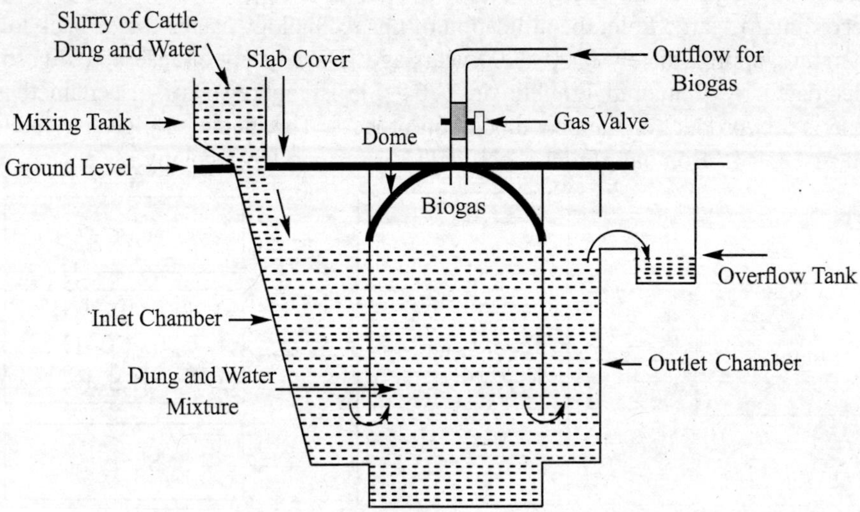

Figure 13.1 Fixed dome type Biogas plant

Working: The process involves combining different types of raw materials to create a slurry. Slurry is a mixture of an equal amount of water and raw materials in the mixing tank. The mixture is introduced into the digester via the input chamber. The slurry remains inactive for around sixty days. Over the course of two months, the anaerobic bacteria in the slurry break down the biomass through fermentation in the presence of water, resulting in the production of biogas. Biogas accumulates in the dome of the tank. As the accumulation of biogas increases, the pressure exerted by the biogas compels the spent slurry to move into the output chamber. The discharged slurry flows from the output chamber and enters the overflow tank. The depleted sediment is manually extracted from the overflow tank and utilized as fertilizer for plants. The gas valve, which is linked to a network of pipelines, is activated when there is a need for a supply of biogas. In order to maintain a consistent flow of biogas, a functional plant can be continuously supplied with the produced slurry.

Advantages: The building of a fixed dome biogas plant merely necessitates the use of readily accessible local materials. The expenses associated with a plant are comparatively inexpensive. It is simple to build.

13.5.2.2 Floating-drum Plant

A floating-drum plant comprises a mixing tank positioned above ground level, which is connected to a cylindrical or dome-shaped digester. Additionally, it includes a mobile gas-holder, also known as a drum that floats on top of the digester. The subterranean digester, like a well, is partitioned into two chambers by a dividing wall. The structure consists of two elongated cement pipes.

1. One of these pipes serves as the inlet pipe, which allows the slurry to be introduced into the intake chamber.
2. The outlet pipe is connected to the overflow tank to facilitate the removal of the used slurry.

A gas holder, also referred to as an inverted steel drum, is positioned above the digester. The drum has the ability to remain buoyant above the digester. The upward movement of the drum indicates the production of biogas, while the sinking of the gas-holder signifies the consumption of gas. The overflow tank is situated at an elevated position above the ground level.

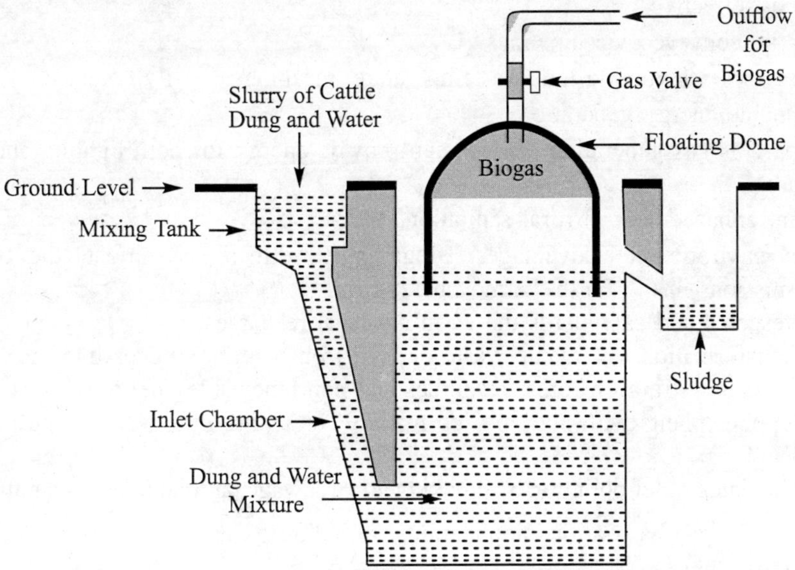

Figure 13.2 Floating drum type Biogas plant.

Working: The process involves preparing a slurry in a mixing tank. This slurry is then introduced into the digester's input chamber through the inlet pipe. The facility remains idle for around sixty days, and thc addition of further slurry is halted and get anaerobic fermented to produce biogas in the digester. The lighter biogas ascends and accumulates in the gas holder. The gas holder is currently ascending. The gas holder has a maximum limit beyond which it cannot ascend. As the accumulation of gas increases, the slurry experiences a corresponding increase in pressure. The exhausted mixture is now compelled into the exit outlet. Once the outlet chamber becomes filled with the used slurry, any surplus is expelled through the exit pipe and directed into the overflow tank. Subsequently, it is employed as fertilizer for plants. The gas valve of the gas outlet is turned on in order to have a steady flow of biogas. To maintain a consistent supply of biogas, it is necessary to regularly remove the used slurry and add fresh slurry during the production process.

Disadvantages: The cost of this particular biogas plant is high. The steel drum is susceptible to corrosion and necessitates frequent require to repair.

1. The primary drawback is the deprivation of the organic waste for composting or fertilization purposes.
2. It has a highly restricted capacity to generate electricity on a worldwide level.
3. The pace of gas production lacks significant control, however the gas can be stored and utilized to a certain degree as needed.

Advantages: Biogas is regarded as a favorable fuel due to its beneficial properties.

1. It is a form of gasoline that is free from impurities or pollutants. It is smokeless and residue-free, making it environmentally friendly.
2. It produces high calorific value.
3. It is cost-effective.
4. It can be conveyed via pipelines.
5. It possesses a favorable ignition temperature, meaning it easily combusts.
6. Sustainable energy generation.
7. Compact biogas generating devices can provide energy for both lighting and culinary purposes.
8. Evident enhancement in rural sanitation.
9. Global environmental advantages: Biogas plants effectively mitigate the greenhouse gas emissions in the Earth's atmosphere.
   (a) Preserves and safeguards the earth's natural resources.
   (b) Transformation of organic waste derived from natural sources into bio-fertilizer.
   (c) Alleviates strain on forest resources and non-renewable energy sources.
   (d) Supplies plants with nutrient-rich manure containing nitrogen (N) and phosphorus (P).
   (e) Alleviates water pollution by breaking down sewage, animal feces, and human waste.

Uses of Biogas: Biogas has various applications, including its use as:

1. Household fuel
2. Street lighting
3. Electric power generating
4. When subjected to compression, it can serve as a substitute for compressed natural gas in vehicular applications.

Limitations of Biogas Plants: Biogas plants have the following limitations:

1. The initial installation cost of the plant is expensive.
2. The quantity of livestock possessed by a typical agricultural household is insufficient to sustain a biogas facility.

13.6 CALORIFIC VALUE

The calorific value of a fuel is the amount of heat released when a certain amount (unit weight) of the fuel is completely burned in the presence of air or oxygen, and the resulting combustion products are cooled back to the original temperature of the fuel. This value is typically measured in calories or kilocalories.

Units: The energy content of solid or liquid fuels is typically measured in calories per gram (cal/g), kilocalories per kilogram (kcal/kg), or joules per kilogram (J/kg). Gaseous fuels are measured in kilocalories per cubic meter (kcal/m^3) or joules per cubic meter (J/m^3).

13.6.1 Gross and Net Calorific Value

When it comes to fuels that contain hydrogen, there are two distinct calorific values: the gross calorific value and the net calorific value.

13.6.1.1 Gross or Higher Calorific Value (GCV)

The gross calorific value, also known as the higher calorific value, refers to the amount of heat released when a unit weight of fuel undergoes complete combustion and the resulting combustion products are subsequently cooled to the original temperature of the fuel. The term "heat of combustion" refers to the quantity of heat released when a specific amount of fuel is entirely burned in air and the resulting combustion products are cooled to room temperature.

13.6.1.2 Net or Lower Calorific Value (NCV)

Typically, water vapors generated during combustion are not condensed and are released together with the hot gases under regular operating conditions. Therefore, a reduced quantity of heat is present, known as the lower or net calorific value.

The net calorific value refers to the amount of heat generated when a specific mass of fuel is entirely burned and the resulting combustion products are allowed to leave. The term refers to "the quantity of thermal energy generated when a specific mass or volume of fuel undergoes complete combustion in the presence of air, with the resulting combustion products being released into the atmosphere."

The net calorific value (NCV) can be calculated by subtracting the weight of hydrogen multiplied by 9 and the latent heat of water vapors from the gross calorific value (GCV).

$$\text{NCV} = \text{GCV} - \text{weight of hydrogen} \times 9 \times \text{Latent heat of water vapour}$$

The latent heat of water vapor is 587 kcal/kg.

Calculation of Net Calorific Value: The hydrogen in the fuel undergoes a chemical reaction with oxygen, resulting in the formation of water.

The chemical equation for the reaction between hydrogen gas (H_2) and oxygen gas (O_2) to form water (H_2O) is represented as follows:

$$H_2 + \frac{1}{2}O_2 \longrightarrow H_2O$$

$$2H = \frac{1}{2}O_2 = H_2O$$

$$2 \text{ parts} = 16 \text{ parts} = 18 \text{ parts}$$

$$1 \text{ parts} = 8 \text{ parts} = 9 \text{ parts}$$

Let H be the hydrogen concentration in the fuel, expressed as a percentage.

The water yield from burning one unit mass of fuel is 9H/100 g.

The latent heat of steam is 587 calories per gram.

The heat generated by the condensation of steam is equal to 9H/100 multiplied by 587 calories.

Amount of heat produced by condensation of steam = 9H/100 × 587 cal/g

GCV = NCV + Latent heat of condensation of steam

NCV = GCV – 9 × Mass of hydrogen × Latent heat of steam

NCV = [GCV – 0.09 × % of hydrogen × Latent heat of steam] cal/g

The calorific value of a fuel is determined by its composition and the ratio of its constituents, with higher levels of hydrogen leading to an increase in calorific value. The presence of moisture significantly decreases the calorific value of a fuel. The calorific value can be theoretically determined based on the chemical composition of the fuel.

The **Dulong formula** is used to calculate the calorific value of a fuel that contains carbon, hydrogen, oxygen, and sulphur. The formulae is as follows:

$$\text{Calorific value} = \frac{1}{100}\left[8080 \times \%C + 34500\left(\%H - \frac{\%O}{8}\right) + 2240 \times \%S\right] \text{KCal/ Kg}$$

The calorific value is determined by the percentages of carbon (C), hydrogen (H), oxygen (O), and sulphur (S) present.

13.7 BOMB CALORIMETER

The term "calorimeter" is derived from the Latin words "*calor*" meaning "heat" and "*metron*" meaning "measure". A bomb calorimeter is employed to quantify the thermal energy generated by the combustion of a fuel (in liquid or solid form) in the presence of oxygen within a small enclosed container (referred to as a bomb). This container is immersed in measured amount of water, which absorbs the energy released during the reaction.

The Process of Combustion: The unit mass of solid or liquid materials, also known as fuel, is measured by weighing it in a crucible. The fuel is then placed into a stainless-steel container, referred to as the Bomb, where supply of oxygen at a pressure of 30 bar (435 psi) is employed. The sample is ignited using a cotton thread that is linked to an ignition wire or a battery-operated electrode fuse wire. This ignition process takes place inside the decomposition vessel, causing the sample to burn and undergo combustion.

During the process of combustion, the temperature within the crucible increases and the pressure experiences a temporary rise. Under these circumstances, fuel undergoes combustion and is oxidized. The thermal energy generated during the combustion process can be calculated. In order to obtain accurate heat measurements, it is necessary to determine or ascertain the heat capacity of the bomb calorimeter.

The heat capacity considers all the components of the calorimeter that are capable of losing or gaining heat.

$$C_{total} = C_{Water} + C_{Thermometer} + C_{Stirrer} + C_{Container}$$

Given that the masses of the other components remain constant, the inclusion of mass units in the heat capacity value is unnecessary.

Thus, the equation $q_{cal} = C\Delta T$ represents the relationship between the heat transferred (q_{cal}) and the change in temperature (ΔT) of the system. Here, C represents the heat capacity of the calorimeter.

The value of the reaction (q_{rxn}) is equal to the negative value of the heat transferred (q_{cal}). Mathematically,

$$q_{rxn} = -q_{cal}$$

The total heat released, q_{cal}, is equal to the sum of the heat released by the bomb, q bomb, the heat released by the water, q water, the heat released by the wires, q wires, and so on.

$$q_{cal} = q_{bomb} + q_{water} + q_{wires} + \cdots$$

The heat capacity of the calorimeter is the quantity of heat energy needed to increase the temperature of the calorimeter by one degree Celsius.

The equation $q_{cal} = \Sigma m_i c_i \Delta T - C_{cal}\Delta T$ represents the relationship between the heat absorbed or released by a system (q_{cal}), the mass of each component (m_i), the specific heat capacity of each component (c_i), and the change in temperature (ΔT).

13.7.1 Calculation of Calorific Value of Fuel with the Help of Bomb Calorimeter

Principle: The fuel sample is combusted entirely with an excess of oxygen. The released thermal energy is absorbed by both the water and the calorimeter. The thermal energy released from the combustion of fuel is transferred to both the water and the calorimeter, resulting in a heat exchange. The fuel's calorific value is determined based on the collected data. The diagram of working Bomb Calorimeter is displayed in Figure 13.3.

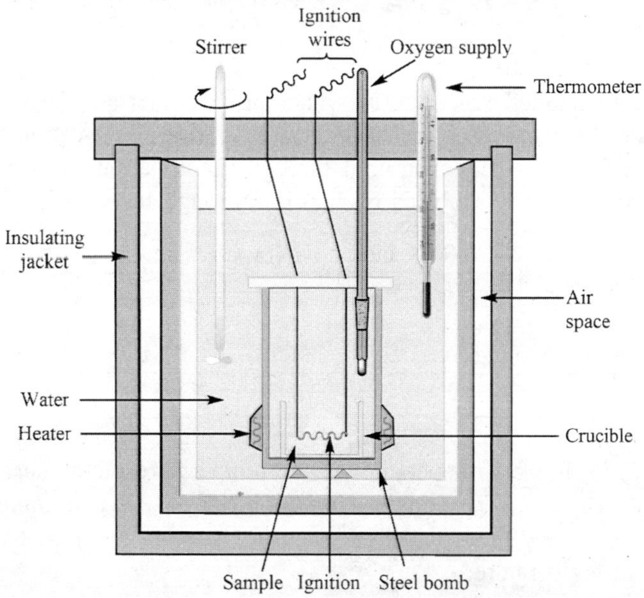

Figure 13.3 Bomb calorimeter.

Observations and Calculations

The mass of the sample fuel = m g

The mass of water in the copper calorimeter = W g

Water equivalent of calorimeter = w g

Initial temperature of water = t_1°C

Final temperature of water = t_2°C

Heat released by complete combustion of fuel = Heat absorbed by water and calorimeter

$$m \times \text{GCV} = (W + w)(t_2 - t_1)$$

$$\text{GCV} = \frac{(W + w)(t_2 - t_1)}{m} \text{ Cal/g}$$

$$\text{GCV} = \frac{(W + w)(t_2 - t_1)}{m} \times 4.184 \times 10^3 \text{ J/Kg}$$

Corrections

1. **Cotton Thread Correction:** It is determined by the weight of the dry cotton thread utilized for ignition.
2. **Fuse Wire Correction:** The calorific value estimation process involves the release of heat, which includes the heat emitted upon igniting the fuse wire. Therefore, it is necessary to deduct it from the overall value.
3. **Acid Correction:** Fuels that contain sulphur (S) and nitrogen (N) undergo oxidation at high pressure and temperature during ignition, resulting in the formation of sulphuric acid and nitric acid, respectively.

$$S + H_2 + 2O_2 \longrightarrow H_2SO_4 + \text{Heat}$$

$$2N + H_2 + 3O_2 \longrightarrow 2HNO_3 + \text{Heat}$$

The aforementioned reactions are inherently exothermic. Hence, the quantified quantity of heat encompasses the heat released during the creation of the acid. The quantity of H_2SO_4 is solely measured through precipitation as $BaSO_4$, whereas the quantity of these acids is analyzed by titration through the washing of the bomb.

For every ml of $\frac{N}{10} H_2SO_4$ and $\frac{N}{10} HNO_3$, 3.6 and 1.43 calories, respectively, should be subtracted.

$$\text{GCV} = \frac{[(W + w)(t_2 - t_1 + \text{cooling correction})] - (\text{Acid} + \text{Fuse wire} + \text{cotton thread correction})}{m}$$

4. **Cooling Correction:** The time required for the water in the copper calorimeter to decrease from its highest temperature to the temperature of the surrounding room is measured and recorded. The cooling correction ($dT \times t$) is determined by the rate of cooling (dT°/min.) and the actual time taken for cooling (t min.). It is then added to the increase in temperature.

13.8 CHARACTERISTICS OF GOOD (IDEAL) FUEL

1. **Ignition Temperature:** It refers to the minimum temperature at which fuel must be heated in order to initiate combustion without any difficulty. When the ignition temperature is low, the fuel readily undergoes combustion. The presence of a low ignition temperature poses a significant risk during the storage and transportation of gasoline. Elevated temperature hinders the process of igniting. Therefore, an ideal fuel should possess a relatively low ignition temperature.
2. **Low Moisture Content:** An ideal fuel should possess little moisture content, as moisture content has a detrimental effect on the fuel's calorific value.
3. **Low Non-combustible Matter Content:** An ideal fuel should possess minimal amounts of non-combustible substances, as these substances are converted into ash during combustion, hence reducing the fuel's calorific value.
4. **Moderate Rate of Combustion:** The temperature at which fuel combusts is contingent upon the rate of combustion. If the rate of combustion is slow, then the necessary high temperature may not be achieved promptly. Conversely, a too rapid combustion rate results in a rapid increase in temperature.
5. **Minimum Smoke and Non-poisonous Gases:** When fuel is burned, it should not release noxious and toxic fumes. Put simply, it is imperative that gaseous byproducts do not contaminate the atmosphere. Carbon monoxide (CO), sulphur dioxide (SO_2), and hydrogen sulfide (H_2S) are examples of noxious gases.
6. **Economical:** An ideal fuel should possess the qualities of being cost-effective and easily accessible.
7. **Efficient Transportation:** An ideal fuel should possess characteristics that facilitate convenient handling and transportation while maintaining a reasonable cost.
8. **Controllable Combustion:** Initiating or halting the combustion of fuel should be readily achievable as needed.
9. **Non-spontaneous Combustion:** The combustion of fuel must not occur spontaneously, as this might lead to the occurrence of fire dangers.
10. **Low Storage Cost:** An ideal fuel should possess the characteristic of being conveniently storable at a minimal expense.

13.9 OTHER SOURCES OF ENERGY

1. **Biomass:** Biomass refers to any organic material, such as trees, plants, plant fibers, and animal excrement. Biomass is an inexhaustible energy resource.
2. **Bioenergy:** Bioenergy refers to the energy obtained from recently deceased organic matter, such as wood, crops, or animal manure, as opposed to fossil fuels, which are generated from decomposed organic matter. Bioenergy refers to the energy that is stored in biomass. Biomass can be combusted for thermal energy or transformed into biofuels like biodiesel or ethanol. Bioenergy exhibits remarkable versatility since it may

be harnessed for many purposes such as heat provision, fuel production, and electricity generation. Bioenergy sources can be renewed by cultivating rapidly growing grasses, trees, and agricultural leftovers such as vegetable oils, wheat straw, or corn.

3. **Biofuels:** Biomass is a limited but renewable option for producing fuels used in transportation. Biofuels emit lower levels of pollutants compared to petroleum fuels. There are two primary categories of biofuels: ethanol and biodiesel.
   (a) ***Ethanol***: Ethanol, commonly referred to as ethyl alcohol or grain alcohol, is derived from the starch found in specific grains such as wheat, corn, or corn derivatives. The process of ethanol manufacturing often commences by pulverizing biomass, such as wheat or corn. After being pulverized, the starch or cellulose undergoes a transformation into sugar. The sugar is subsequently supplied to microorganisms that metabolize it as a nutrient, resulting in the production of ethanol.
   (b) ***Biodiesel:*** Biodiesel is a renewable fuel derived from sources like vegetable oils or animal fats, which can be utilized as an alternative to conventional diesel fuel. Biodiesel is an environmentally friendly fuel that can be produced from vegetable oils, animal fats, or recycled restaurant greases, all of which are obtained inside the country.

   History of biodiesel: The diesel engine, created by Dr. Rudolf Diesel, was specifically engineered to function on a wide range of fuels, including vegetable oil. In 1911, he asserted that the diesel engine possesses the capacity to be powered by vegetable oils, so making a significant contribution to the progress of agriculture in the nations that embrace it. He asserted that while the current significance of using vegetable oils as engine fuels may seem insignificant, in the long run, these oils have the potential to become just as essential as petroleum and the existing coal tar products.

   Biodiesel today: Biodiesel, similar to ethanol, is predominantly utilized as a gasoline additive. Diesel blends typically consist of 20 percent biodiesel and 80 percent petroleum diesel. No more alterations are required for a vehicle that operates on this biodiesel blend. Furthermore, engines manufactured after 1994 may readily accommodate greater blends, even 100% biodiesel, with minimal or no adjustments.

   How is biodiesel made: Fats and oils undergo a chemical reaction with methanol to generate chemical molecules referred to as fatty acid methyl esters. Biodiesel refers to these esters when they are specifically intended for use as a fuel.

4. **Bio-power:** Bio-power refers to the utilization of biomass, which encompasses plant and organic matter, for the purpose of producing electricity. Since the inception of fire, biomass has been utilized for illumination, culinary purposes, and thermal energy generation. Biomass represents the most significant contributor of non-hydro renewable electricity. Most of the electricity generated from biomass is utilized as base load power inside the current electrical distribution system. Approximately two-third of this power is generated and utilized by the wood and paper sectors. Biomass constitutes the most significant origin of non-hydro renewable electricity.
   (a) ***Advantages of Bio-power for the Environment:*** While bio-power production does lead to higher levels of particle emissions, they are nonetheless considered cleaner

than those from coal burned power plants due to the absence of sulphur dioxide release. In addition, bioenergy systems are carbon dioxide neutral due to the plant material's ability to absorb an equal amount of carbon dioxide over its lifespan as is emitted when it is combusted for electricity generation.

13.10 INTRODUCTION TO ENVIRONMENT

The environment refers to the entirety of external factors, situations, and forces that encompass and impact the progression and maturation of an individual organism or a collective group of organisms. Environmental health encompasses the examination and control of environmental factors that impact human health and overall welfare. The environment encompasses biotic elements such as animals, plants, forests, and fungi, as well as abiotic components including continents, oceans, clouds, soil, rocks, and our constructed environment comprising buildings and human settlements. Additionally, it includes social ties and institutions. It consists of the following contents:

1. Physical factors encompass elements such as air, water, soil, housing, climate, geography, heat, light, noise, debris, radiation, and so on.
2. Biological factors include humans, viruses, microorganisms, insects, rodents, animals, and plants.
3. Psychosocial factors encompass cultural values, conventions, beliefs, habits, attitudes, morals, religion, education, lifestyles, community life, health services, and social and political organization.

Humans coexist within the ecosystem and are an integral component of the natural world. The continued existence of humanity is wholly contingent upon the well-being and proper functioning of the Earth. Humans are an integral component of the natural ecosystem, and their interactions with other elements of the ecosystem hold significant importance. Human survival, well-being, prosperity, health, mobility, and leisure time are entirely reliant on the environment. However, natural systems have suffered degradation in the form of pollution, erosion, and the extinction of species. Long-term health and survival are jeopardized by environmental changes.

Environmental science is the scientific examination of the functioning of the natural world and the reciprocal interactions between the environment and human beings. The fundamental objective of the environmental assessment is to identify potential avenues for addressing environmental issues.

Modern Environmental Movement: The modern environmental movement emerged in the early 1970s. At first, Rachel Carson in her book Silent Spring (1962), an American marine biologist raised awareness among the general public on the ecological threat presented by pesticides. The book has sparked discussion and catalyzed a significant paradigm shift in regards to the safety of utilizing pesticides and other noxious substances.

> *We spray our elms, and the following spring, trees are silent of robin song, not because we sprayed the robins directly but because the poison traveled step by step through the now familiar elm-earthworm-robin cycle* —RACHAEL CARSON

David Brower was an American environmentalist. Environmental activists were introduced to the utilization of litigation, regulatory action, and mass media. Barry Commoner was an American biologist and activist scientist who courageously raised awareness about environmental hazards.

Rachel Carson

David Brower

Barry Commoner

The human population relies on several components of the Earth, including the hydrosphere (water), atmosphere (air), lithosphere (soil), and biosphere (living world).

13.11 POLLUTION

Pollution encompasses the harmful changes in our environment that have negative effects on plants, animals, and humans. This phenomenon arises when society prioritizes immediate economic advantages while disregarding the long-term ecological advantages that are beneficial to humanity. Human actions have resulted in more substantial ecological modifications than any other natural occurrence. In recent decades, humans have contaminated the air, water, and land, which are vital for supporting life, with a wide variety of waste materials. Pollutants are anthropogenic compounds that exist at higher concentrations than those occurring naturally, and can be in the form of solids, liquids, or gasses. They are generated via anthropogenic activities and exert a detrimental influence on our environment.

The effect on human health is dependent by the specific attributes and concentration of a pollutant. An average human typically inhales roughly 12 kg of air per day, a quantity that is nearly 12 to 15 times greater than our food consumption. Hence, even a minuscule presence of pollutants in the air becomes more significant when compared to equal quantities found in food. Waterborne pollutants has the ability to spread across great distances, especially within the marine ecosystem.

Ecologically, contaminants can be categorized as:

1. Degradable or non-persistent pollutants refer to substances that can be swiftly decomposed by natural processes. For example, this includes waste from households, such as sewage and leftover vegetables.
2. Slowly degradable or persistent pollutants are substances that remain in the environment for extended periods without undergoing significant degradation, often taking decades or longer to break down. For example, substances like DDT and the majority of plastics.
3. Non-degradable pollutants are substances that cannot be broken down by natural processes. Once they are introduced into the environment, they pose a significant challenge to eliminate and persistently accumulate. For example, harmful substances such as lead or mercury.

13.12 AIR POLLUTION

13.12.1 History of Air Pollution

Air pollution was referenced by Hippocrates circa 400 BC. The issue of smoke pollution was acknowledged 700 years ago in London. This led King Edward I to enact the initial legislation against pollution in 1273, prohibiting the use of coal for household heating.

In 1300, a new law was passed to forbid the use of coal. The act of violating the law led to the enforcement of capital punishment. During the industrial revolution, the use of coal in industrial activity led to the emergence of air pollution as a significant problem in London. The initial recorded major disaster was the 'London Smog' in 1952, resulting in almost 4000 deaths due to the extended occurrence of air pollution in the metropolis for a period of five days.

In the mid-19th century, a melanistic form of the Peppered moth was identified in European countries characterized by significant industrialization. Usually, the common Peppered moth is well hidden on a clean tree that is covered with lichen. The introduction of transportation networks and the widespread use of petrol and diesel in the early 20th century led to an increase in air pollution. The first instance of notable air pollution induced by the production of photochemical haze from the emissions of diesel and petrol engines was observed in Los Angeles (Figure 13.4).

Figure 13.4 Photochemical haze in Los Angeles.

Auto-exhaust pollution remains a prominent environmental issue in many developed and developing countries, including India. The Air Pollution Control Act was implemented in India in 1981, whereas the Motor Vehicle Act, which focuses on the regulation of air pollution, was enacted more recently. The objective of this act is to alleviate air pollution.

The Bhopal disaster, which took place on the evening of December 3rd, 1984, involved the accidental release of a highly toxic gas called methyl isocyanide from Union Carbide's pesticide production facility. This incident caused a major industrial calamity and resulted in serious contamination of the atmosphere (Figure 13.5). The long-lasting consequences of this disaster on both human health and the soil continue to exist up until the present time.

Figure 13.5 Contamination of the atmosphere occur due to Bhopal gas tragedy.

13.12.2 Levels of Atmosphere

The composition of the atmosphere typically consists of 79 percent nitrogen, 20 percent oxygen, and one percent as a blend of carbon dioxide, water vapor, and small quantities of several other gases like neon, helium, methane, krypton, hydrogen, and xenon. The atmospheric composition exhibits various significant characteristics that are pertinent to environmental issues.

The atmosphere is segmented into multiple layers, as follows.

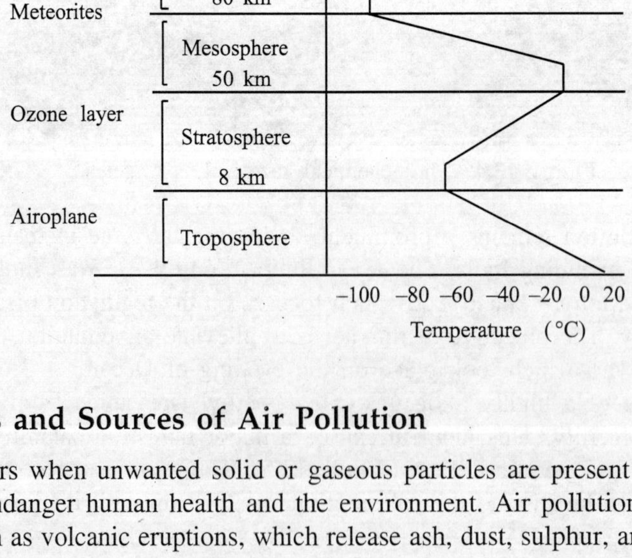

13.12.3 Types and Sources of Air Pollution

Air pollution occurs when unwanted solid or gaseous particles are present in the atmosphere at amounts that endanger human health and the environment. Air pollution can occur due to natural events such as volcanic eruptions, which release ash, dust, sulphur, and other pollutants, or periodic forest fires that are triggered by lightning. In contrast to pollutants produced by human activities, naturally occurring pollutants have a short duration in the atmosphere and do not result in long-term changes to the atmosphere. Pollutants are produced by both natural phenomena, including dust storms and volcanic eruptions, and human activities, such as emissions from vehicles and industries. These compounds are commonly known as primary pollutants.

Five primary pollutants jointly contribute to around 90 percent of air pollution worldwide. The substances present are carbon oxides (CO and CO_2), nitrogen oxides, sulphur oxides, volatile organic compounds (VOCs) (mostly hydrocarbons), and suspended particulate matter (SPM). Chemical reactions between primary pollutants give rise to secondary pollutants in the environment. Examples include sulphuric acid, nitric acid, carbonic acid, and others.

Table 13.1 Sources of Indoor Air Pollution

Pollutant	Sources	Threat
Chloroform	Chlorine-treated water in hot showers	Cancer
Para-dichlorobenzene	Air fresheners, mothball crystals	Cancer
Tetrachloroethylene	Dry-cleaning fluid fumes on clothes	Nerve disorders, damage to liver and kidneys, possible cancer
Formaldehyde	Furniture stuffing, paneling, particle-board, foam insulation	Irritation of eyes, throat, skin, and lungs; nausea; dizziness
Styrene	Carpets, plastic products	Kidney and liver damage
Benzo-α-pyrene	Tobacco smoke, woodstoves	Lung cancer
Radon-222	Radioactive soil and rock surrounding foundation, water supply	Lung cancer
Tobacco smoke	Cigarettes	Lung cancer, respiratory ailments, heart disease
Carbon monoxide	Faulty furnaces, unvented gas stoves and kerosene heaters, woodstoves	Headaches, drowsiness, irregular heartbeat, death
Asbestos	Pipe insulation, vinyl ceiling and floor tiles	Lung disease, lung cancer
Particulates	Pollen, pet dander, dust mites, cooking smoke particles	Irritated lungs, asthma attacks, itchy eyes, runny nose, lung disease
Nitrogen oxides	Unvented gas stoves and kerosene heaters, woodstoves	Irritated lungs, children's colds, headaches
1,1,1-Trichloroethane	Aerosol sprays	Dizziness, irregular breathing
Methylene chloride	Paint strippers and thinners	Nerve disorders, diabetes

13.12.3.1 Major Outdoor Air Pollutants

1. **Ozone (O_3):** Ozone is present in both the high and lower atmosphere of Earth. The ozone layer in the upper atmosphere shields the Earth from detrimental solar radiation like UV, whereas in the lower atmosphere, it poses significant risks to human health.

$$\underset{\text{Nitrogen oxides}}{NO_x} + \underset{\text{Volatile organic compounds}}{VOC} \longrightarrow \underset{\text{Ozone}}{O_3}$$

2. **Carbon Monoxide:** Carbon monoxide is a colorless, odorless, and poisonous gas that is generated during the incomplete combustion of organic substances like natural gas, coal, or wood. Vehicle emissions are the primary contributor of carbon monoxide. The vehicles are also inadequately maintained and many of them lack sufficient pollution control technology, leading to the emission of higher levels of carbon monoxide.

$$\underset{\text{(gasoline)}}{\text{Fuel}} + O_2 \longrightarrow \underset{\substack{\text{Primary}\\\text{pollutants}}}{CO/CO_2 + H_2O}$$

3. **Sulphur Dioxide (SO_2):** Generated from the combustion of coal that contains pyrite, which reacts with oxygen in the environment. Sulphuric acid (acid rain) is formed when moisture in the atmosphere comes into contact with sulphur compounds.

$$\underset{\substack{\text{in coal}\\\text{and oil}}}{S} \xrightarrow{\text{Burning}} \underset{\substack{\text{Sulphur}\\\text{dioxide}}}{SO_2} \xrightarrow{O_2} \underset{\substack{\text{Sulphur}\\\text{trioxide}}}{SO_3} \xrightarrow[\substack{\text{Water}\\\text{vapour}}]{H_2O} \underset{\substack{\text{Sulphuric}\\\text{acid}}}{H_2SO_4} \xrightarrow{NH_3} \underset{\substack{\text{Ammonioum}\\\text{sulphate}}}{(NH_4)_2SO_4}$$

The acid generated causes deterioration in the quality of monuments, architecture, and statuary (Figure 13.6).

Figure 13.6 Deterioration of statue due to acid rain.

4. **Nitrogen Dioxide (NO_2):** Nitrogen Dioxide (NO_2) is a chemical compound generated from the combustion of fuels. A prevalent constituent of smog, which is a combination of smoke and fog. Typically found in industrial zones and locations with a dense presence of motor vehicles.

$$\underset{\substack{\text{in coal}\\\text{and oil}}}{N} \xrightarrow{\text{Burning}} \underset{\substack{\text{Nitric}\\\text{oxide}}}{NO} \xrightarrow{O_2} \underset{\substack{\text{Nitrogen}\\\text{dioxide}}}{NO_3} \xrightarrow[H_2O]{hv} \underset{\substack{\text{Nitric}\\\text{oxide}}}{NO} + \underset{\substack{\text{Oxygen}\\\text{atom}}}{O} \xrightarrow{O_2} \underset{\text{Ozone}}{O_3} \xrightarrow{\text{VOCs}} \text{PANs and other pollutants}$$

$$NO_3 \xrightarrow{\text{Hydrocarbons}} \text{Peroxyacyl nitrates (PANS)}$$

5. **Hydrocarbons:** Hydrocarbons are a class of chemical compounds composed of carbon and hydrogen atoms. They can either evaporate from fuel supply or be residual remains of fuel that underwent incomplete combustion. Hydrocarbons are removed from the atmosphere by precipitation and then flow into bodies of water. They create a greasy layer on the surface and do not pose a significant problem until they undergo a reaction to produce secondary pollutants. Modifications such as increasing the oxygen concentrations in the fuel-air combination and installing valves to prevent gas leakage, as well as equipping automobiles with catalytic converters, can effectively decrease the emission of hydrocarbons into the atmosphere.
6. **Lead:** Lead, a significant airborne contaminant, is predominantly unmonitored and emitted by cars. Elevated levels of lead have been documented in the atmospheric air in urban areas. Leaded petrol is the main contributor to the release of lead particles into the air in Indian cities.
7. **Particulate Matter:** Particulate matter denotes solid particles present in the atmosphere. Common examples of particulate matter consist of soil, construction dust, shards of tire rubber, asbestos derived from brake linings, and emissions from cars.

Types of Particulates

Term	Meaning	Examples
Aerosol	A broad term used to describe particles that are suspended in the air.	Sprays from pressurized cans
Mist	An aerosol composed of small liquid droplets.	Sulphuric acid mist
Dust	An aerosol composed of solid particles that are either blown into the air or created by grinding bigger particles into smaller ones.	Dust storm
Smoke	An aerosol that is composed of solid particles or a combination of solid and liquid particles. It is typically generated through chemical reactions, such as those occurring during fires.	Cigarette smoke, smoke from burning garbage
Fume	The emission of aerosols resulting from the condensation of hot metal vapors, however it is generally synonymous with smoke.	Zinc/lead fumes
Plume	The geometric configuration or structure of the smoke emitted from a chimney.	
Fog	An aerosol composed of small water droplets.	
Smog	A term that refers to a combination of smoke and fog.	

13.12.4 Effects of Air Pollution

13.12.4.1 Effects on Living Organisms

The respiratory system employs multiple mechanisms to protect us from air pollution. Nasal hair functions as a filtration mechanism to eliminate large particles. The thick mucus in the epithelial lining of the upper respiratory tract traps tiny particles and chemically degrades certain gaseous pollutants. Sneezing and coughing are physiological responses via which the upper respiratory system expels contaminated air and mucus when it is exposed to contaminants.

Prolonged exposure to air pollution or long-term smoking can overwhelm or damage the body's natural defenses, resulting in or worsening illnesses such as lung cancer, asthma, chronic bronchitis, and emphysema. Vulnerable populations, such as the elderly, infants, pregnant women, and individuals with heart disease, asthma, or other respiratory disorders, are especially prone to the detrimental impacts of air pollution.

Ozone in the lower atmosphere can cause chronic respiratory tract infections, eye discomfort, coughing, and shortness of breath, nausea, wheezing, and headaches.

Exposure to carbon monoxide in heavy traffic results in symptoms such as migraines, drowsiness, and compromised eyesight. Carbon monoxide exposure can lead to a range of symptoms. At lower levels, individuals may experience symptoms such as headaches, fatigue, and flu-like symptoms. At elevated levels, symptoms may encompass compromised visual and auditory abilities, challenges in executing precise motor movements, and diminished ability to focus. Carbon monoxide is a very poisonous gas that has a greater tendency to bind to iron (haemoglobin) in red blood cells than oxygen. Carbon monoxide accumulates in the bloodstream and reduces the blood's capacity to carry oxygen due to its long-lasting attachment to hemoglobin. These consequences encompass diminished cognitive function and logical thinking, reduced reflexes, as well as the manifestation of headaches, drowsiness, dizziness, nausea, and maybe even death.

Exposure to sulphur dioxide can lead to respiratory infections, diminished lung function, and worsening of preexisting heart disease. Extended exposure leads to a disease that closely resembles bronchitis. Furthermore, it reacts with water, oxygen, and other molecules found in the environment, leading to the creation of sulphur-containing acids. Upon inhalation, the acids have the ability to attach to particles, resulting in significant corrosion within the lungs.

Nitrogen oxides, specifically NO_2, has the capacity to induce lung irritation, exacerbate symptoms of asthma or chronic bronchitis, and increase susceptibility to respiratory infections such as influenza or common colds. Exposure to nitrogen dioxide can lead to symptoms such as eye irritation, nasal congestion, coughing, throat pain, lung inflammation, and various respiratory illnesses. Exposure to particulate matter can lead to irritation of the respiratory system, illness, and specific forms of cancer.

Cigarette smoking is the main factor responsible for the most significant degree of carbon monoxide exposure. Exposure to as little as 0.001 percent of carbon monoxide in the air for several hours can lead to collapse, coma, and potentially deadly outcomes. Smoking possesses the capacity to inflict harm onto numerous biological systems, resulting in the emergence of cancer and chronic illnesses.

Particulate pollution worsens the symptoms of bronchitis and asthma. Extended exposure to these particles leads to the degradation of lung tissue and contributes to the development of chronic respiratory illness and cancer.

Several volatile organic compounds, such as benzene and formaldehyde, along with toxic particles like lead and cadmium, have the capacity to cause mutations, reproductive problems, or cancer. Inhaling ozone, a component of photochemical smog, causes symptoms like coughing, chest pain, breathing difficulties, and irritation of the eyes, nose, and throat.

13.12.4.2 Impact on Vegetation

Gaseous contaminants that infiltrate leaf stomata cause harm to agricultural plant foliage. Prolonged exposure of the leaves to air pollution can degrade the hydrophobic layer that serves as a barrier against excessive transpiration, resulting in susceptibility to diseases, pests, drought, and frost damage. This level of exposure disrupts the process of photosynthesis and stunts plant growth, impairs the absorption of nutrients, and leads to the discoloration, browning, or complete shedding of leaves (Figure 13.7). At an elevated concentration of sulphur dioxide, the majority of flower buds undergo a transformation, becoming rigid and inflexible. Ultimately, they drop from the plants due to their inability to bloom. Extended exposure to elevated concentrations of various air pollutants emitted by smelters, coal-fired power plants, industrial facilities, as well as vehicles, can cause harm to trees and other vegetation.

Figure 13.7 Impact of polluted environment on vegetation.

13.12.4.3 Impacts of Air Pollution on the Stratosphere

The upper stratosphere harbors substantial amounts of ozone, acting as a highly effective shield against UV rays. The ozone layer, referred to as this stratum, extends vertically for a distance of 60 kilometers from the Earth's surface. Ozone is present at altitudes of up to 60 kilometers, with its most abundant levels occurring between 20 to 25 kilometers. The ozone layer is not solely comprised of ozone, but rather a mixture of other abundant atmospheric molecules. In the densest part of the ozone layer, the proportion is such that there is a single ozone molecule for every 100,000 gas molecules. Therefore, even slight modifications in the ozone concentration can have significant effects on life on Earth. Total column ozone is the aggregated amount of ozone in the atmosphere, extending from the Earth's surface to an altitude of 50 km. The measurement is denoted in Dobson Units (DU), which represents the thickness of the ozone layer as an analogous layer of pure ozone gas under standard temperature and pressure conditions at sea level. Thus, it can be deduced that a measurement of 100 Dobson Units (DU) equates to a thickness of 1 millimeter of pure ozone gas, assuming standard temperature and pressure conditions at sea level. Ozone is a molecular form of oxygen that contains three atoms instead of the typical two. Ozone is generated via the natural phenomenon of sunlight dissociating oxygen gas molecules in the atmosphere. The ozone produced is constantly broken down by naturally occurring processes that maintain its balance in the ozone layer.

Depletion of Ozone: Changes in the ozone layer have important consequences for humanity. In the absence of contaminants, the production and breakdown of ozone are exclusively controlled by natural systems. Nevertheless, the presence of some contaminants might accelerate the breakdown of ozone. Prior understanding recognized that ozone levels vary and occasionally undergo slight reduction. However, it was not until 1985 that the substantial degradation of the ozone layer, often known as the Ozone Hole, attracted widespread notice. This event transpired when British researchers disseminated measurements pertaining to the ozone layer.

Subsequent to these findings, scientific investigation on the ozone layer experienced substantial acceleration, presenting solid proof that CFCs were responsible for its depletion. Chlorofluorocarbons (CFCs) are highly stable compounds that exhibit both non-flammability and non-toxicity, rendering them safe for handling. These qualities make them very suitable for various industrial applications, including aerosols, air conditioners, freezers, and fire extinguishers. Various aerosol cans employ chlorofluorocarbons (CFCs) for the purpose of generating foam and spray, such as in scents and air fresheners. CFCs are also utilized in the manufacturing of foams used in mattresses and cushions, disposable Styrofoam cups, glasses, packaging material for insulation, and cold storage, among other uses.

However, their stability also grants them an extended lifespan in the environment. Halons have a similar structure to CFCs, however they replace chlorine atoms with bromine ones. They present a more significant hazard to the ozone layer compared to CFCs. Halons are employed as fire suppressants because they do not pose a threat to people and equipment in close vicinity during firefighting operations. Halons, a kind of CFCs, must be transported by air currents until they reach a position just above the lower atmosphere due to their greater density relative to air. Subsequently, they gradually scatter over the higher regions of the atmosphere.

This is a lengthy process that can span from five to fifteen years. Unfiltered UV radiation in the stratosphere induces the dissociation of chemical bonds, leading to the release of chlorine from the remaining CFC molecules. This process entails the deliberate attack on the ozone molecule, resulting in its fragmentation into an oxygen molecule and an oxygen atom. While CFCs are widely dispersed worldwide, the depletion of ozone is most prominent in the vicinity of the South Pole as a result of the harsh weather conditions in the Antarctic region.

The ice crystals enhance the Cl-O bonding. The ozone layer over countries such as Australia, New Zealand, South Africa, and specific areas of South America is also undergoing reduction. In 1992, India formally approved the Montreal Protocol, an international agreement aimed at controlling the production and utilization of chemicals that cause ozone depletion.

Mechanism of Ozone Depletion

Step 1: Ultraviolet radiation induces the dissociation of a chlorine atom from the chlorofluorocarbon (CFC) molecule.

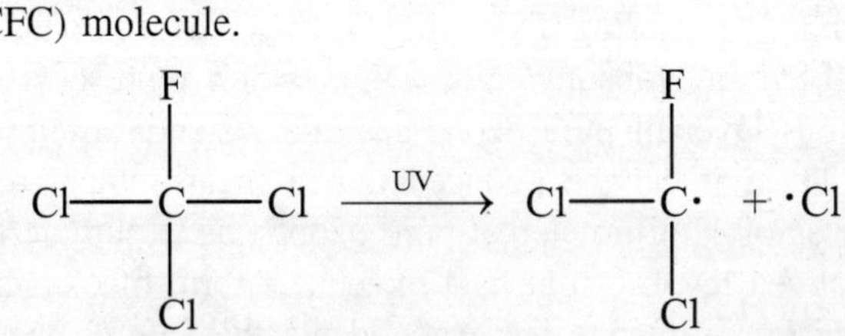

Step 2: The unbound chlorine atom collides with an ozone molecule. The chlorine atom displaces one oxygen atom.

$$\cdot Cl + O_3 \longrightarrow \underset{\text{Chlorine monoxide}}{ClO} + O_2$$

Step 3: A free oxygen atom collides with the chlorine monoxide molecule. The outcome yields an additional free chlorine atom. The presence of free radicals will persistently diminish the ozone layer in the stratosphere.

$$\underset{\text{Free oxygen from stratosphere}}{ClO + O_3} \longrightarrow \cdot Cl + O_2$$

Effects on Human Health: Human health is negatively impacted by increased levels of ultraviolet radiation, leading to sunburn, cataracts, premature skin aging, and an increased risk of skin cancer. It compromises the immune system by inhibiting the body's overall resistance to specific infections, including as measles, chickenpox, and other viral diseases that cause rashes, as well as parasite diseases like malaria that enter through the skin.

Food Production: Ultraviolet radiation hampers plants' capacity to harness light energy in the process of photosynthesis. This diminishes the nutritional composition and stunts the growth of plants. This phenomenon is particularly evident in beans and cabbage. Ultraviolet light causes harm to both plant and animal plankton. Changes in radiation lead to a reduction in the breeding time of zooplanktons, which are minute creatures. Planktons, being the foundation of the marine food chain, have a direct impact on the production of fish and shellfish due to changes in their population and species composition.

Impact on Materials: Elevated UV radiation leads to the deterioration of paints and fabrics, resulting in accelerated fading.

Impact on Climate: Pollution-induced atmospheric changes contribute to global warming, a phenomena resulting from the elevated concentration of certain gases such as carbon dioxide, nitrogen oxides, methane, and CFCs.

Empirical evidence has unequivocally demonstrated that atmospheric elements, including water vapor, carbon dioxide, methane, nitrogen oxides, and Chloroflurocarbons, effectively retain heat in the form of infrared radiation in close proximity to the Earth's surface. This phenomenon is commonly referred to as the 'Greenhouse Effect'.

The phenomenon exhibits similarities to the processes occurring within a greenhouse. The glass in a greenhouse permits the entry of solar radiation, which is then absorbed by the items within. These objects emit heat in the form of terrestrial radiation, which is unable to travel through the glass. Consequently, the greenhouse effectively retains the heat, resulting in elevated temperatures within and promoting the flourishing growth of plants.

A portion of solar energy is reflected by both the Earth and the atmosphere. The majority of radiation is absorbed by the Earth's surface, resulting in its warming. A portion of the infrared light is able to penetrate the atmosphere. Greenhouse gas molecules absorb and subsequently re-emit a portion of the energy in various directions. This phenomenon results in the elevation of the Earth's surface temperature and the lower portion of the atmosphere. The Earth's surface emits infrared light.

Global Warming can have Numerous Detrimental Consequences: The increase in global temperature will lead to the melting of the polar ice caps, resulting in a rise in sea levels and subsequent flooding of coastal regions. This would have disastrous consequences in countries such as Bangladesh or the Maldives. In the event of a 3-meter rise in sea level, the Maldives would be entirely submerged beneath the ocean. An increase in temperature will result in a decrease in agricultural yield. Alterations in the allocation of solar energy can lead to modifications in habitats. An agricultural region that was once productive would experience significant droughts, while areas that were previously desert-like will receive rainfall.

This has the potential to induce alterations in the genetic makeup of natural flora, cultivated crops, insects, cattle, and microorganisms. The Polar Regions would experience catastrophic consequences due to the temperature increases caused by global warming. Alaska harbors immense amounts of methane that are confined beneath the frozen earth. The melting of permafrost can trigger the release of methane, which has the potential to expedite the progression of global warming.

13.12.5 Control Measures for Air Pollution

There are two primary methods for managing air pollution:

13.12.5.1 Methods for Preventing and Controlling Effluent

Installing appropriate equipment is an efficient method for mitigating air pollution. This encompasses various equipment such as scrubbers, closed collection recovery systems, dry and wet collectors, filters, and electrostatic precipitators, which are employed to eliminate pollutants from flue gases and prevent their escape. Elevating the stacks to a larger height can effectively promote the dispersion of pollutants, minimizing their proximity to the ground. Industries ought to be situated in locations that minimize pollution impacts, taking into account the topography and prevailing wind directions. It is possible to replace raw materials that contribute to higher levels of pollution with ones that result in lower levels of pollution.

13.12.5.2 Air Quality Monitoring

Air quality monitoring refers to the process of measuring and assessing the level of pollutants and contaminants present in the air. Currently, India lacks a robust framework for monitoring air pollution. During the late 1960s, when air quality monitoring was initiated in India, the planners primarily concentrated on monitoring a limited number of pollutants, including sulphur dioxide, nitrogen oxides, and suspended particle matter. Carbon monoxide and lead were only monitored to a limited extent as additional contaminants. The potential risks posed by other airborne pollutants, such as benzene, ozone, and other fine particles, remain uncertain due to the lack of monitoring for these substances. The National Environmental Engineering Research Institute (NEERI), Nagpur, has compiled a database on ambient air quality in Indian cities through its monitoring networks. In 1985, the Central Pollution Control Board (CPCB) launched its own nationwide program called the National Ambient Air Quality Monitoring (NAAQM) program. Ambient air quality standards in India developed by the CPCB is tabulated in Table 13.2.

13.12.6 Legal Aspects of Air Pollution Control in India

The Air (Prevention and Control of Pollution) Act was enacted in 1981. The Act was enacted to ensure the prevention, control, and reduction of air pollution. In regions designated under this legislation, no industrial activity that causes pollution may commence without obtaining permission from the relevant State Pollution Control Board. However, this Act lacked the necessary strength to effectively serve as a preventive or remedial measure. In the aftermath of the Bhopal tragedy, the government enacted a more extensive Environment Protection Act (EPA) in 1986.

This legislation granted enforcement agencies the authority to use punitive measures for the purpose of limiting any activities that may cause harm to the environment. The Central Motor Vehicles Act of 1939 was modified in 1989 in order to reduce vehicular pollution. After this modification, the regulations regarding exhaust emissions for vehicle owners were officially announced in 1990, and the requirements for mass emissions set for car manufacturers were first implemented in 1991. The emission standards for mass have been further changed for the year 2000. India has not yet developed a clearly defined program for managing air quality. In order to improve air quality, it is crucial to enhance air quality management through the use of continuous air monitoring. An integrated approach, along with stringent air pollution control legislation, would be necessary. Real-time Air Quality Index (AQI) of top 10 Indian cities are tabulated in Table 13.3.

Several recommendations for accomplishing this task include:

1. Prioritizing pollution prevention over control measures.
2. Decreasing the reliance on fossil fuels.
3. Enhancing the caliber of automotive gasoline.
4. Enhancing the utilization of sustainable energy sources.

Table 13.2 Ambient air quality standards in India developed by the CPCB

Area Category	**SPM μg/m^3**	**SO_2 μg/m^3**	**CO μg/m^3**	**NO_x μg/m^3**
Industrial and mixed use	500	120	5000	120
Residential and rural	200	80	2000	80
Sensitive	100	3	1000	30

Table 13.3 Real-time Air Quality Index of top 10 cities of India on 23 Dec. 2023

Rank	**City**	**AQI**	**Status**
1.	Faridabad	466	Hazardous
2.	Bhiwadi	462	Hazardous
3.	Ghaziabad	461	Hazardous
4.	Sonipat	448	Hazardous
5.	New Delhi	447	Hazardous
6.	Noida	436	Hazardous
7.	Rohtak	435	Hazardous
8.	Meerut	435	Hazardous
9.	Gurgaon	425	Hazardous
10.	Bulandshahr	405	Hazardous

Source: https://www.aqi.in/winter-air-pollution-ranking

SOLVED PROBLEMS

PROBLEM 13.1 Calculate the gross and net calorific values of coal having the following compositions, carbon = 85%, hydrogen = 8%, sulphur = 1%, nitrogen = 2%, ash = 4%, latent heat of steam = 587 cal/gm.

Solution

Higher Calorific Value (HCV)

$$HCV = \frac{1}{100}\left[8,080 \times \%C + 34,500\left(\%H - \frac{O\%}{8}\right) + 2,240 \times \%S\right] \text{ kcal/kg}$$

$$HCV = \frac{1}{100}\left[8,080 \times 85 + 34,500\left(8 - \frac{0}{8}\right) + 2,240 \times 1\right] \text{ kcal/kg}$$

$$= \frac{1}{100}[6,86,800 + 276000 + 2,240] \text{ kcal/kg}$$

$$= \frac{1}{100}[9,65,040] \text{ kcal/kg}$$

$$= 9650.4 \text{ kcal/kg}$$

Net Calorific Value (NCV):

$$NCV = GCV - \frac{9}{100} H \times 587 \text{ kcal/kg}$$

$$= 9650.4 - \frac{9}{100} \times 8 \times 587 \text{ kcal/kg}$$

$$= 9650.4 - 422.64 \text{ kcal/kg}$$

$$= 9227.76 \text{ kcal/kg}$$

PROBLEM 13.2 Calculate the net and gross calorific value of a coal sample having following composition: C = 82%, H = 8%, O = 5%, N = 1.4% and ash = 3.6%.

Solution

Higher Calorific Value (HCV):

$$HCV = \frac{1}{100}\left[8{,}080 \times \%C + 34{,}500\left(\%H - \frac{O\%}{8}\right) + 2{,}240 \times \%S\right] \text{ kcal/kg}$$

$$HCV = \frac{1}{100}\left[8{,}080 \times 82 + 34{,}500\left(8 - \frac{5}{8}\right) + 0\right] \text{ kcal/kg}$$

$$= \frac{1}{100}[662560 + 254437.3] \text{ kcal/kg}$$

$$= 9169.8 \text{ kcal/kg}$$

Net Calorific Value (NCV)

$$NCV = GCV - \frac{9}{100} H \times 587 \text{ kcal/kg}$$

$$= 9169 - \frac{9}{100} \times 8 \times 587 \text{ kcal/kg}$$

$$= 8747.34 \text{ kcal/kg}$$

PROBLEM 13.3 Calculate the gross and net calorific valve of a fuel having following composition: 82%, C, 8% H, 5% O, 2.5% S, 1.4% N and 2.1% ash.

Solution

Higher Calorific Value (HCV)

$$HCV = \frac{1}{100}\left[8{,}080 \times \%C + 34{,}500\left(\%H - \frac{O\%}{8}\right) + 2{,}240 \times \%S\right] \text{kcal/kg}$$

$$HCV = \frac{1}{100}\left[8{,}080 \times 82 + 34{,}500\left(8 - \frac{5}{8}\right) + 2240 \times 2.5\right] \text{kcal/kg}$$

$$= 9225.97 \text{ kcal/kg}$$

Net Calorific Value (NCV)

$$NCV = GCV - \frac{9}{100} H \times 587 \text{ kcal/kg}$$

$$= 9225.97 - \frac{9}{100} \times 8 \times 587 \text{ kcal/kg}$$

$$= 8803.3 \text{ kcal/kg}$$

PROBLEM 13.4 The following data is obtained in a bomb calorimeter experiment:

Weight of crucible = 3.649 gm

Weight of crucible + fuel = 4.678 gm

Water equivalent of calorimeter = 570 gm

Water taken in the calorimeter = 2200 gm

Observed rise in temperature = 2.3 °C

Cooling correction = 0.047 °C

Acid correction = 62.6 °C

Fuse wire correction = 3.8 °C

Cotton thread correction = 1.6 calories

Calculate the gross calorific value of the fuel sample. If the fuel contain 6.5% H, determine the net calorific value.

Solution

Weight of fuel = 4.678 − 3.649 = 1.029 g

Gross Calorific Value:

$$\text{GCV} = \frac{[(W + w)(t_2 - t_1 + \text{cooling correction})] - (\text{Acid + Fuse wire + cotton thread correction})}{\text{Weight of the fuel } (x)}$$

$$\text{GCV} = \frac{[(2200 + 570)(2.3 + 0.047)] - (62.6 + 3.8 + 1.6)}{1.029}$$

$$= 6251.88 \text{ cal/g}$$

Now, Net calorific value = [GCV – (0.09 × H × 587)] cal/g
= (6251.88 – 0.09 × 6.5 × 587) cal/g
= (6251.88 – 343.99) cal/g = 5907.89 cal/g.

PROBLEM 13.5 A sample of coal contains: C =83%, H = 6% and ash 1%. The following data were obtained when the above coal was tested in a bomb calorimeter:
Weight of coal burnt = 0.92 g
Weight of water taken = 550 g
Water equivalent of calorimeter = 2,200 g
Rise in temperature = 2.42°C
Fuse wire correction = 10.0 cal
Acid correction = 50.0 cal
Calculate the gross and net calorific values of coal, assuming that the latent heat of condensation of steam ss 580 cal/g.

Solution

Weight of coal sample (x) = 0.92 g
Weight of water (W) = 550 g
Water equation of calorimeter (w) = 2200 g
Rise in temperature ($T_2 - T_1$) = 2.42°C
Acid correction = 50.0 cal
Fuse wire correction 10.0 cal
Latent heat of steam 580 cal/g
% of H = 6%

Gross Calorific Value:

$$\text{GCV} = \frac{[(W + w)(t_2 - t_1)] - (\text{Acid + Fuse wire correction})}{\text{Weight of the fuel } (x)}$$

$$\text{GCV} = \frac{(550 + 2200) \times 2.42 - (50 + 10)}{0.92}$$

$$= 7{,}168.5 \text{ cal/g}$$

Now, Net calorific value =

$$NCV = [GCV - 0.09H \times \text{Latent heat of steam}] = (7{,}168.5 - 0.09 \times 6 \times 580) \text{ cal/g}$$
$$= 6{,}855.3 \text{ cal/g}$$

QUESTIONS

13.1 Define GCV and NCV? Calculate the gross and net calorific value of coal sample having the following composition: C = 80%, H = 7%, O = 3%, S = 3.5%, N = 2.1% and ash = 4.4%.

13.2 Give the composition of biogas. With the help of diagram, explain a biogas plant.

13.3 Explain GCV and NCV. A sample of coal contains C = 83%, H = 6% and ash = 1%. The following data were obtained when the above coal was tested in a bomb calorimeter:

Weight of coal burnt: 0.92 g

Weight of water taken: 550 g

Water equivalent of calorimeter: 2,200 g

Rise in temperature: 2.42 °C

Fuse wire correction: 10 Cal

Acid correction: 50.0 Cal

Calculate the gross and net calorific value of coal, assuming that the latent heat of condensation of steam is 580 cal/g.

13.4 Discuss the principle of the bomb calorimeter. How is it used to determine the GCV and LCV of the Fuel?

13.5 How does Gross Calorific Value differ from Net Calorific Value?

13.6 Explain the different parameters that are determined in the proximate analysis of coal. On burning 0.92 g of a solid fuel in a bomb calorimeter, the temperature of 3300 g of water increased by 2.42°C. Water equivalent of calorimeter and latent heat of steam are 385.0 g and 587.0 cal/g, respectively. If the fuel contains 0.7% hydrogen, calculate its GCV and NCV.

13.7 Explain the construction and working principle of bomb calorimeter. A sample of coal contains 80% C, 15% H, and 5% Ash. The following data were obtained when the above coal sample was tested in bomb calorimeter:

Weight of coal burnt = 0.98 g

Weight of water taken = 1000 g

Water equivalent of bomb calorimeter = 2500 g

Observed rise in temperature = 2.5°C

Fuse wire correction = 8 cal

Acid correction = 50 cal

Cooling correction = 0.02°C

Calculate gross and net calorific value of coal if the latent heat of condensation of water is 580 cal/g.

13..8 What is the difference between Gross calorific value and Net calorific value of a fuel?

13.9 With the help of a neat diagram, explain the construction and working of bomb calorimeter.

13.10 What is the basic principle of bomb calorimeter? A 0.80 g sample of solid fuel was completely combusted in the excess of oxygen using bomb calorimeter. The rise in temperature of water in calorimeter was 2.5°C. Calculate the High calorific value of the fuel. If water taken in calorimeter is 2000 g and water equivalent of calorimeter is 2200 g. Also calculate low Calorific value. (Given: % Hydrogen in fuel = 2.2).

13.11 Describe proximate analysis of fuels. A coal sample has the following composition by weight: C = 90%, O = 3%, S = 0.5%, N = 0.5% and Ash = 2.5%. Net calorific value of the coal was found to be 8,490.5 kcal/kg. Calculate the percentage of hydrogen and Gross calorific value.

13.12 A fuel sample contains 92% C, 5% H and 2% ash gave following analysis on combustion by bomb Calorimeter: Weight of coal burnt = 0.95g. Weight of water taken = 750 g, Water equivalent of calorimeter = 2000 g, Rise in temperature-2.8°C, Cooling correction 0.20°C, Fuse wire correction = 10.0 cal, Acid correction= 40.0 cal. Calculate GCV and NCV of the fuel.

13.13 Write the relationship between high and low calorific values. If HCV of a coal sample is 7500 cal/g and % H is 5 then calculate the low calorific value of the fuel. (Given that latent heat of condensation of steam is 580 cal/g.).

13.14 Define determination of calorific value of fuel by bomb calorimeter. The Following data is obtained in a bomb calorimeter experiment: Wt. of Fuel = 2.02 gm, Water equivalent of calorimeter = 585 gm, Water taken in calorimeter = 2500 gm, Observed rise in temperature =2.5°C, Calculate the gross calorific value of fuel sample. If the fuel contain 8.5% Hydrogen, determine the LCV of fuel. (Latent heat of Steam = 587 cal/gm).

CHAPTER 14

Lubricants

14.1 INTRODUCTION

Lubricants are chemicals that are used to reduce the amount of friction that exists between the moving parts of the machine that are in contact with one another. The resistance to relative motion of the surfaces is created whenever there is movement on the surface of machinery that is moving across the surface. With the human eye, the solid surface appears to be smooth; nevertheless, when viewed under a high-powered microscope, this smooth surface reveals imperfections in the form of projections and voids.

When such a surface is positioned on top of another, the projections of the first surface fall into the cavities of the second surface, causing them to become entangled (Figure 14.1). Since the surfaces are interlocking with one another, there is resistance to the relative motion of the surfaces. The frictional forces or frictional resistance of friction are the names given to this phenomenon. The previous projections eventually break, and malformations appear as a result of the natural sequence of events.

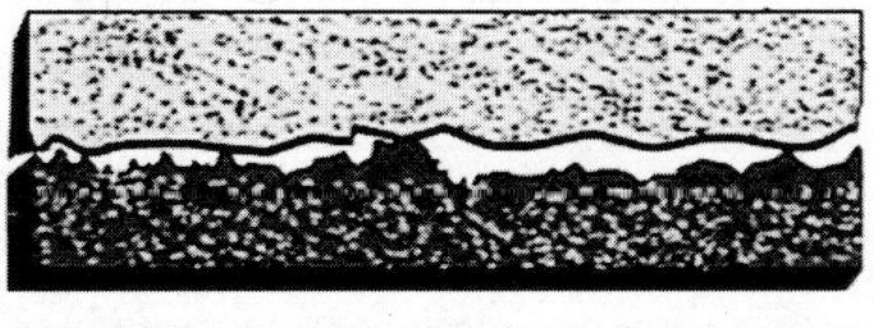

Figure 14.1 Irregular project of two surcaces.

14.2 FRICTION

The opposing force that is created between the surfaces of contact when one body moves across the surface of another body is referred to as friction. Friction can be defined as the force that is created.

The relative motion that occurs between the moving parts of a machine is something that is opposed by the frictional forces (Figure 14.2). Therefore, additional energy must be expended in order to overcome the friction, which results in a rise in the costs associated with energy. In addition, heat is produced as a result of the friction that occurs between the moving elements of machines, which in turn causes damage to the machinery. The moving elements of equipment that are in contact with one another are subjected to wear and tear as a result of friction, which ultimately results in the machines losing their efficiency and becoming obsolete.

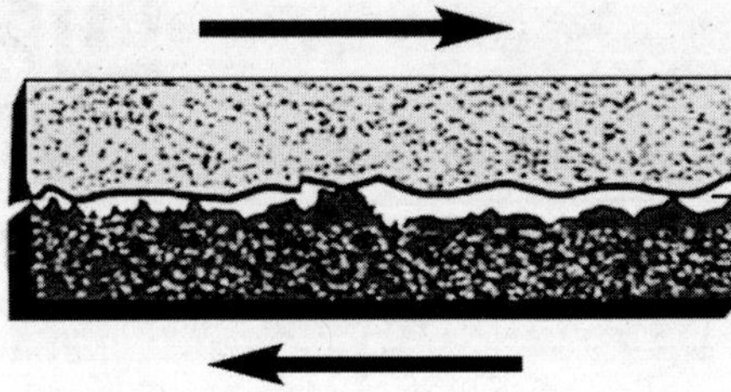

Figure 14.2 Movement two irregular surfaces.

14.3 FUNCTIONS OF LUBRICANTS

1. Lubricants reduce the amount of heat that is produced, which lessens the likelihood that the moving parts of machines will be damaged.
2. Lubricants are what keep the moving parts of machines apart, which helps lessen the amount of wear and tear that machines experience.
3. The use of lubricants lowers the costs associated with the machine's operation and maintenance.
4. Lubricants perform the function of coolant because they reduce the amount of heat that is produced by the moving parts of the machine that are in contact with one another.
5. Lubricants cut down the amount of energy that is wasted, which in turn boosts the efficiency of the machine.
6. The use of lubricants causes the relative motion of the machine's moving parts to become smooth, which thus reduces the noise level the machine made while it was running.
7. Lubricants also serve the purpose of preventing corrosion in most cases.
8. As with the piston, lubricants also perform the function of a seal. Lubricant that is utilized between the piston and the walls of the container (cylinder) acts as a seal, preventing hot gases that are created by the internal combustion from escaping while the piston is in motion.

14.4 MECHANISM OF LUBRICANTS (TYPES OF LUBRICATION)

In order to lubricate a machine effectively, it is essential to consider the specific working circumstances under which the equipment functions optimally. There exist three distinct ways by which lubrication is achieved.

1. Hydrodynamic or Fluid Film or Thick Film Lubrication
2. Boundary Lubrication or Thin Film Lubrication
3. Extreme Pressure Lubrication

14.4.1 Hydrodynamic or Fluid Film or Thick Film Lubrication

Thick film lubrication refers to the lubrication process in which a thick layer of fluid is used to separate two surfaces in relative motion, reducing friction and wear. This sort of lubrication is employed when low-viscosity liquid lubricants are utilized under operating conditions, while simultaneously ensuring that the lubricant remains in position and effectively prevents contact with the machine's moving or sliding components (Figure 14.3).

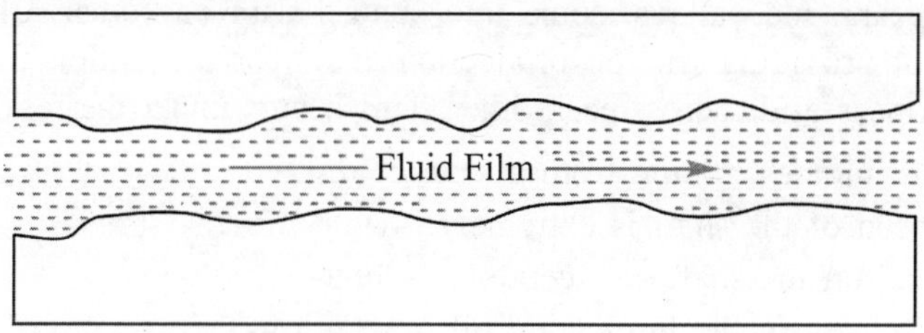

Figure 14.3 Working of fluid film lubrication between two surfaces.

Fluid film lubrication involves the filling of grooves or irregularities on sliding surfaces with a liquid lubricant, creating a substantial layer that prevents direct contact between the surfaces. The lubricant minimizes wear and tear by lowering the frictional resistance generated by the sliding motion of parts. Many machines use lubricants to separate sliding and rolling materials. This is achieved by creating pressure between the surfaces, which holds a thin coating of lubricant in place.

The friction encountered by the sliding components is solely caused by the internal resistance between the particles of the lubricant. The frictional resistance is not affected by the load, but it is influenced by the surface area and relative velocity of the sliding components, as well as the viscosity and thickness of the lubricants. Light machinery such as watches, clocks, weapons, scientific equipment, etc., as well as sensitive devices, are equipped with fluid film lubrication for this purpose.

14.4.2 Boundary Lubrication or Thin Film Lubrication

Boundary Lubrication, also known as Thin Film Lubrication, refers to the lubrication mechanism where a thin layer of lubricant separates two surfaces in contact. Thin film lubrication is used when thick film lubrication is not effective. Thin film or border lubrication is employed when the presence of a continuous lubricating film is not sustainable, allowing for the potential occurrence of direct contact between metal surfaces (Figure 14.4).

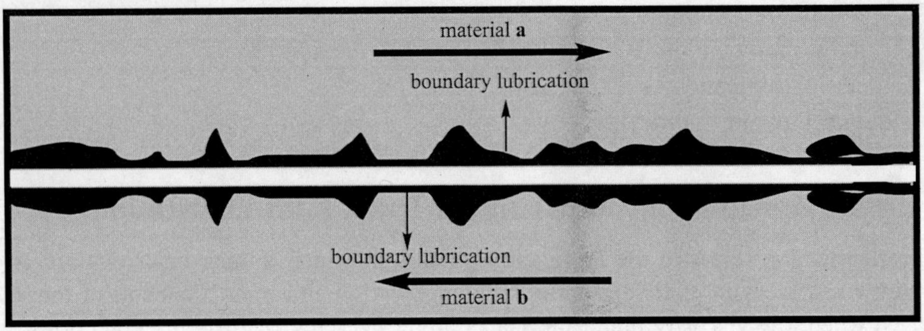

Figure 14.4 Working of thin film lubrication between two surfaces.

The lubricant coating is insufficient to achieve complete surface separation. Surface asperities, or small peaks and valleys, come into contact with one other. Chemical additives are the sole means of providing friction production and wear prevention. High velocity and heavy load when surfaces glide across each other. This occurs under the following conditions:

1. When a shaft initiates motion from a state of rest.
2. When the speed of the shaft is extremely low.
3. When the load on the shaft is exceptionally high.
4. When the viscosity of the lubricating oil is very low.

Boundary lubrication involves the application of an oily lubricant between two metallic surfaces that are sliding or moving. The lubricant is absorbed onto both surfaces through physical or chemical processes and is not easily removable. The lubricant adheres to the sliding metallic surfaces, forming two thin layers, which prevent direct contact between the metal surfaces. Consequently, the frictional resistance between the sliding metallic surfaces is diminished. An ideal thin lubricant should possess a high viscosity index, excellent heat and oxidation resistance, good oiliness, and a low pour point. Vegetable and animal oils, along with their soaps, have a remarkable ability to adsorb and are utilized for lubricating in thin films. Graphite and molybdenum disulphide suspended in oil are employed for the purpose of boundary lubrication. These lubricants provide a thin layer on the metal surfaces, which decreases friction and can withstand compression and high temperatures.

The lubricant molecules must possess these qualities for boundary lubrication. Ideally, the hydrocarbon chains should be lengthy and exhibit strong lateral attraction to one another. In order to facilitate spreading and establish a connection with metallic surfaces under high pressure, it is necessary for the molecules to possess active polar groups or atoms.

14.4.3 Extreme Pressure Lubrication

Extreme Pressure Lubrication refers to the lubrication mechanism that is specifically designed to handle high pressure conditions. When metallic surfaces move or slide against one other with intense pressure, a significant amount of heat is generated, resulting in a very high temperature. At elevated temperatures, conventional liquid lubricants undergo decomposition or vaporization, resulting in their inability to adhere to metallic surfaces. In order to withstand such circumstances, mineral oils are supplemented with extreme pressure additives.

Extreme pressure additives are organic chemicals that include active radicals or groups such as chlorine, sulphur, phosphorus, etc. At elevated temperatures, these compounds undergo reactions with metallic surfaces, resulting in the formation of metallic chlorides, sulphides, or phosphides. These compounds can then create surface layers on the moving or sliding metallic surfaces, effectively functioning as high-quality lubricants.

14.5 FLASH POINT AND FIRE POINT

The flash point refers to the lowest temperature at which the vapors emitted by a fuel, when heated, may ignite momentarily upon contact with a test flame positioned above the fuel's surface, without the fuel itself catching fire. It is crucial in assessing the flammability of the fuel. The flash point provides insight into the quantity of low boiling fraction inside the liquid fuel, as well as the volatility of the fuel and its potential for explosive risks. The flash points of the diesel-biodiesel blend increase as the concentration of biodiesel increases. The majority of International Standards require a flash point that exceeds 120°C. The rise in the flash point of the diesel-biodiesel blends is attributed to the higher quantity of *Sterculia foetida* oil present in the blends.

The fire point is the lowest temperature at which flammable vapors will consistently generate and burn when ignited. The flash and fire points of the fuel blends were determined using the Pensky-Martens flash point closed apparatus. The flash point and fire point of *Sterculia foetida* oil are approximately three times higher than those of diesel fuel. It guarantees the secure transportation and storage of biodiesel fuels. The flash point of tri-compound fuel blends exhibited significant divergence from that of diesel, with an exceptionally low range of 13 to 17°C. All ethanol-containing mixes exhibited strong flammability, as their flash point temperature was lower than the surrounding temperature. The flash point of the tri-compound fuel blend has a direct impact on the categorization of fuels for shipping and storage purposes. Therefore, it is crucial to exercise caution when handling and transporting this gasoline. Typically, when measuring flash points, the fuel component with the lowest flash point tends to have the most influence. Ethanol mostly governs the flash point of a combination consisting of three compounds. In order to prevent explosions, it is necessary to carefully control the storage, handling, and transportation of blended gasoline.

14.6 CLOUD POINT AND POUR POINT

The cloud point of a petroleum fuel oil is the temperature at which the oil forms a visible cloud of wax crystals upon cooling at a certain rate. As the proportion of biodiesel in the mix is raised, the cloud point of the blend also increases. The cloud point of neat *Sterculia foetida* oil is 6°C. The pour point of an oil is the minimum temperature at which the oil will flow or pour, without any agitation, under standard conditions.

The terms cloud point and pour point are frequently employed to indicate the suitability of fuel oils for usage in low temperatures. All the mixtures of *Sterculia foetida* oil and diesel exhibited a pour point of 10°C. Tri-compound fuel blends have much lower pour point in comparison to biodiesel-diesel fuel blends. Ethanol exhibits a significantly low pour point, whereas biodiesel typically possesses a pour point that surpasses that of traditional diesel fuel.

14.7 ANILINE POINT

The aniline point is primarily a characteristic of diesel fuel. The term "miscibility point" refers to the temperature at which the oil being tested becomes fully soluble in an equal amount of aniline, as determined using a standardized apparatus. Aromatics readily dissolve other aromatics. Therefore, a higher concentration of aromatics in the oil will result in a lower aniline point. However, an increase in the paraffin content of the oil will result in a higher aniline point.

14.8 VISCOSITY INDEX

The viscosity index is a characteristic of petroleum fractions that quantifies the rate at which viscosity changes in response to temperature variations. The viscosity index is represented by a numerical value on a scale of 0 to 100. A higher number indicates a higher viscosity index. A viscosity index of 100 indicates that the oil maintains its thickness at high temperatures and does not become too viscous at low temperatures. Paraffinic base lubricating lubricants typically have a viscosity index close to 100.

14.8.1 Viscosity

Viscosity is a numerical assessment of a fluid's opposition to movement. The SI unit of viscosity is the pascal-second (Pa·s), which is equivalent to (N·s)/m^2 or kg/(m·s).

The viscosity coefficient (η): The definition of shear stress is the force parallel to a surface divided by the area, necessary to maintain a constant velocity gradient. The unit is expressed as Newton per square meter (Nm^{-2}).

14.8.2 Factor Affecting Viscosity

(i) The viscosity of a liquid reduces as the temperature increases. The increase in kinetic energy of the liquid leads to a decrease in intermolecular attraction.

(ii) The viscosity of a substance increases as the pressure applied to it increases. As pressure increases, the volume decreases, causing the molecules to move closer together.

(iii) The viscosity of a liquid is predicted to be higher if it has larger intermolecular forces of attraction. Alcohol exhibits a greater viscosity compared to benzene.

QUESTIONS

14.1 What are lubricants? Explain the theories of lubrication.

14.2 Name one solid lubricant and explain its lubricating action.

14.3 Define Lubrication. Explain the different mechanisms of lubrication.

14.4 What is Viscosity index?

14.5 Explain flash and fire point of lubricant. Discuss its significance.

14.6 What are lubricants? How are they classified?

14.7 Explain extreme high pressure lubrication mechanism.

UNIT 5

Spectroscopic Technique and Applications, Organic Reaction and Synthesis of a Drug

CHAPTER 15

Spectroscopic Technique and Applications

15.1 INTRODUCTION

Spectroscopy is the scientific study of how radiation and matter interact in order to get information about a sample. Spectrochemical methods are extensively utilized for determining molecular structure and quantitatively and qualitatively analyzing inorganic and organic substances. Spectroscopic analytical procedures rely on quantifying the radiation emitted or absorbed by specific molecular or atomic species of interest.

Atoms and molecules engage with electromagnetic radiation (EMR) through a diverse range of interactions. Atoms and molecules have the ability to absorb and/or emit electromagnetic radiation (EMR). Electromagnetic radiation (EMR) induces various forms of motion in atoms and/or molecules through absorption. The spectral characteristics refer to the specific wavelengths and intensities of light absorbed or emitted by a substance. Spectroscopy is the study of **spectra** and how they relate to the structure of atoms and molecules, as well as their surroundings.

Upon exposure to an external electromagnetic radiation source, the sample can undergo many processes. A portion of the incident radiation can be absorbed, causing certain analyte species to transition to an excited state. Absorption spectroscopy involves measuring the extent to which light is absorbed at different wavelengths.

James Clerk Maxwell, born on June 13, 1831, and deceased on November 5, 1879, was a Scottish physicist renowned for his extensive research in various fields. He formulated the classical theory of electromagnetic radiation, which provided the initial structure for understanding the separate yet interrelated properties of electricity, magnetism, and light.

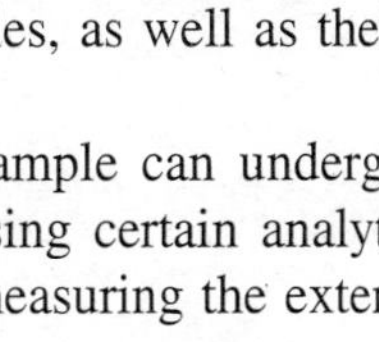

James Clerk Maxwell

15.2 ELECTROMAGNETIC RADIATION

Electromagnetic (EM) radiation is a type of energy that is propagated through space at tremendous speeds. EM radiation can be defined as a wave characterized by its wavelength, frequency, velocity, and amplitude. EM radiation is commonly represented as waves that consist of electric and magnetic fields oscillating perpendicular to one other. This is illustrated in Figure 15.1 and 15.2. An oscillation in the electric field occurs perpendicular to the path of transmission at a 90° angle. The magnetic field oscillation is perpendicular to both the direction of propagation and the electric field oscillation, at a 90° angle. Unlike sound waves, light does not need a medium to travel through, hence it can easily pass through a vacuum.

Electromagnetic radiation can be seen as discrete units of energy or particles known as **photons** or **quanta**. The simultaneous consideration of radiation as both particles and waves is not contradictory, but rather mutually reinforcing.

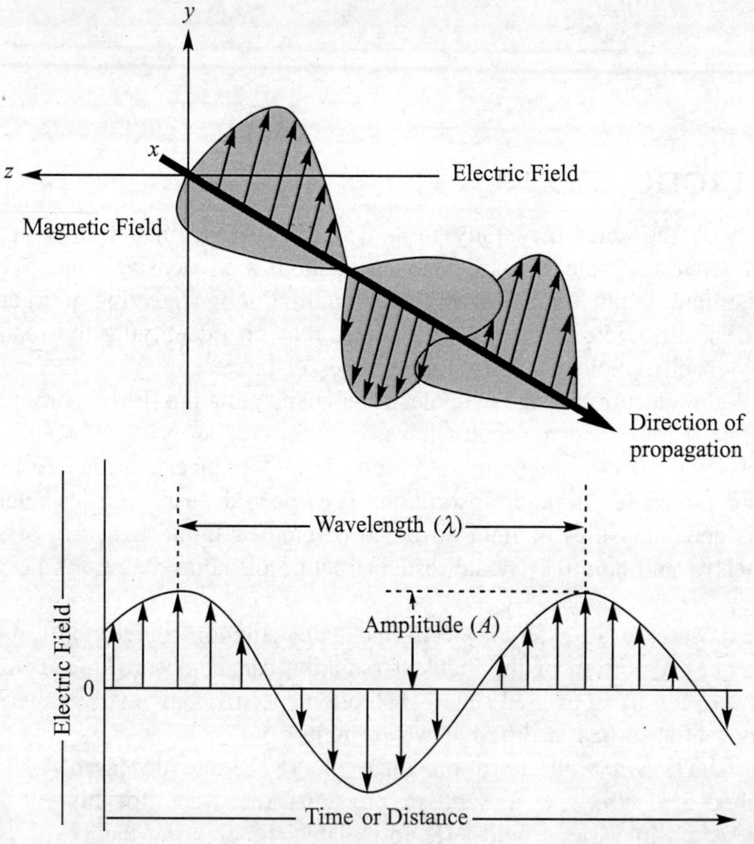

Figure 15.1 Oscillation of electromagnetic radiation with electric and magnetic fields.

15.3 INTERACTION OF ELECTROMAGNETIC RADIATION WITH MATTER

Radiation-matter interaction can result in the redirection of radiation and/or the shifting of energy levels in atoms or molecules. During interaction three phenomenon occur, i.e., absorption, emission and scattering.

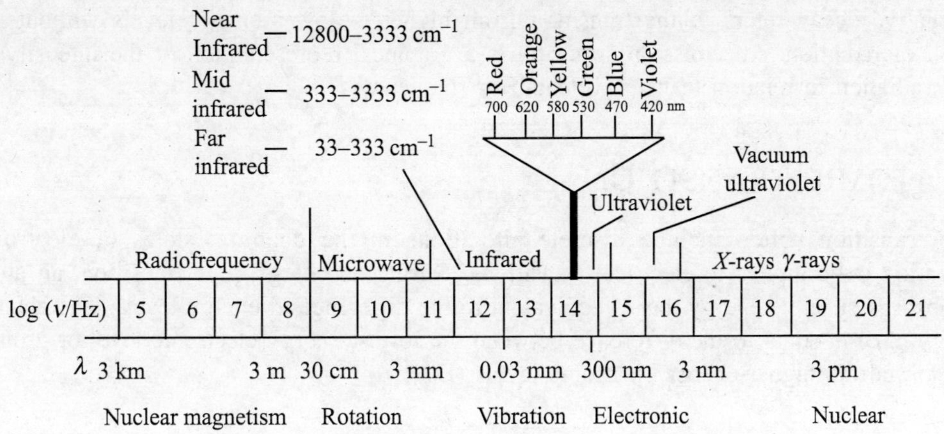

Figure 15.2 Electromagnetic radiation categorization.

1. Absorption refers to the process in which energy is transferred from the radiation source to molecule, resulting in a change from a lower energy level to a higher energy level.
2. The process of energy transfer from a higher level to a lower level is considered as emission. In this case, the energy is transmitted to the radiation field. The emission can be non-radiative decay when no radiation is released.
3. Scattering refers to the redirection of light caused by its interaction with materials. This phenomenon can occur with or without the transfer of energy. In other words, the dispersed radiation may have a slightly different wavelength or the same wavelength as the original light.

Types of Spectra: An absorption spectrum is detected when energy is transferred from the radiation source to a substance, resulting in a change from a lower energy level to a higher energy level. An absorption spectrum is a graphical representation that shows the relationship between the amount of light absorbed by a substance and the wavelength or frequency of that radiation (Figure 15.3).

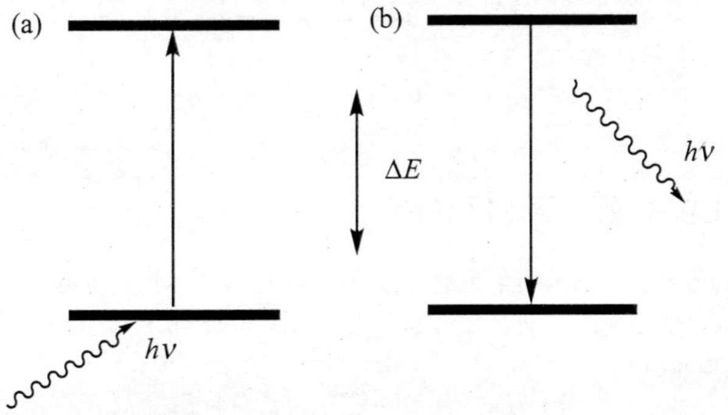

Figure 15.3 Types of spectra.

Observation of an emission spectrum occurs when energy is transferred from the emitter to the radiation field during a transition from a higher energy level to a lower energy level.

Non-radiative decay refers to the transition from higher to lower energy levels without the emission of radiation. An emission spectrum is a graphical representation of the intensity of emitted radiation in relation to its wavelength or frequency.

15.4 ATOMIC TRANSITION

Atomic transitions often include discrete alterations in the quantum states of electrons. Quantization only applies to electrical transitions. During an energy state transition, an atom either absorbs or emits energy that is equivalent to the difference in energy levels. The change in energy (ΔE) is equal to the difference between the lowest energy electronic level or ground state (E_0) and the higher-energy electronic levels (E_1).

$$E = E_1 - E_0$$

The wavelength or frequency of radiation received or emitted during a transition is directly proportional to the change in energy (ΔE). Line spectra are generated by transitions between electronic energy levels (Fig. 15.4).

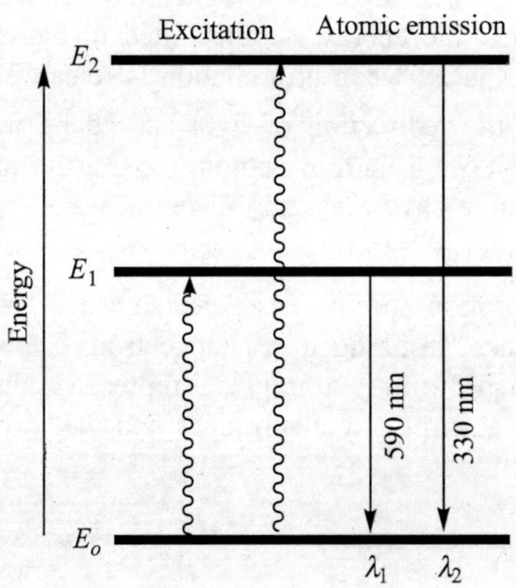

Figure 15.4 Atomic transition.

15.5 MOLECULAR TRANSITION

In molecules, the electronic states are further separated into vibrational states (Fig. 15.5). The energy of a band in a chemical absorption spectrum is determined by the combination of three distinct energy components.

$$E = E_{\text{electronic}} + E_{\text{vibrational}} + E_{\text{rotational}}$$

Spectra that exhibit bands are a result of transitions occurring between electronic-vibrational-rotational states.

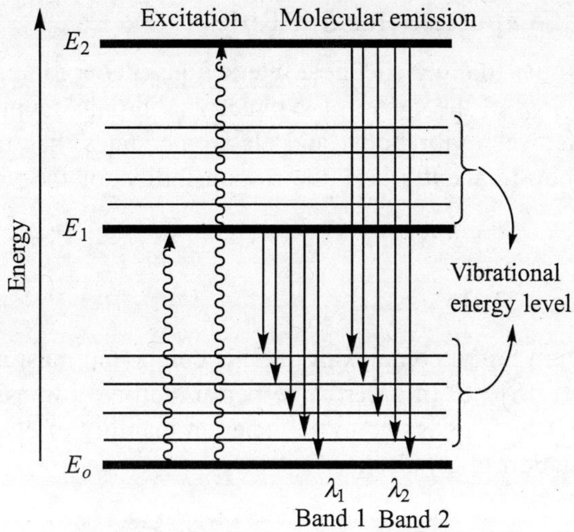

Figure 15.5 Molecular transition.

15.6 SELECTION RULES

The statement "some transitions are not allowed" indicates that while such a transition may still occur, it is less probable and the intensity (molar absorption coefficient) of the corresponding absorption band is very low. Not all potential transitions are observed. The transition of an electron is subject to specific quantum mechanical constraints known as "selection rules". As an illustration, it is impossible for an electron to alter its spin quantum number while undergoing a transition. Such changes are considered "forbidden". The quantity of electrons that can undergo excitation simultaneously is contingent upon the molecular's symmetry qualities and the symmetry of its electronic states. In addition, there are instances where "forbidden" transitions are occasionally detected (albeit with low strength) as a result of several other circumstances. The permissibility of transitions, as well as the extent to which they may be prohibited, is contingent upon selection rules.

15.6.1 Laporte Selection Rule

A molecule with a centre of symmetry cannot transition between states of the same parity. That means the transition is treated as forbidden. Laporte-allowed transitions occur when there is a change in the orbital angular momentum quantum number ($\Delta l = \pm 1$). The conditions are as follows:

1. Transitions such as '$s \leftrightarrow p$', '$p \leftrightarrow d$', '$d \leftrightarrow f$', etc. are allowed, with a change in the principal quantum number (Δl) of ± 1.
2. Transitions between the '$s \leftrightarrow d$', '$p \leftrightarrow f$', and so on, are not allowed. The difference in the position of the letters should be either +2 or –2.
3. Transitions '$s \leftrightarrow s$', '$p \leftrightarrow p$' , '$d \leftrightarrow d$', '$f \leftrightarrow f$', etc. with a change in angular momentum (Δl) equal to zero are not allowed.

15.6.1.1 Relaxation of Laporte Selection Rule

Octahedral complexes, similar to other chemical interactions, undergo unsymmetrical vibrations that result in a temporary loss of the center of symmetry. Vibronic coupling, a phenomena that involves the interaction between vibrational and electronic states, has the ability to relax the Laporte selection rule. They frequently bear the responsibility for the vibrant colors exhibited by these structures.

15.6.2 Spin Selection Rule

The total angular momentum S of a complex or organic compound must remain constant during an electronic transition, so $\Delta S = 0$. In order to be permissibility, a transition must not involve any alteration in the spin state. This is due to the inherent inability of electromagnetic radiation to alter the relative orientation of an electron's spin.

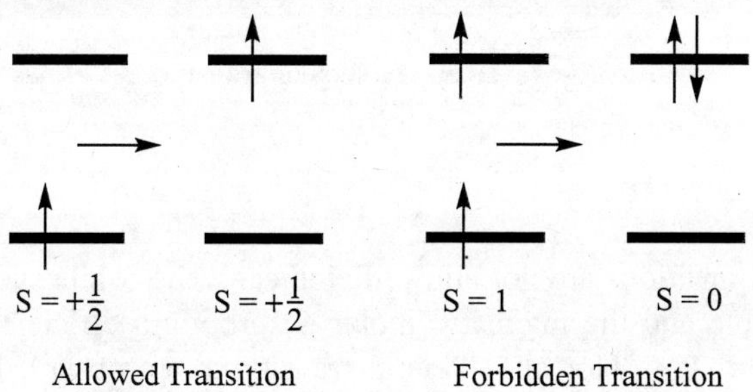

15.7 ELECTRONIC OR ULTRAVIOLET AND VISIBLE SPECTROSCOPY

UV spectroscopy involves exposing the sample to a wide range of UV rays. If a specific electronic transition aligns with the energy level of a specific ultraviolet (UV) band, it will undergo absorption. The residual UV light passes the material and is subsequently observed. An *absorption spectrum* is obtained from the residual radiation, which exhibits "gaps" at specific discrete energies. Identification of chromophores are performed by this spectroscopy in the process of determining molecular structures. Chromophores are unsaturated system (dienes), polyenes, aromatic compounds, and conjugated ketones. Additionally, it is highly valuable in the quantitative study of substances containing chromophores.

15.7.1 Observed Electronic Transitions or Origin of Electronic Spectra

Electronic transitions typically arise from valence electrons located in a chromophore. Transition types can be allocated based on the chemical orbital involved. In molecule σ, π, n, σ^* and π^* exists. As a result of their symmetrical nature in molecular orbitals, transitions like $p \rightarrow \pi^*$ and $n \rightarrow \pi^*$ can either be allowed or forbidden in alkene and keto group respectively. Absorptions with high ε values represent permissible transitions, while absorptions with low ε values indicate prohibited transitions. The UV Spectral origin occurs when a photon of light causes an electron to move from the Highest Occupied Molecular Orbital (HOMO) to the Lowest Unoccupied Molecular Orbital (LUMO) (Figure 15.6).

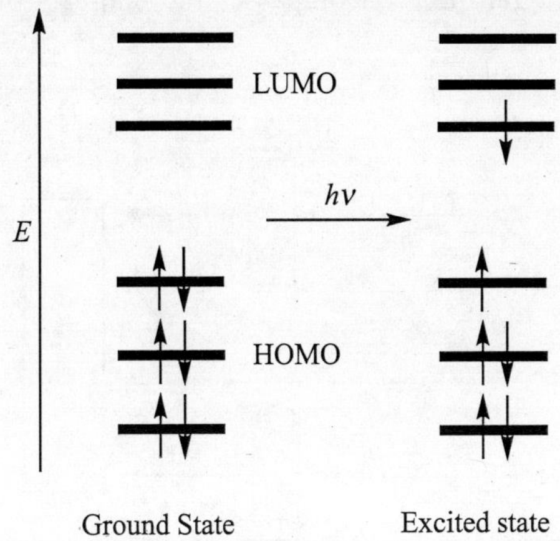

Figure 15.6 HOMO to LUMO transition by light photon.

15.7.2 UV-visible Absorption Measurement

Electromagnetic radiation can be characterized by its wavelength λ (measured in nanometers), frequency ν (measured in inverse seconds), or wavenumber (measured in inverse centimeters). These quantities are related to each other by the energy difference as expressed in the following equations:

$$\lambda \nu = c;\ \bar{\nu} = \frac{1}{\lambda}$$

Where λ is wavelength in nm, ν is frequency in cycle/s, $\bar{\nu}$ is wave number in 1/cm

$$\Delta E = h\nu = \frac{hc}{\lambda} = \frac{28636}{\lambda}\ \text{kcal/mol} = \frac{119809}{\lambda}\ \text{kJ/mol}$$

15.7.3 UV Regions

The energy associated with electromagnetic radiation in the ultraviolet (UV) area is separated into two segments:

1. The near UV region, which spans from 190 to 400 nm.
2. The far UV or vacuum UV extends less than 190 nm.

In the spectrometer, the interference in the far UV region caused by the absorption of oxygen needs to be eliminated or cleared by flushing it with nitrogen in order to acquire the sample's spectra. To cover the complete UV-VIS band, two sources are necessary for scanning. The deuterium lamp emits ultraviolet (UV) light in the range of 200–330 nm. The tungsten lamp emits light in the wavelength range of 330 to 700 nm.

15.7.4 Instrumentation of UV-Visible Spectroscopy

The design of a conventional UV-Vis spectrometer closely resembles that of an IR spectrometer, as it necessitates identical functionalities such as sample handling, irradiation, detection, and output. Here is a concise diagram that encompasses the majority of modern UV spectrometers (Figure 15.7).

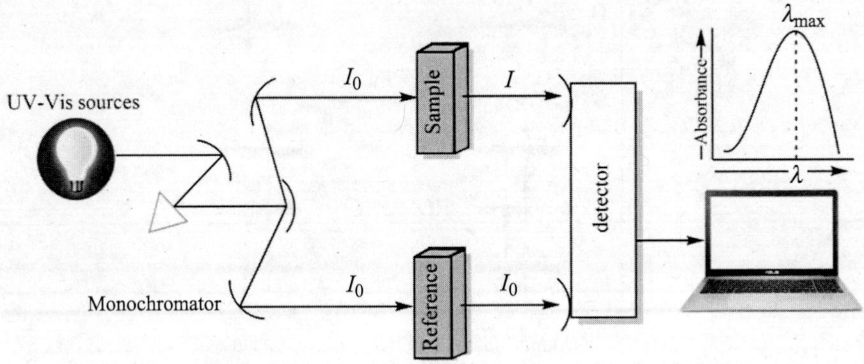

Figure 15.7 Systematic elucidation of spectrophotometer.

The lamps emit light across the complete spectrum of UV-Vis light, while the monochromator systematically alters the narrow radiation directed towards the beam splitter. It directs a distinct wavelength range to a reference and sample cell. Detector identify the light that passes through the sample (I) and the initial light (I_0), and transmits this data to the recorder.

The diode-array spectrophotometer is a recent advancement that utilizes a prism as a dispersion device to separate the complete spectrum that passes through the sample. Almost all ultraviolet spectra are typically recorded in a solution phase. Cells can consist of materials such as plastic, glass, or quartz. Quartz is the only material that allows light to pass through completely throughout the entire range of 200–700 nm. Plastic and glass, on the other hand, are only capable of transmitting light in the visible spectrum. Concentration is determined by empirical means. The transparency of solvents is essential for observation, and the wavelength at which a solvent becomes opaque is known as the cutoff. Given that spectra are limited to a range of 200 nm, fluids generally just need to be free of conjugated π complexes or carbonyls.

In addition, solvents must maintain the integrity of the fine structure, particularly in the UV region where it is directly observed. The presence of hydrogen bonding adds complexity to the influence of vibrational and rotational energy levels on electronic transitions. In contrast, dipole-dipole interactions have a lesser effect, particularly in solvents with lower polarity. Therefore, solvents that are more non-polar are often more favorable, although this may not always be feasible.

15.7.5 Measurement of Absorbance by Beer-Bouger-Lambert Law

The absorbance (A) or molar absorptivity (ε) of a UV band is determined using the Beer-Bouger-Lambert Law.

$$I = I_0 e^{-kl}$$

$$I = I_0 10^{-\varepsilon cl}$$

$$\log \frac{I_0}{I} = \varepsilon cl$$

The variable "I_0" represents the intensity of the light that is incident. The variable "I" represents the magnitude of light that is able to pass through a medium. l represents the length of the path, measured in centimeters. c represents the concentration of a substance in units of moles per liter (mol L^{-1}). The absorption coefficient, denoted as k, refers to the measure of a material's ability to absorb electromagnetic radiation. The molar extinction coefficient, also known as molar absorptivity, is a measure of the ability of a substance to absorb light. It is expressed in units of cm^2 mol^{-1} or L mol^{-1} cm^{-1}.

Absorbance

The relationship between k and ε is $k \approx 20303\ \varepsilon c$. The equation that links the actual measured quantity, absorbance $\left(A = \log\frac{I_o}{I}\right)$, with the molar absorptivity (ε) is: $A = \varepsilon cl$. For i absorbing species: $A = l\sum_i \varepsilon_i c_i$. The measured quantity is the relative intensities of the light beam transmitted through a reference cell containing pure solvent and a matching cell holding the sample solution.

The absence of any fine structure, spectra are rarely displayed in their unprocessed state. Instead, the highest points of the peaks are typically presented as a numerical compilation of "lambda max" values or λ_{max}. The vertical axis of the spectrum is measured in absorbance units, denoted as A (Figure 15.8).

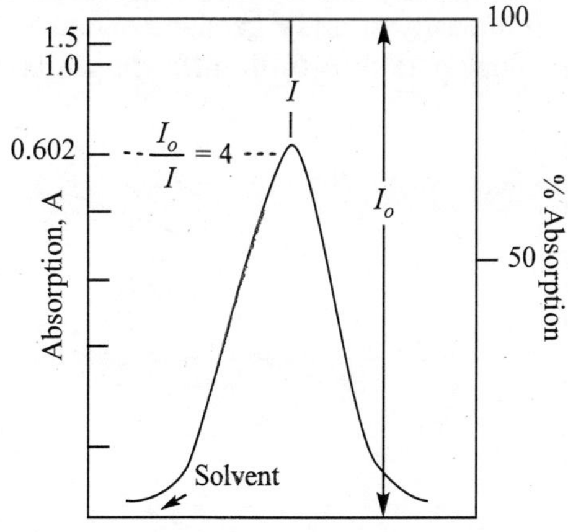

Figure 15.8 Measurement of solute absorbance A by a double-beam spectrophotometer.

From an experimental standpoint, three additional factors must be taken into account:

1. A greater path length, denoted as l, across the sample will result in increased absorption of UV radiation. This relationship is linear in nature.
2. The absorption of UV light increases linearly with the concentration, c, of the sample. Certain electronic transitions exhibit more efficiency in absorbing photons compared to others, as indicated by their molar absorptivity. This variation can span several orders of magnitude.

In the majority of UV spectrometers, the value of l, representing the path length, remains constant. It is worth noting that standard cells normally have a path length of 1 cm. Concentration is usually adjusted based on the intensity of absorption that is observed or anticipated, normally being weak, with a value below 0.001 M. Molar absorptivities exhibit significant variations across many magnitudes.

1. Absorptions with values ranging from 10^4 to 10^6 are referred to have high intensity.
2. Low intensity absorptions are defined as values ranging from 10^3 to 10^4.
3. The absorptions of forbidden transitions range from 0 to 10^3 in terms of values.

A is dimensionless, hence the values for ε are $cm^{-1} \cdot M^{-1}$ and are infrequently stated. By eliminating the influence of path length and concentration effects, absorbance can be directly proportional to the molar absorptivity (ε). Consequently, the y-axis can be represented as either ε itself or the logarithm of ε.

15.7.6 Frank-Condon Principle

The broadened UV bands suggest a wide range of energy distribution, resulting from the overlapping of multiple vibrational modes on the electrical level. According to the potential energy diagram, the ground state has a lower energy and a shorter equilibrium bond length, indicating a stronger bond. On the other hand, the excited state has a higher energy and a longer equilibrium bond length, indicating a weaker bond. Each electrical state contains several vibrational states. The shape and intensity of a UV band are determined by the transitions from the ground state's lowest vibrational state ($v = 0$) to different vibrational levels in the excited state (Figure 15.9–15.10).

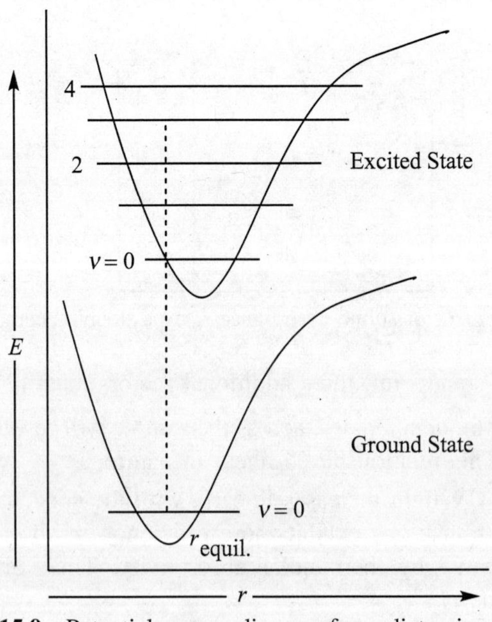

Figure 15.9 Potential energy diagram for a diatomic molecule.

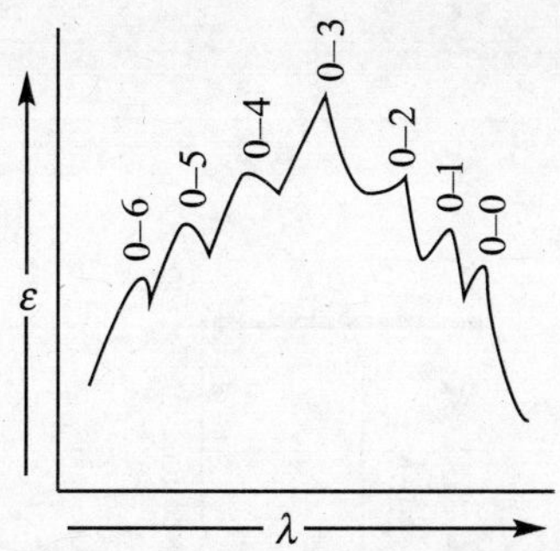

Figure 15.10 Intensity distribution among vibronic bands.

The Frank-Condon Principle can be summarized as follows:

1. The movement of atomic nuclei (which occurs on a timescale of 10^{-13} seconds) is significantly slower than the movement of electrons during a transition (which occurs on a timescale of 10^{-16} seconds). As a result, the motion of atomic nuclei can be negligible when considering electronic excitation.
2. Due to the immobility of the nucleus during stimulation, the internuclear distance remains constant, resulting in electronic transitions primarily occurring in a vertical direction.
3. The transition from the $v = 0$ to $v = 3$ is a vertical transition because it occurs at the peak of the electron probability curve for $v = 3$ in the excited state. The probabilities of transition for other vertical transitions ($0 \rightarrow 2$, $0 \rightarrow 1$, ..., $0 \rightarrow 4$, $0 \rightarrow 5$,...) are reduced, as indicated by the composite fine structure of the vibronic broad band.

Electronic Transition: UV electronic transitions quantify the probability and amount of energy required to stimulate a molecule. Both singlet and triplet states are available for each energy state. In the singlet state, the spins of an electron pair are oriented in opposite directions, which is known as being antiparallel. On the other hand, if the spins are parallel in the same direction, there are three potential states that are collectively referred to as the triplet state. The equation is $M = 2S + 1$, where M represents the multiplicity. S is overall angular momentum (total spin).

Selection Rules: The selection rules permit self-interactions ($S \rightarrow S$) and self-transitions ($T \rightarrow T$), but prohibit interactions between different entities ($S \rightarrow T$ and $T \rightarrow S$). Ground states typically exist as singlets, resulting in the majority of excitations occurring to singlet excited states, such as $S_0 \rightarrow S_1$, $S_0 \rightarrow S_2$, and so on. Triplet states typically arise through intersystem crossing from an excited singlet state, such as S_1, rather than through direct excitation from the ground state S_0 (Figure 15.11).

15.7.7 Classification of Electronic Transitions

The electronic transition is determined by the disparity in energy levels between the ground state and the excited state. The determination of the λ value can be approximated by considering the difference in energy (ΔE) between the molecular orbitals (MO) occupied by the electrons at the ground state and the higher excited MO at the excited state. This approximation proves to be useful. The order of the molecular orbitals is as follows: $\sigma > \pi > n > \pi > \sigma^*$. Different kinds of transition are illustrated in Figure 15.12.

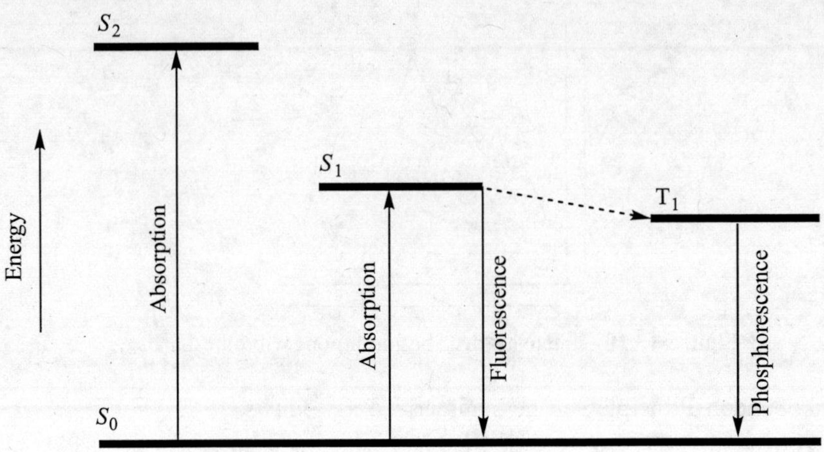

Figure 15.11 Different kind of emissions.

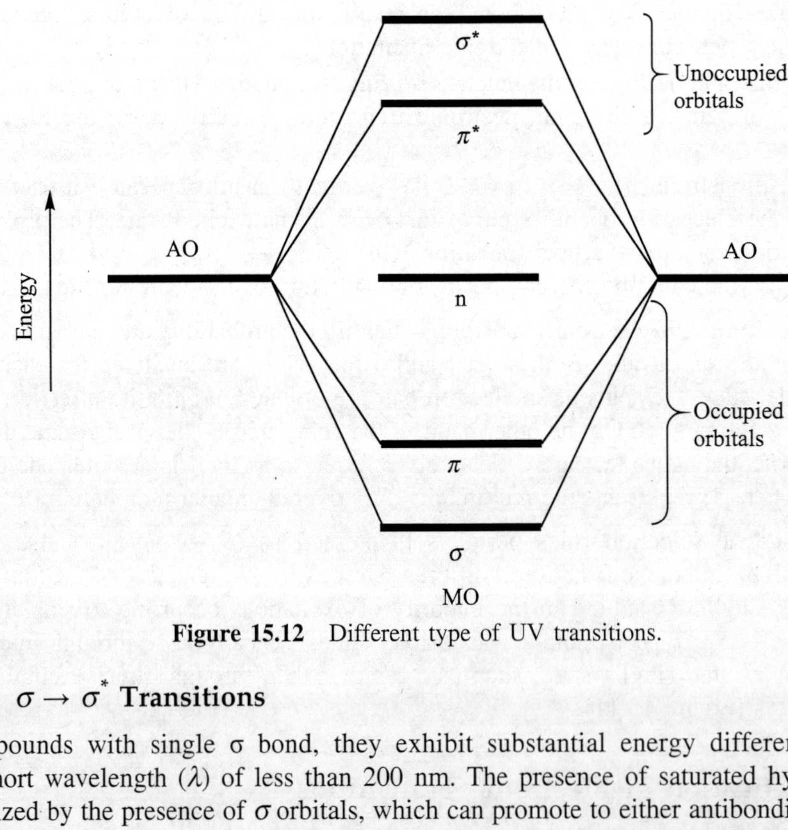

Figure 15.12 Different type of UV transitions.

15.7.7.1 $\sigma \rightarrow \sigma^*$ Transitions

For compounds with single σ bond, they exhibit substantial energy difference (ΔE) and have a short wavelength (λ) of less than 200 nm. The presence of saturated hydrocarbons is characterized by the presence of σ orbitals, which can promote to either antibonding σ^* orbitals located in higher valence shell orbitals such as 3*s*, 3*p*, 4*s*, and so on. This transition entails a significant energy difference (ΔE) and results in a tiny maximum wavelength (λ_{max}) that is observed in the far-ultraviolet (FUV) area. For cyclopropane, the maximum wavelength (λ_{max}) is 190 nm, while for cycloalkane, the maximum wavelength is 135 nm (Vacuum ultraviolet (VUV)).

15.7.7.2 $n \rightarrow \pi^*$ Transitions

The process involves the stimulation of an electron in a nonbonding orbital, namely unshared electron pairs on oxygen, nitrogen, sulphur, etc., to an antibonding π^* orbital. This typically occurs in an unsaturated bond with heteroatoms such as carbon-oxygen (C=O), carbon-sulphur (C=S), nitrogen-oxygen (N=O), etc. It is an electrically prohibited and weakly intense transition. The example provided states that saturated aldehydes and ketones have a maximum absorption wavelength (λ_{max}) ranging from 185 to 300 nm.

15.7.7.3 $\pi \rightarrow \pi^*$ Transitions

For compounds that have double, triple bonds, or aromatic rings, the excitation of a π electron occurs, moving it to an antibonding π^* orbital. This transition is typically permitted and characterized by great intensity. Ethylene exhibits strong absorption at a wavelength of 162 nm (10000) in the vacuum ultraviolet range. Extending conjugation decreases the energy difference, ΔE, and increases the wavelength of maximum absorption, λ_{max}, if prolonged beyond 5 double bonds, it enters the visible range.

15.7.7.4 $n \rightarrow \sigma^*$ Transitions

Excitation occurs when an electron moves from a nonbonding orbital (n) to an antibonding σ^* orbital. The compound CH_3OH in vapor phase absorbs light at a wavelength of 183 nm with an extinction coefficient of 150. The compound NEt_3 in vapor phase absorbs light at a wavelength of 227 nm with an extinction coefficient of 900. The compound MeI in hexane solution absorbs light at a wavelength of 258 nm with an extinction coefficient of 380.

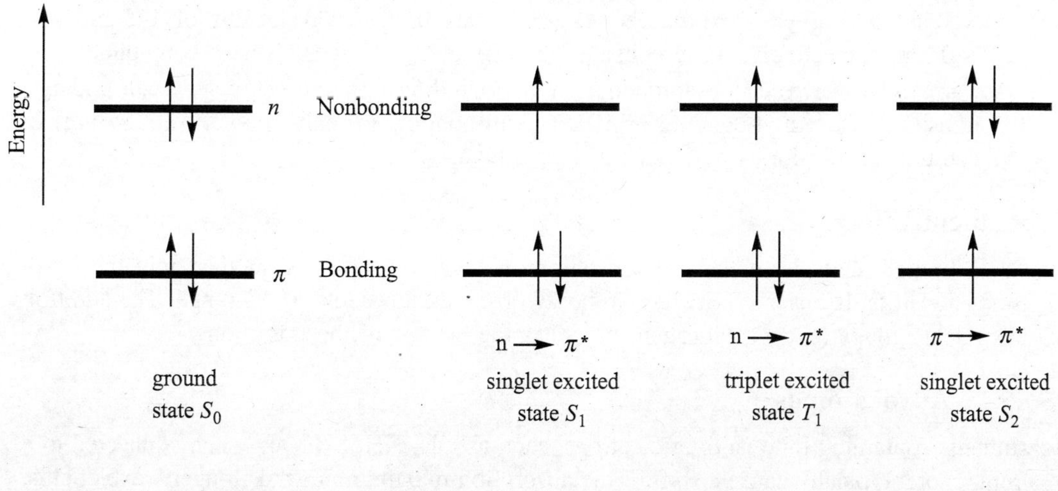

15.7.8 Application of UV Spectroscopy

1. Ultraviolet (UV) spectroscopy was the initial organic spectral technique, however it is seldom employed as the principal approach for determining molecular structure.
2. It is particularly beneficial when used alongside NMR and IR data to clarify distinctive electronic characteristics that may be unclear in those approaches.
3. It can be utilized to determine the appropriate wavelengths of irradiation for photochemical investigations or the development of UV resistant paints and coatings, through the assessment of λ_{max} and molar absorptivity.
4. UV is commonly used as a detection tool in HPLC. It is easily integrated into the design of liquid chromatography (LC) systems since it is used to analyze samples in solution phase compared to a reference solvent.
5. UV plays a similar role in HPLC as mass spectrometry (MS) does in gas chromatography (GC).

15.7.9 Terms Describing UV Absorptions

15.7.9.1 Chromophores

A chromophore, also known as a color-loving functional group, refers to a group that exhibits characteristic electronic transitions. The chromophore's structural or electronic alterations can be measured and utilized to forecast variations in the detected electronic transitions.

Functional Groups that give Electronic Transitions

Organic Chromophores

1. Alkanes only contain sigma-bonds and lack any lone pairs of electrons, resulting in the observation of solely the high energy $\sigma \rightarrow \sigma^*$ transition in the far ultraviolet region.
2. In the case of simple, aliphatic examples of alcohols, ethers, amines, and sulphur compounds, the n to σ^* transition is the most frequently observed. Similar to the alkane σ to σ^* transition, it typically occurs at wavelengths shorter than 200 nm.
3. Alkenes and alkynes exhibit the $\pi \rightarrow \pi^*$ transition at wavelengths of 175 nm and 170 nm, respectively, when considering individual instances of these compounds.
4. Carbonyls, which are unsaturated systems including nitrogen or oxygen, can undergo transitions from non-bonding orbitals to anti-bonding orbitals ($n \rightarrow \pi^*$) (~285 nm) in addition to transitions between bonding orbitals ($\pi \rightarrow \pi^*$).

Substituent Effects

Based on limited analysis of these broad chromophores, we have shown that only the weak ($n \rightarrow \pi^*$) transition is observed in the commonly observed ultraviolet (UV) range. The addition of substituent groups (except hydrogen) can alter the energy of the transition.

15.7.9.2 Auxochromes

Substituents containing lone pairs of electrons, such as OH, NH, SH, etc., when connected to a π chromophore, typically cause a shift of the absorption maximum towards longer wavelengths.

Auxochromes are substituents that enhance the intensity and frequently alter the wavelength of an absorption. Typical auxochromes consist of alkyl, hydroxyl, alkoxy, amino groups, and halogens.

Substituents can exert one of four effects on a chromophore (Figure 15.13).

1. **Bathochromic shift (red shift)** refers to a shift towards longer wavelengths and lower energy levels. Transition to longer wavelengths, commonly known as red shift.
2. **Hypsochromic shift**, often known as a **blue shift**, refers to a change towards shorter wavelengths and higher energy. Transition to a smaller wavelength, commonly known as blue shift.
3. The **hyperchromic effect** refers to an augmentation in intensity. Enhancement in the ε (epsilon) value of a band.
4. **Hypochromic effect** refers to a reduction in intensity. Reduction in the epsilon (ε) value of a band.

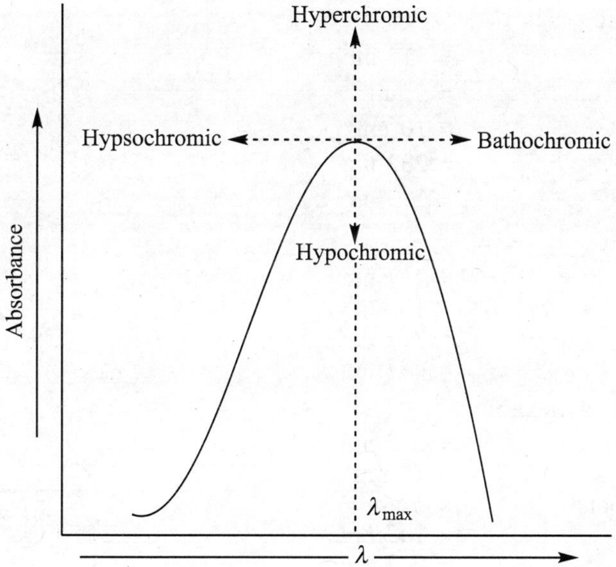

Figure 15.13 Shifts in the position of an absorption band

15.7.10 Woodward-Fieser Rules

Woodward and the Fiesers conducted thorough investigations on terpene and steroidal alkenes. They observed that the presence of identical substituents and structural characteristics may reliably determine the wavelength at which the lowest energy $\pi \rightarrow \pi^*$ electronic transition occurs. Woodward (1942) proposed a set of empirical guidelines for determining the absorption wavelengths of dienes based on factors such as their substitution and conjugation.

The chromophore under observation's base value for max is where the rules start. This base value is increased by the incremental contribution of substituents derived from the group tables:

Basic λ_{max} value of conjugated dienes		217 nm	
Group	**Increment**	**Group**	**Increment**
Extended conjugation	+30	–OR	+6
Each exocyclic C=C	+5	–SR	+30
Alkyl	+5	–Cl, –Br	+5
$-OCOCH_3$	+0	$-NR_2$	+60

For example:

Isoprene

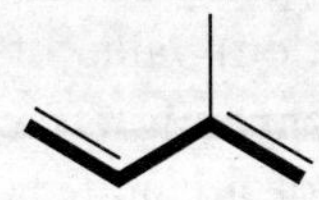

acyclic butadiene = 217 nm
one alkyl subs = 5 nm
Total = 222 nm
Experimental value = 220 nm

Allylidenecyclohexane

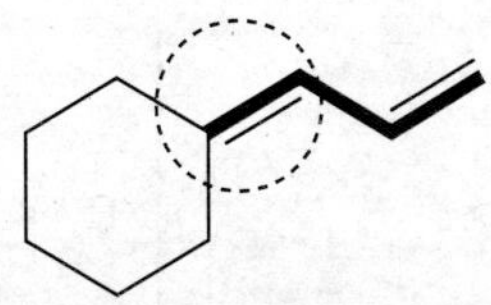

acyclic butadiene = 217 nm
one exocyclic C=C = 5 nm
2 alkyl subs. = 10 nm
Total = 232 nm
Experimental value = 237 nm

Dimethyl butadiene

acyclic butadiene = 217 nm
2 alkyl substituents (2 × 5) = 10 nm
Total = 227 nm
Experimental value = 227 nm

Pent-3-en-2-ylidenecyclohexane

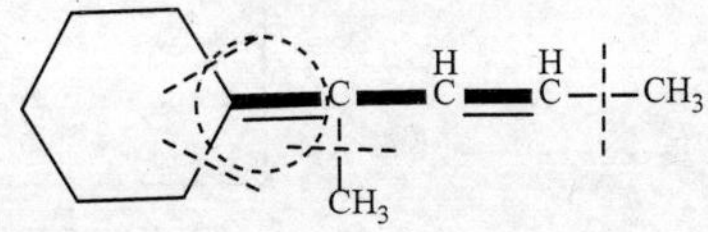

acyclic butadiene = 217 nm
2 alkyl substituents (2 × 5) = 10 nm
1 Exocyclic double bond = 5 nm
Total = 242 nm
Experimental value = 242 nm

Cyclic Dienes

Cyclic dienes come in two main varieties, each with a different base value.

Heteroannular (transoid)	**Homoannular (cisoid)**
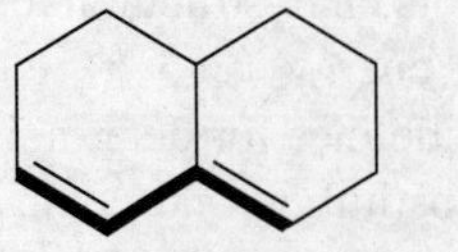	
e = 5,000 – 15,000 Base λ_{max} = 214	e = 12,000 – 28,000 Base λ_{max} = 253

With a few additions, the increment table is identical to that for acyclic butadienes:

Group	Increment
Additional homoannular (Where both types of diene are present, the one with the longer l becomes the base)	+30
Exocyclic double bond	+5

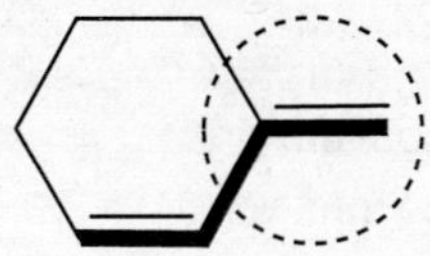

For example:

1. Base value of Transoid = 214 nm
   3 alkyl substituents (3 × 5) = 15 nm
   1 Exocyclic double bond = 5 nm
   Total = 234 nm
   Experimental value = 234 nm

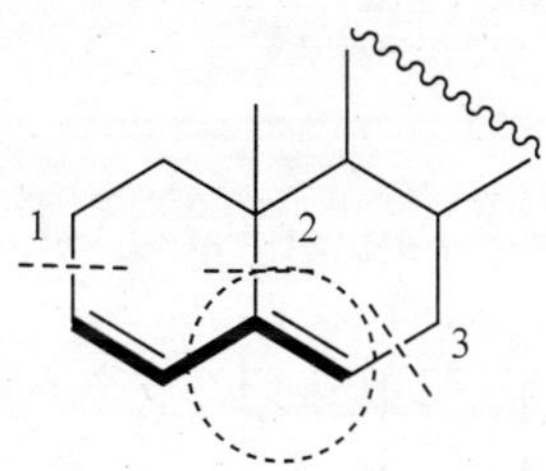

2. Base value of Transoid = 214 nm
   3 alkyl substituents (3 × 5) = 15 nm
   1 Exocyclic double bond = 5 nm
   –OR = 6 nm
   Total = 240 nm
   Experimental value = 241 nm

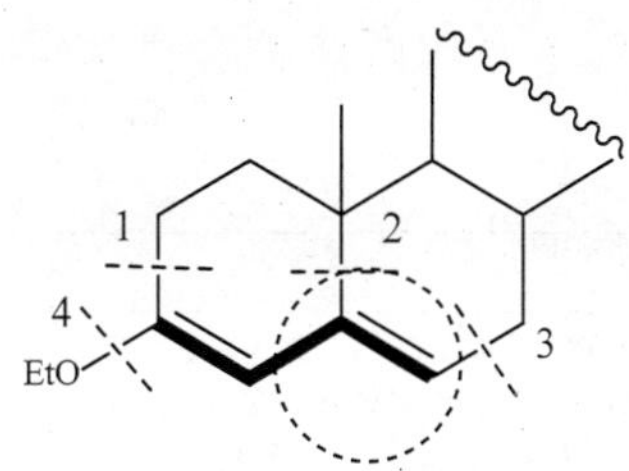

3. Base value of Cisoid = 253 nm
   3 alkyl substituents (3 × 5) = 15 nm
   1 Exocyclic double bond = 5 nm
   –OR = 6 nm
   Total = 279 nm
   Experimental value = 275 nm

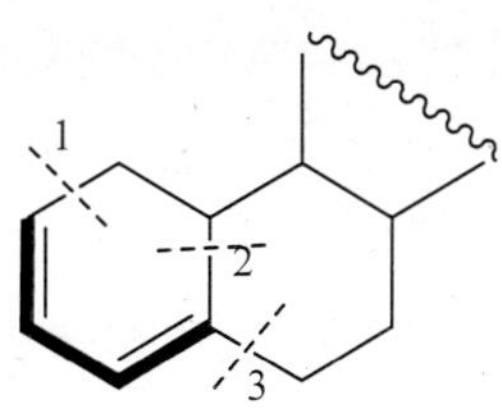

4. Base value of Cisoid = 253 nm
   3 alkyl substituents (3 × 5) = 15 nm
   1 Exocyclic double bond = 5 nm
   Double-bond extending conjugation = 30 nm
   Total = 303 nm
   Experimental value = 304 nm

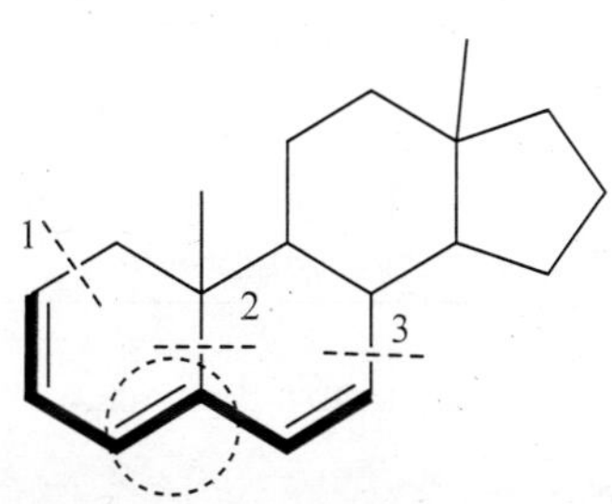

5. Base value of Transoid = 214 nm
   5 alkyl substituents (5 × 5) = 25 nm
   3 Exocyclic double bond = 15 nm
   Double-bond extending conjugation = 30 nm
   Total = 284 nm
   Experimental value = 283 nm

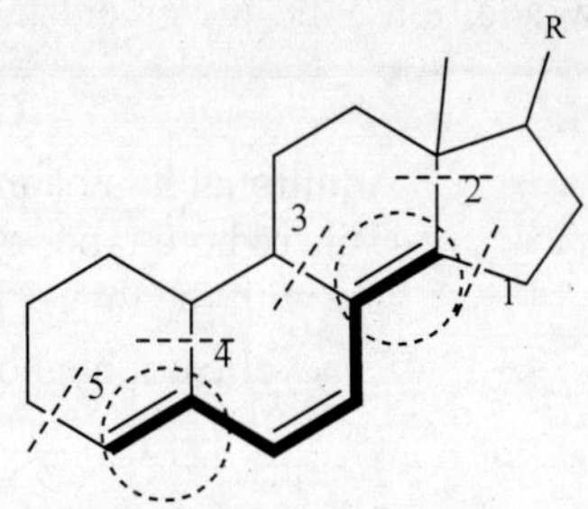

6. Base value of Cisoid = 253 nm
   5 alkyl substituents (5 × 5) = 25 nm
   3 Exocyclic double bond = 15 nm
   2 Double-bond extending conjugation = 60 nm
   Total = 353 nm
   Experimental value = 355 nm

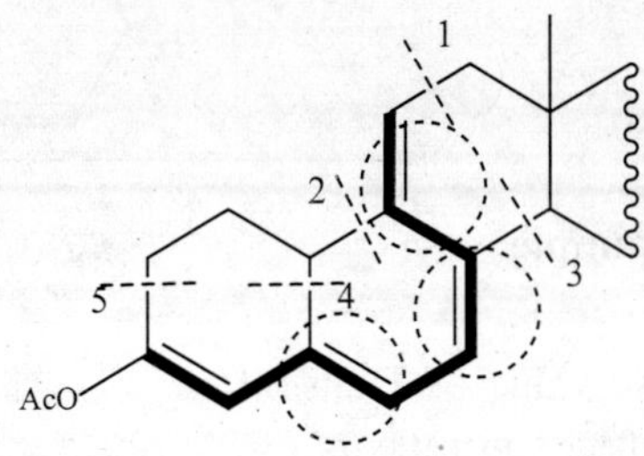

Enones cyclic α, β-unsaturated ketones

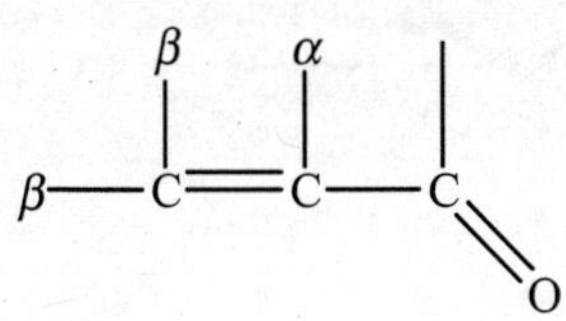

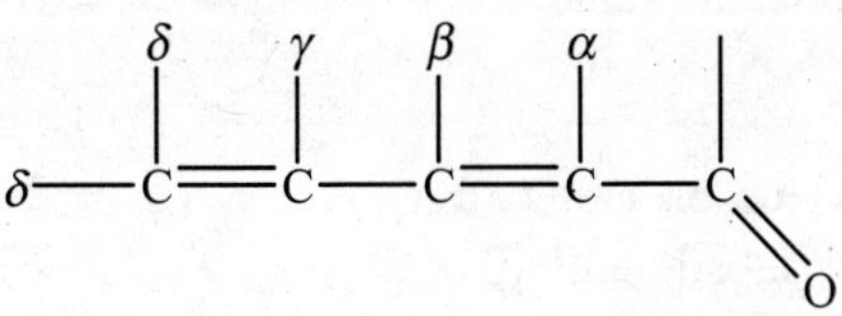

Group		Increment
6-membered ring or acyclic enone		Base 215 nm
5-membered ring parent enone		Base 202 nm
Acyclic dienone		Base 245 nm
Double bond extending conjugation		30
Alkyl group or ring residue	α, β, γ and higher	10, 12, 18
–OH	α, β, γ and higher	35, 30, 18
–OR	α, β, γ, δ	35, 30, 17, 31
–O(C=O)R	α, β, γ,	6
–Cl	α, β	15, 12
–Br	α, β	25, 30
–NR_2	β	95
Exocyclic double bond		5
Homocyclic diene component		39

The base values of aldehydes, esters, and carboxylic acids differ from those of ketones.

Unsaturated system	Base Value
Aldehyde	208
With α or β alkyl groups	220
With α, β or β, β alkyl groups	230
With α, β, β alkyl groups	242
Acid or ester	
With α or β alkyl groups	208
With α, β or β, β alkyl groups	217
Group value – exocyclic α, β double bond	5
Group value – endocyclic α, β bond in 5 or 7 membered ring	5

In contrast to conjugated alkenes, solvents do affect λ_{max}. The Woodward-Fieser rules also describe these consequences.

Solvent correction	Increment	Solvent correction	Increment
Water	+8	Dioxane	–5
Ethanol, methanol	0	Ether	–7
Chloroform	–1	Hydrocarbon	–11

For example:

1. Base value = 215 nm
   α substituent = 10 nm
   β substituent = 12 nm
   Total = 237 nm
   = 232 nm

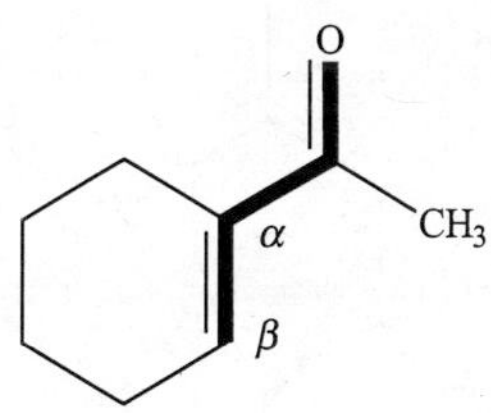

2. Base value = 215 nm
   2β substituent = 24 nm
   1 Exocyclic double bond = 5 nm
   Total = 244 nm
   Experimental value = 245 nm

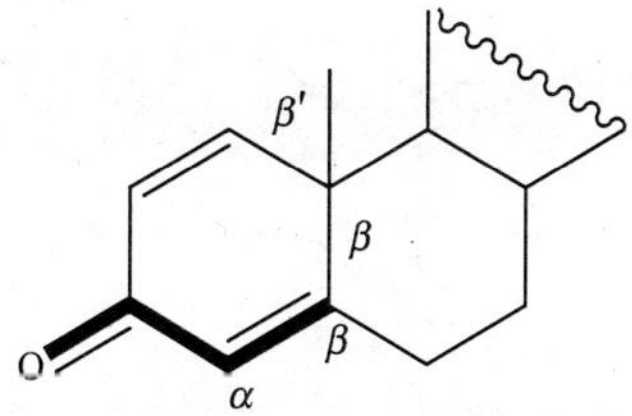

3. Base value = 202 nm
   β substituent = 12 nm
   α–OH = 35 nm
   Total = 249 nm
   Experimental value = 247 nm

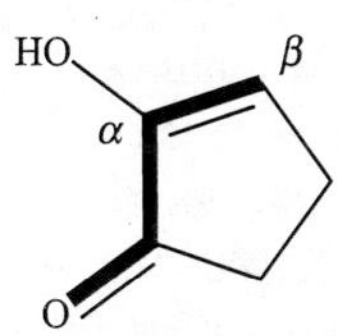

4. Base value = 215 nm
   2β substituent = 24 nm
   α-OH = 35 nm
   Total = 274 nm
   Experimental value = 270 nm

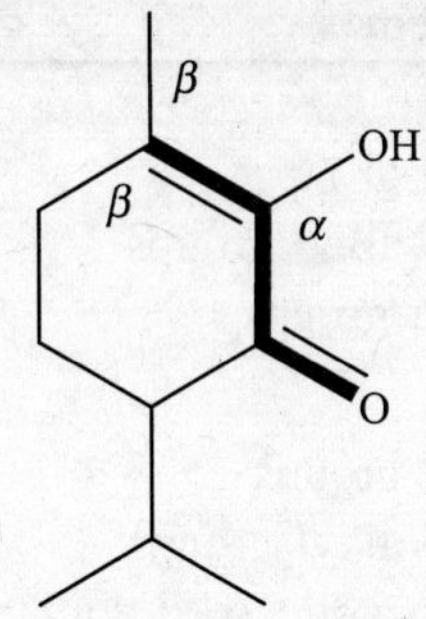

15.7.10.1 Violations of Woodward Rules

1. Biphenyls: Biophenyls exhibit a non-planar conformation with a dihedral angle of 45 degrees. Substituting elements leads to a decrease in the alignment of orbitals in the same plane. Decreased overlap resulting in a shift towards shorter wavelengths with a decrease in intensity.

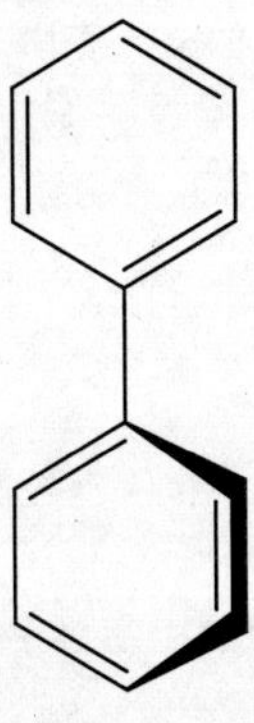

250 nm

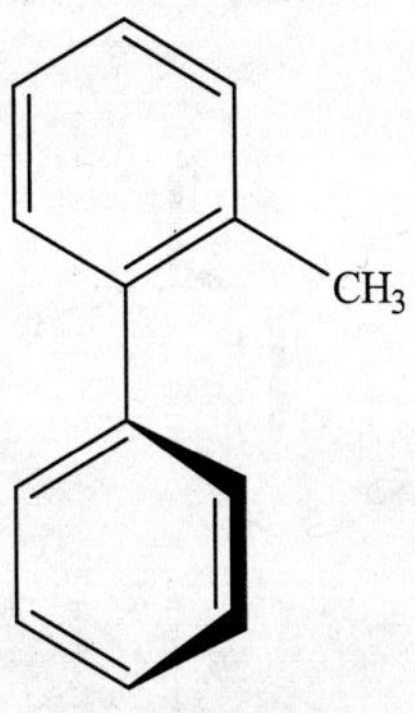

237 nm

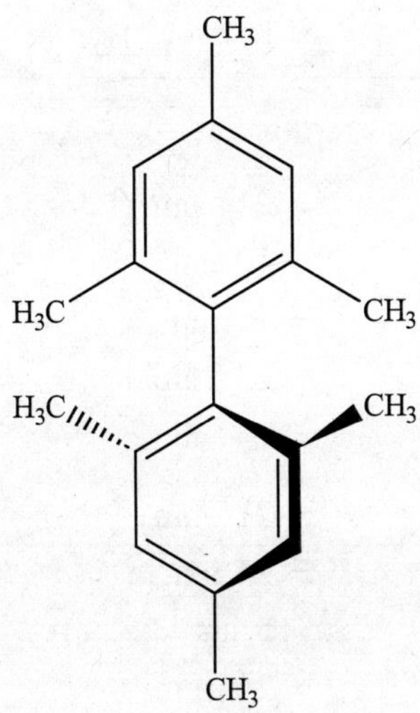

266 nm

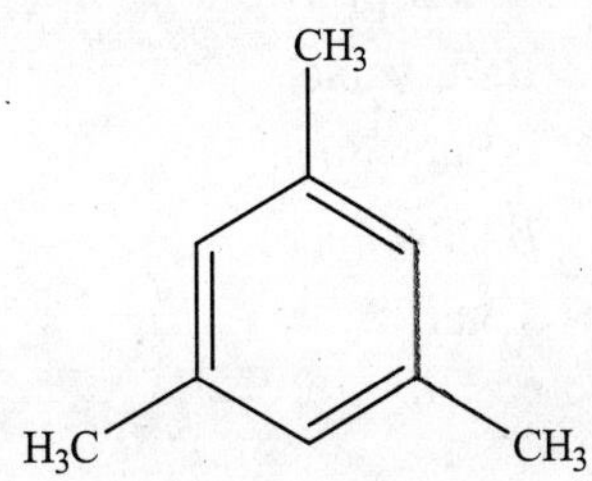

266 nm

2. Steric Inhibition of Resonance: Trans isomers in alkenes display absorption at longer wavelengths with a higher intensity. The Woodward criteria are only relevant in the absence of strain surrounding the chromophore. If the estimated and experimental values of λ_{max} do not coincide, it can be inferred that the molecule is under strain. Therefore, ultraviolet (UV) radiation can be employed as an indirect method to ascertain the presence of strain in a molecule.

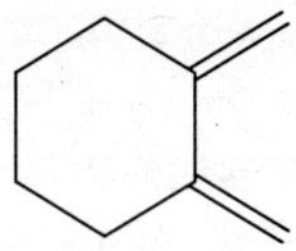

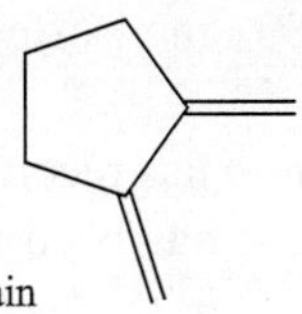

Ring strain

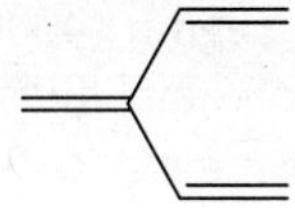

Cross conjugation

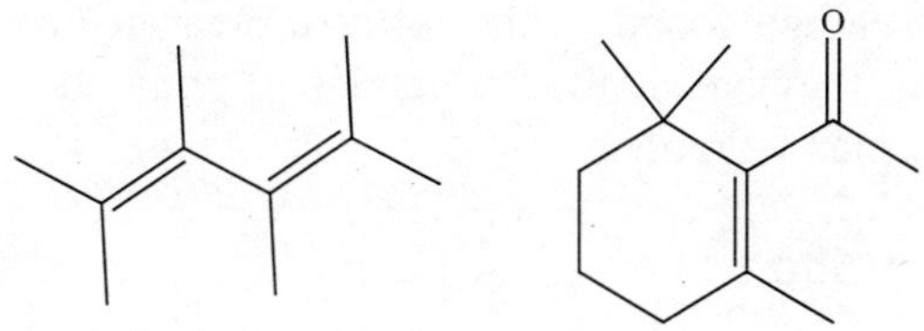

Steric inhibition of resonance

15.8 THE VIBRATIONAL OR IR SPECTROSCOPY

Infrared spectroscopy (IR) is a highly powerful technique that employs electromagnetic radiation within the infrared range to ascertain and recognize the molecular composition of substances. It is crucial to acknowledge that atoms have the ability to assimilate energy from electromagnetic radiation, and this assimilated energy modifies the state of the atoms within the molecule.

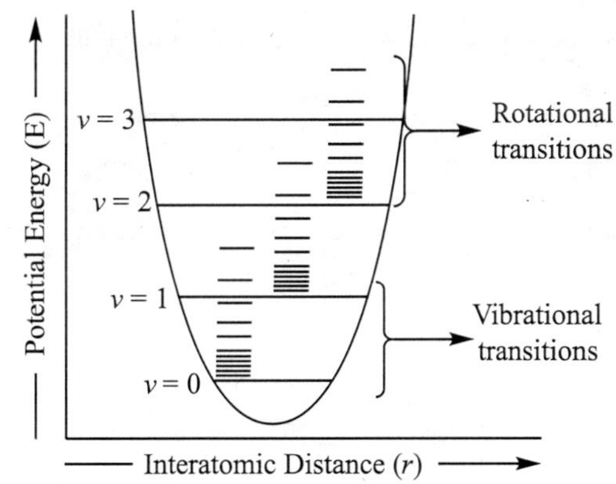

Figure 15.14 Potential energy resembles classic Harmonic Oscillator.

Infrared spectrometers employ optical instruments to disperse and concentrate electromagnetic radiation of infrared frequency. This radiation is then transmitted through the sample, and any alterations in absorbance are quantified by comparing it to a reference beam. The IR absorption of molecules occurs during their vibrational and rotational transitions. When a typical organic molecule absorbs radiation in this range, it becomes excited in its vibrational, rotational, and bending modes, while still remaining in its electronic ground state.

The IR spectrum is divided into three distinct regions: near, mid, and far. The demarcation lines between these entities lack clear definition and continue to be a subject of ongoing discussion. However, they can be broadly categorized as follows:

1. **The near infrared range** (12820–4000 cm^{-1}) lacks distinct absorptions and primarily consists of overtones and combination bands that arise from vibrations in the mid-infrared area of the spectrum.

2. **The mid-infrared region** (4000–400 cm^{-1}) offers valuable structural insights into the majority of organic compounds.
3. **The far Infrared range** (400–33 cm^{-1}) has received less research attention compared to the other two regions. Nevertheless, it has been utilized in studies involving inorganic compounds.

The relatively low energy found in the infrared area are inadequate to induce electronic transitions. Nevertheless, they are significant enough to bring about alterations in the frequency and amplitude of molecular vibrations.

Theory of IR Absorption

Molecular Vibrations: Infrared radiation mostly consists of thermal energy. It amplifies the intensity of molecular vibrations in covalent connections, which can be conceptualized as the tension of springs connecting two masses, or atoms. Particular bonds exhibit a particular response to (absorb) distinct frequencies.

Harmonic Oscillator Model: An approximate depiction of the stretching of an atom, when two masses are connected by a spring (Figure 15.15).

$$E = \frac{1}{2} ky^2$$

where, y is the spring displacement and k is the spring constant.

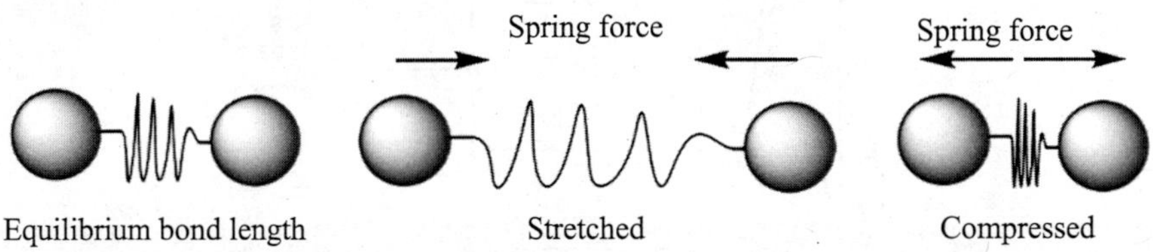

Vibrational frequency given by: $\nu = \frac{1}{2}\pi\sqrt{k/\mu}$

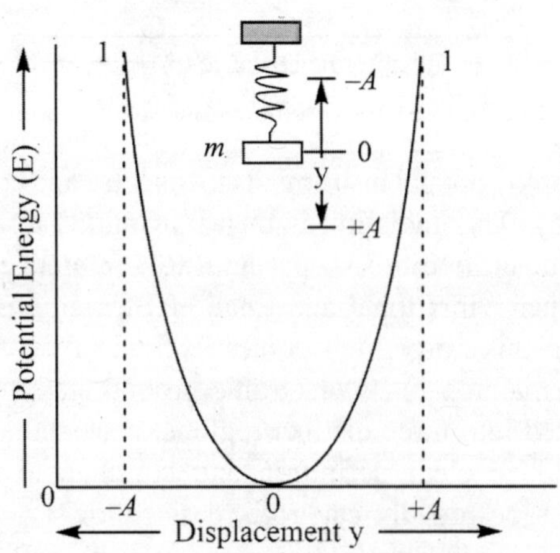

Figure 15.15 Harmonic Oscillator Model.

where ν is frequency, k is force constant (measure of bond stiffness), μ is reduced mass which is equal to $m_1m_2/m_1 + m_2$. If know ν and atoms in bond, can get κ. For single bonds the value of $\kappa \sim 3 \times 10^2$ to 8×10^2 N/m (Avg ~ 5×10^2), double and triple bonds ~2x and 3x k for single bond.

$$\nu \propto k$$

So, vibration n occur in order: i.e. single < double < triple

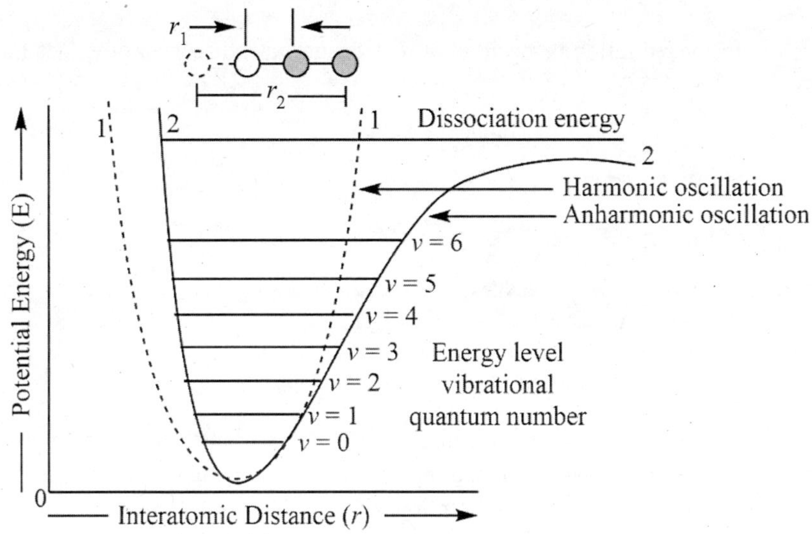

Figure 15.16 Harmonic and Anharmonic Oscillator Model.

(ii) Anharmonic Oscillation: The harmonic oscillation model is effective for low energy levels (ν_0, ν_1, ν_2, ...) but in high energy levels it is attributed to the interplay of atomic repulsion and attraction (Figure 15.16).. When atoms come closer, the repulsive force between them adds to the bonding force, resulting in a higher increase in energy compared to a harmonic function. Conversely, when atoms move apart, the dissociation energy is approached and the harmonic function rises more rapidly.

The presence of anharmonics leads to the observation of overtone lines at frequencies approximately 2–3 times higher than the fundamental frequency, due to the occurrence of energy differences (ΔE) and frequency differences (Δv) of ±2 and ±3 at low levels. Typically, the change in frequency (Δv) is either positive or negative one.

(iii) Types of Molecular Vibrations: The energy levels observed in IR spectroscopy correspond to the vibrational states of molecules as described by quantum mechanics. The sensation of heat is the result of our perception of this vibration. When we refer to the length of a covalent connection between two atoms, we are actually referring to an average value. This is because the bond functions in a manner similar to a vibrating spring that connects the two atoms.

Two varieties of bond vibration exist:

1. Stretch: It refers to the vibration or oscillation that occurs along the line of the bond.

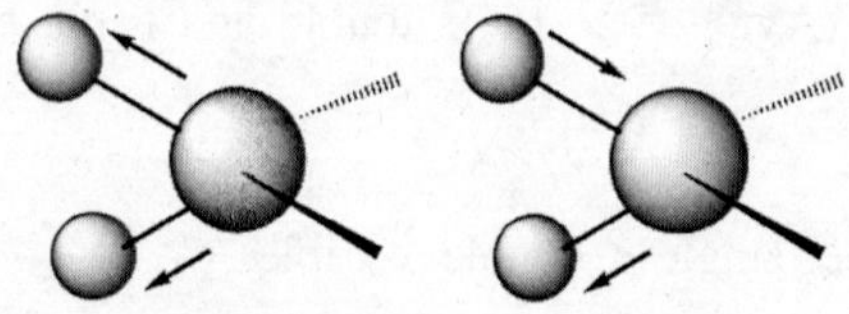

Symmetric Streching **Asymmetric Streching**

2. Bend: It refers to the vibration or oscillation that occurs in a direction other than along the line of the bond.

In-plane Bending

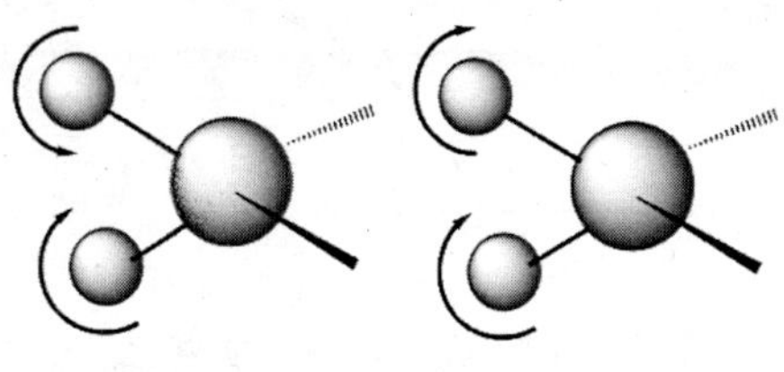

Scissoring **Rocking**

When a covalent bond oscillates, it causes the dipole of the molecule to oscillate as well, resulting in the production of a fluctuating electromagnetic field. The magnitude of the electromagnetic field created is directly proportional to the extent of change in dipole moment during the vibration. When an oscillating electromagnetic field formed by an oscillating dipole of the same frequency intersects with a wave of infrared light, the two waves interact and the infrared light is absorbed. The wave that is now connected oscillates with double the magnitude.

Out of plane Bending

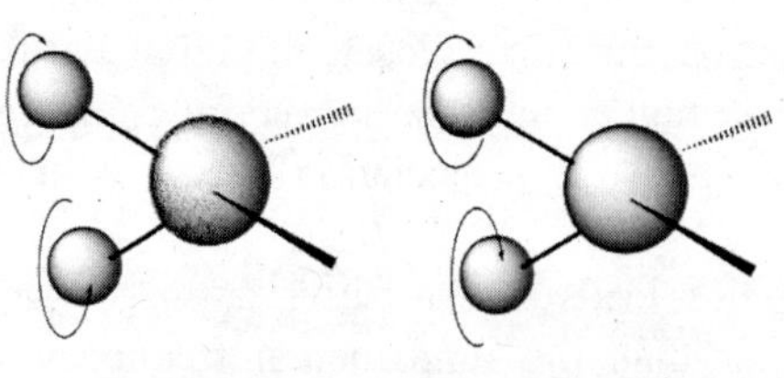

Twisting **Wagging**

(iv) Number of Vibrational Modes: The number of vibrational modes can be determined as follows:

For **non-linear** molecules, the number of different types of vibrations is equal to 3N-6. For **linear** molecules, the number of different types of vibrations is equal to 3N-5.

The observed vibrations may be lower in frequency than predicted due to factors such as symmetry (resulting in no change in dipole moment), similar vibration energies, low absorption intensity, or frequencies that are beyond the range of the instrument.

Examples:

1. HCl: Since HCl is linear molecule so 3N-5 is the applicable formulae. The number of atom is 2 so, 3(2) – 5 = 1. Hence HCl has only 1 mode of vibration.
2. CO_2: Since CO_2 is linear molecule so 3N-5 is the applicable formulae. The number of atom is 3 so, 3(3) – 5 = 4 modes, moving in-out of plane.

(v) IR Active Vibrations: For a molecule to absorb infrared (IR) radiation, it must exhibit a vibration that occurs at the same frequency as the light. Additionally, the molecule must also undergo a change in its overall dipole moment as a direct consequence of this vibration.

Examples:

1. CO_2: 3(3) – 5 = 4 modes

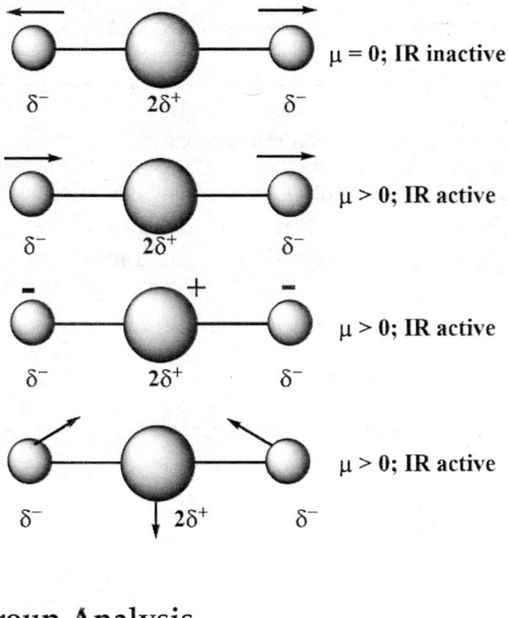

15.8.1 Infrared Group Analysis

1. Qualitative Analysis (Compound Identification): The utilization of infrared spectroscopy (IR), in conjunction with nuclear magnetic resonance (NMR) and mass spectrometry (MS), during the late 1950s brought about a significant transformation in the field of organic chemistry. It reduced the time required to verify compound identification by a factor of 10 to 1000.

The primary use of the IR is to ***detect functional groups***

(i) **General Scheme:** IR analyze the presence of functional groups by observing the frequency region associated with each group. The range is from 3600 cm^{-1} to 1200 cm^{-1}.

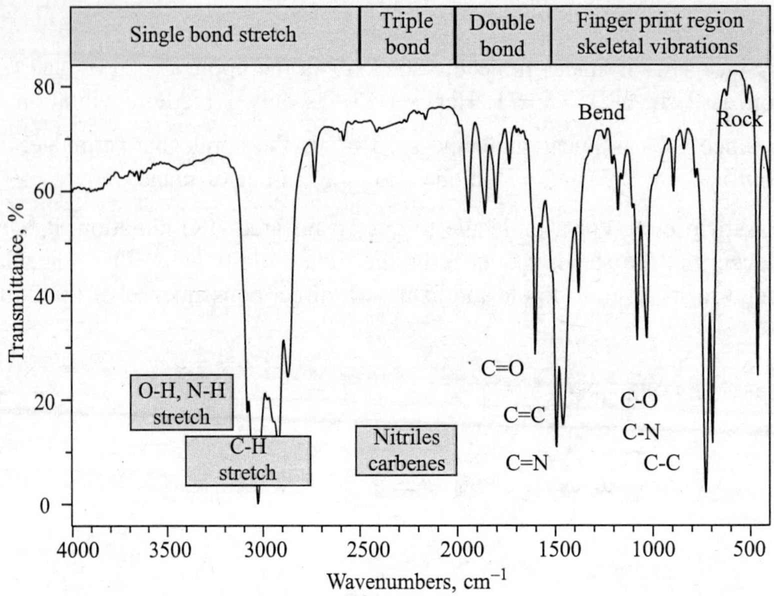

(ii) Frequency Region of the Group: The frequency of various functional groups (such as C=O, C=C, C–H, and O–H) can be estimated by utilizing atomic masses and force constants. Atomic positions exhibit slight variations in proximity to nearby atoms, typically within the same general vicinity. Provides a reliable preliminary indication of compound identity, but does not provide definitive evidence.

Bond	Type of Compound	Frequency Range, cm^{-1}	Intensity
C–H	Alkanes	2850–2970	Strong
C–H	Alkenes	3010–3095	Medium strong
		675–995	
C–H	Alkynes	3300	Strong
C–H	Aromatic rings	3010–3100	Medium strong
		690–900	
O–H	Monomeric alcohols, phenols	3590–3650	Variable
	Hydrogen–bonded alcohols, phenols	3200–3600	Variable,
	Monomeric carboxylic acids	3500–3650	sometimes broad
	Hydrogen–bonded carboxylic acids	2500–2700	Medium broad
N–H	Amines, amides	3300–3500	medium
C=C	Alkenes	1610–1680	Variable
C=C	Aromatic rings	1500–1600	Variable
C=C	Alkynes	2100–2260	Variable
C–N	Amines, amides	1180–1360	Strong
C N	Nitriles	2210–2280	Strong
C–O	Alcohols, ethers, carboxylic acids, esters	1050–1300	Strong
C=O	Aldehydes, ketones, carboxylic acids, esters	1690–1760	Strong
NO_2	Nitro compounds	1500–1570	Strong
		1300–1370	

(iii) **Fingerprint Region (1200–700 cm^{-1}):** This region primarily contains signals from single bonds. Many compounds have similar frequency, which results in mutual influence and contributes to the distinctive pattern characteristics of their overall skeletal structure. The complexity of this region of spectra often hinders precise interpretation. Complexity can be equated to uniqueness.

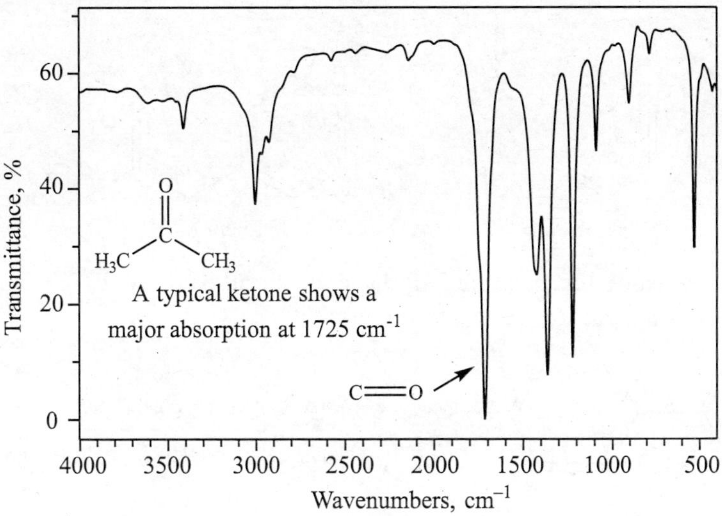

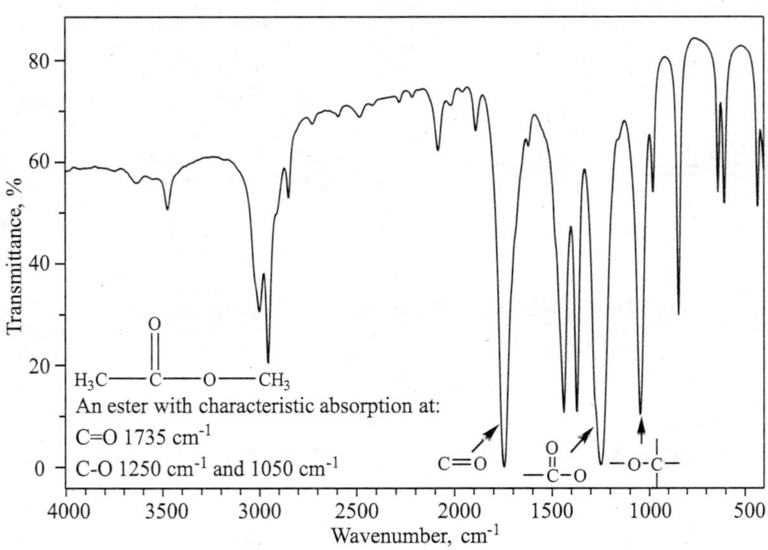

2. Quantitative Analysis: UV/Vis is superior to it in terms of accuracy and precision. Advanced spectra with greater complexity. Reduced spectral width (deviation from Beer's Law). IR equipment have limitations in terms of their light throughput and the sensitivity of their detectors. Elevated infrared radiation levels in the background. Challenging to align reference and sample cells. Modifications in the variable ε (where $A = \varepsilon bc$) are frequent. An inherent benefit is the high selectivity, as numerous substances exhibit distinct infrared spectra. An example of a typical use case is the identification of airborne pollutants.

15.8.2 Effects on IR Bands

15.8.2.1 Conjugation Effects

Conjugation is a phenomenon where the energy of a double or triple bond is reduced by resonance effects. This phenomenon is easily visible in the infrared spectrum.

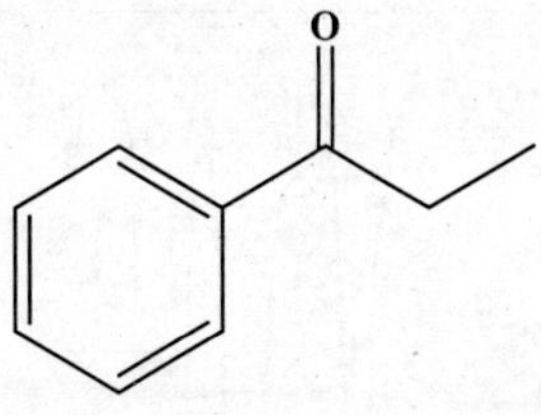

C=O; 1684 cm^{-1}

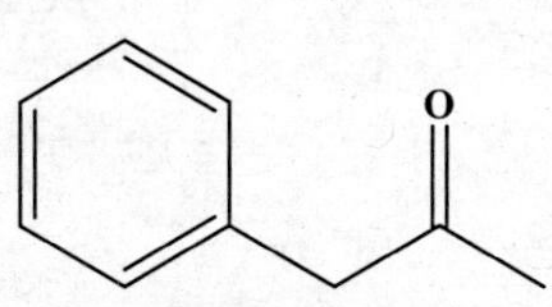

C=O; 1715 cm^{-1}

Conjugation decreases the measured IR band of a carbonyl by 20–40 cm^{-1}, given that conjugation produces a significant resonance contributor.

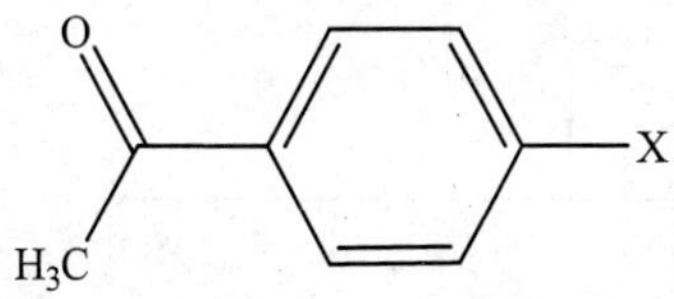

X =	NH_2	CH_3	Cl	NO_2	
	1677	1687	1692	1700	cm^{-1}

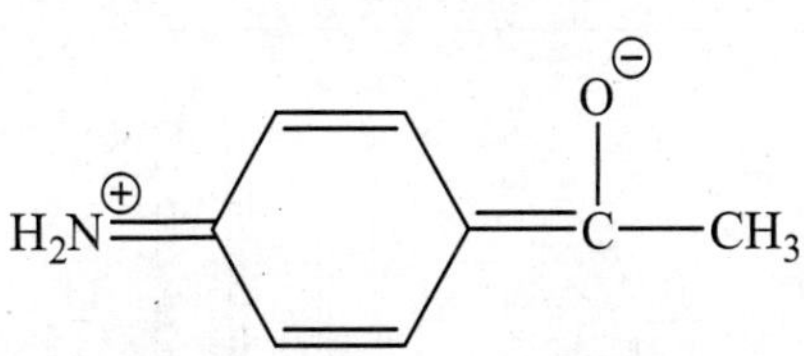

Strong resonance contributor

vs.

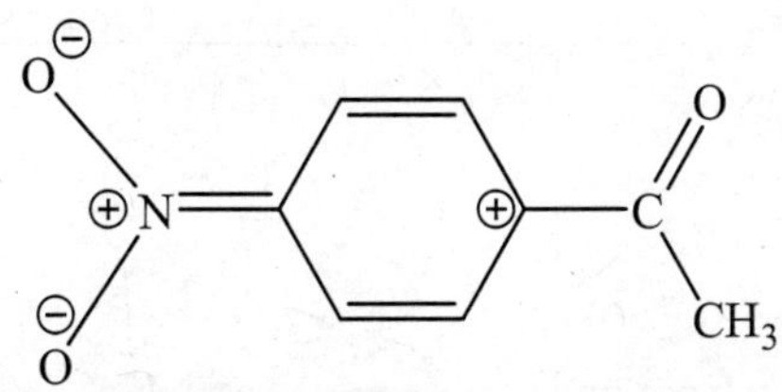

Poor resonance contributor
(cannot resonate with C=O)

Inductive effects are often negligible, unless they are combined with a resonance contributor (such as $-CH_3$ and –Cl above).

15.8.2.2 Steric Effects

Steric effects are often insignificant in IR spectroscopy, unless they diminish the bond strength (typically π) by obstructing appropriate orbital overlap.

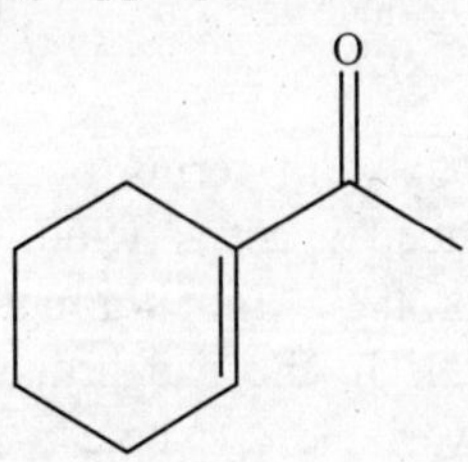

C=O; 1686 cm^{-1}

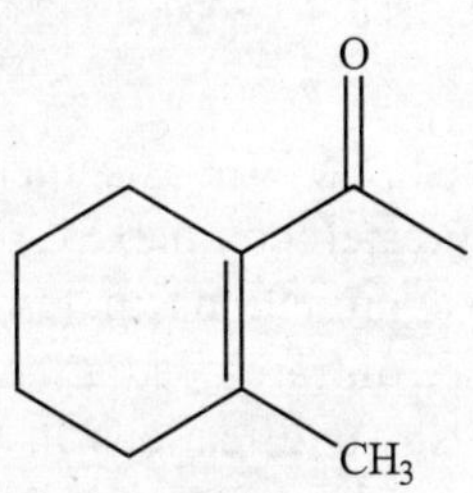

C=O; 1693 cm^{-1}

The presence of the methyl group in the structure on the right induces a tiny deviation of the carbonyl group from the plane, so disrupting resonance.

15.8.2.3 Strain Effects

The restrictions of a ring will induce alterations in bond angle, resulting in a little modification in hybridization and consequently, bond strength.

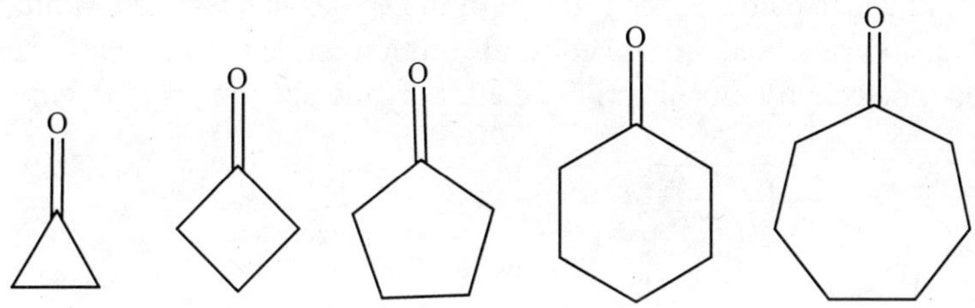

1815 cm^{-1} 1775 cm^{-1} 1750 cm^{-1} 1715 cm^{-1} 1705 cm^{-1}

As the bond angle falls, the electronegativity of carbon increases, and its sp^2 hybridization diminishes (when the bond angle is less than 120°).

15.8.2.4 Hydrogen Bonding Effects

Hydrogen bonding leads to the widening of the band as it generates a range of bond energies that form a continuous spectrum.

The effects mentioned are clearly observable in the solution phase, but they disappear in the gas phase. In the absence of steric effects, the band looks as crisp as any other infrared (IR) bands.

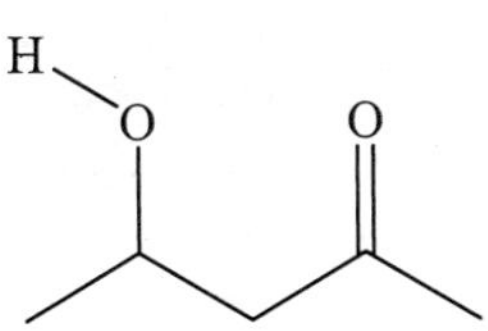

C=O; 1701 cm^{-1}

Gas phase spectrum of 1-butanol

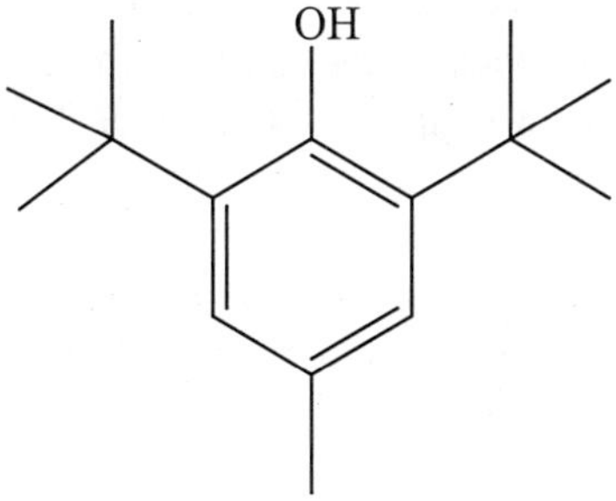

Steric hindrance to H-bonding in a di-tert-butylphenol

H-bonding has the ability to interact with other functional groups in order to decrease frequencies.

15.9 MICROWAVE SPECTROSCOPY OR ROTATIONAL SPECTROSCOPY

Electromagnetic waves can stimulate the rotational levels of molecules if the molecules possess an electric dipole moment. The molecule experiences a torque due to the electromagnetic field.

The rotational transitions of molecules commonly occur within the microwave range of the electromagnetic spectrum. The quantum energy of microwave photons, which falls within the range of 0.00001–0.001 eV, aligns with the energy ranges that differentiate quantum states of molecule rotations and torsion. The rotational frequencies span a range of approximately $8 \times 10^{10} - 4 \times 10^{11}$ Hz, corresponding to wavelengths of around 0.75 – 3.75 mm. These wavelengths are situated within the microwave range of the electromagnetic spectrum.

It is important to understand that the rotation of molecules is quantized, similar to electronic and vibrational transitions, which are linked to absorption and emission lines. The heating is a result of enhanced molecule rotational activity caused by the absorption of microwave radiation.

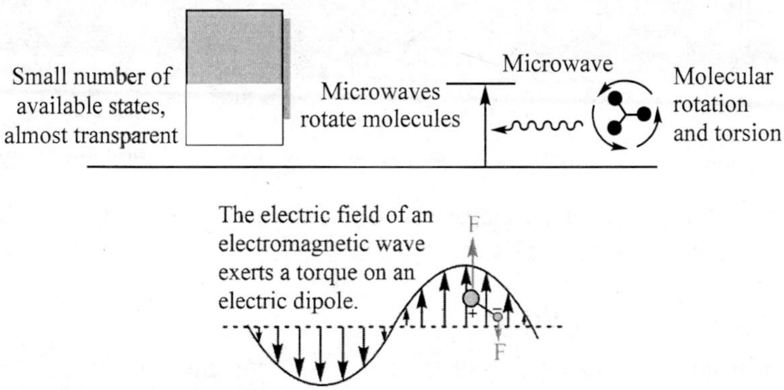

Molecular rotational energies exhibit quantization, meaning they possess only discrete energy levels $E = h\nu$. E represents the energy in joules (J), h represents Planck's constant in Joule-seconds (Js), and ν represents the rotational frequency in hertz (Hz).

Homonuclear diatomic molecules, such as H_2, O_2, N_2, and Cl_2 possess a dipole moment of zero, making them nonpolar. Homonuclear diatomic molecules are considered microwave inactive due to their lack of dipole shift during rotation, resulting in no interaction with radiation.

Heteronuclear diatomic molecules, such as HCl, HF, and CO, possess a permanent dipole moment, making them polar compounds. A dipole undergoes a change in orientation during rotation, leading to an interaction with radiation. Hence, heteronuclear diatomic compounds exhibit microwave activity.

15.9.1 Rigid Rotor

We can focus on the rotational motion of a rigid diatomic molecule. A diatomic molecule is capable of rotational motion along a vertical axis. Rotational energy exhibits quantization. A diatomic molecule can be conceptualized as a pair of atoms connected by a massless and rigid rod (*rigid rotator model*) (Figure 15.17).

Let's examine a diatomic molecule composed of two atoms with distinct masses, m_1 and m_2. The distances of these atoms from the center of mass are r_1 and r_2, respectively. The system's moment of inertia with respect to the center of mass is: $I = m_1 r_1^2 + m_2 r_2^2$.

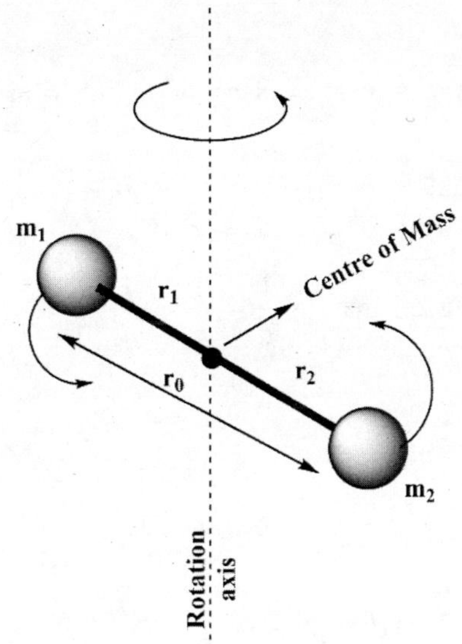

Figure 15.17 Rigid rotator model.

15.9.1.1 Rotational Levels

The classical expression for rotational energy is: In the harmonic oscillator model, the total energy was in the form of potential energy. In the rigid rotor, the entire energy is in the form of kinetic energy.

$$E_r = \frac{I\omega^2}{2} = \frac{L^2}{2I} \quad \text{where } L = \text{angular momentum}$$

$$L^2 \rightarrow \left(\frac{h}{2\pi}\right)^2 J(J+1)$$

where J is the rotational quantum number

$$E(J) = \frac{1}{2I}\left(\frac{h}{2\pi}\right)^2 J(J+1) = hcB_v\, J(J+1)$$

$$B_v = \frac{h}{8\pi^2 cI} \quad \text{the rotational constant}$$

Assume a rigid (not elastic) bond

$$r_0 = r_1 + r_2$$

For rotation about center of gravity, C:

$$m_1 r_1 = m_2 r_2 = m_2(r_0 - r_1)$$

$$r_1 = \frac{m_2 r_0}{m_1 + m_2} \quad r_2 = \frac{m_1 r_0}{m_1 + m_2}$$

Moment of inertia about C:

$$I_C = m_1 r_1^2 + m_2 r_2^2 = m_2 r_2 r_1 + m_1 r_1 r_2 = r_1 r_2 (m_1 + m_2)$$

$$\Rightarrow \qquad I = \frac{m_1 m_2}{m_1 + m_2} r_0^2 = \mu r_0^2$$

$$\mu = \text{reduced mass}, \quad \frac{1}{\mu} = \frac{1}{m_1} + \frac{1}{m_2}$$

More detailed derivation:

$$\begin{aligned} I_C &= m_1 r_1^2 + m_2 r_0^2 \\ &= m_1 r_1 \left[\frac{m_2 r_2}{m_1}\right] + m_2 r_2 \left[\frac{m_1 r_1}{m_2}\right] \\ &= r_1 r_2 [m_1 + m_2] \\ &= \left[\frac{m_1 r_0}{m_1 + m_2}\right]\left[\frac{m_2 r_0}{m_1 + m_2}\right][m_1 + m_2] \\ &= \frac{m_1 m_2}{m_1 + m_2} r_0^2 \end{aligned}$$

A diatomic molecule exhibits rotational motion about a vertical axis. Rotational energy exhibits quantization. The rotational energy levels permitted for a rigid diatomic molecule can be determined using the Schrödinger equation.

$$E_J = \frac{h^2}{8\pi^2 I} J(J+1) \text{ joules} \qquad \text{where } J = 0, 1, 2, \ldots$$

Rotational energy level in *wavenumbers* (cm^{-1})

$$E_J = \frac{h}{8\pi^2 I_c}$$

The rotational quantum number, denoted as J, takes on integer values starting from 0 and increasing by 1. It should be noted that when $J = 0$, the molecule is at its lowest energy level and does not undergo rotational motion. I represents the moment of inertia, which is equal to the product of the mass (m) and the square of the radius (r). μ represents the decreased mass, which is calculated as the product of the masses (m_1 and m_2) divided by their sum ($m_1 + m_2$) (μ = reduced mass = $m_1 m_2/(m_1 + m_2)$). The variable "r" represents the distance between two atomic nuclei.

The rotational frequency is equivalent to the frequency of the microwave radiation required to induce rotation (Figure 15.18). The equation $\nu/\text{Hz} = \Delta E/h$ can be rewritten as $J(J + 1)\ cm^{-1} = BJ(J + 1)\ cm^{-1}$. The term "rotational constant" refers to the value of B for a specific molecule. The units of J are cm^{-1}, as it represents a quantum number (label).

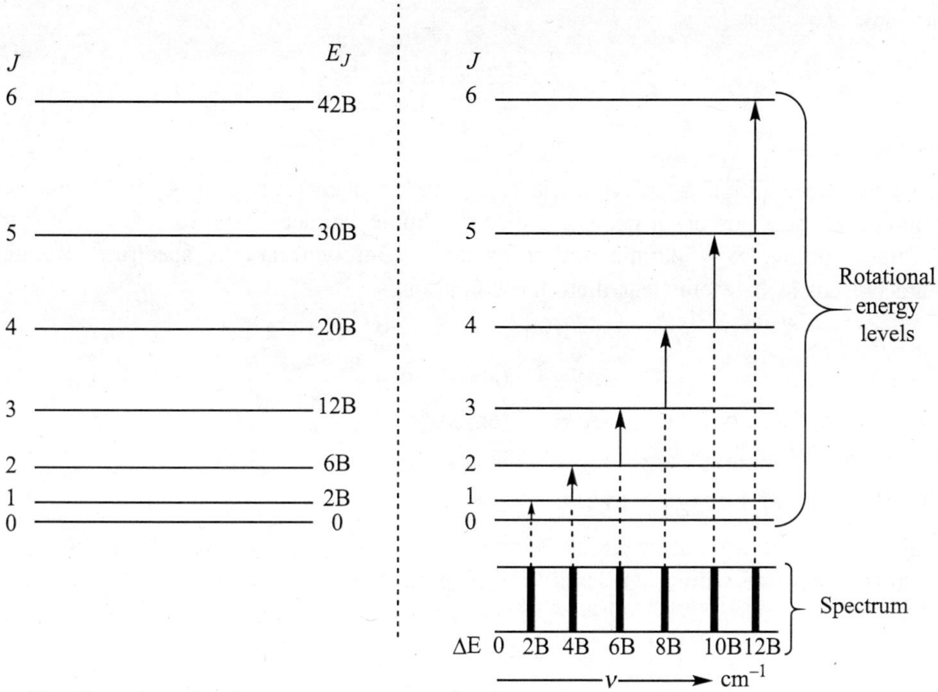

The allowed rotational energy levels of a rigid diatomic molecule

Allowed transitions between the energy levels of a rigid diatomic molecule and the spectrum

Figure 15.18 Allowed rotational energy levels.

15.9.1.2 Rotational Spectra of Rigid Diatomic Molecule

Rotational Energy Levels for rigid rotor:

$$E_J = \frac{h^2}{8\pi^2 I} J(J+1) \text{ joules} \qquad \text{where } J = 0, 1, 2, \ldots$$

$$\varepsilon_J = \frac{E_J}{hc} = \frac{h^2}{8\pi^2 I_c} J(J+1) \text{ cm}^{-1} \qquad (J = 0, 1, 2, \ldots)$$

$$\varepsilon_J = BJ(J+1) \text{cm}^{-1} \qquad (J = 0, 1, 2, \ldots)$$

where B is the rotational constant and is given by

$$B = \frac{h^2}{8\pi^2 I_B c} \text{ cm}^{-1}$$

where

$$I = \frac{m_1 m_2}{m_1 + m_2} r_0^2 = \mu_0^2$$

For rigid rotor, $J \rightarrow J + 1$,

$$\bar{\nu}_{J \rightarrow J+1} = 2B(J+1)\ \text{cm}^{-1}$$

Selection rule: $\Delta J = \pm 1$

Selection Rule: In addition to the Specific rule, another need is the Gross rule, which states that the molecule must possess a persistent electric dipole moment, represented by the symbol μ. Therefore, homonuclear diatomic molecules lack a purely rotational spectrum. Rotational spectra are present in heteronuclear diatomic compounds.

$$\Delta j = \pm 1$$
$$\Delta j = +1 \text{ (absorption)}$$
$$\Delta j = -1 \text{ (emission)}$$

15.9.1.3 Rotational Energy Levels

Rotational energy levels after absorption of electromagnetic radiation are depicted in Figure 15.9. which indicates that one vibrational level may contain many rotational levels.

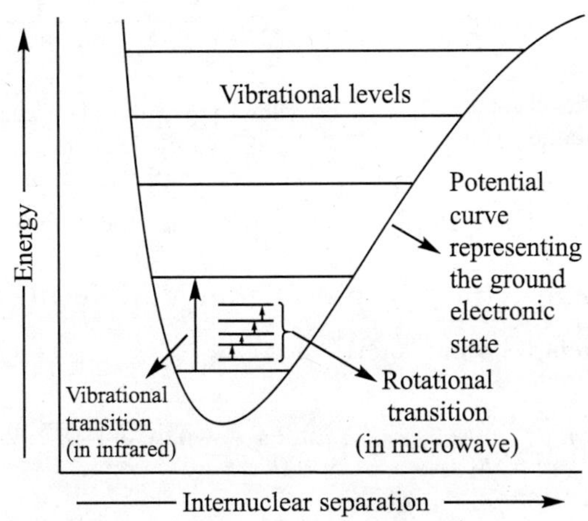

Figure 15.19 Rotational Energy levels.

15.9.1.4 Relative Intensities of Rotation Spectral Lines

With our comprehension of the places of lines in the microwave spectrum, we can now identify the lines that exhibit the greatest intensity.

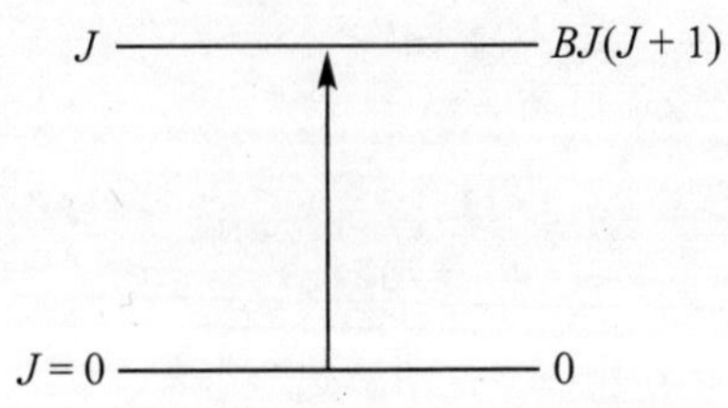

Intensity depends upon two factors:

1. A larger starting population of atoms results in more intense spectral lines. The size of this population is contingent upon the temperature, T.

$$k = \text{Boltzmann's constant, } 1.380658 \times 10^{-23} \text{ J K}^{-1}$$

$$(k = R/N) \quad \frac{N_J}{N_0} \propto \exp\left(-\frac{E_J}{kT}\right) = \exp\left(-\frac{hc\bar{\nu}}{kT}\right)$$

We conclude that the population is smaller for higher J states.

$$\frac{N_J}{N_o} \propto e\left(-\frac{1.52034\bar{\nu}}{T}\right) \quad \frac{hc}{k} = 1.52034 \text{ cmK}$$

2. Degeneracy of initial state regulates intensity. The term "degeneracy" refers to the presence of two or more energy states that possess identical energy levels (Figure 15.20). The degeneracy of each level J is equal to $(2J + 1)$. The population is larger for states with higher J values.

The overall relative population at energy E_J can be summarized by the equation $E_J \propto (2J + 1) \exp(-E_J/kT)$. The highest population occurs at the nearest integral J value. To analyze the values of N_J/N_0, refer to the Figure 15.20 which shows the plot of population of rotating energy levels vs the value of J.

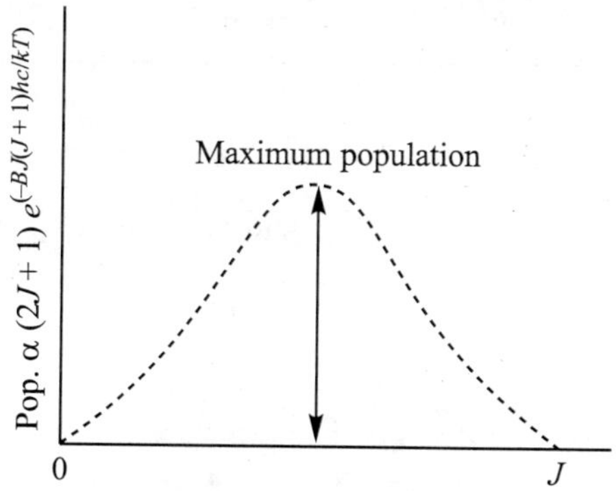

Figure 15.20 Overall relative population to energy.

15.10 NUCLEAR MAGNETIC RESONANCE

This phenomenon takes place when the nuclei of specific atoms are placed within a stationary magnetic field and subjected to a fluctuating electromagnetic field. NMR spectroscopy is utilized in various scientific fields. Chemists commonly employ NMR spectroscopy to investigate chemical structure through the use of straightforward one-dimensional techniques. Complex molecules can be analyzed using two-dimensional techniques to ascertain their structure. The wide range of applications of NMR makes it prevalent in various scientific disciplines.

What is the Concept of Spin?

Spin is an intrinsic characteristic of nature, comparable to electrical charge or mass. Spin is a quantification of the angular momentum, which refers to the rotation around an axis. Spin is quantized in increments of 1/2 and can have positive or negative values. Protons, electrons, and neutrons exhibit spin. Each individual unpaired electron, proton, and neutron has a spin value of 1/2. Nuclei consisting of neutrons and protons can also exhibit spin. Atoms possessing an uneven quantity of protons and neutrons will exhibit spin. Atoms possessing an equal amount of protons and neutrons, both of which are even, will not exhibit spin. The sign I represents the nuclear spin, which is the quantum number that represents the value of nuclear spin. The sequence is defined as $I = 0, 1/2, 1, 3/2, 2, 5/2$..... A nucleus possessing a spin value of I can exist in a total of $(2I + 1)$ distinct spin states.

Common Nuclei with Spin

Nuclei	Unpaired Protons	Unpaired Neutrons	Net Spin
^{1}H	1	0	1/2
^{2}H	1	1	1
^{31}P	1	0	1/2
^{23}Na	1	2	3/2
^{14}N	1	1	1
^{13}C	0	1	1/2
^{19}F	1	0	1/2

15.10.1 Magnetic Behavior of Atomic Nuclei

A rotating electric charge produces a magnetic field. The atomic nucleus possesses an intrinsic charge as a result of the presence of protons. Therefore, an atomic nucleus with a non-zero spin ($I > 0$) will possess an associated magnetic field.

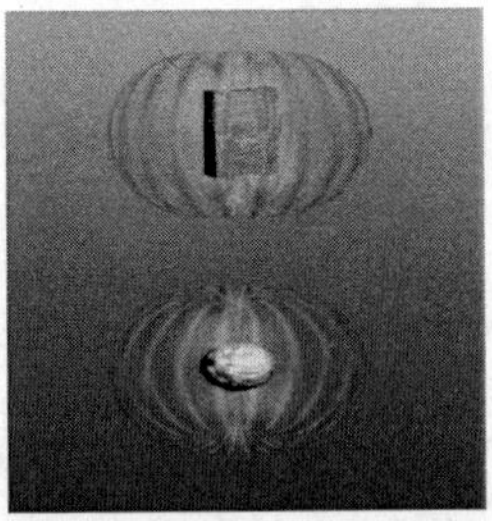

The nucleus produces a magnetic dipole aligned with its spin axis, and the inherent strength of this dipole is a fundamental characteristic of the nucleus known as the nuclear magnetic moment, μ. The nuclear magnetic moment is determined by the nuclear angular momentum quantum number I, as described by the following equation:

$$\mu = \gamma_N m_N \sqrt{[I(I + 1)]}$$

Both γN and mN are constants specific to a certain nucleus N.

A spinning atomic nucleus exhibits characteristics similar to those of a tiny bar magnet.

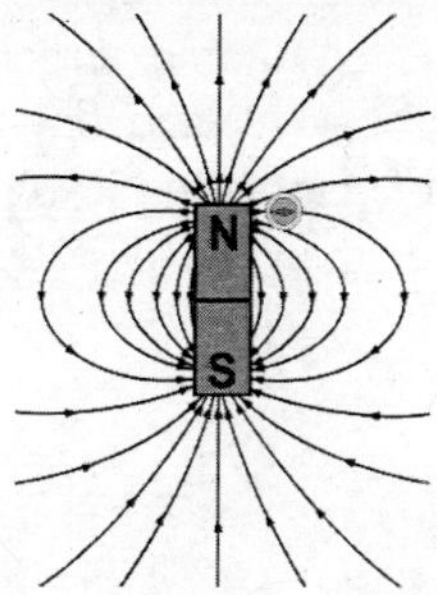

15.10.2 Magnetic Properties of Atomic Nuclei

Let's examine the behaviour of a hydrogen atom with a single proton. The magnetic moment, denoted as μ, is a vector that possesses both magnitude and direction. The magnetic moments of a group of many hydrogen atoms orient themselves randomly in the absence of an external magnetic field. In other words, no one orientation is favoured.

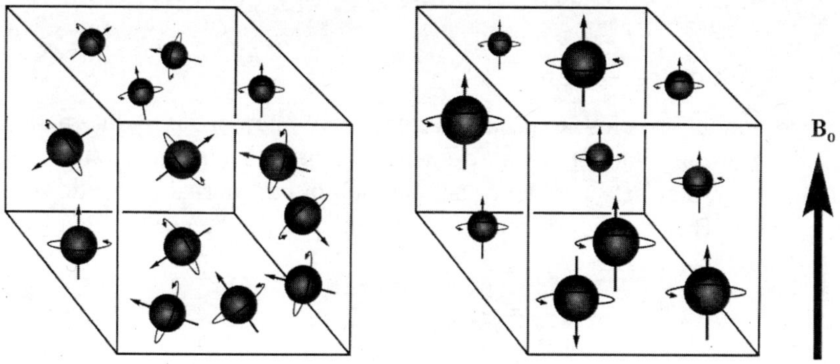

Random orientation | Orientation under applied external magnetic field effect

According to quantum mechanics, there are only $2I + 1$ conceivable orientations in which the nuclear magnetic moment of a nucleus can align with an externally supplied magnetic field of strength B_o. The magnetic moment is aligned parallel to the applied field (spin + 1/2) and is commonly denoted as α. On the other hand, the higher energy anti-parallel orientation (spin −1/2) is referred to as β.

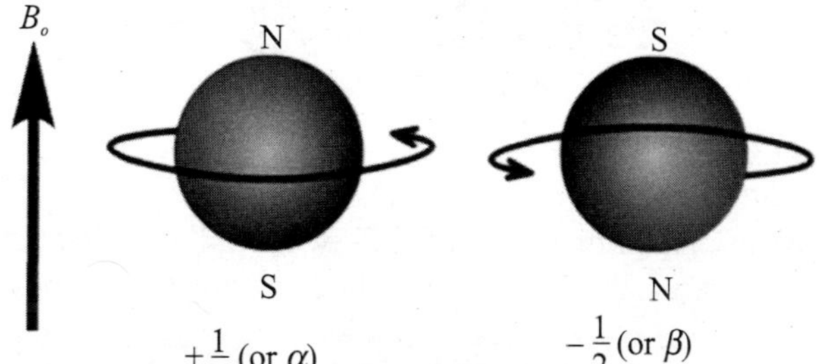

Spin Energy States

15.10.3 Quantized Spin

When a hydrogen atom ($I = 1/2$) is subjected to an external magnetic field, only two spin states are observed. The two states labelled as α (+1/2) and β (–1/2) and they possess distinct energy levels. Due to the inherent preference of nature for lower energy states, a greater number of hydrogens will opt for the α state rather than the β state, albeit with a modest difference in population.

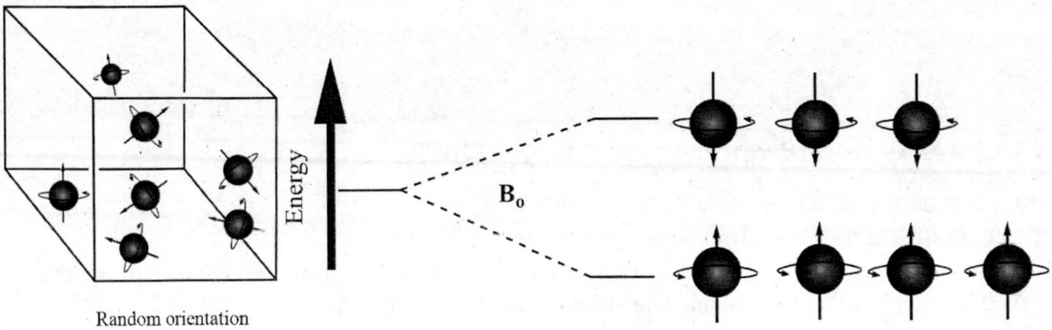

15.10.3.1 Gyroscopic Precession

When magnets are introduced close to each other, they will align precisely in a parallel manner and remain in this position without any movement. Magnetic nuclei, constrained by quantum mechanics, do not align precisely parallel or antiparallel to the external magnetic field. Instead, they align at an angle. This has a significant implication that can be exemplified by examining a gyroscope (Figure 15.21).

Figure 15.21 A Spinning Gyroscope in a Gravity Field.

When a gyroscope is set in a particular position, it will tend to maintain that position despite the influence of external forces such as gravity. When a gyroscope is positioned vertically in a gravitational field, where gravity pulls directly downward, it will remain in this orientation without any movement. When a force is exerted on the gyroscope in a direction perpendicular to the gravitational field, it will undergo rotational motion around an axis that is parallel to the gravitational field. This phenomenon is known as precession. The frequency of this precession is contingent upon two factors: the gravitational force exerted and the gyroscope's force exertion.

15.10.3.2 Nuclear Precession

The gyroscope will undergo precession when subjected to a gravitational field, similarly, the magnetic moment μ of a rotating spherical charge will undergo precession when exposed to an external magnetic field. In this instance, the rotating nucleus is positioned at the starting point of a cartesian coordinate system, and the external field is aligned with the z-axis.

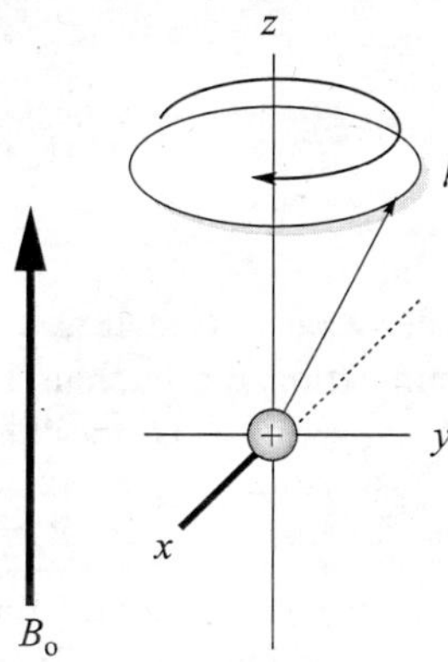

The angular frequency at which this precession occurs is determined by the equation $\nu = \gamma B_o / 2\pi$ and is referred to as the Larmor frequency. The symbol γ represents the magnetogyric ratio, which is a unique property of each nucleus type. It pertains to the intensity of the magnetic field of the nucleus. B_o represents the magnitude of the magnetic field that is applied externally. For instance, when a hydrogen (^{1}H) atom is subjected to a magnetic field of 1.41 Tesla, its Larmor frequency is precisely 60 megahertz (MHz).

15.10.3.3 Spin State Energy and B_o

The energy levels associated with the orientations of a magnetic nucleus in relation to an external magnetic field are not equal. Spin states aligned parallel to the external field have lower energy compared to when there is no external field. Conversely, spin states that are more closely aligned against the external field have a higher energy compared to when there is no external field present (Figure 15.22).

The energy difference between the two spin states is dependent upon the intensity of the external magnetic field and is consistently small. The diagram demonstrates that the two spin states possess equivalent energy levels in the absence of an external field, but deviate when the field intensity escalates.

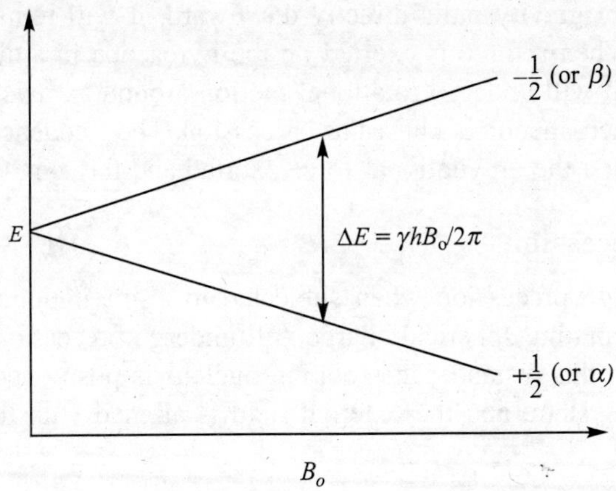

Figure 15.22 Energy levels.

15.10.3.4 Spin State Transitions

When there is a difference in energy between two states, it is possible to cause a transition between the different spin states. By subjecting the nucleus to electromagnetic radiation with the appropriate energy (defined by its frequency), it is possible to stimulate a nucleus in a low energy state to transition to a higher energy state. The NMR approach relies on the absorption of energy during this transition.

If radio frequency (R_f) with a frequency that matches the Larmor frequency the rotating nucleus will take in energy and the magnetic moment will change to its $I = -1/2$ state. The R_f is applied perpendicular (i.e. along the x-axis) to the external magnetic field. The graphic below illustrates this excitation.

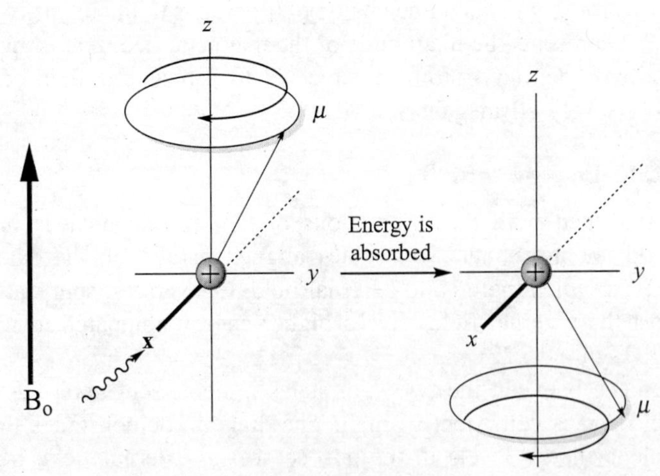

15.10.3.5 Spin Inversion

When magnetic nuclei are exposed to a stationary external magnetic field, there will consistently be a small surplus of atoms in the lower energy α state. By exposing the nucleus to electromagnetic radiation with the appropriate energy, specified by its frequency, a nucleus in a low energy α state can be prompted to transition to a higher energy β state. The NMR approach relies on the absorption of energy during this transition.

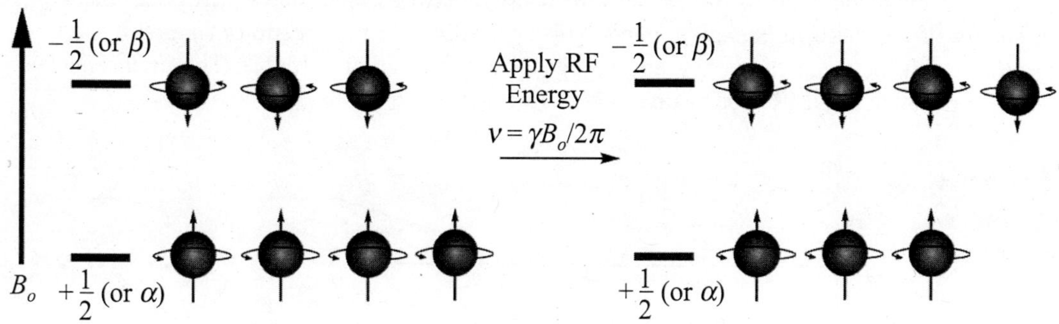

15.10.3.6 Irradiation Frequency vs. Field Strength

NMR spectroscopy requires the presence of powerful magnetic fields. The tesla (T) is the standard unit of measurement for magnetic flux on an international scale. The magnitude of the Earth's magnetic field at ground level is roughly 10^{-4} Tesla. In the context of nuclear magnetic resonance (NMR), the little variation in energy (ΔE) is typically expressed as a frequency measured in megahertz (MHz), which corresponds to 10^6 Hz. This frequency range can vary from 20 to 900 MHz, depending on the intensity of the magnetic field. Exposing a sample to radio frequency (R_f) energy that matches the difference in spin states of a particular group of nuclei will result in the stimulation of those nuclei in the +1/2 state, causing them to transition to the higher –1/2 spin state (Figure 15.23).

Energy of a photon, $E = h\nu$; Spin state energy difference, $\Delta E = \gamma h B_o/2\pi$; when $E = \Delta E$, spin flip occurs $h\nu = \gamma h B_o/2\pi$, the necessary frequency is: $\nu = \gamma B_o/2\pi$.

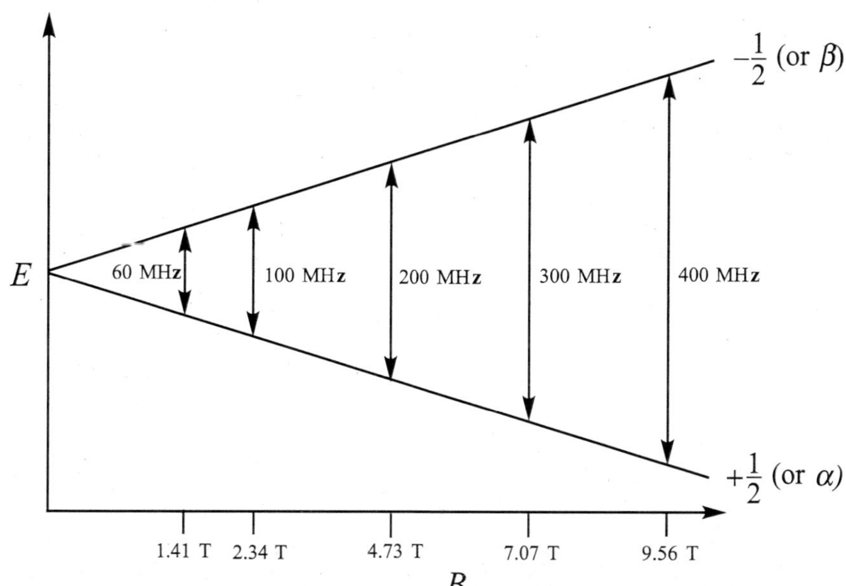

Figure 15.23 Proton spin energy differences.

15.10.4 Experiment of NMR

Throughout the experiment, a targeted frequency of electromagnetic radiation is utilized. By varying the magnetic field, a discrepancy in energy between spin states will arise, matching the energy of the radio frequency applied. This will allow for the creation of a graph showing the relationship between frequency and energy absorption (Figure 15.24). This is the nuclear magnetic resonance (NMR) spectrum.

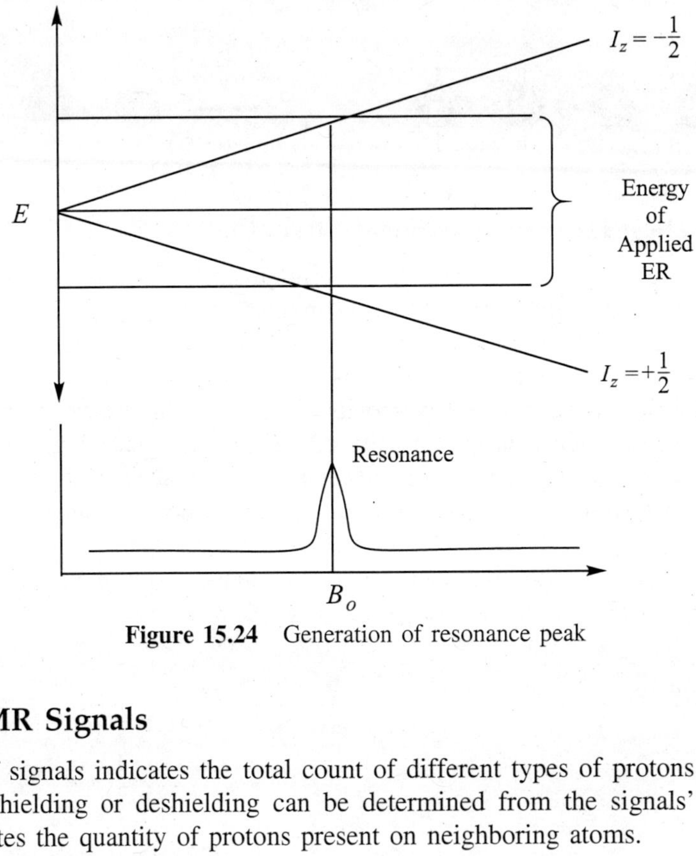

Figure 15.24 Generation of resonance peak

15.10.5 NMR Signals

The number of signals indicates the total count of different types of protons that are present. The proton's shielding or deshielding can be determined from the signals' location. Signal splitting indicates the quantity of protons present on neighboring atoms.

15.10.5.1 Number of Signals

The protons inside a molecule encounter different magnetic surroundings, resulting in distinct signals in the NMR spectrum. Protons that are in the same magnetic environment and have the same chemical properties are referred to as chemically equivalent. They exhibit identical chemical shifts, regarding the distinction between equivalent and non-equivalent protons.

15.10.5.2 Position of Signals

The position of the signal on the horizontal frequency scale where the analogous proton signals are observed referred to chemical shift. The chemical shift is influenced by the fluctuating local magnetic fields generated by the adjacent protons.

The Chemical Shift

Protons possess identical magnetic moments, it is reasonable to anticipate that all ^{1}H nuclei will exhibit resonance signals at equivalent field. This statement is false when it comes to chemistry applications. There are many exemplary proton signals exhibited across a same range of magnetic fields. Examining solitary protons in the spectrometer is not feasible.

"The chemical shift refers to the range of field/frequency values shown by hydrogen atoms in a chemical environment." The resonance frequencies, or bandwidth, for the majority of hydrogen atoms in organic molecules under an external magnetic field (B_o = 1.41T) typically span a narrow range of 600–1000 Hz (Figure 15.25).

PPM Scale of Chemical Shift

The chemical shift of a nucleus is quantified in parts per million (ppm) on a scale known as the delta (δ) scale. The ratio of the change in frequency from the reference compound tetramethylsilane (TMS) to the overall frequency of the spectrometer, expressed in hertz (Hz) and megahertz (MHz) respectively. The machine operates at the same frequency of 60, 100, or 300 MHz. The value of Tau (τ) is equal to 10-δ. δ is the shift in Hz from the TMS/spectrometer frequency. On a 60 MHz spectrometer, a chemical shift of 60 Hz is equivalent to a shift of 1 ppm (parts per million), calculated as (60/60,000,000).

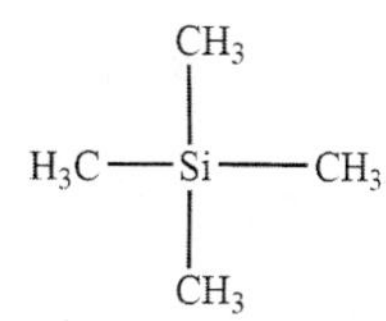

Tetramethylsilane (TMS)

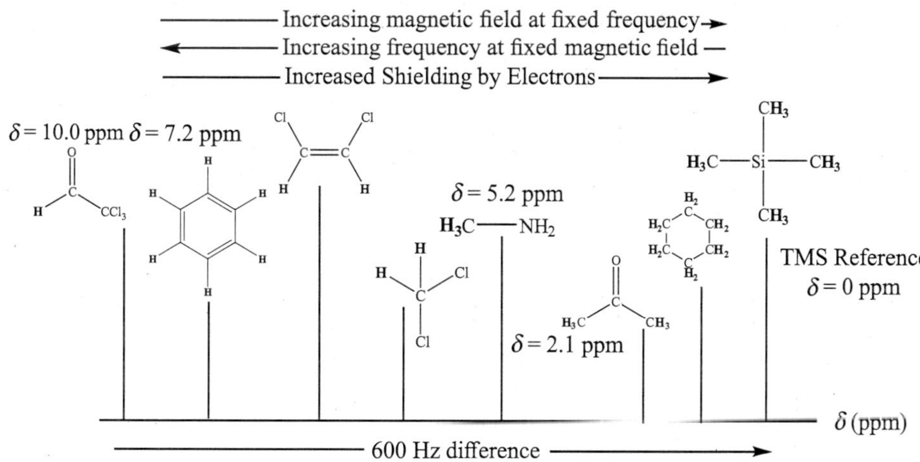

Figure 15.25 ^{1}H NMR resonance signals for some different compounds.

In the field of NMR spectroscopy, the commonly used reference compound is tetramethylsilane, which is represented by the chemical formula $Si(CH_3)_4$ and is commonly abbreviated as TMS. The chemical shift is a very accurate measure of the chemical surroundings surrounding a nucleus. The sample is supplemented with TMS. Due of silicon's lower electronegativity compared to carbon, TMS protons exhibit a greater degree of shielding. The signal is defined as having a value of zero. Protons in organic compounds experience a shift towards lower magnetic field values (to the left) relative to the signal of tetramethylsilane (TMS).

Diamagnetic Shielding

The behaviour of proton nuclei in various substances varies in the NMR experiment. The reason is that the electrons orbiting the proton. Electrons, being charged particles, exhibit movement in reaction to the external magnetic field in order to produce a secondary field that counteracts the far stronger applied field. The presence of this secondary field acts as a protective barrier for the nucleus against the external field (Figure 15.26). Consequently, in order to accomplish resonance (which involves the absorption of R_f energy), the strength of B_o must be heightened. Therefore, the value of B_o must be augmented in order to counterbalance the generated shielding field.

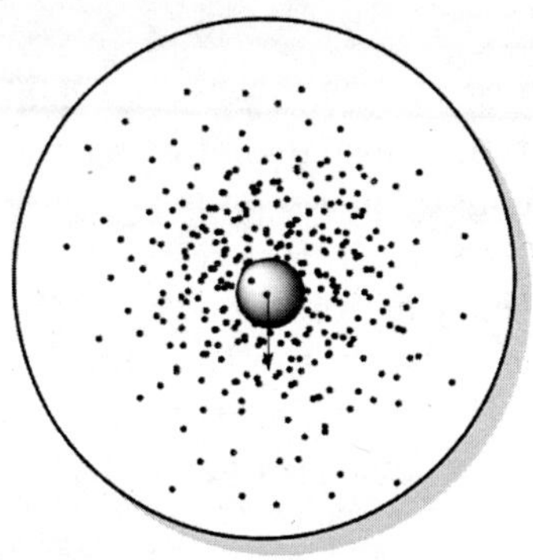

Figure 15.26 Field effect of nucleus.

An atom's electrons orbit in the direction of the magnetic field when it is subjected to one. Particle circulation creates a tiny magnetic field at the nucleus that opposes the field applied from the outside. The effective field, or magnetic field that the nucleus experiences, is usually smaller than the applied field by a factor of σ.

$$B = B_o(1 - \sigma)$$

Shielding and Deshielding

The electron cloud consistently envelops the nucleus of a molecule. Due to the electron's magnetic field exhibiting an opposing effect to the external magnetic field, the nucleus is effectively protected or "shielded" from the complete impact of the external magnetic field. The value of H_{eff} (B_{eff}) is often lower than that of $H_o(B_o)$.

Factors that decrease the electron density are referred to as deshielding the nucleus, as they cause an increase in the effective nuclear charge (H_{eff}) at the nucleus. Likewise, variables that enhance the electron density are referred to as "shielding" the nucleus, resulting in a drop in H_{eff} (Figure 15.27).

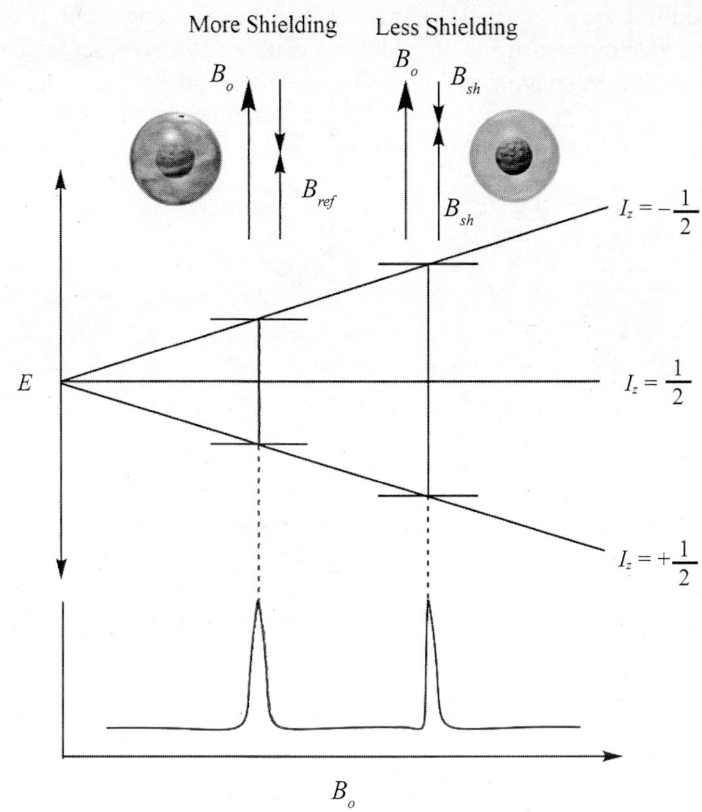

Figure 15.27 Shielding and Deshielding effect of nucleus.

15.10.5.3 Location of Signals: Factor Affecting Chemical Shift

Electronegativity

A nucleus experiences shielding or deshielding when it is impacted by the magnetic fields produced by electrons. The electrons that make chemical bonds with nearby atoms are those that are closest to the nucleus.

Electronegative atoms possess a strong attraction towards electrons. The strength of this affinity increases as the electronegativity of the atom increases. Let's now consider two cases; first case: where hydrogen is attached to a carbon; and second case: hydrogen atom is attached to an oxygen.

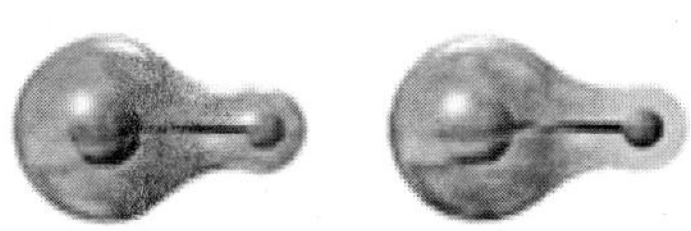

C–H O–H

Oxygen has a higher electronegativity than carbon. In an oxygen-hydrogen bond, the oxygen atom will exert a stronger pull on the bonding electrons. The electron density surrounding the

H will be diminished in contrast to the H connected to a C. Regarding an O–H bond, hydrogen experiences reduced electron density in its vicinity, resulting in decreased shielding. Electron withdrawing groups like oxygen are classified as deshielding, and electron donating groups like carbon that donate electrons are regarded as having a shielding effect.

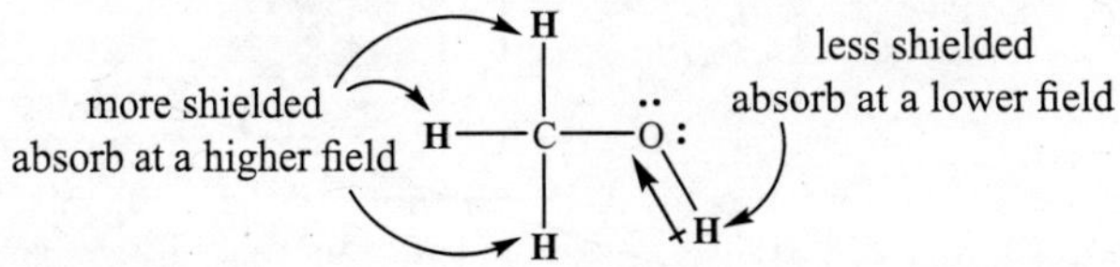

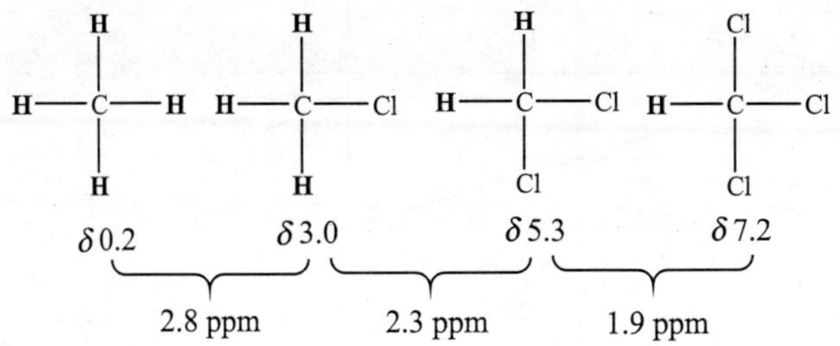

Proton Chemical Shifts

Types of proton	Approximate chemical shift (ppm)	Types of proton	Approximate chemical shift (ppm)
$(CH_3)_4Si$	**0**	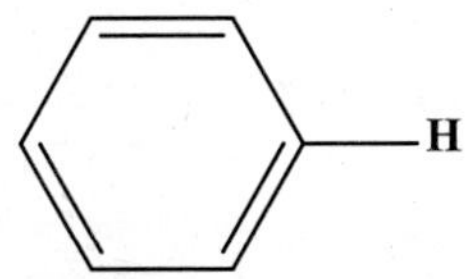	**6.5–8.0**
—CH_3	**0.9**	—C(=O)—H	**9.0–10**
—CH_2—	**1.3**	I—C—H	**2.5–4.0**
—CH—	**1.4**	Br—C—H	**2.5–4.0**

Types of proton	Approximate chemical shift (ppm)	Types of proton	Approximate chemical shift (ppm)
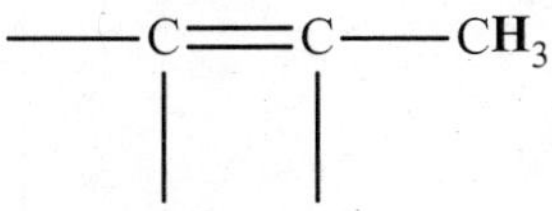	1.7	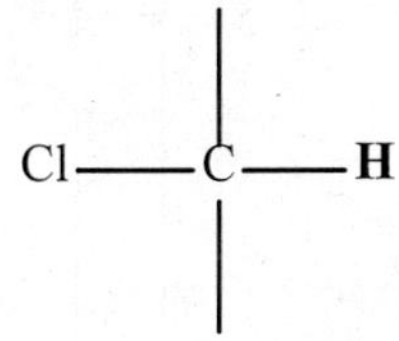	3.0–4.0
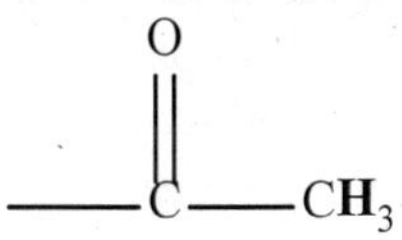	2.1	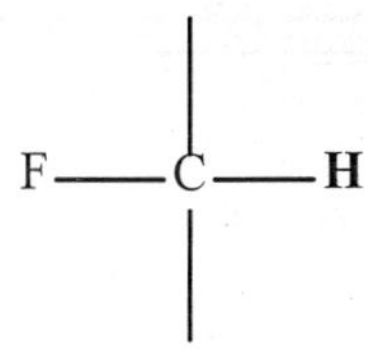	4.0–4.5
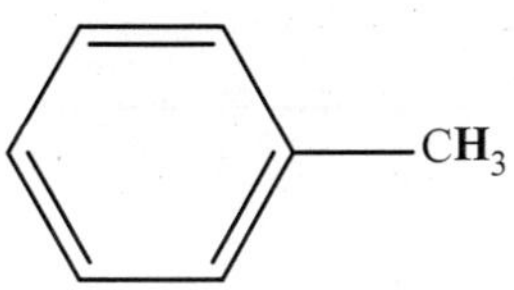	2.3	RNH_2	Variable, 1.5–4.0
—C≡C—H	2.4	ROH	Variable, 2.0–5.0
R—O—CH_3	3.3	ArOH	Variable, 4..0–7.0
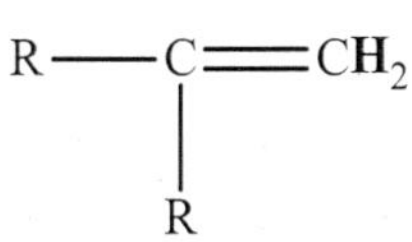	4.7	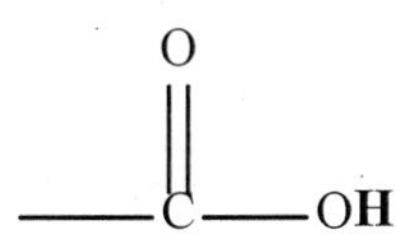	Variable, 10.0–12.0
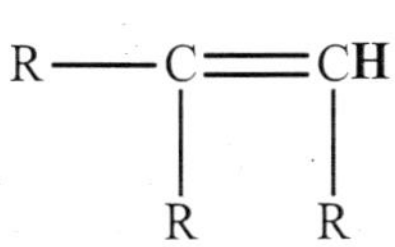	5.3		

15.10.5.4 Intensity of Signals

The magnitude of each peak is directly proportional to the quantity of protons. Demonstrated with the use of integral trace.

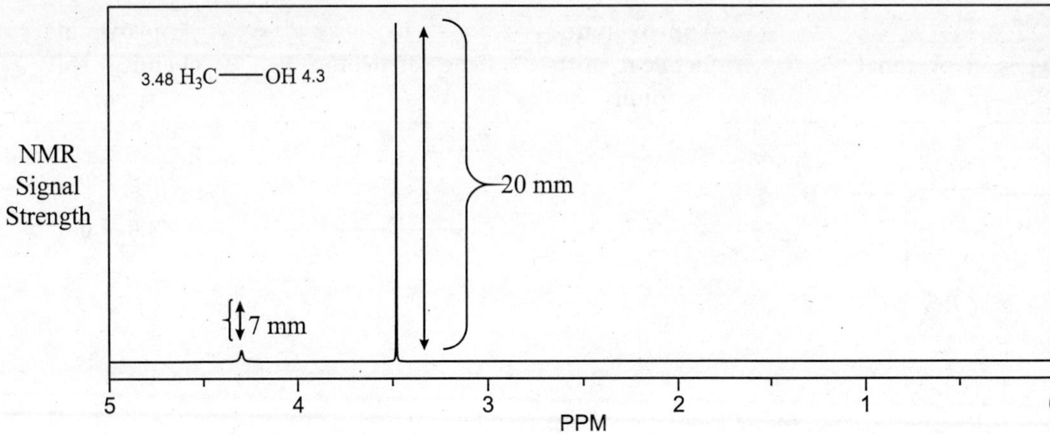

Integration

In the molecular formula, each integral peak can be attributed to a certain number of hydrogen atoms.

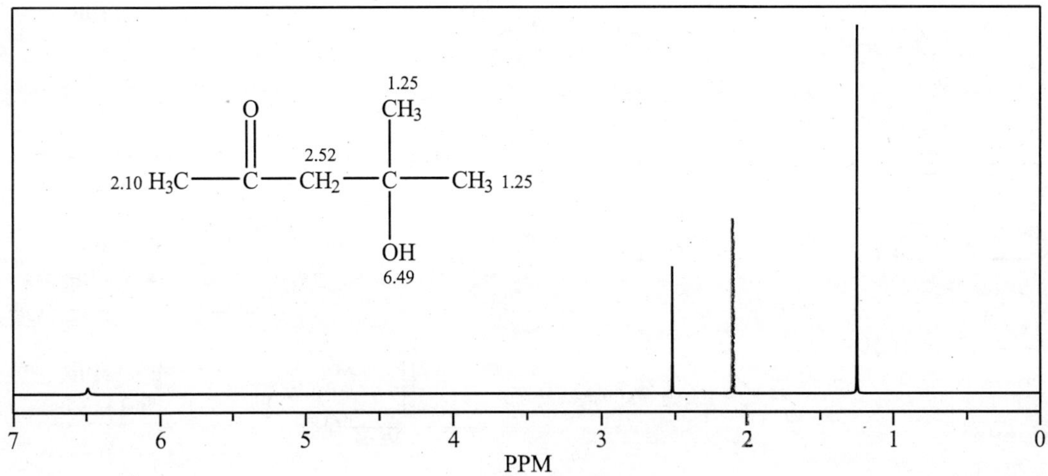

15.10.5.5 Spin-Spin Splitting

Adjacent carbons possess nonequivalent protons that exhibit magnetic fields capable of either aligning with or opposing the external field. The proton experiences a magnetic connection that results in a slight downfield absorption when the external field is strengthened and a slight upfield absorption when the external field is opposed. Every conceivable outcome exists, hence the signal is divided and the division produces patterns (Multiplicity) (Figure 15.28). Relative peak intensities of symmetric multiplets is tabulated in Table 15.1.

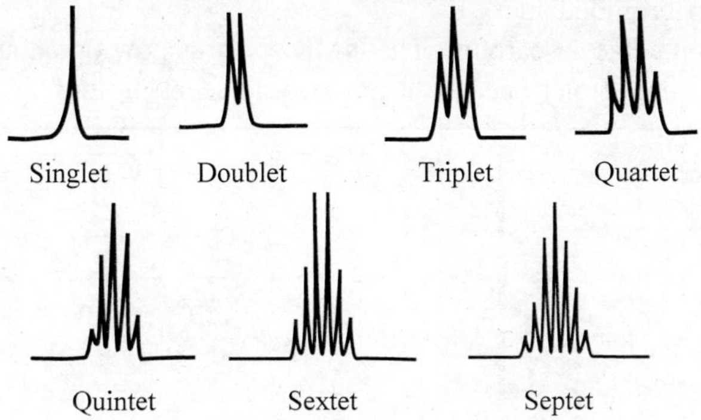

Figure 15.28 Common NMR peaks.

N + 1 Rule: If a signal is split by N equivalent protons, it is split into N + 1 peaks.

Table 15.1 Relative peak intensities of symmetric multiplets

Number of Equivalent protons Causing Splitting	Number of Peaks (Multiplicity)	Area Ratios (Pascal's Triangle)
0	1 (singlet)	1
1	2 (doublet)	1 1
2	3 (triplet)	1 2 1
3	4 (quartet)	1 3 3 1
4	5 (quintet)	1 4 6 4 1
5	6 (sextet)	1 5 10 10 5 1
6	7 (septet)	1 6 15 20 15 6 1

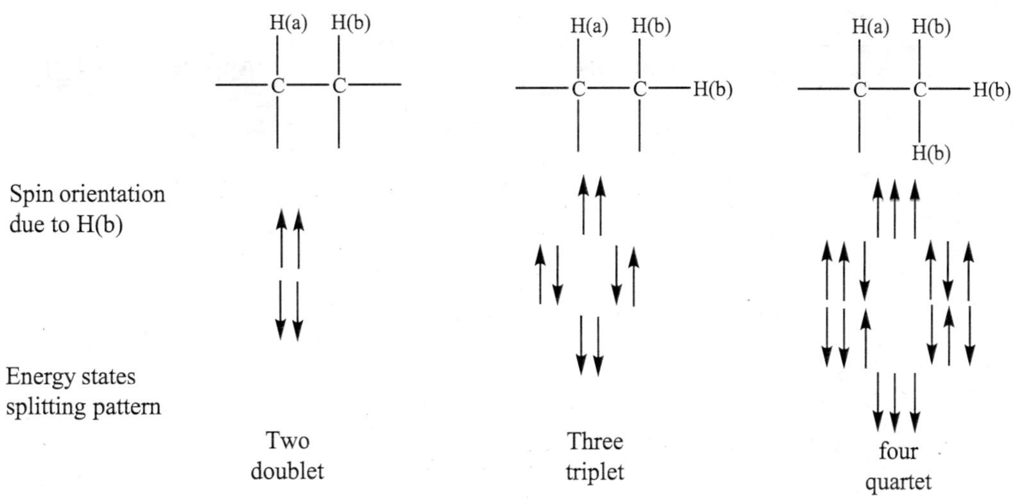

For exemple: 1, 1, 2-Tribromoethane

Nonequivalent protons on adjacent carbons. The NMR spectra show one doublet because it has one adjacent proton and one triplet because it has two adjacent protons.

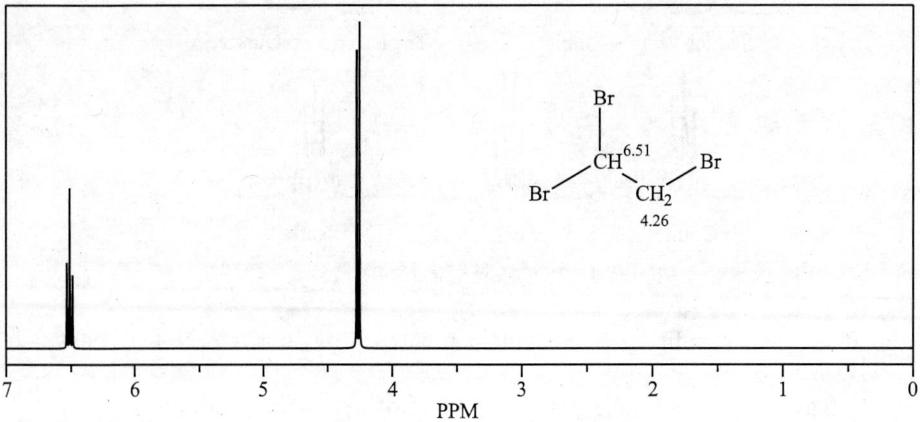

15.10.5.6 Range of Magnetic Coupling

Protons that are equivalent do not cause splitting in one another. Protons that are connected to the same carbon atom will undergo splitting only if they are not identical. Adjacent carbons typically exhibit proton coupling. Protons that are at least four bonds apart will not exhibit coupling.

Coupling Constants

Peak-to-peak distance of a multiplet is expressed in hertz. Coupling constant is independent of the magnitude of the external field. Adjacent groups of protons can divide each other, resulting in multiplets with identical coupling constants.

Values for Coupling Constants

	Approx. J (in Hz)		Approx. J (in Hz)
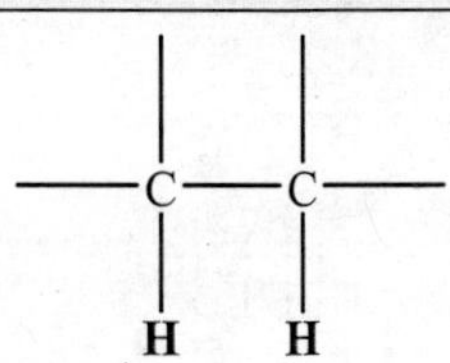 (free rotation)	7	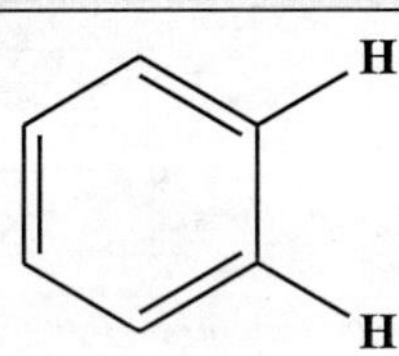 (ortho)	8
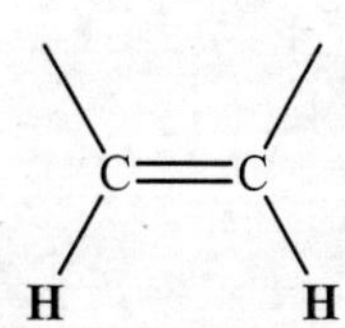 (cis)	10	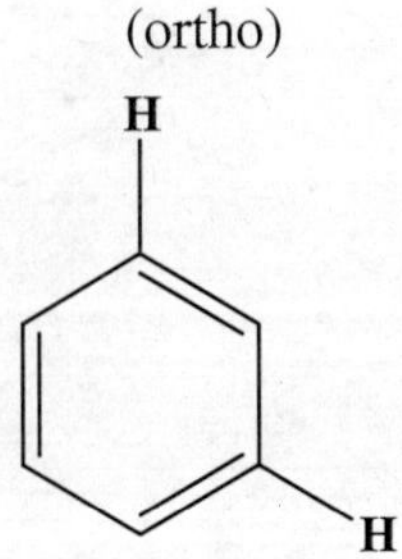 (meta)	2

	Approx. J (in Hz)		Approx. J (in Hz)
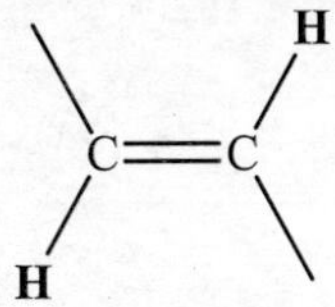 (trans)	15	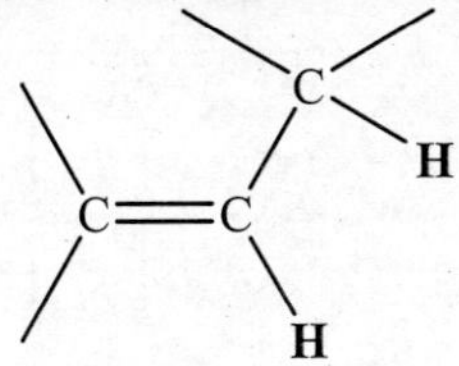 (allylic)	6
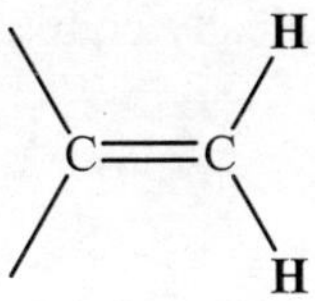 (geminal)	2		

To determine the degree of unsaturation in a molecule, we can compute the double bond equivalence when provided with the molecular formula:

$$DBE = C + 1 - \frac{H}{2} - \frac{X}{2} + \frac{N}{2}$$

1. **O–H and N–H Signals:** The chemical shift is influenced by the concentration and the kind of the solvent. In concentrated solutions, the presence of hydrogen bonding causes a shift in the proton signals, resulting in a chemical shift of approximately δ3.5 for N–H and δ4.5 for O–H. Proton transfers between the molecules can cause the peak to become wider.

2. **Identifying the O–H or N–H Peak:** In order to confirm whether a certain peak is caused by O–H or N–H, agitate the sample with heavy water (D_2O). Deuterium can undergo an exchange with the protons of O–H or N–H. In a subsequent NMR spectrum, the peak will either be absent or much reduced in intensity.

3. **Time Dependence:** Nuclear Magnetic Resonance (NMR) produces a spectrum that represents the average of all molecular orientations when the molecules rotate in relation to the magnetic field. The axial and equatorial protons on cyclohexane undergo rapid interconversion, resulting in a singular signal. Proton transfers involving OH and NH groups can happen rapidly enough to prevent the splitting of the proton by neighboring protons inside the molecule.

PROBLEM Calculate the number of signals in a, b and c.

(a) 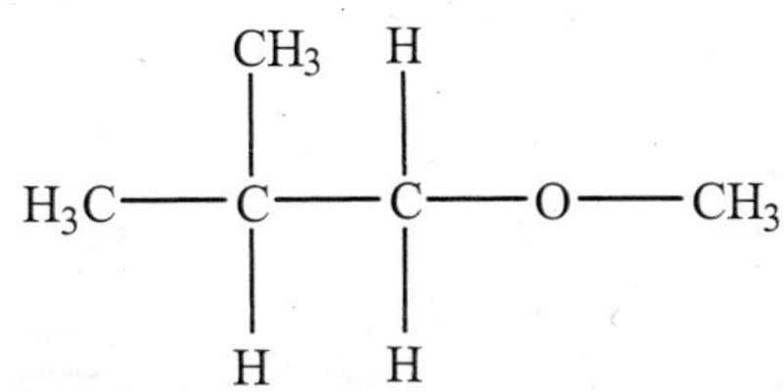 (b) 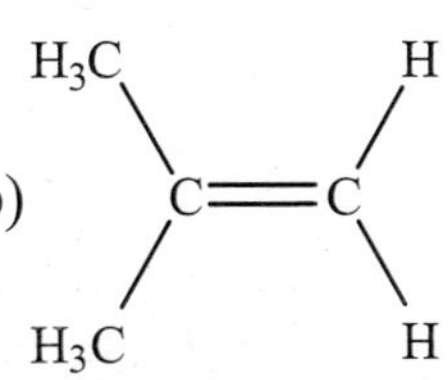 (c) 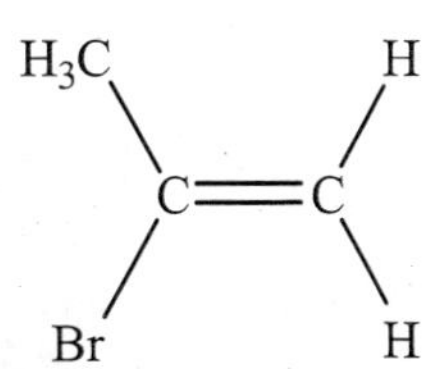

Solution (a) There are four equivalent protons indicated by a, b, c, and d. Hence it has four signals.

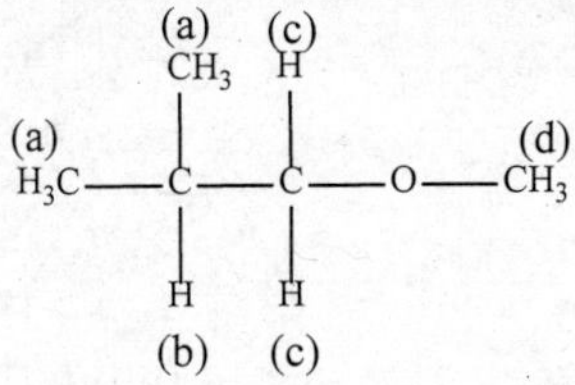

(b) Protons need to be stereochemically equivalent in order to be chemically equivalent. So this molecule has two signals.

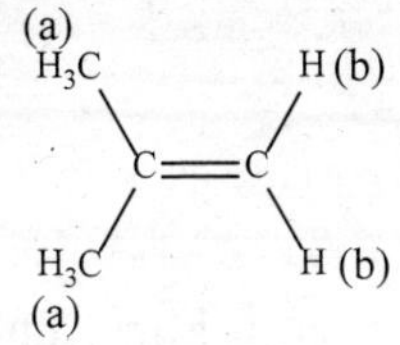

(c) b and c, the two vinylic protons, are not interchangeable. To the bromine, one is *trans* and the other is *cis*.

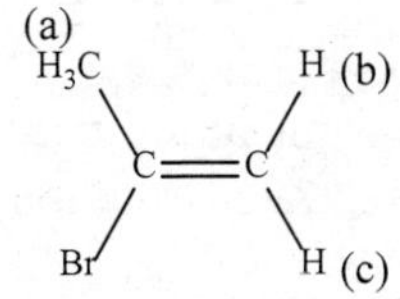

PROBLEM Give a structure consistent with the following NMR data:
Molecular formula = $C_{10}H_{14}$

(i) Singlet at δ 1.30; 9H
(ii) Singlet at δ 7.28; 5H

Solution

Step 1: First determine the double bond equivalent (DBE) using the formula

$$DBE = C - (H/2) + (N/2) + 1 = 10 - 7 + 1 = 4$$

DBE is 4, so the possibility to have benzene ring or 4 double bonds in given molecule.

Step 2: In the given data the δ value is 7.28, which indicates that the benzene ring should present in given molecule. Which also supported by value of DBE.

The singlet at δ 1.30 of 9H is only possible when single carbon is attached with three $-CH_3$ groups.

So, based on above findings the structure of compound is follows:

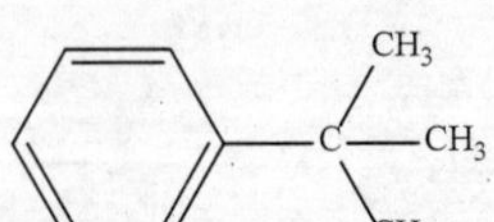

tert-butylbenzene

15.11 MAGNETIC RESONANCE IMAGING (MRI)

15.11.1 Introduction

MRI was developed in 1973. The parents are Paul Lauterbur, who was affiliated with Stony Brooks at the time, and Sir Peter Mansfield, who was affiliated with Nottingham but had not yet been knighted. In 2003, they received the Nobel Prize in Physiology or Medicine. MRI is a non-invasive imaging modality that generates three-dimensional, high-resolution anatomical images without the need for harmful radiation. It is frequently employed for the purpose of disease detection, diagnosis, and monitoring of treatment. The technology is based on advanced principles that stimulate and detect alterations in the rotational axis of protons present in the water comprising biological tissues.

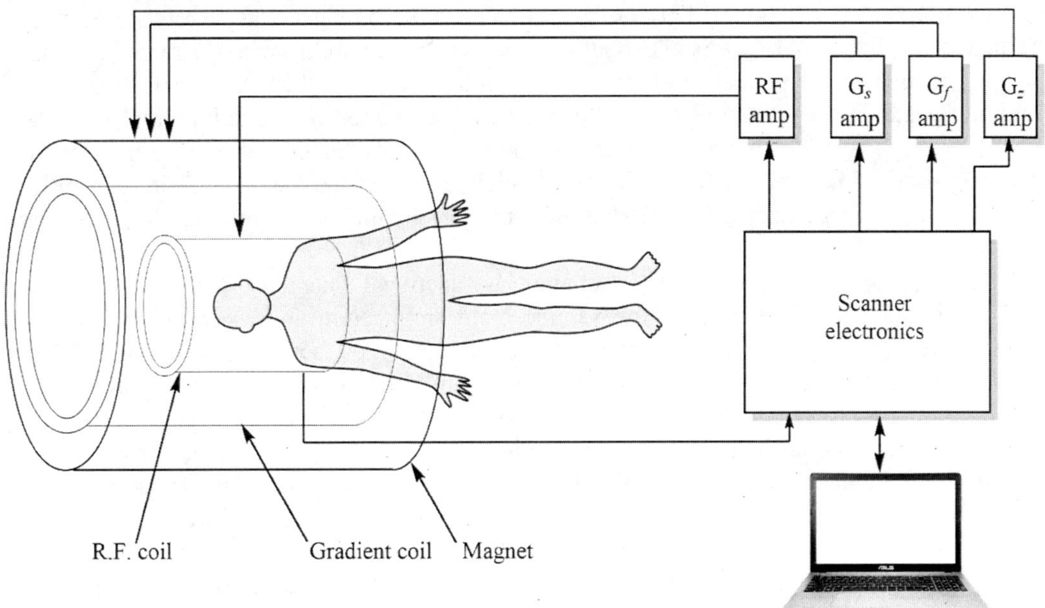

Figure 15.29 Working of MRI.

15.11.2 Working of MRI

Magnetic resonance imaging (MRI) uses strong magnets to force protons inside the body to align with the strong magnetic field. Then, a radiofrequency current is pulsed into the patient, exciting and driving the protons out of balance and pushing against the force of the magnetic field. When the radiofrequency field is turned off, the energy released as the protons realign with the magnetic field is picked up by the MRI sensors. The amount of energy generated and the time it takes for the protons to realign with the magnetic field depend on the environment and the composition of the molecules. These magnetic characteristics allow medical professionals to discriminate between various tissue kinds (Figure 15.29).

MRI scanners are highly adept at visualizing the non-osseous or soft tissues of the body. Contrary to computed tomography (CT), they abstain from utilizing the harmful ionizing radiation emitted by X-rays. Magnetic resonance imaging (MRI) provides a much enhanced visualization of the brain, spinal cord, nerves, muscles, ligaments, and tendons compared to conventional X-rays and CT scans. Consequently, MRI is frequently employed for the examination of knee and shoulder injuries. Magnetic resonance imaging (MRI) can discern between white matter and grey matter in the brain, as well as serve as a diagnostic tool for aneurysms and tumors. MRI is the preferred imaging method for frequent diagnosis or therapy, particularly in the brain, as it does not involve the use of X-rays or other forms of radiation. Nevertheless, magnetic resonance imaging (MRI) incurs higher costs compared to X-ray imaging or computed tomography (CT) scanning.

15.11.3 Applications of MRI

Magnetic Resonance Imaging (MRI) utilizes potent magnets to generate a robust magnetic field that compels protons within the body to align with said field. When a radiofrequency current is applied to the patient, it stimulates the protons, causing them to deviate from their equilibrium state and resist the magnetic field. Upon deactivation of the radiofrequency field, the MRI sensors may discern the energy emitted during the reorientation of protons in response to the magnetic field. The duration for proton realignment with the magnetic field, as well as the magnitude of energy released, varies based on the surrounding conditions and the molecular composition.

An example of a customized MRI technique is functional Magnetic Resonance Imaging (fMRI). This technique is employed to examine brain architecture and ascertain the specific regions of the brain that exhibit increased oxygen consumption, or "activation," during different cognitive tasks. It is utilized to enhance comprehension of brain organization and presents a promising new benchmark for evaluating neurological condition and neurosurgical peril.

In order to acquire an MRI image, a patient is positioned within a substantial magnet and is required to maintain absolute stillness throughout the imaging procedure to prevent any distortion of the image. Contrast agents, typically containing Gadolinium, can be administered intravenously to a patient prior to or during the MRI procedure in order to expedite the reorientation of protons in response to the magnetic field. The speed at which the protons reorient determines the intensity of the picture.

QUESTIONS

15.1 Calculate λ_{max} for the following compounds?

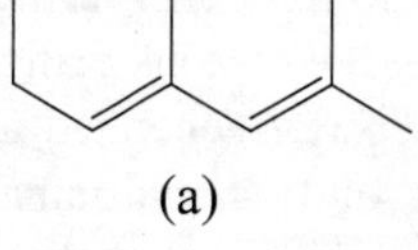

(a)

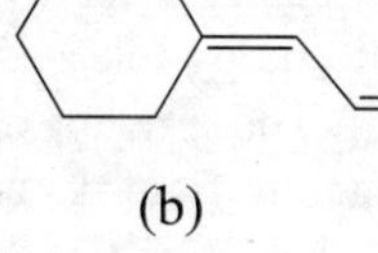

(b)

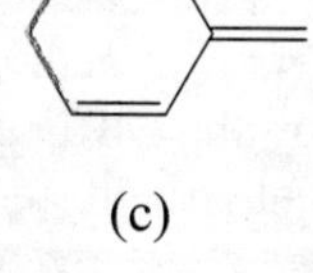

(c)

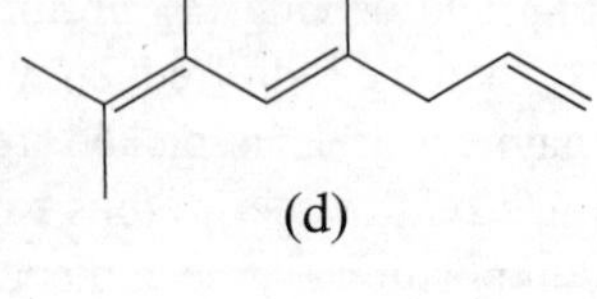

(d)

Answer: (a) 229 nm, (b) 229 nm, (c) 229 nm, (d) 234 nm.

15.2 Calculate λ_{max} for the following compounds:

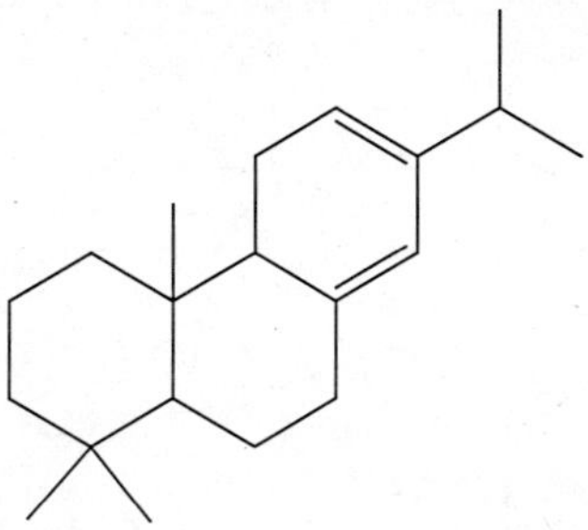
(a)

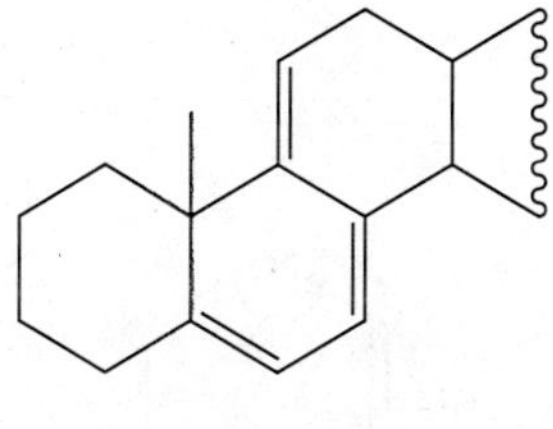
(b)

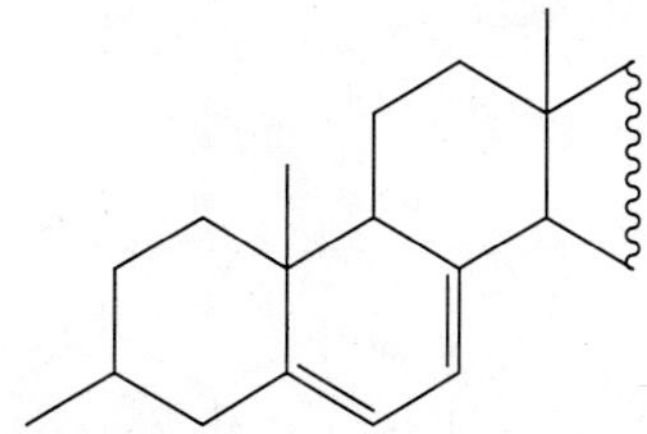
(c)

Answer: (a) 278 nm, (b) 323 nm, (c) 283 nm

15.3 Assign equivalent and non-equivalent protons to each of the following compounds to determine the number of peaks.

(a) 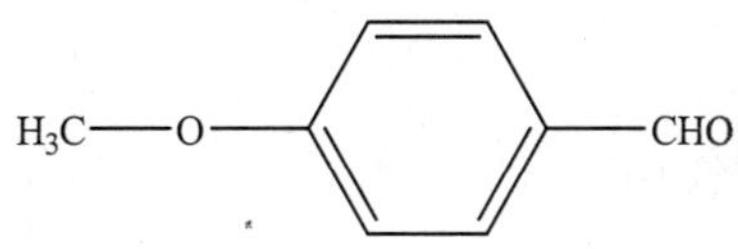 (b) CH_3–COO–CH_2–CH_3 (c) 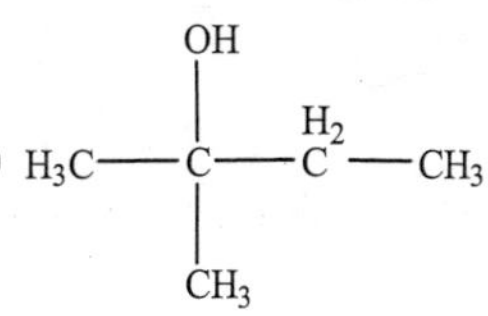

(d) 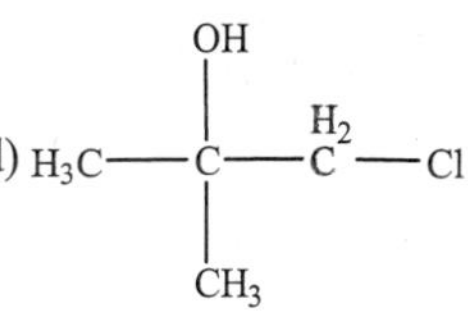 (e) 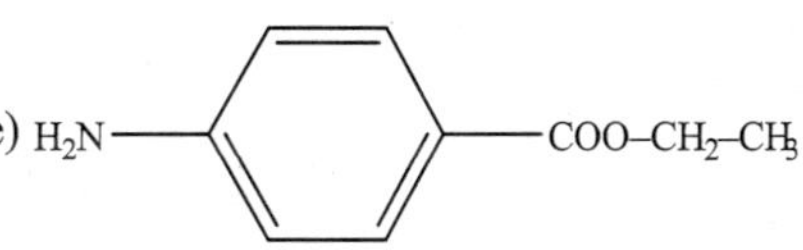

15.4 Write down splitting pattern of the following compounds:

(a) 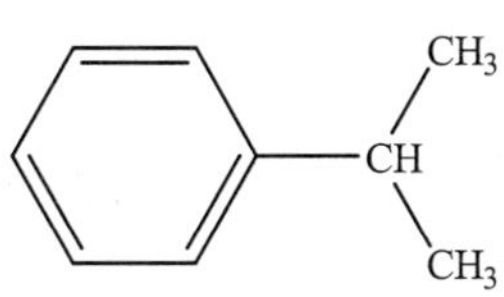 (b) 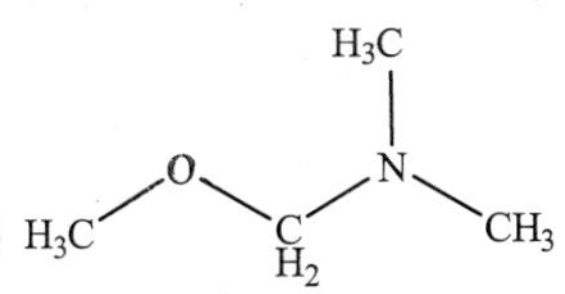 (c) 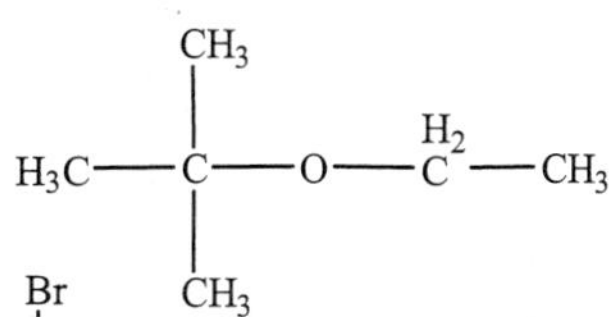

(d) 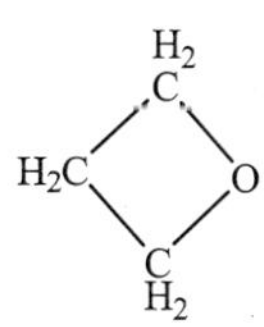 (e) 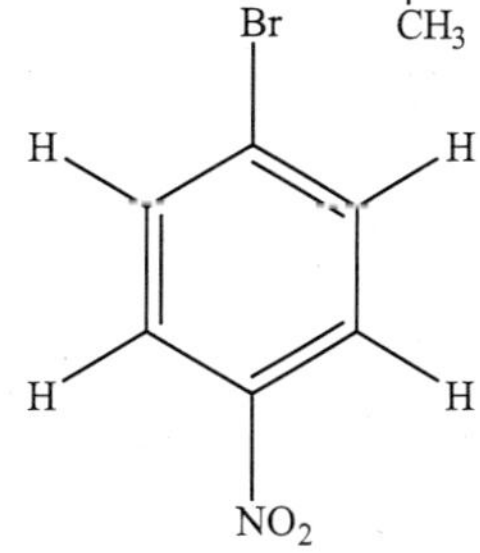

15.5 Give a structure or structures consistent with the following data of a molecule with Molecular Formula: $C_4H_8O_2$

IR (cm^{-1}): 3012 (m),1725 (s), 1423 (m), 1262 (m), 1058 (m).

^1H NMR (δ ppm): 0.98 (t, J = 6.7 Hz, 3H), 2.67 (q, J = 6.7 Hz, 2H), 3.52 (s, 3H).

15.6 Calculate the double bond equivalent and Number of signals of the following compounds: (i) C_6H_6, (ii) C_3H_7N, (iii) $C_{10}H_7Cl$, (iv) $C_7H_6O_2$.

15.7 Calculate the λ_{max} for the following using Woodward-Fieser rule:

(i) 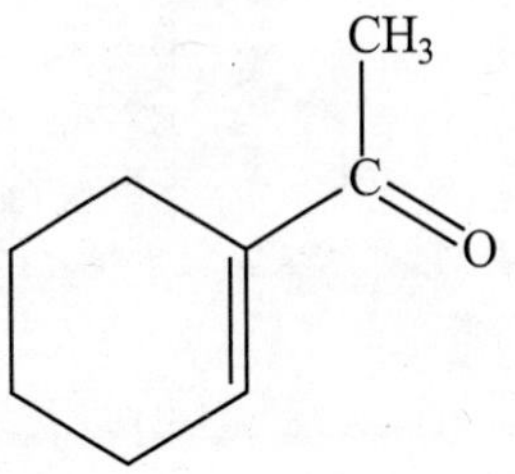 (ii) 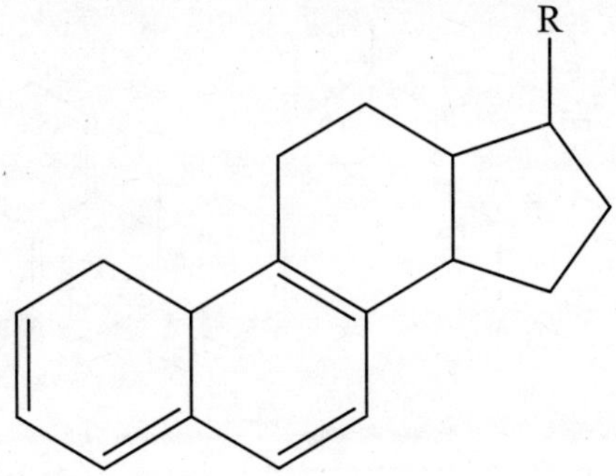

15.8 Calculate the number of NMR signals and splitting of each proton in the given molecules.

(i) 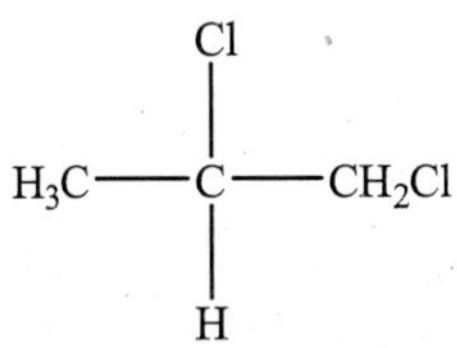 (ii) $CH_3CH_2COCH_2CH_2CHO$ (iii) 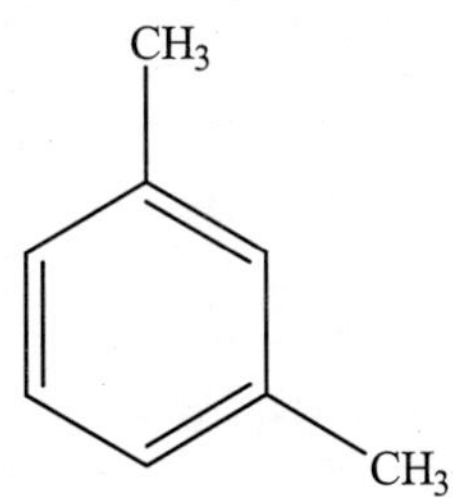

15.9 Calculate the λ_{max} for the following compounds:

(i) 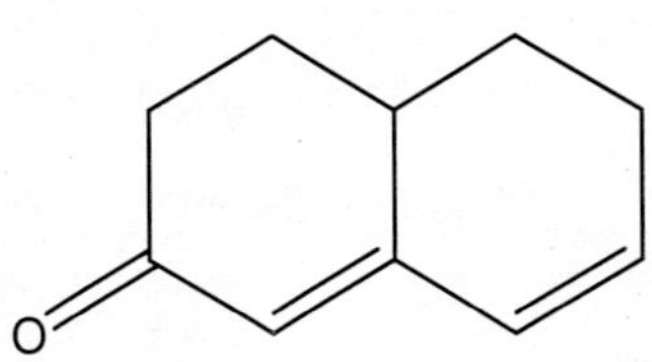 (ii) 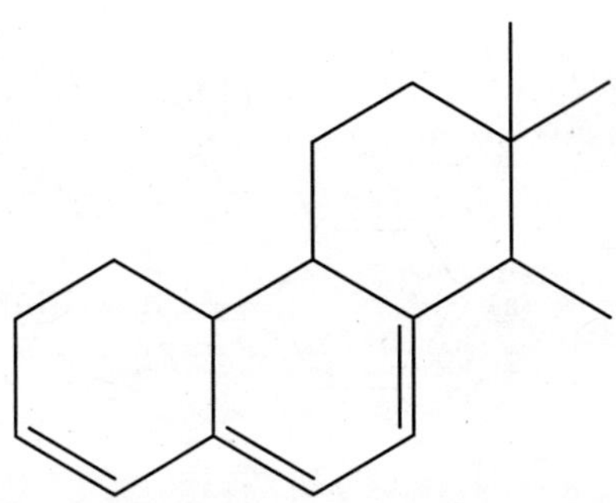

15.10 Calculate number of peaks, splitting pattern and chemical shifts of the following compounds:

(i) 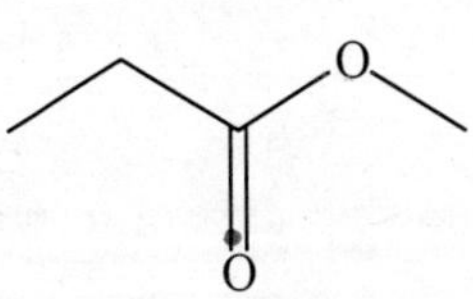

methyl propionate

(ii) 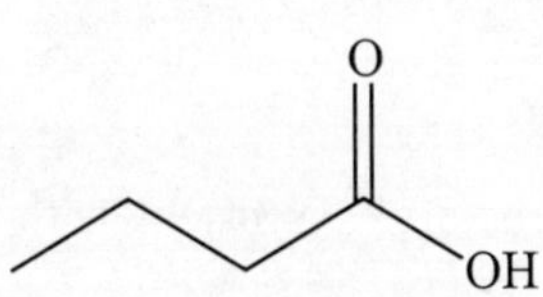

butyric acid

(iii) 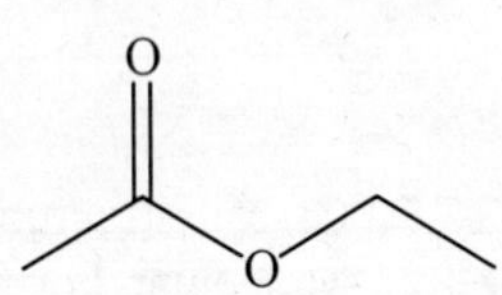

ethyl acetate

15.11 Comment on the IR activity and microwave activity of following compound along with reason: COS, NH_3, CO_2, CH_4.

15.12 Discuss the various vibrational modes in IR spectroscopy of polyatomic molecules. Draw all possible vibrational mode for CS_2 molecule.

15.13 Write note on Auxochromic and Chromophore. Discuss application of UV spectroscopy.

15.14 Drive relation for rigid rotator.

15.15 With the help of examples differentiate between Microwave active and Microwave inactive compounds.

15.16 Illustrate Finger print region in IR spectroscopy. Two Isomers I and II of the molecular formula C_3H_6O give I.R. absorption band near 3550 cm^{-1} and 1717 cm^{-1} respectively. Assign structural formula to A and B consistent with their IR absorption bands.

15.17 Discuss electronic transitions involved in UV-visible spectroscopy. Illustrate, the effect of polar solvent on $n \rightarrow \pi^*$ transition in acetone. Also describe Absorption and Intensity shift in the UV spectroscopy with the help of examples.

15.18 Which of the following molecules will show rotational spectrum: H_2, HCl, CH_4, CH_3Cl, CH_2Cl_2, H_2O and SF_6?

15.19 Write informative notes on:

(i) Importance of fingerprint region in IR spectroscopy

(ii) Applications of UV-visible spectroscopy

15.20 Give the number of signals in the following molecules:

(i) CH_3CH_2Cl, (ii) C_6H_6, (iii) $C_6H_6CH_3$, (iv) CH_3OCH_3, (v) CH_3CH_2OH.

15.21 Show the number of signals in NMR spectra of following compounds:

CH_2CH_2COOH, CH_2CH_2OH, CH_3CHO, $CH_2OCH(CH_3)_2$

15.22 Distinguish between: Fluorescence and phosphorescence.

15.23 Draw and explain the splitting pattern of $CHCl_2CH_2Cl$.

15.24 Describe and discuss Jablonski diagram for depicting various photo processes.

15.25 Explain photosensitization and quenching by taking suitable examples.

15.26 State and illustrate with suitable potential energy curves the Franc-Condon principle in the vibronic spectrum of a diatomic molecule.

15.27 Calculate the number of signals present in the following compound:

$CH_3CH_2CH_2CH_3$

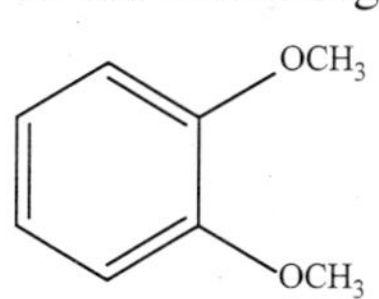

15.28 In IR spectroscopy why the region below 1500 cm^{-1} is called as fingerprint region?

15.29 Explain the selection rules of UV-vis Spectroscopy.

15.30 Which of the following will absorb at higher wave number for C=O stretching?

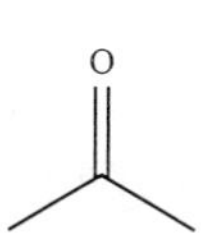

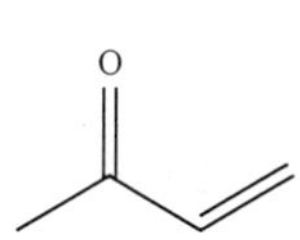

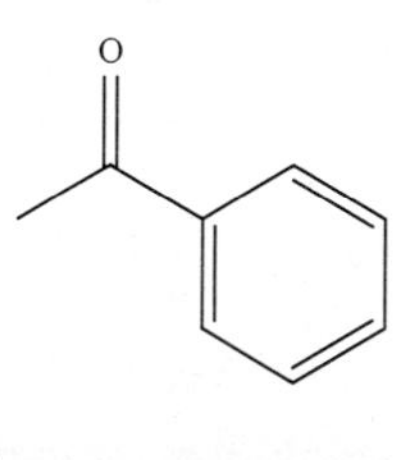

CHAPTER 16

Organic Reactions

16.1 INTRODUCTION

The organic molecule undergoes a reaction with a suitable nucleophilic or electrophilic reagent in an organic reaction. The process results in the generation of one or more intermediate species, which subsequently undergo further transformations to provide the end product(s). The process of organic reaction entails the cleavage and formation of covalent bonds. The process of breaking and forming covalent bonds often takes place in a series of distinct stages prior to the conversion into a final product.

This phenomenon is commonly referred to as a reaction mechanism. A reaction mechanism provides a comprehensive account of the processes involved in the breaking and formation of chemical bonds during the conversion of starting materials into products. This text elucidates the intricate mechanisms underlying electron movement, the energy aspects involved in bond cleavage and bond creation, as well as the kinetics and thermodynamics governing the conversion of reactants into products.

16.2 THE PROCESS OF BOND FORMATION AND BOND DISSOCIATION

Chemical reactions can take place through either a single step or a sequence of multiple steps. A reaction that occurs in a single step is referred to as a **concerted reaction**. Concerted reactions involve the simultaneous breaking or formation of many bonds, resulting in the direct conversion of a starting material into a product. A stepwise response is characterized by the involvement of multiple sequential steps. The initial step in the process involves the conversion of a substance into an intermediary or transitional state that is characterized by its lack of stability. Subsequently, this transitional state proceeds to undergo further transformations, ultimately resulting in the formation of the desired end product.

A molecular-level chemical reaction refers to an occurrence wherein two molecules interact by colliding in a manner that results in the rupture of one or more of their existing bonds, subsequently leading to the formation of one or more novel bonds and, consequently, new molecules. The precise order and temporal coordination of the bond cleavage and bond formation events will play a crucial role in enhancing our comprehension of the underlying mechanisms. In order to ascertain the intermolecular interactions and anticipate the outcomes of chemical reactions, a comprehensive understanding of the reaction mechanism is required.

$$\text{Reactant} + \text{Reagent} \xrightarrow[\text{Temperature/Pressure}]{\text{Solvent/Light}} \begin{matrix}\text{[Intermediate]} \\ \text{or} \\ \text{[Transition State]}\end{matrix} \longrightarrow \text{Product}$$

The distribution of electrons inside a chemical bond can occur in either an equal or unequal manner between the two participating atoms. There exist two distinct methods for breaking or cleaving a chemical bond, namely homolytic and heterolytic pathways.

16.3 KINDS OF ORGANIC REACTIONS

A chemical reaction refers to the process by which one or more chemical substances undergo a transition, resulting in the production of different chemical substances. A chemical reaction encompasses the formation of new chemical bonds and the dissociation of old chemical bonds.

The categorization of reactions involving organic compounds can be broadly classified into two aspects: the nature of the reactions and the mechanisms by which these reactions take place. Organic compounds can be classified into four distinct categories based on the types of reactions they undergo. These categories include:

1. Addition reactions—the combination of two molecules
2. Elimination reactions—one molecule splits into two
3. Substitution—the exchange of components between two molecules
4. Rearrangement reactions—the alteration of atomic connectivity within a molecule
5. Cyclisation—The Diels-Alder Cycloaddition

16.3.1 Addition Reactions

Addition reactions are characterized by the combination of two initial reactants, resulting in the formation of a singular product without any remaining atoms. During an addition reaction, the initial substance undergoes a chemical transformation resulting in the incorporation of new groups X and Y. A π-bond is cleaved and two sigma bonds are generated. The replacement of one π-bond and one σ-bond with two σ-bonds results in a negative change in enthalpy (ΔH).

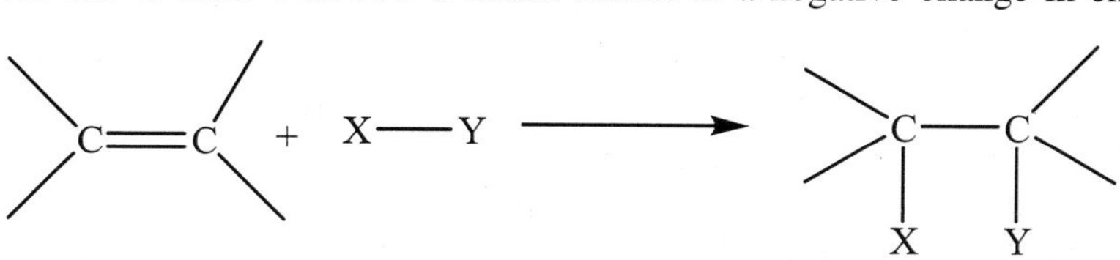

In the process of addition, all components of the reagent are incorporated into the resulting product, resulting in the fusion of two individual molecules into a single entity. For example,

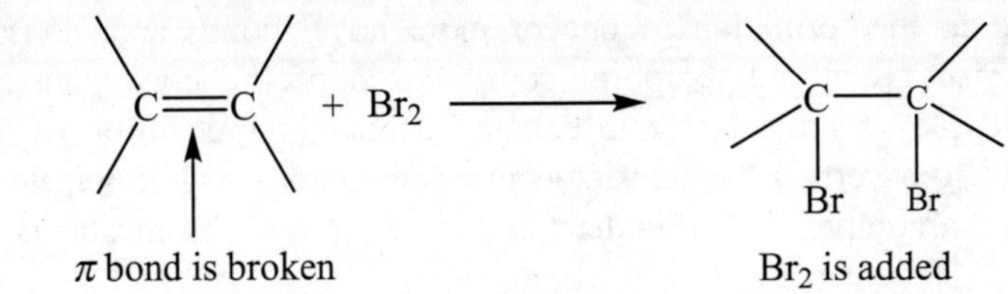

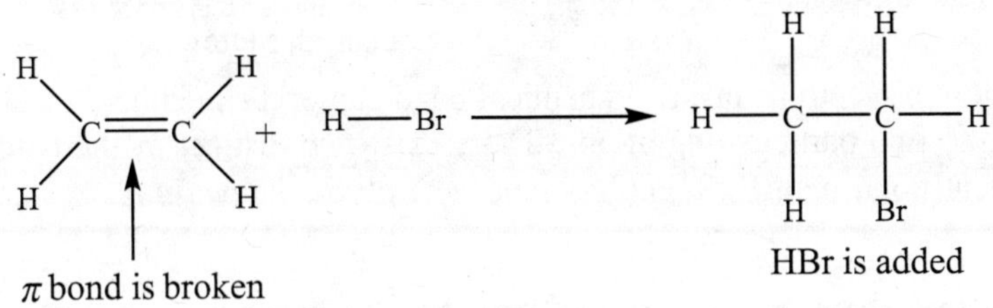

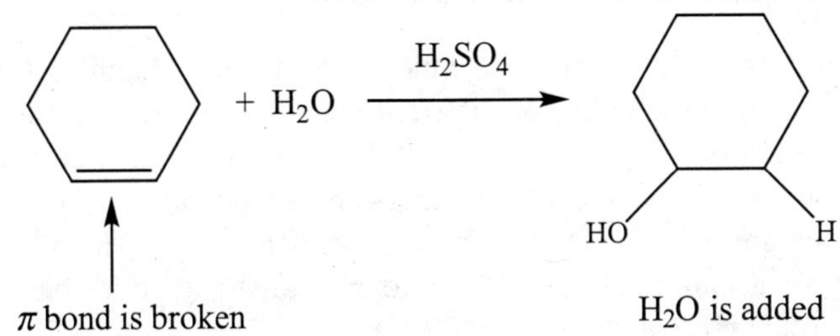

16.3.1.1 Polar Reactions as Addition Reaction

Differences in electronegativities can lead to the presence of local unsymmetrical electron distributions within molecules. This phenomenon results in the development of a partial negative charge on one atom, while a nearby atom experiences a corresponding partial positive charge. The atom with a higher electronegativity exhibits a bigger electron density. Elements such as oxygen (O), fluorine (F), nitrogen (N), and chlorine (Cl) exhibit greater electronegativity compared to carbon (C). One illustrative instance of a polar reaction: The process of adding hydrogen bromide (HBr) to ethylene (C_2H_4).

Hydrogen bromide (HBr) undergoes addition to the pi (π) bond of a carbon-carbon (C–C) double bond. The π bond exhibits a surplus of electrons, enabling it to operate as a nucleophile. Hydrogen bromide (HBr) might be considered electron deficient at the hydrogen atom due to the higher electronegativity of bromine (Br). This characteristic renders HBr an electrophile.

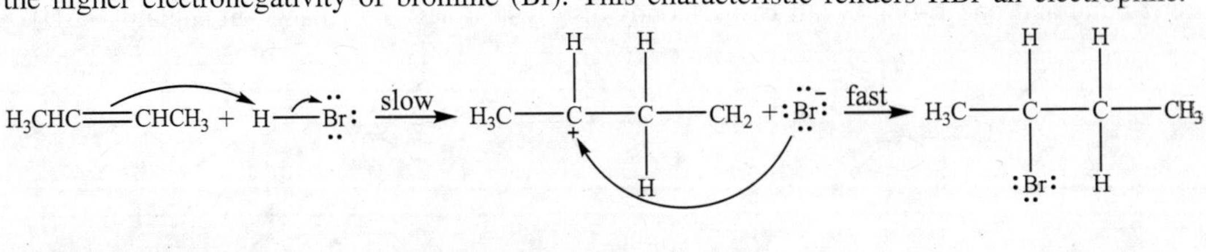

16.3.1.2 Mechanism of Addition Reactions

The process of alkene addition can be described by a two-step mechanism. The mechanism underlying addition reactions as follows:

Step I: In the initial stage, the π-bond undergoes cleavage, resulting in the formation of a new C—H bond and the generation of a cation. The initial stage is characterized by a slow, rate-determining step.

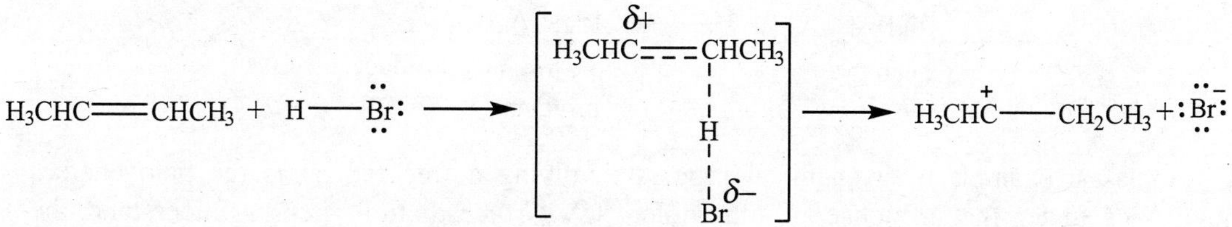

Step II: In the second stage, a chemical connection is established between the negatively charged bromide ion and the positively charged carbon atom. The second phase is executed rapidly.

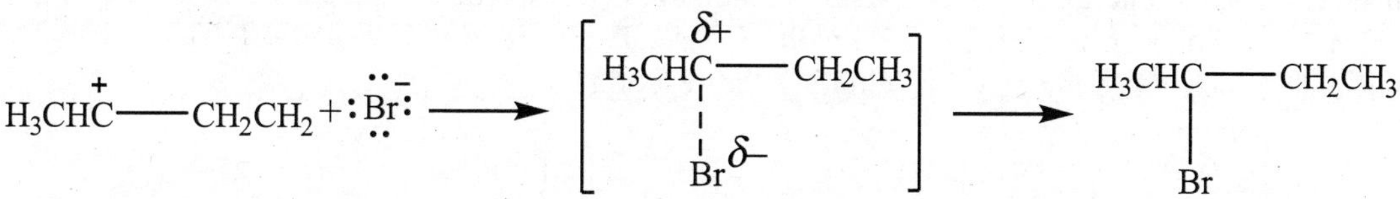

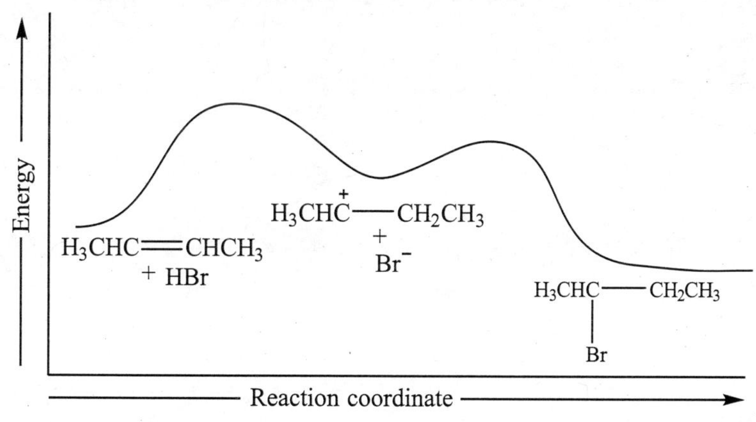

16.3.1.3 Free Radical Reactions

The term "free radical" pertains to an atom or a collection of atoms that possesses an odd number of electrons. Consequently, the electrons within a free radical are unable to form complete pairs. Despite lacking a positive or negative charge, free radicals exhibit a remarkably high level of reactivity.

Radical species exhibit reactivity in order to get a fully filled octet of valence electrons. Radicals possess the ability to cleave a chemical bond within a separate molecule and subsequently extract an electron from a partner, resulting in substitution within the initial molecule. The addition of a radical to an alkene results in the formation of a new radical, so initiating an addition reaction.

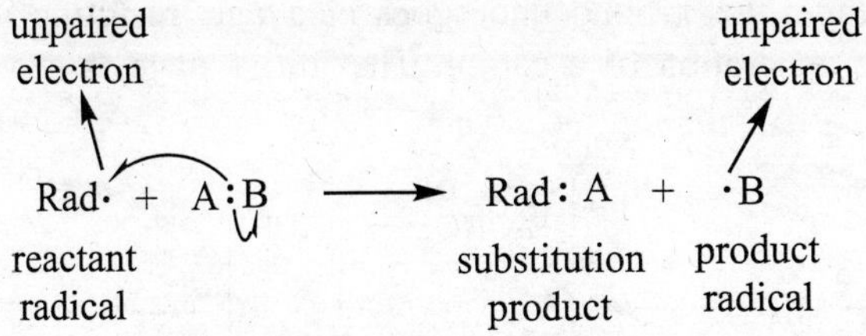

A classic example of a chemical process involving a free radical is the halogenation of alkanes under the influence of ultraviolet (UV, $h\upsilon$) radiation. Let us understand the monochlorination process of methane. In order for the reaction to proceed, it is necessary to cleave a carbon-hydrogen (C–H) link and a chlorine-chlorine (Cl–Cl) bond, while simultaneously establishing a carbon-chlorine (C–Cl) bond and a hydrogen-chlorine (H–Cl) bond. The Cl–Cl link, with a bond energy of 58 kcal/mole, exhibits lower strength compared to the C–H bond, which has a bond energy of 104 kcal/mole. Consequently, it is the weaker Cl–Cl bond that will undergo cleavage initially, so initiating the reaction.

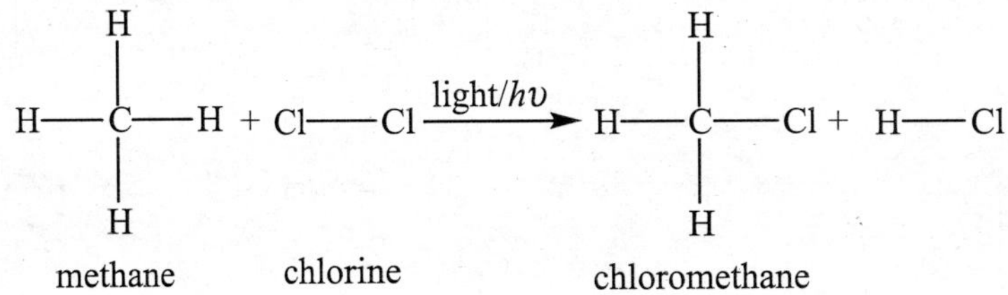

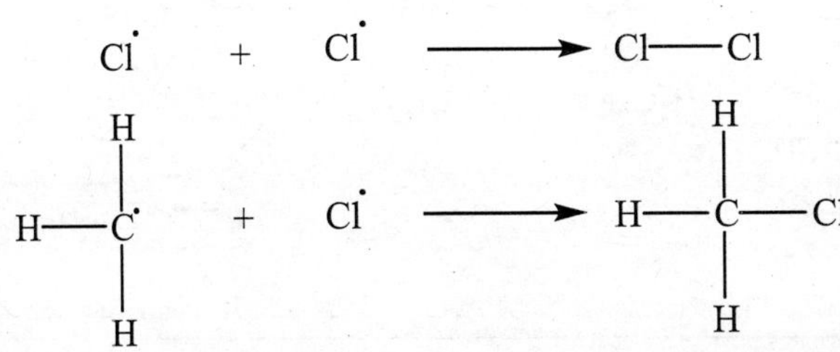

The above reaction involves three steps:

Step I: Initiation The initiation process involves the homolytic production of two reactive species, each possessing unpaired electrons. The synthesis of chlorine atoms results in the production of chlorine molecules (Cl_2) and the emission of light.

The initial stage in the chlorination process of methane involves the homolytic cleavage of Cl_2, resulting in the formation of two chlorine free radicals. The energy necessary for the cleavage of Cl_2, which amounts to 58 kcal/mole, can be supplied through either UV light or heat. The process of halogenation of alkanes does not occur under conditions of darkness.

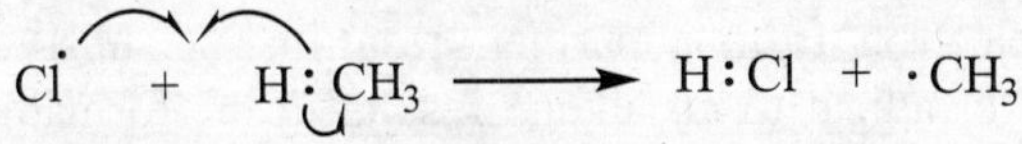

Step II: Propagation Propagation refers to the chemical process in which a molecule reacts with another molecule, resulting in the generation of a radical species. The interaction between a chlorine atom and methane results in the formation of hydrogen chloride (HCl) and methyl radical (CH_3).

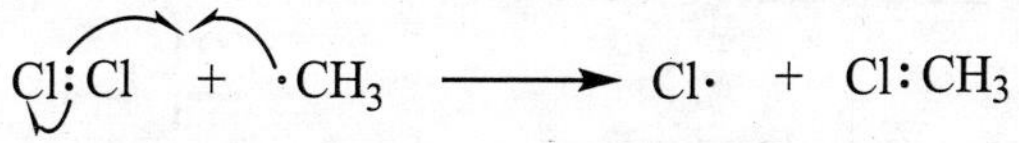

Step III: Termination Termination is the process by which two radicals combine to produce a stable product. The reaction between two methyl radicals (CH_3.) yields ethane (CH_3CH_3).

16.3.2 Substitution Reaction

A substitution reaction involves the replacement of an atom or a group of atoms with another atom or group of atoms. In a typical substitution reaction, the carbon atom undergoes a replacement of Z with Y.

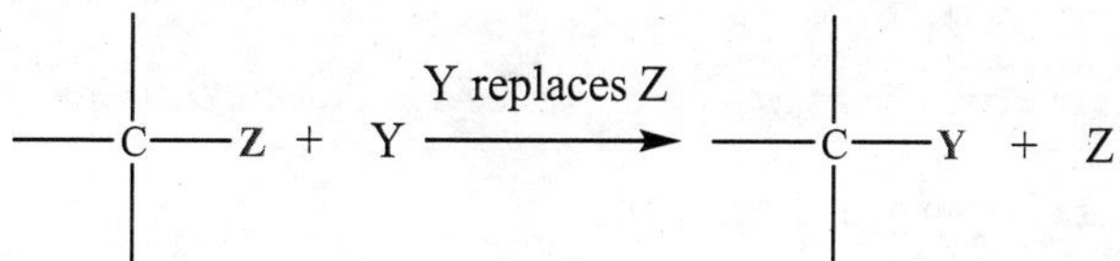

Y/Z = H or a heteroatom

Substitution reactions encompass the breaking of a sigma (σ) bond, while simultaneously forming another sigma bond at the carbon atom. Substitution reactions predominantly manifest when Z represents either hydrogen or a heteroatom possessing more electronegativity than carbon. Two examples are displayed below:

$$H_3C{-}I + Cl^- \xrightarrow{\text{Cl replaces I}} H_3C{-}Cl + I^-$$

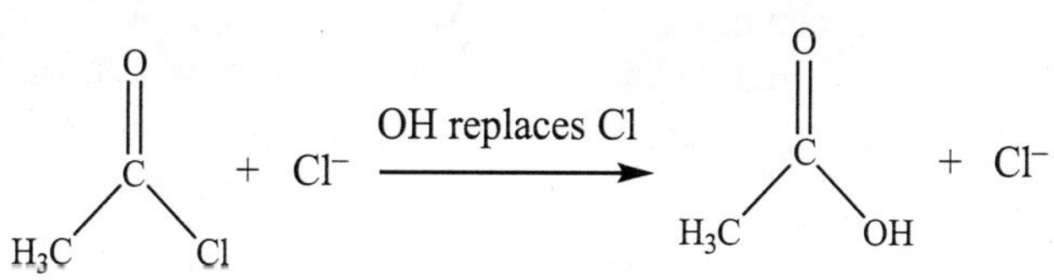

16.3.2.1 Nucleophilic Substitution Reactions (S_N)

In the context of nucleophilic substitution, the reagent employed is a nucleophile. In a chemical reaction, a nucleophile displaces a leaving group. Competing reaction refers to a phenomenon that encompasses the simultaneous occurrence of elimination reactions and rearrangements. The typical reaction involves substitution occurring on a carbon atom that is sp^3 hybridized, resulting in a tetrahedral configuration.

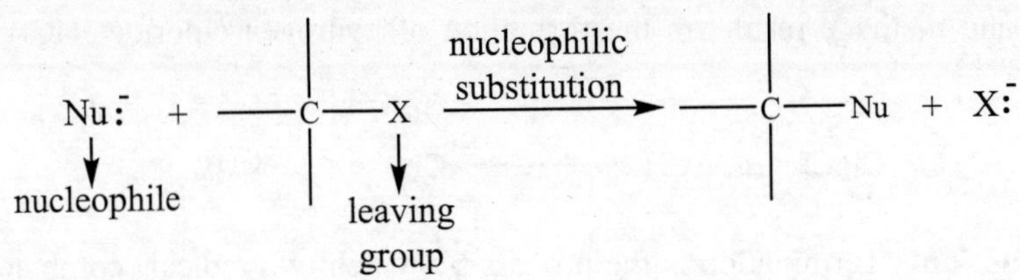

Some nucleophilic substitution reactions

$$Nu:^- + CH_3X \longrightarrow CH_3Nu + X:^-$$

Based on the different nucleophile the product is different.

$HO:^- + CH_3X \longrightarrow CH_3OH + X:^-$	An alcohol
$RO:^- + CH_3X \longrightarrow CH_3OR + X:^-$	An ether
$HS:^- + CH_3X \longrightarrow CH_3SH + X:^-$	A thiol (a mercaptan)
$RS:^- + CH_3X \longrightarrow CH_3SR + X:^-$	A sulphide (a thioether)
$I:^- + CH_3X \longrightarrow CH_3I + X:^-$	An alkyl iodide
$:NH_3 + CH_3X \longrightarrow CH_3\overset{+}{N}H_3 + X:^-$	An alkyl ammonium ion
$H\ddot{O}H + CH_3X \longrightarrow H_3C-\overset{+}{\ddot{O}}(H)-H + X:^-$	An alcohol (after proton transfer)

Mechanism

Chemists have put forth two distinct mechanisms that serve as limits for nucleophilic displacement. One essential distinction between the two entities is in the temporal sequence of bond cleavage and bond formation processes. At the far end of the spectrum, the two reactions occur concurrently, referred to as S_N2, where S represents substitution, N denotes nucleophilic, and 2 signifies bimolecular, indicating the involvement of two species in the rate-determining phase.

$$\text{Rate} = k\,[\text{haloalkane}]\,[\text{nucleophile}]$$

In the alternative limiting mechanism, the process of bond cleavage between carbon and the departing group is fully accomplished prior to the initiation of bond formation with the nucleophile. The mechanism is denoted as S_N1, where the letter S represents substitution, N represents nucleophilic, and the numeral 1 signifies that the reaction is unimolecular, with only one species participating in the rate-determining step.

$$\text{Rate} = k\,[\text{haloalkane}]$$

Bimolecular Nucleophilic Substitution (S_N2 Reaction)

In the context of this chemical reaction, it is seen that both the reactants participate in the transition state of the rate-determining step. The nucleophile exhibits a tendency to engage with the reactive center in a manner that is diametrically opposed to the position occupied by the leaving group. The transition state (T.S.) involves the simultaneous formation and breaking of bonds.

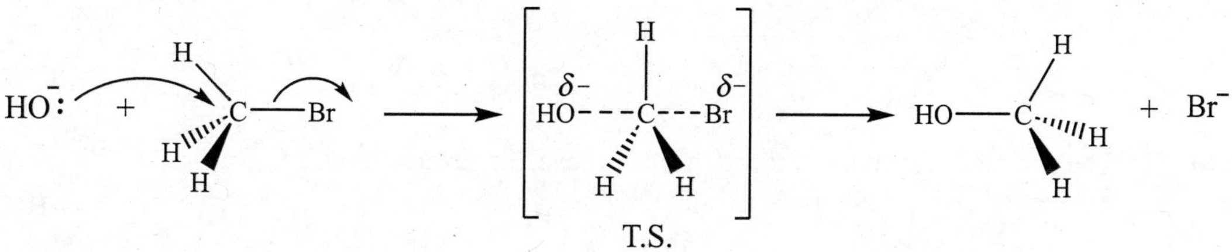

An Energy Diagram for an S_N2 Reaction

This diagram consist one transition state and no reactive intermediate:

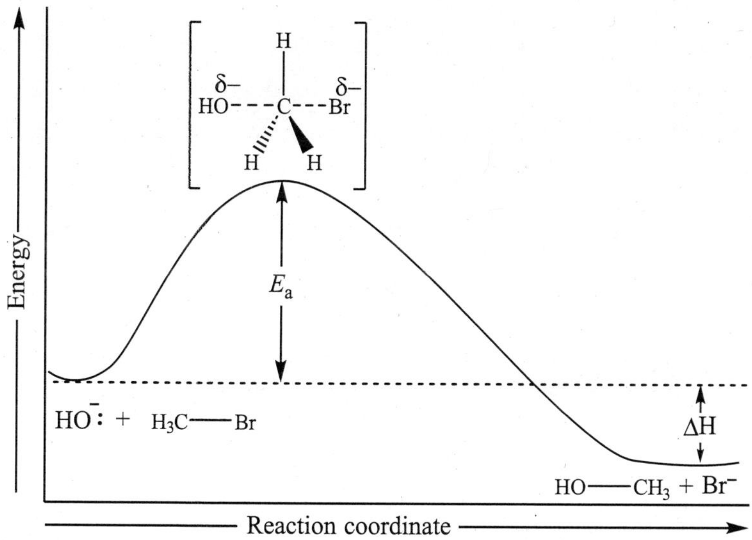

Unimolecular Nucleophilic Substitution (S_N1 Reaction)

The S_N1 mechanism can be exemplified using the solvolysis reaction of tert-butyl bromide. It proceeds through three steps.

Step I: The process of ionization of the C–X bond results in the formation of a carbocation intermediate (carbon is trigonal planar).

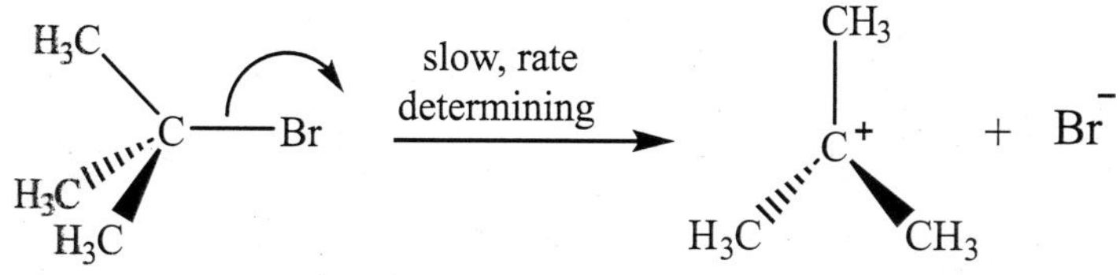

Step II: The carbocation, which acts as an electrophile, undergoes a reaction with methanol, a nucleophile, resulting in the formation of an oxonium ion.

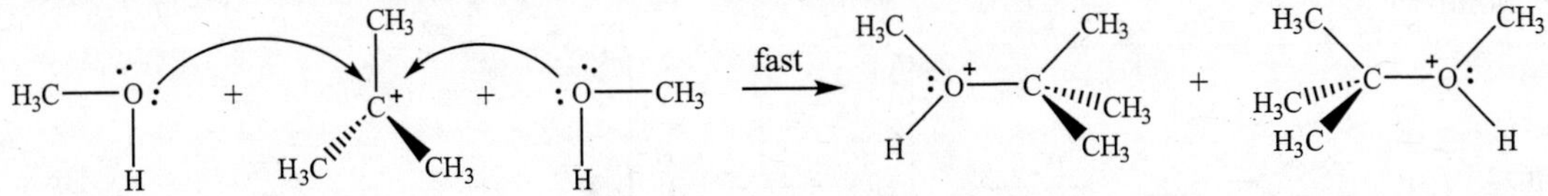

Step III: The reaction is finalized through the process of proton transfer.

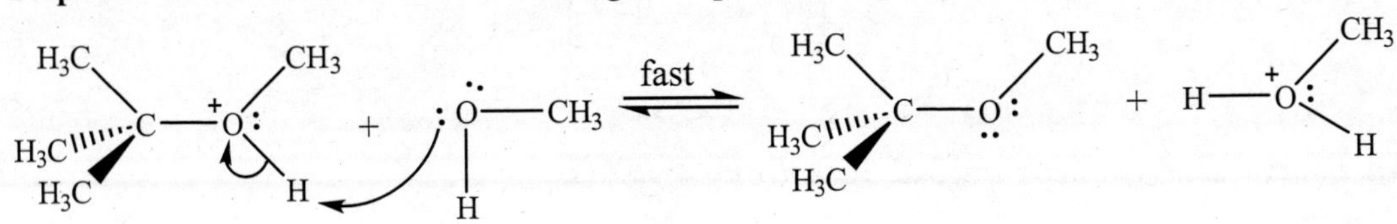

An energy diagram for an S_N1 reaction

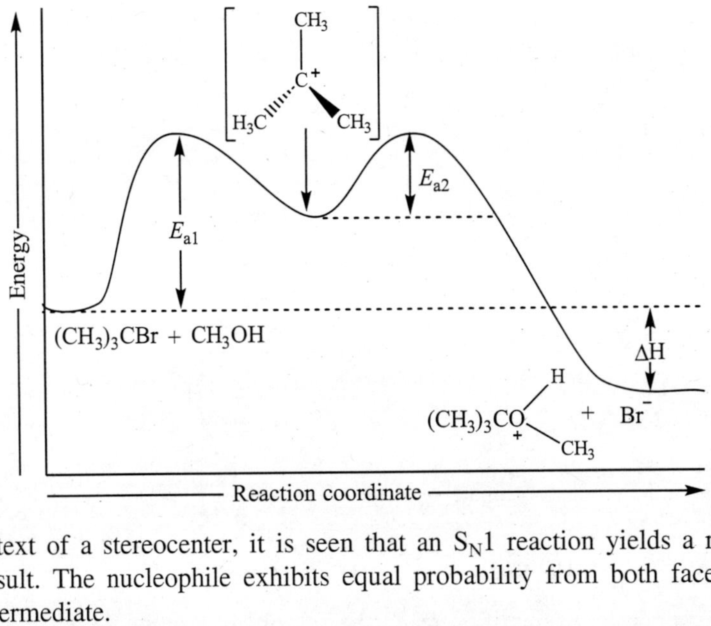

In the context of a stereocenter, it is seen that an S_N1 reaction yields a racemic mixture as the final result. The nucleophile exhibits equal probability from both faces of the planar carbocation intermediate.

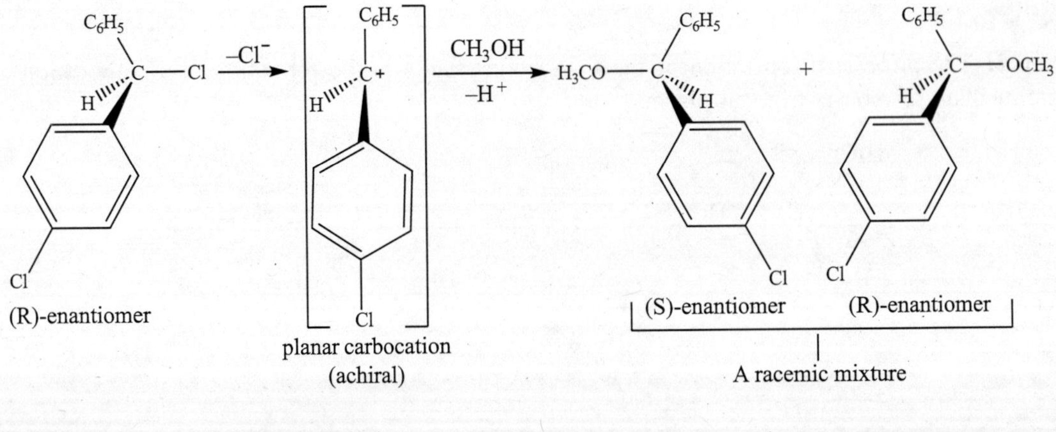

Effect of Variables on S_N Reactions

The nucleophilic substitution reaction is influenced by the properties of the substituents attached to the atom that is being targeted by the nucleophile. Hence, the reactivity of the reaction is influenced by the characteristics by both the nucleophile and the leaving group. The impact of the solvent on reactivity is of comparable importance.

1. Effect of Substituents on S_N Reactions

Effect on S_N1 Reactions: The S_N1 reaction is primarily influenced by electronic variables, namely the relative stabilities of the carbocation intermediates. The relative rates of reaction exhibit the following order: tertiary > secondary > primary > methyl. Methyl and primary alkyl halides do not typically undergo S_N1 reactions, but secondary and tertiary alkyl halides readily undergo S_N1 reactions. The occurrence of a stable 3° carbocation formation as an intermediate is attributed to electronic factors in reactions.

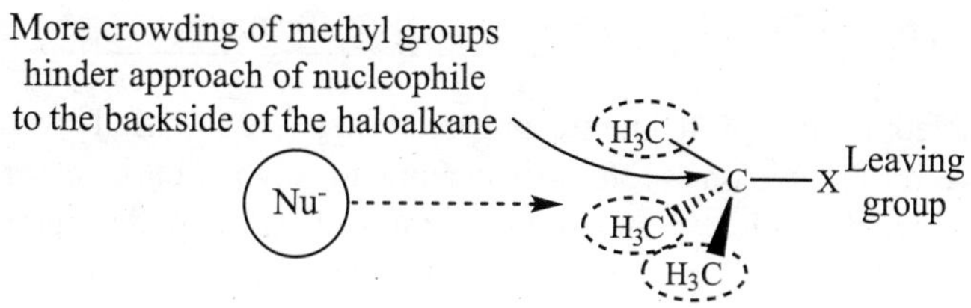

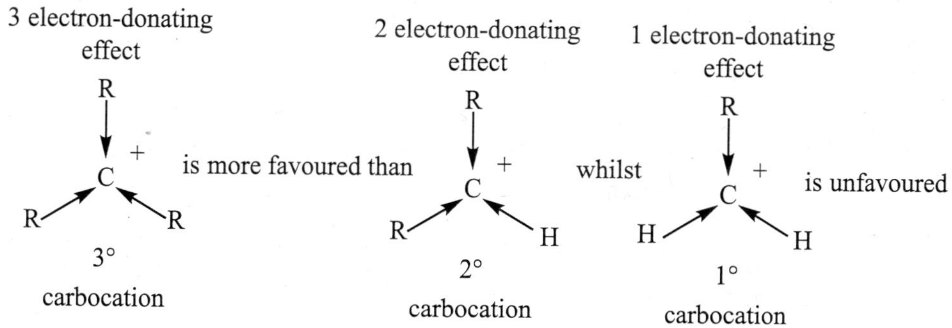

Halides	Relative rate of reaction	Class of Halide
CH_3–	1	Primary
CH_3CH_2–	1	Primary
$(CH_3)_2CH$–	12	Secondary
$(CH_3)_3C$–	1.2×10^6	Tertiary

Effect of Substituents S_N2 Reactions: The S_N2 reaction is influenced by steric considerations, namely the relative accessibility of the nucleophile to the reaction site. The relative rates of reaction can be observed to follow the following order: methyl > primary > secondary > tertiary.

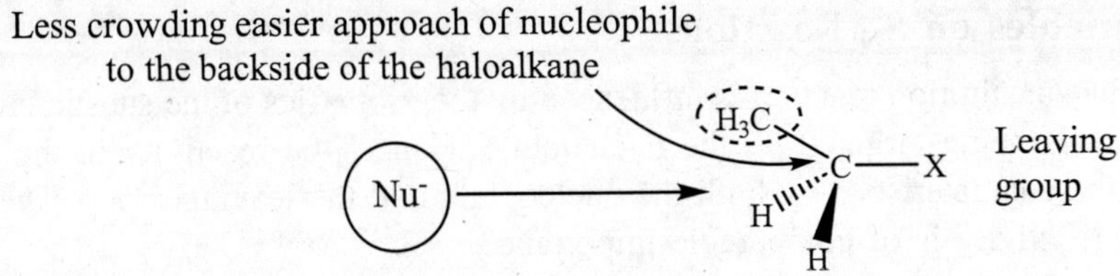

Effect of Structure on rates of S_N2 reactions

Halides	Relative rate of reaction	Class of Halide
CH_3–	30	Primary
CH_3CH_2–	1	Primary
$(CH_3)_2CH$–	0.03	Secondary
$(CH_3)_3C$–	0	Tertiary

Nucleophilicity: Nucleophilicity is a kinetic property that is quantified by the rate at which a nucleophile initiates an assault on a reference chemical under a defined set of experimental parameters. For instance, the velocity at which a group of nucleophiles replaces the bromide ion in bromoethane.

$$CH_3CH_2Br + NH_3 \longrightarrow CH_3CH_2\overset{+}{N}H_3 + Br^-$$

Nucleophilicity has two significant characteristics. An anion has superior nucleophilic behavior compared to an uncharged conjugated acid, while strong bases demonstrate favorable nucleophilic properties.

Nucleophilicity of Common Reagents

	Nucleophile	Name
Very good nucleophiles	NC^- HS^- I^-	Cyanide Thiolate Iodide
Good nucleophiles	OH^- Br^- NH_3	Hydroxide Bromide Ammonia
Reasonable nucleophiles	Cl^- $CH_3O_2^-$ F^- CH_3OH H_2O	Chloride Acetate Fluoride Methanol Water

Leaving Group

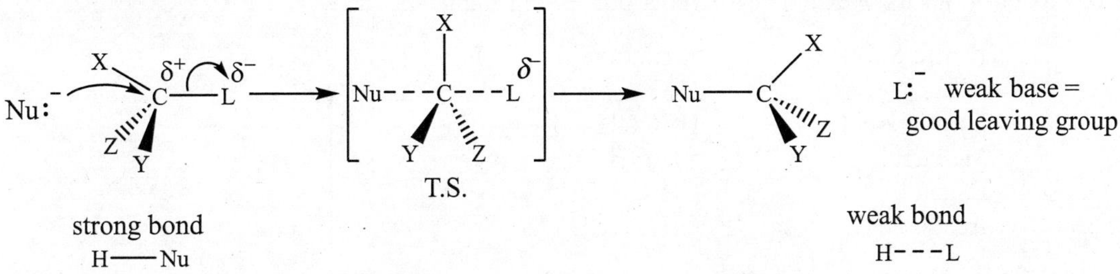

$$CH_3Br + OH^- \longrightarrow CH_3OH + Br^-$$

strong base, good nucleophile ⟶ weak base, good leaving group

$$CH_3OH + Br^- \not\longrightarrow CH_3Br + OH^-$$

weak base, poor nucleophile ⟶ strong base, poor leaving group

Leaving group categorization

Good leaving groups	I^-, Br^-, Cl^-, Me_2S, $CH_3SO_3^-$, H_2O, RCO_2^-
Moderate leaving group	HS^-, CN^-, NH_3
Moderate leaving group	RNH_2, RS^-
Poor leaving group	F^-, OH^-, RO^-
Very poor leaving group	NH_2^-, R_2N^-, H^-, R^-

The halogens I^-, Br^-, and Cl^- are considered to be the most favorable leaving groups within this particular series. Anions exhibit enhanced stability and conjugate acids demonstrate greater strength. The anionic species OH^-, RO^-, and NH_2^- exhibit such low reactivity as leaving groups that they are infrequently, if ever, displaced in nucleophilic substitution processes. In nucleophilic substitution and β-elimination processes, they exhibit infrequent reactivity as leaving groups.

Solvent Effect

The phenomenon known as the solvent effect refers to the influence of the solvent on the properties and behavior of a solute in a solution.

Protic Solvent

A protic solvent is defined as a solvent that possesses a hydroxyl (–OH) functional group.

These solvents exhibit a preference for S_N1 reactions, whereby the formation of carbocations is facilitated to a larger extent in solvents possessing higher polarity.

Protic Solvent	Structure	Polarity of Solvents
Water	H_2O	↑
Formic Acid	$HCOOH$	
Methanol	CH_3OH	
Ethanol	CH_3CH_2OH	
Acetic acidc	CH_3COOH	

Aprotic Solvent

An aprotic solvent is characterized by the absence of a hydroxyl (–OH) group. The formation of carbocations is comparatively more challenging in aprotic solvents. Aprotic solvents are known to exhibit a preference for S_N2 reactions.

Aprotic solvent	Structure	Polarity of Solvents
Dimethyl sulfoxide (DMSO)	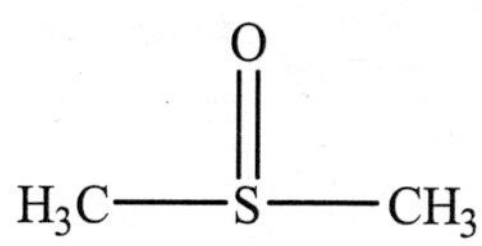	↑
Acetone	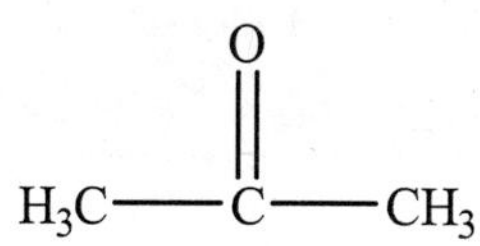	
Dichloromethane	CH_2Cl_2	
Diethyl ether	$H_3CH_2C—O—CH_2CH_3$	

16.3.3 Elimination Reaction

Elimination is a chemical reaction wherein the constituent atoms of the initial substance are effectively removed, resulting in the formation of a π bond. During an elimination reaction, the removal of two groups, X and Y, takes place from a given starting material. Two sigma bonds are dissociated, and a pi bond is generated between neighboring atoms. The instances of elimination reactions are frequently observed when X is equal to hydrogen (H) and Y represents a heteroatom that possesses a higher electronegativity than carbon. When alkyl halides are exposed to a base, they undergo a process known as elimination of hydrogen halide (HX). The substances under consideration are classified as alkenes.

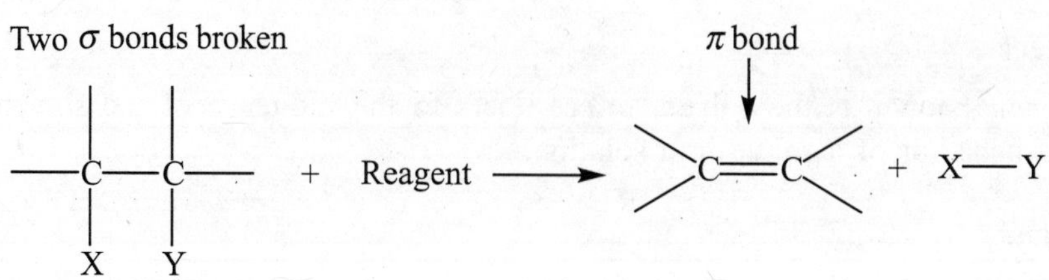

Elimination reactions typically necessitate the use of external energy or the use of reagents to proceed. The conditions required for this reaction include the application of heat and the presence of a strong base. The elimination reactions that alkyl halides undergo are commonly referred to as i2-eliminations or β eliminations.

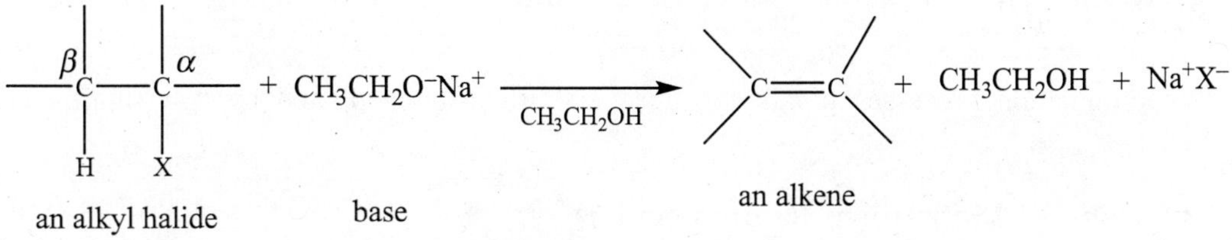

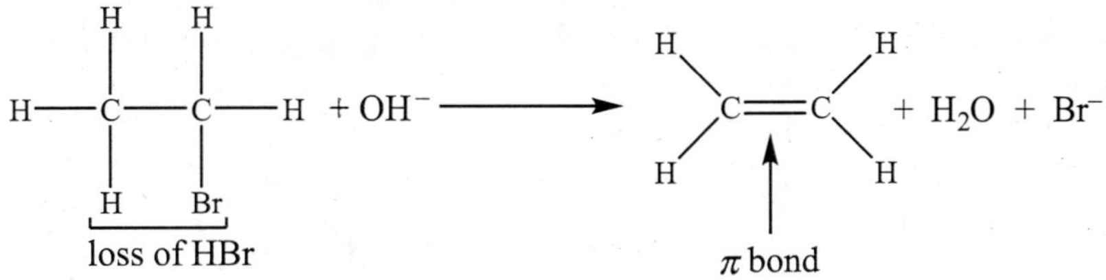

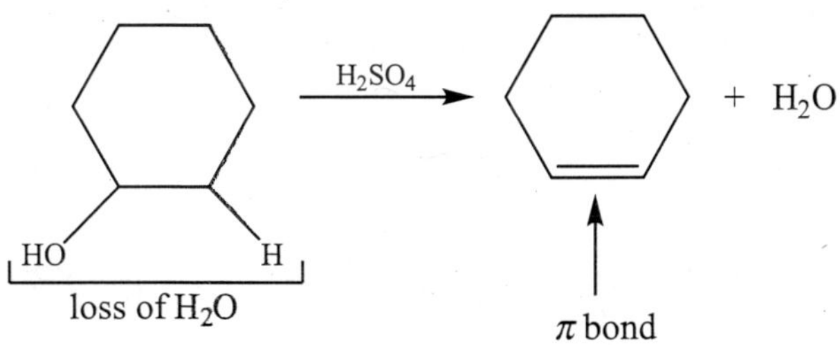

Addition and elimination reactions are considered to be inverse processes. In elimination reactions, the formation of a π bond occurs, while in addition reactions, the breaking of a π bond takes place.

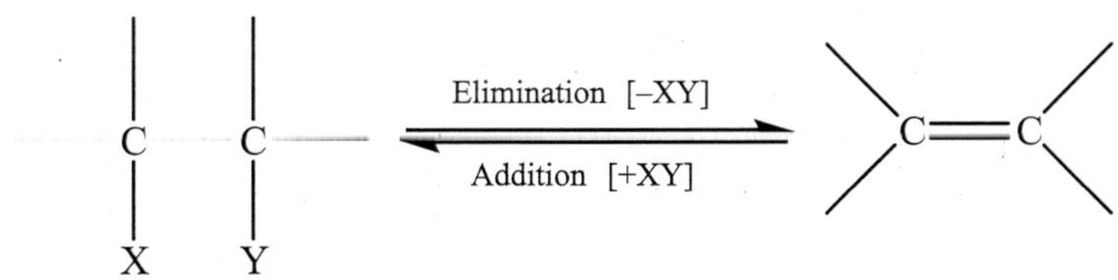

β-Elimination: The process involves the elimination of atoms or groups of atoms from neighboring carbon atoms in order to generate a carbon-carbon double bond. There exist two distinct mechanisms that impose limitations on β-elimination processes. The surrounding carbon atoms undergo a loss of the elements present in H–X, resulting in the formation of a carbon-carbon double bond (C=C). The primary carbon of the alkyl halide is referred to as α ("alpha"), while the adjacent carbon atom(s) are denoted as β ("beta"). The halogen atom is dissociated from the alpha carbon, and one of the β carbons loses a hydrogen atom.

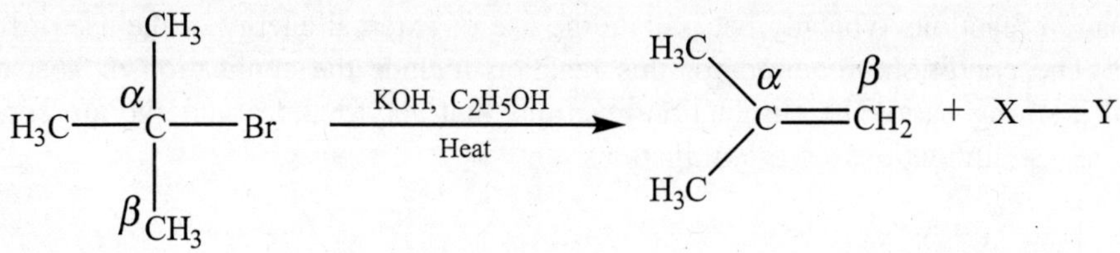

The elimination processes of alkyl halides are primarily facilitated by two significant mechanisms.

16.3.3.1 The E_1 Mechanism (Unimolecular)

The breaking of the C–X bond occurs in its entirety prior to the C–H bond being broken through interaction with a base. The rate-determining phase, which is the slowest step, involves solely one species, R–X (the alkyl halide). The E1 (unimolecular) process gives rise to the formation of a carbocation intermediate when alkyl halides undergo elimination.

The rate of the reaction = k [alkyl halide]

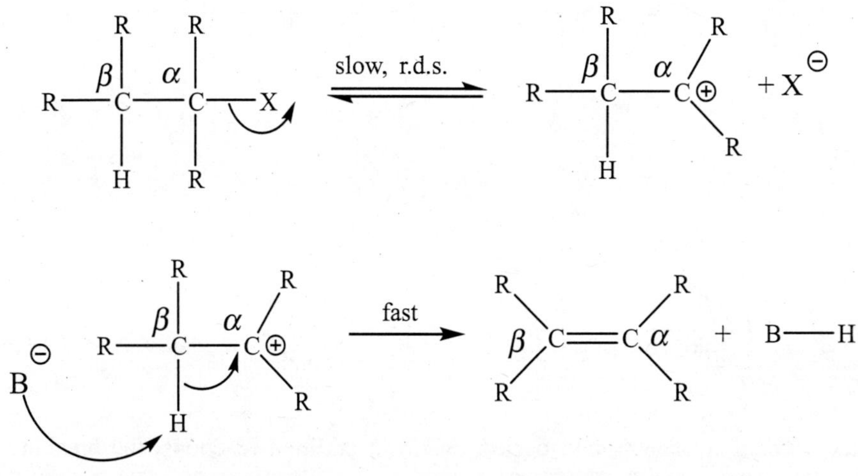

Tertiary alkyl halides are prone to undergo HX elimination by this particular mechanism. The elimination reaction for *t*-butyl bromide in aqueous alcoholic KOH:

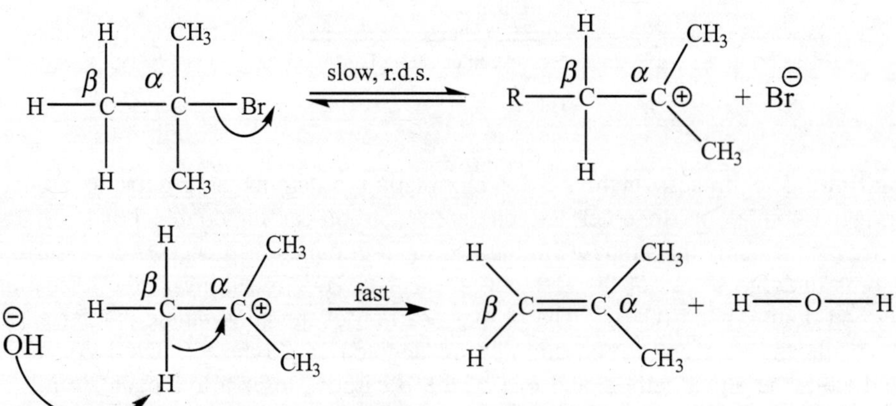

16.3.3.2 The E_2 Mechanism (Bimolecular)

Bond creation and breaking occur simultaneously in E_2 elimination reaction, i.e., the C–X and C–H bonds break simultaneous. This is a concerted reaction. The base and the alkyl halide both are involved in the rate-determining phase

$$\text{Rate} = k\,[\text{alkyl halide}][\text{base}]$$

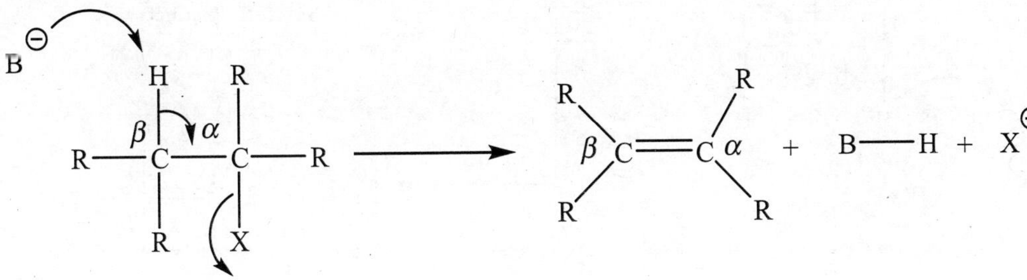

The H and X groups of an alkyl halide must be opposite to one another (i.e. antiperiplanar), in the same plane as the carbon atoms they are connected to, and undergo elimination via the E_2 process.

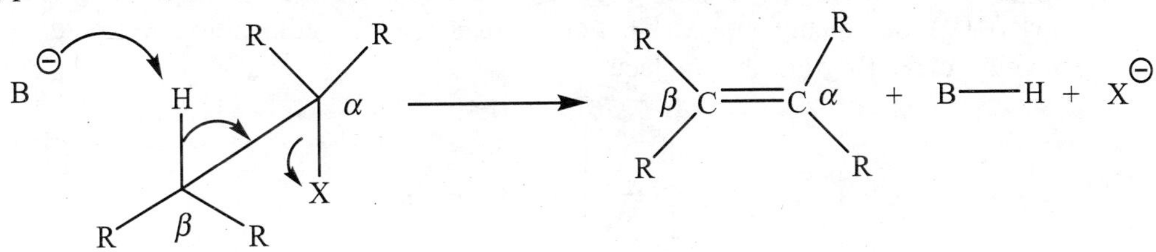

There are typically two probable elimination products for alkyl halides:

1. In 3° alkyl halides which contains three β carbons, if two identical methyl (CH_3) groups, one methylene (CH_2) group is present.

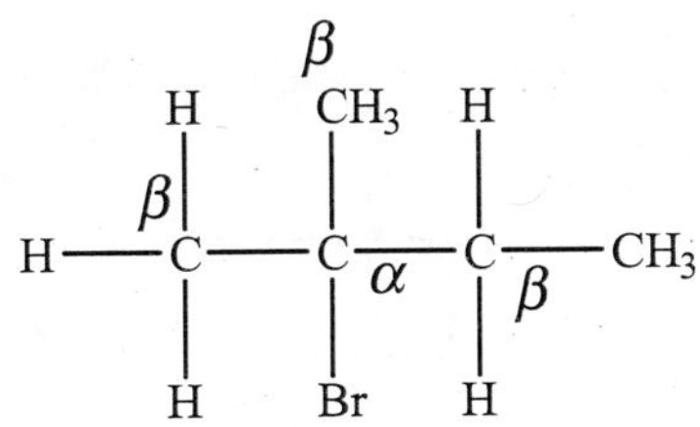

For example, the E_1 mechanism removes H–Br from this molecule.
Two products can result from the loss of H–Br.

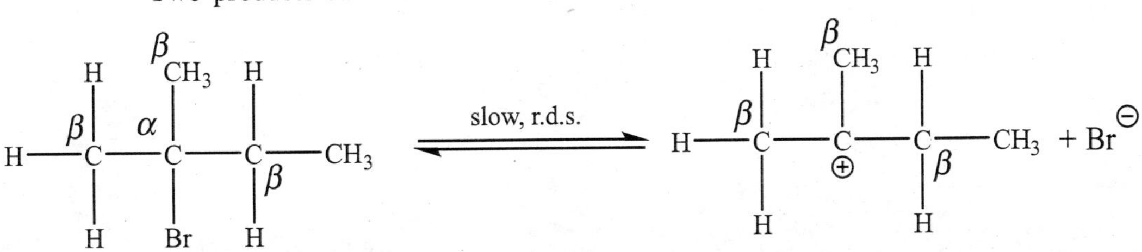

Carbocation

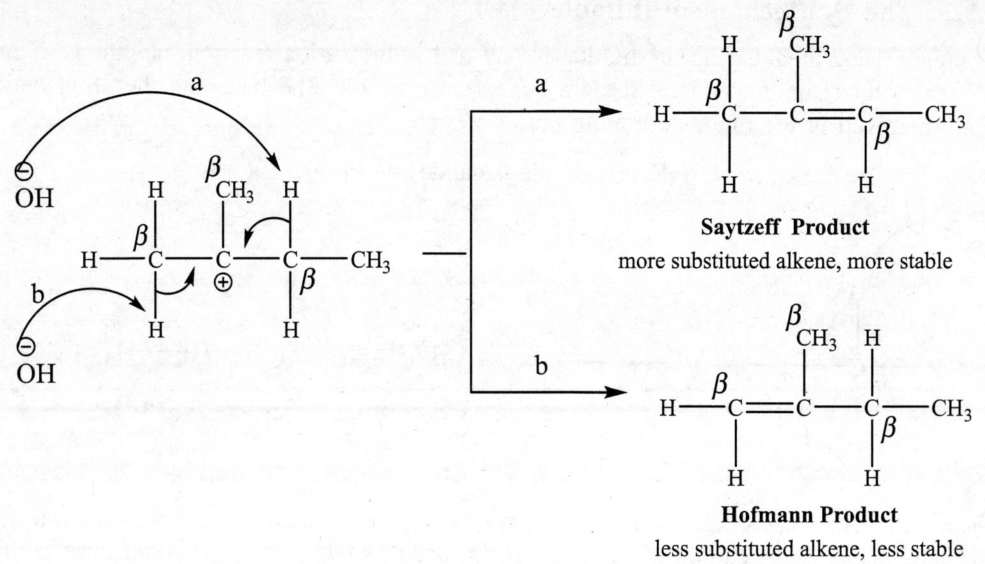

2. In the 2° alkyl halide two carbons are not the same. There are two types of groups: methyl (CH_3) and methylene (CH_2). For example, the E_2 elimination followed the elimination of H–Br from this molecule.

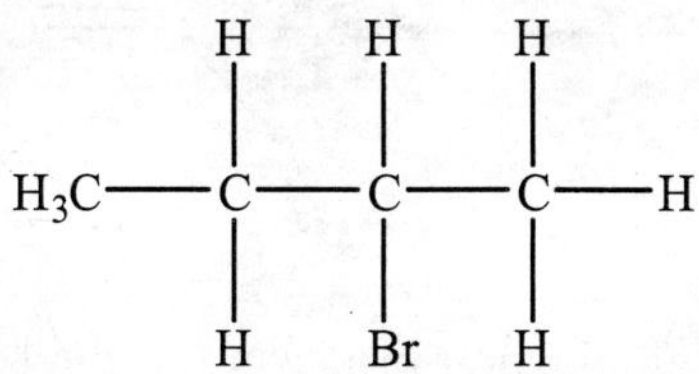

The mechanism of the above reaction via the E_2 mechanism is presented below. Which results into two products. These two results are govern by two rules: Saytzeff rule and Hofmann elimination.

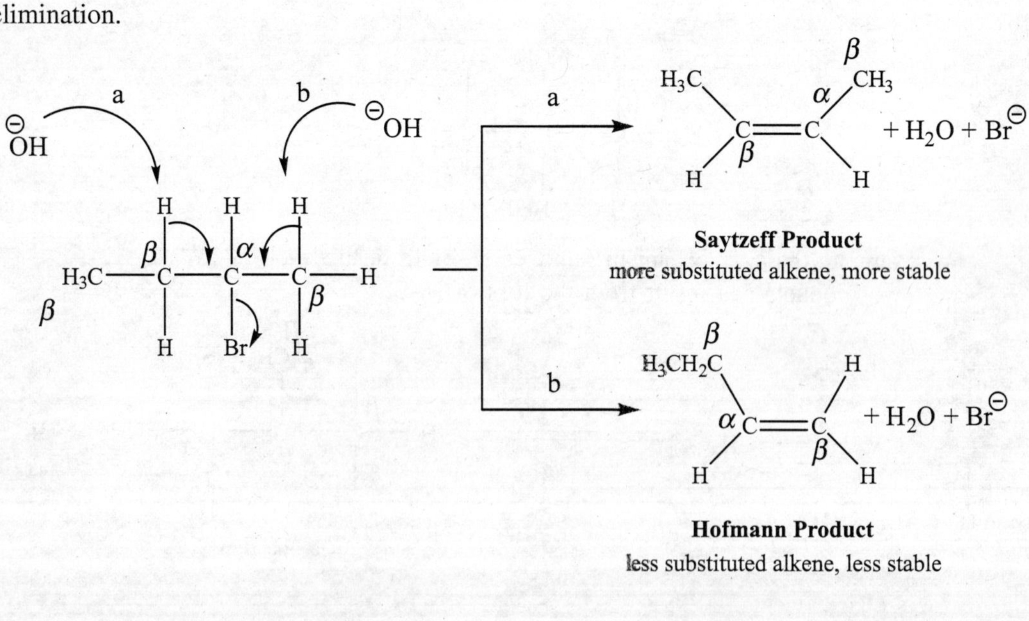

Saytzeff Rule: According to Saytzeff rule the major product of an E_2 elimination is the more highly substituted alkene that is more stable alkene.

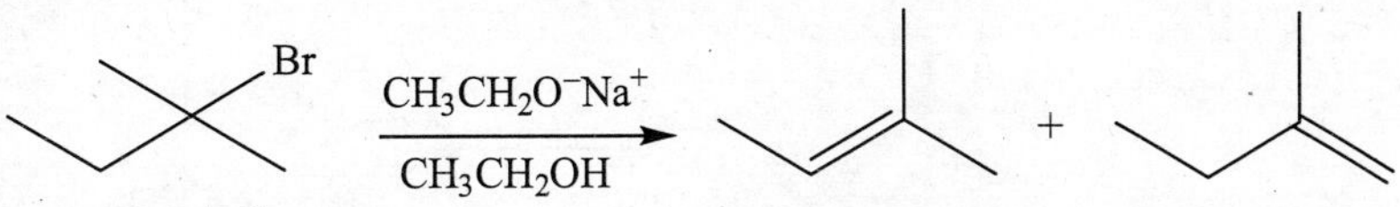

2-bromo-2-methylbutane → 2-methylbut-2-ene (major product) + 2-methylbut-1-ene

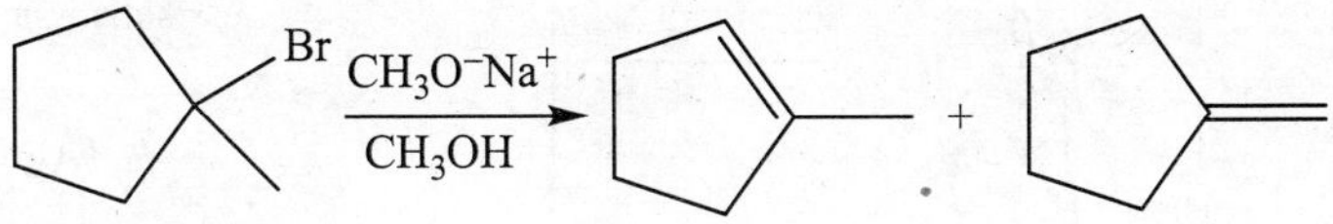

1-bromo-1-methylcyclopentane → 1-methylcyclopent-1-ene (major product) + methylene cyclopentane

Hofmann Elimination: The elimination that does not govern by saytzeff rule will be followed by Hofmann elimination. According to the Hofmann the major product of the elimination reaction is less substituted alkene. The proportion of the less substituted alkene (Hofmann product) can be increased by using a very bulky base. This starts with a quaternary ammonium salt and a tertiary amine acts as the leaving group. Two examples of bulky bases are shown below:

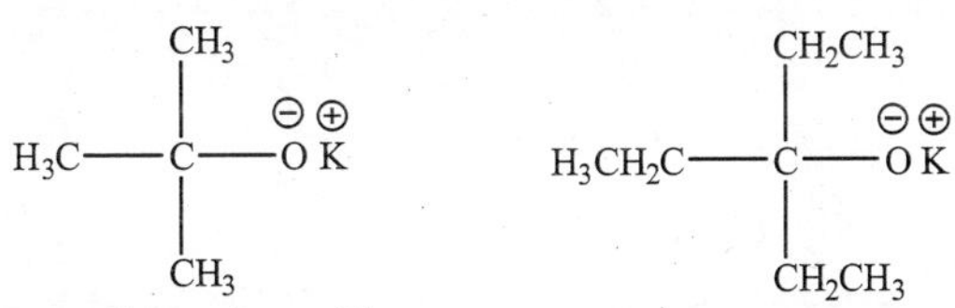

Potassium *t*-butoxide Potassium 3-ethyl-3-pentoxide

Bulky bases increase the proportion of the less substituted alkene (Hofmann product) formed in elimination reactions.

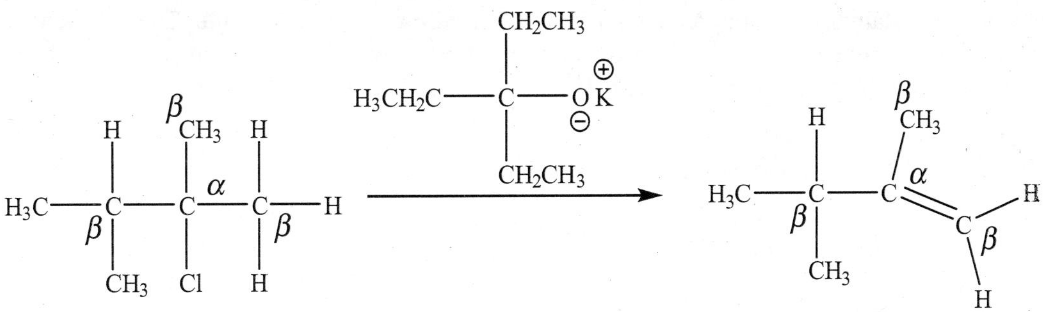

Hofmann Product

90% less substituted alkene

The hydrogens present on less substituted β carbon compared to highly substituted β carbon are more prone to react with bulky base. The approach of the bulky base to less substitute β carbon which is less crowded is easy than the other β carbon, so less substituted alkene product is more stable. This stable product is known a Hofmann product.

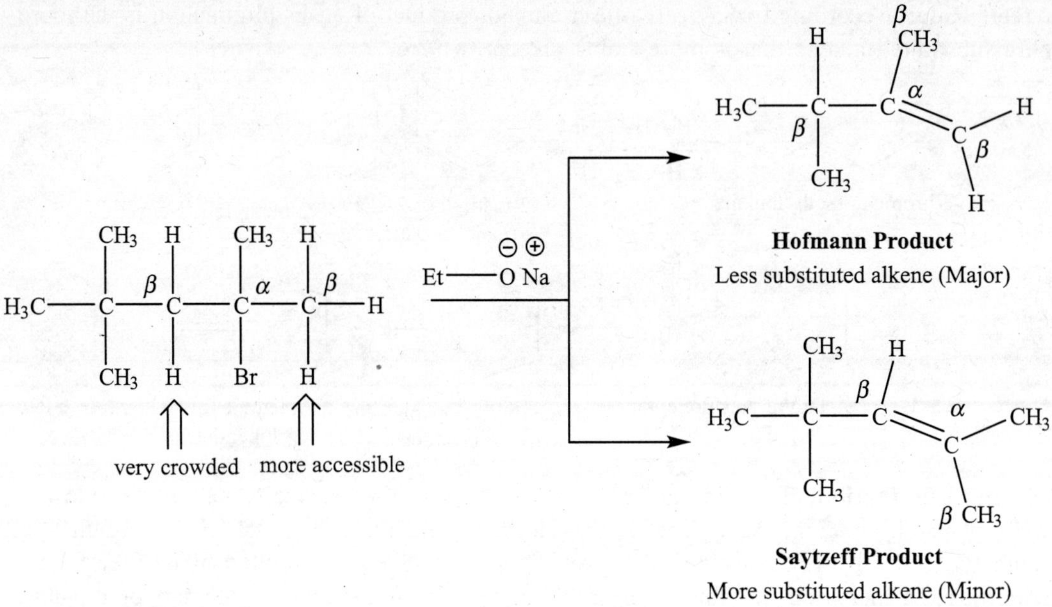

Another example

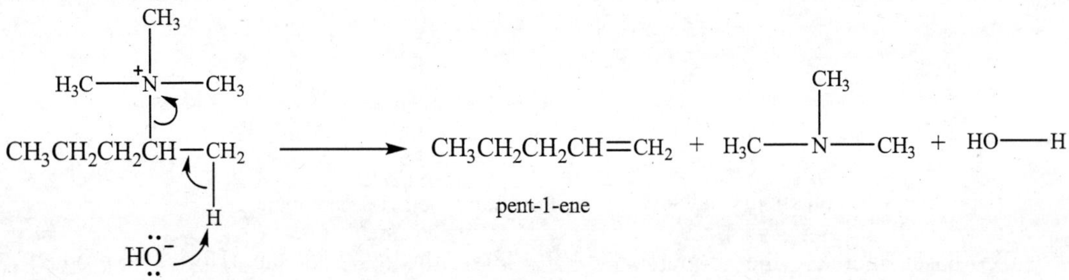

Although 2-pentene and 1-pentene are possible products, 1-pentene, the less substituted, less stable alkene, is almost the only product. This reaction follows Hofmann's rule. The preference for the less substituted product is often quite strong. Formation of the less substituted product is favored by removal of the more acidic H and steric hindrance.

Here is another example.

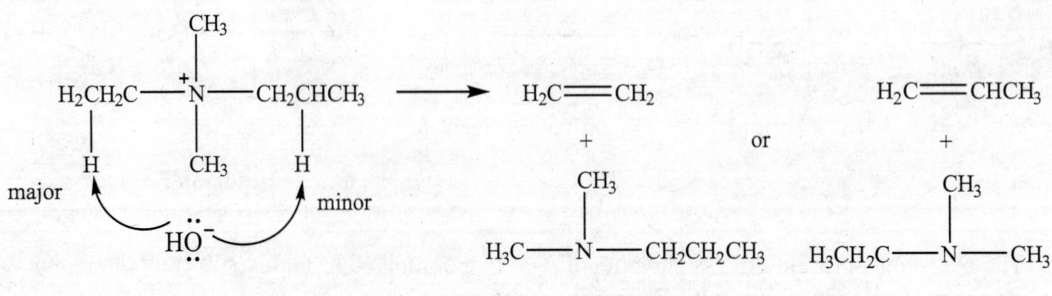

The reasons that the less substituted product is formed are somewhat complex. The leaving group is not very good, so the C–H bond is more broken in the transition state than usual. So removal of the more acidic H is favored. Alkyl groups are weakly electron donating and decrease acidity. Also, the leaving group is bulky (a tertiary amine) so removal of the less hindered H is favoured.

The Competition between Elimination and Substitution Reactions

S_N2 and E_2: S_N1 reaction is more favored by strong base and E_1 is favored by strong nucleophile. S_N2 reaction is influenced by steric hindrance but E_2 does not. For example,

$$CH_3CH_2Br + CH_3CH_2O^- \xrightarrow{CH_3CH_2OH} \underset{99\%}{CH_3CH_2OCH_2CH_3} + \underset{1\%}{H_2C{=}CH_2}$$

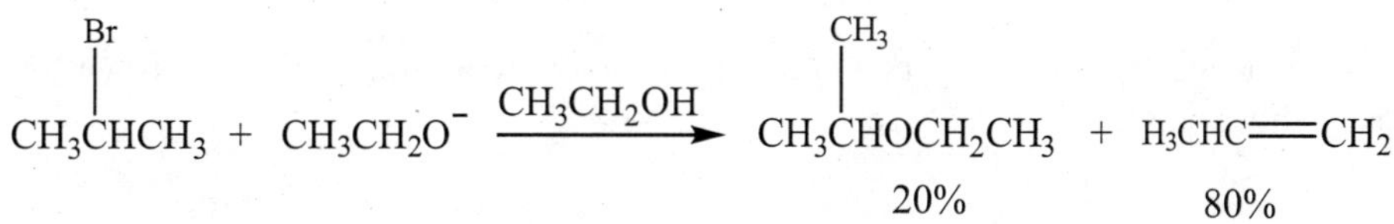

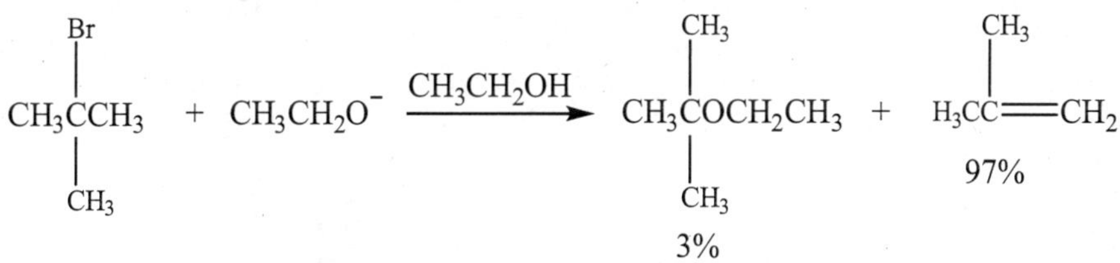

Reactions involving strong bases proceed according to E_2, not S_N1. However, when the base is weaker, S_N1 dominates E_2.

Stronger base

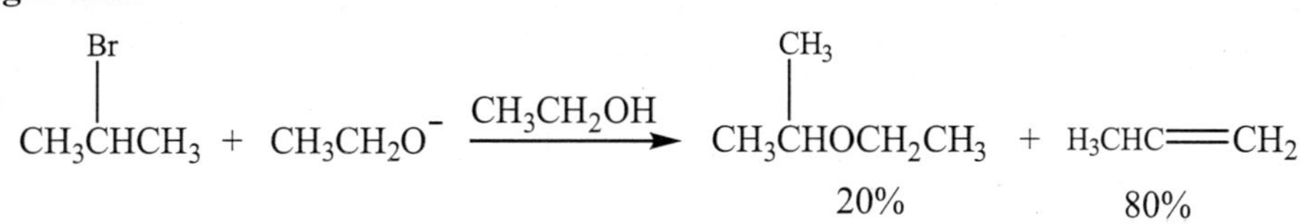

Weaker base

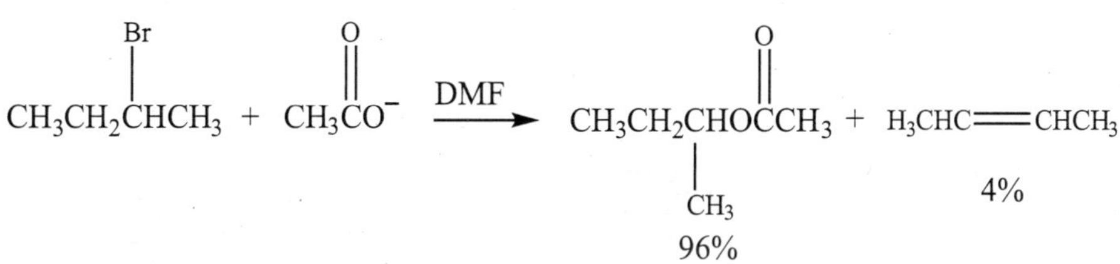

Elimination reaction is more favourable at higher temperatures.

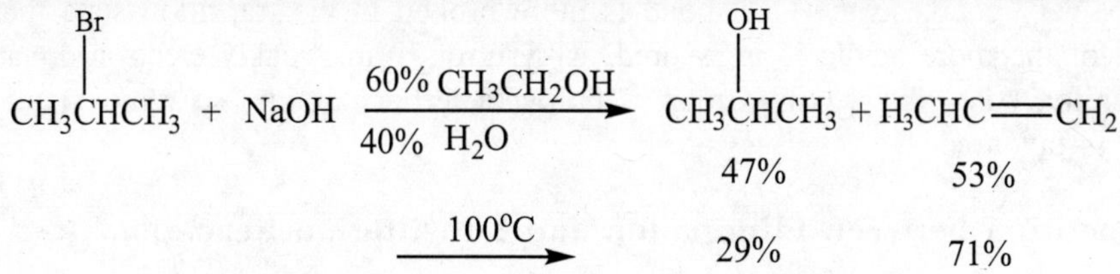

Basic difference between S_N2 and E_2 is as follows:

S_N2	E_2
Weaker bases	Stronger bases
Less steric hindrance	More steric hindrance
Lower temperature	Higher temperature

S_N1 and E_1: S_N1 and E_1 reaction occur in neutral or acidic conditions. Especially with tertiary or secondary substrates in polar solvents.

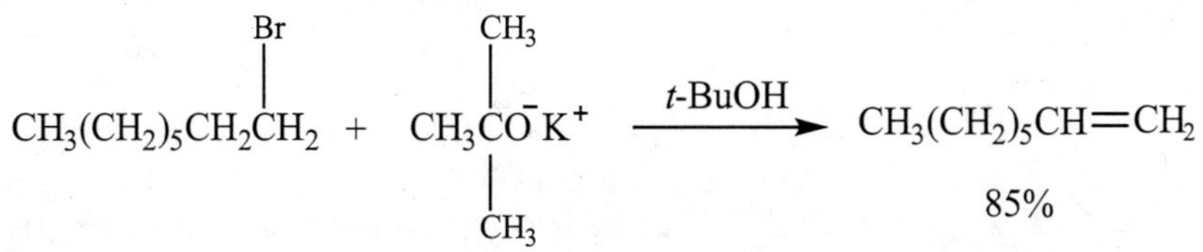

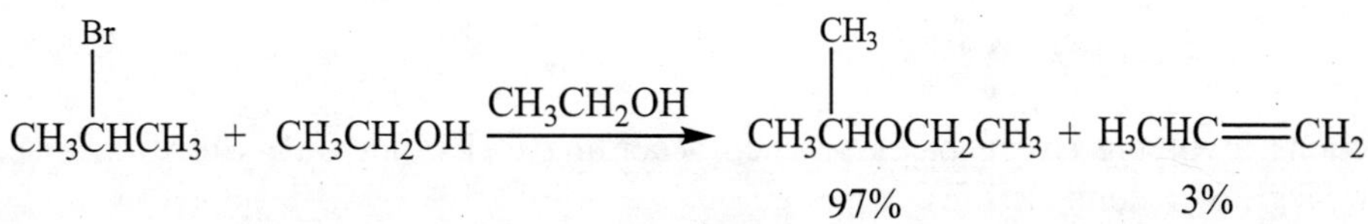

Tertiary alkyl halide does not undergo S_N2, because it is too hindered. E_2 elimination favored with strong bases HO^-, RO^- (NaOH, NaOEt). But S_N1 reaction is favored by weak nucleophile in polar solvent.

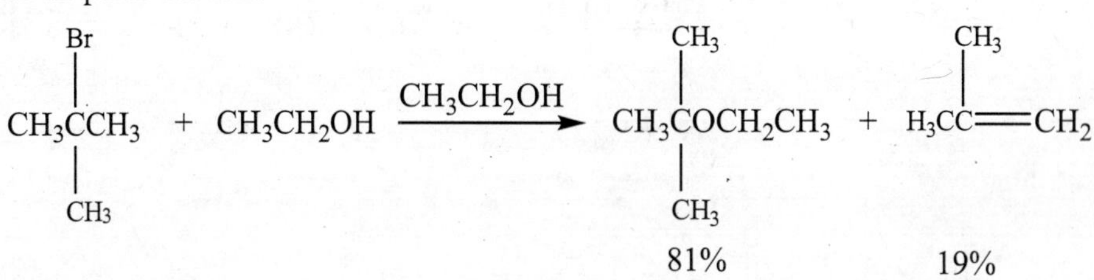

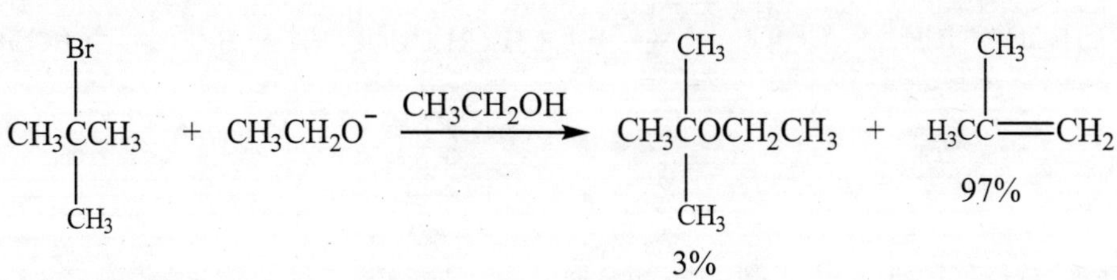

16.3.4 Rearrangement Reactions

Two categories of organic chemical reactions are referred to as "rearrangements." A H atom or a bigger molecule fragment may migrate in one step during a relatively short-lived intermediate during a rearrangement. A wide family of chemical events known as rearrangement reactions occur when a molecule's carbon skeleton is reorganised to produce a structural isomer of the original molecule. A substituent frequently shifts within a molecule, from one atom to another. In organic chemical there are number of reaction mechanism where rearrangement reactions takes place. These are as: Curtius Rearrangement or Curtius Reaction, Claisen Rearrangement, Beckmann Rearrangement, Hofmann Rearrangement, Pericyclic Rearrangement, Photochemical rearrangements, pinacol–pinacolone rearrangement, Benzilic acid rearrangement, and Fries rearrangement.

In the example below there is a rearrangement of methyl group takes place from one carbon atom to another adjacent carbon atom.

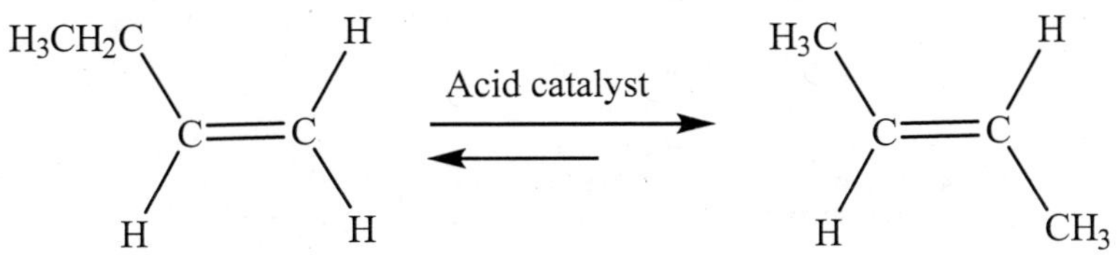

1-Butene 2-Butene

Migratory Aptitude

The relative capacity of a migrating group to move in a rearrangement reaction is known as its migratory aptitude. The leaving group may have an impact on this, depending on whether one produces a more stable carbocation. For example, Hydride > Phenyl = *Tert*-Alkyl > *Sec*-Alkyl > Primary Alkyl > Methyl. The migrating groups adhere to the carbocation's increasing stability sequence. Increased electron density at anionic carbon is the true determining factor. The ability to migrate increases with the number of electron-releasing groups.

A group or an atom migrates within a molecule during rearrangement reactions, moving from one centre (the migration origin) to another (the migration terminal). The type of migrating group or atom can be used to categorise these rearrangements in general. Anionotropic or nucleophilic migration occurs when a group moves together with its electron pair. A group that is migrating without its electron pair is said to be electrophilic or cationotropic. A free radical is a group that migrates with just one electron. Two forms of nucleophilic rearrangements are possible.

Intramolecular: These migrating groups occur within the same molecule and do not totally separate from the migration origin.

Intermolecular: This group of migrants is isolated from its migration source. In this scenario, a group or atom may migrate to a different molecule.

Claisen Rearrangement

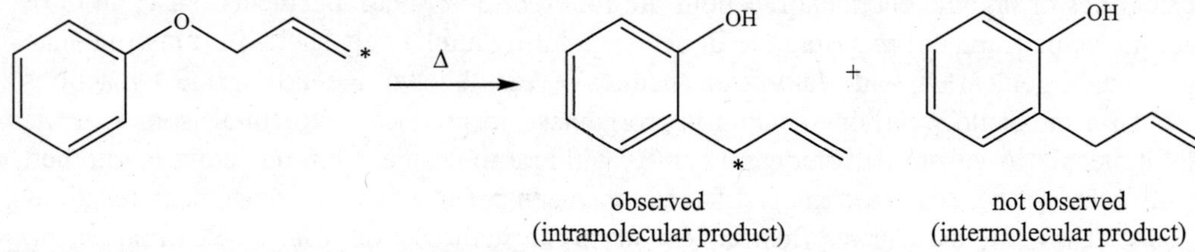

intramolecular mechanism

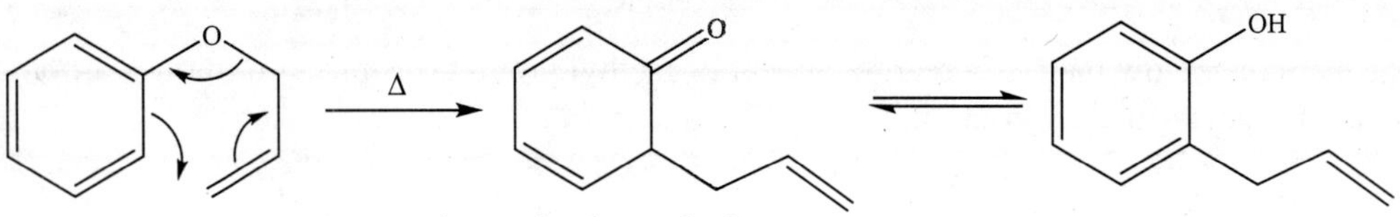

intermolecular mechanism

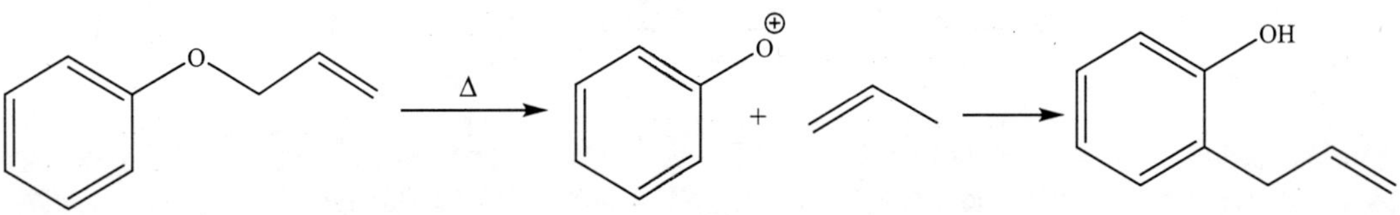

General Features of 1, 2-rearrangement Reactions: There is a species involved that has a valence electron sextet, either a carbocation or a carbenium ion. Thermodynamic driving force for a 1, 2-rearrangement will be important if it produces a more stable carbocation or a structure with an octet on every atom. As an alternative, the pushing force may come from a decrease in angle strain. A carbocation would result from the heterolysis of such a link. The adjacent group then assists the departing group in leaving.

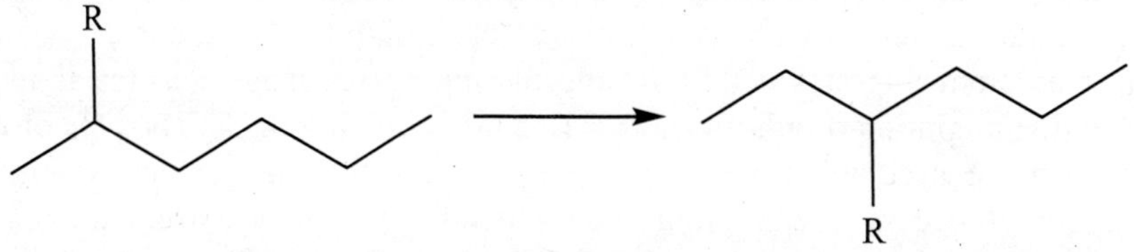

Mechanism of Nucleophilic Rearrangement: Three steps make up broadly nucleophilic rearrangement reactions:

The creation of an electron-deficient centre within the molecule is the first stage. The migration terminus must have an incomplete octet since the migrating group migrates with an electron pair. There are two methods to obtain this by carbocation: There are several ways to form carbohydrate groups. The most typical one is alcohol dehydration. And via Nitrenes: The breakdown of acyl azides can produce nitriles.

The migrating group creates a new electron-deficient centre in the second step by migrating to the electron-deficient centre with its electron pair.

In the third stage, an electron-deficient centre that has just formed gains an octet by either rejecting or accepting a nucleophile. It is frequently noted that two or all three phases happen at the same time. As may be observed in many instances, rearrangement to produce a more stable carbocation typically occurs after the S_N1 type of first step. It is demonstrated by the fact that the rate of reaction is independent of base concentration and rises with solvent ionising power. It has been demonstrated that the degree of electron shortage at the migration terminus rises with the migration rate.

Nature of Migration: Rearrangements are primarily intramolecular. A chiral migrating group can be used to obtain additional evidence. It is highly probable that the rearrangement is intramolecular if the configuration at the migrating group is preserved in the final product.

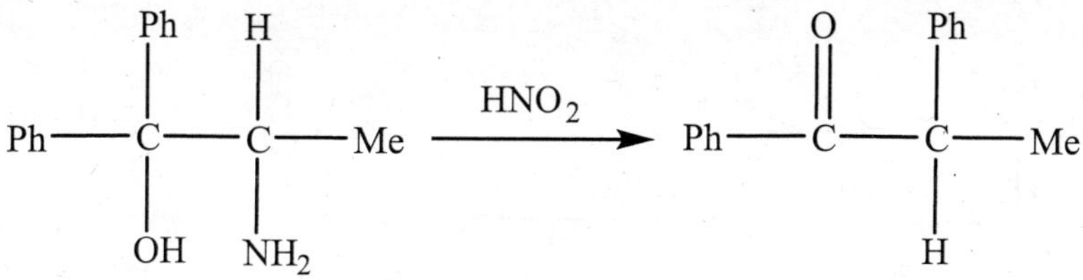

At the migratory terminal, inversion occurs in this scenario. Diazotization and intramolecular 1.2-phenyl migration are involved in the process.

Migratory Aptitude: Many responses, such as those from Hofmann, Curtis, and others, make it very evident who the migrating group is. There are, nevertheless, several options in some other reactions, such as Beckman rearrangement. Which group migrates under such circumstances depends on a number of variables (such as the geometry of the molecule). There are numerous options for Wagner-Meerwein and Pinacol rearrangement since the substrate has multiple groups with comparable migratory propensities. Relative migratory aptitude is studied using such reactions.

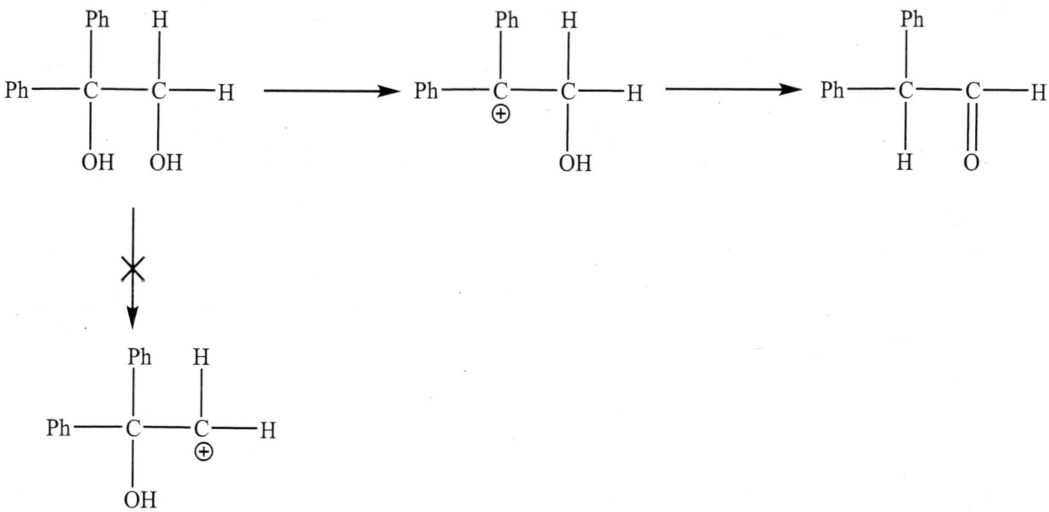

In order to create a more stable carbocation, the hydroxyl group is removed from carbon containing two phenyl groups. Group in the order aryl > alkyl > H improves the stability of the carbocation. The group moved in the subsequent order: tertiary alkyl > aryl > H > secondary alkyl > primary alkyl > methyl.

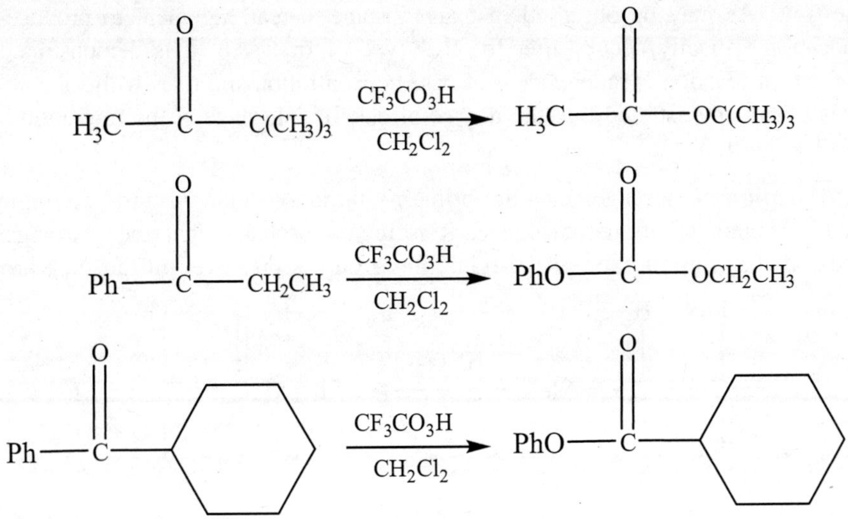

Migratory aptitude is controlled by numerous factors. The two things that can stabilise the developing positive charge are (a) conformational properties, and (b) the relative ability of the groups at the migration origin.

The following example shows that just the phenyl group migrates when tosylate (OTs) is broken down.

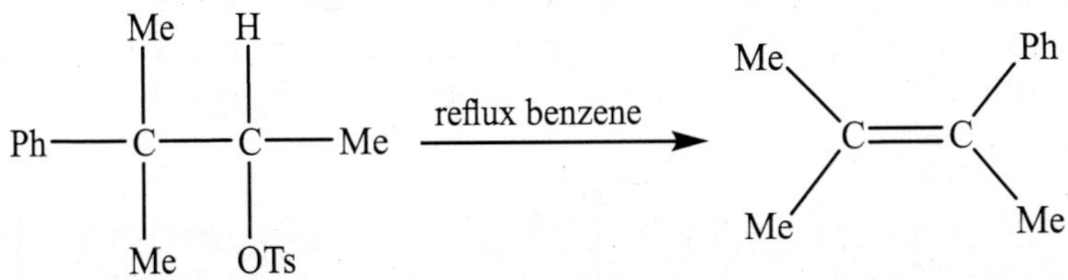

In the case, the phenyl group aids in the tosyl group's exit. Several overarching patterns in the ability of various groups to migrate. The tendency for migration is stronger in alkyl groups than in aryl groups. Hydrogen's ability to migrate is erratic. Thus, a variety of migrating products are produced. When it comes to aryl groups, those that have substituents at the ortho or para locations that donate electrons move more favourably than those that have substituents at the meta or para positions. Reduced migratory aptitude is seen in the aryl group that contains electron withdrawing groups.

16.3.5 The Diels-Alder Cycloaddition

Conjugated dienes cycle along with dienophiles to produce the Diels-Alder reaction. This response is stereotypical and favours endo over exo products. The result of the general Diels-Alder reaction is cylohexene. Two bonds are lost in order to form two new ones. The conjugated diene system consists of four electrons. Dienophiles, or "diene lovers," are two-electron systems. An adduct is the term for the product.

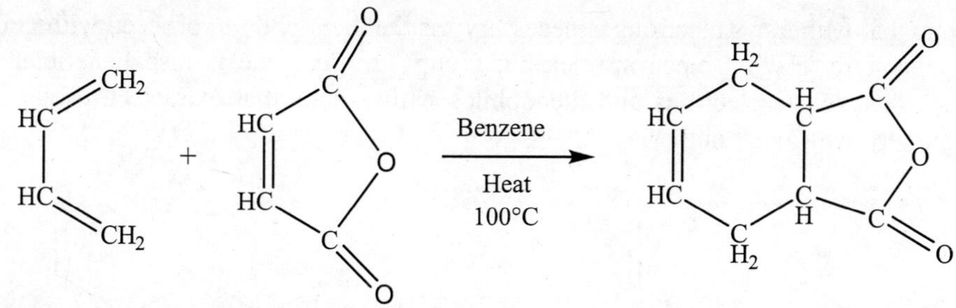

buta-1,3-diene (diene) maleic anhydride (dienophile) Adduct (100%)

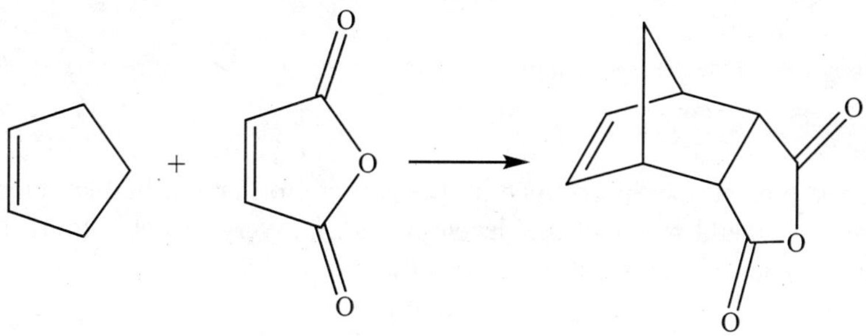

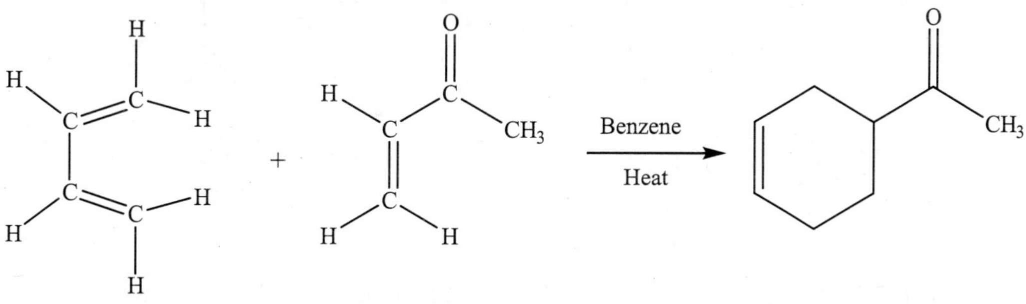

Conjugated diene Dienophile

16.3.5.1 Factors Favoring the DA Reaction

The most basic Diels-Alder process has an extremely low yield and high temperature requirements.

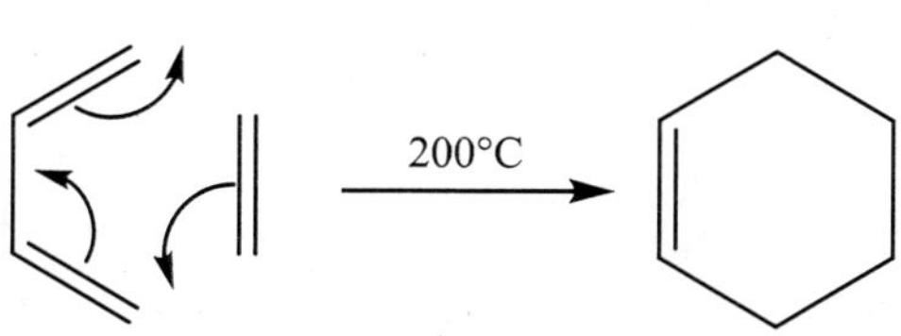

20%

A group that withdraws electrons is necessary for the dienophile to proceed with good yield at low temperature. Having electron-releasing groups in the diene is also beneficial. Dienes with groups that donate electrons and dienophiles with groups that extract electrons can also react favourably with one another.

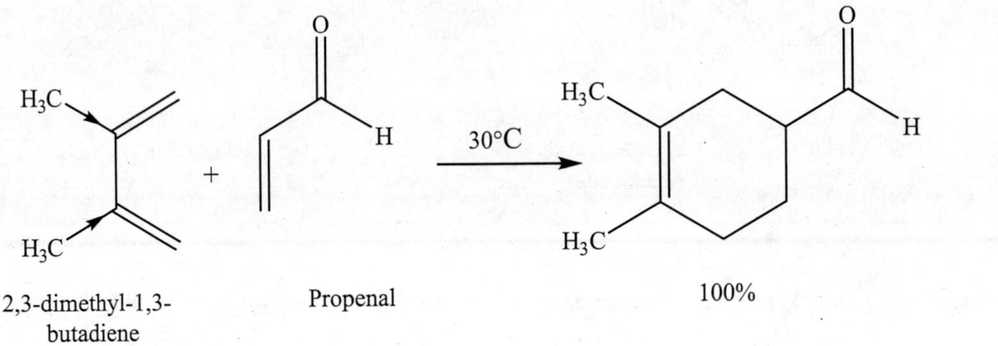

Dienes: For dienes to react, they need to be in the *s-cis* conformation. In a six-membered ring, *S-Trans* conformation would result in the development of a very unstable trans bond. Highly reactive are cyclic dienes, which need the *s-cis* structure.

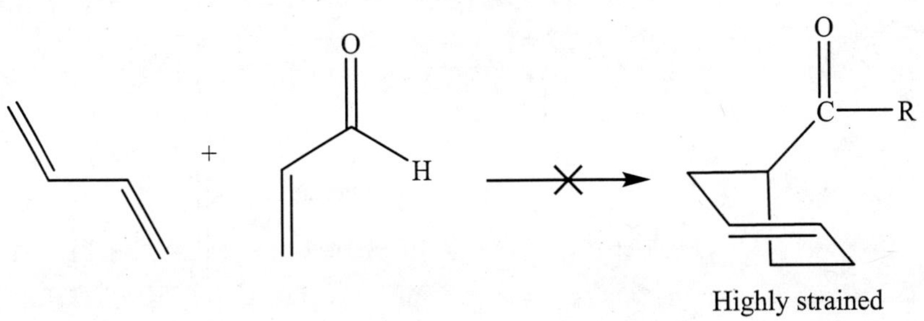

Conjugated Diene: It has single and double bonds that alternate.

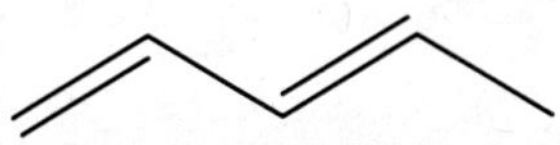

Diene take on an S-cis shape. More stable than diens that are not conjugated.

Conformation of Diens

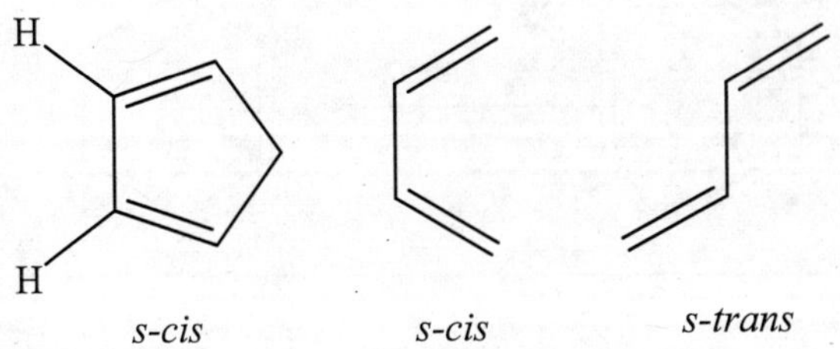

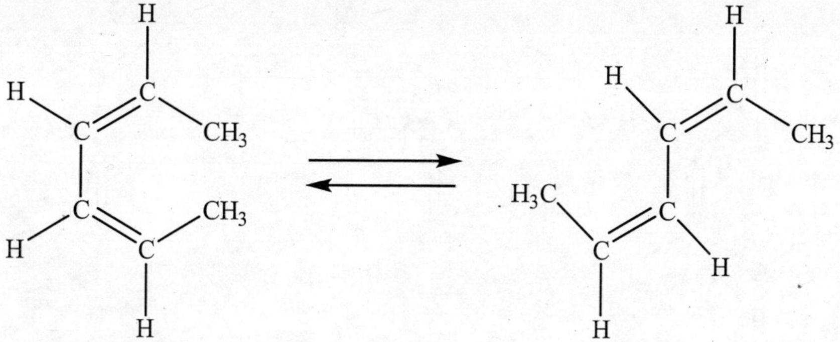

severe steric strain in *s-cis* form

s-trans

At room temperature, cyclopentadiene spontaneously undergoes a Diels-Alder reaction with itself due to its high level of reactivity. Distillation can be used to separate the cyclopentadiene product after heating this dimer and carrying out a retro-Diels-Alder reaction.

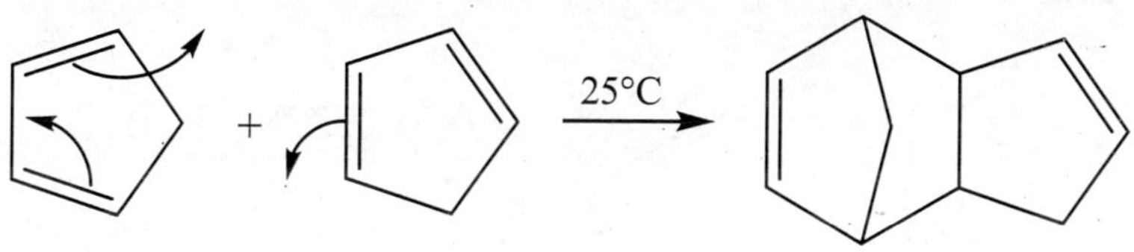

Dicyclopentadiene

Dienophile: It has a double or triple carbon bond next to a substituent group that removes electrons and is positively polarised. Reactive and non-reactive dienophiles are discussed below:

Reactive

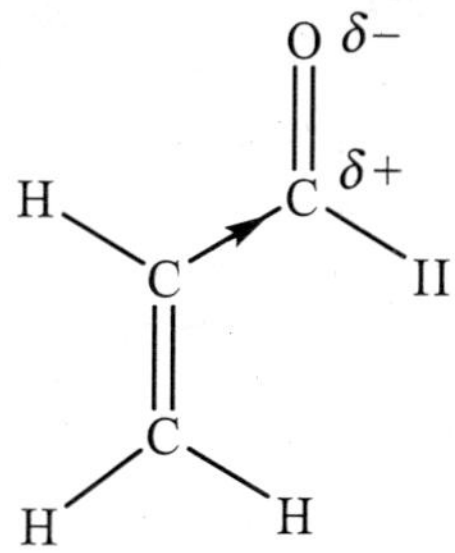

Propenal (Acrolein)

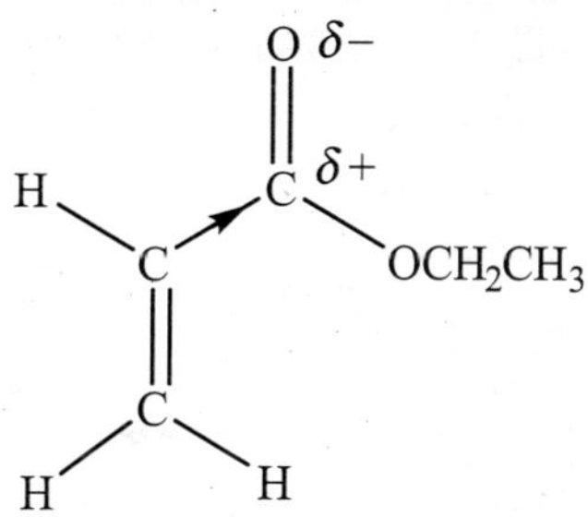

Ethyl propenoate (Ethyl acrylate)

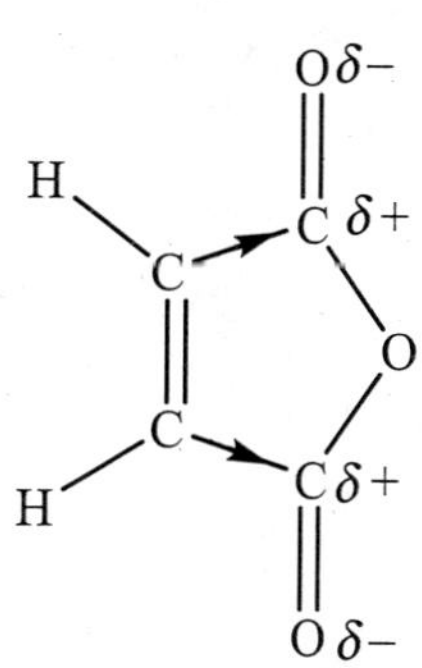

Maleic anhydride

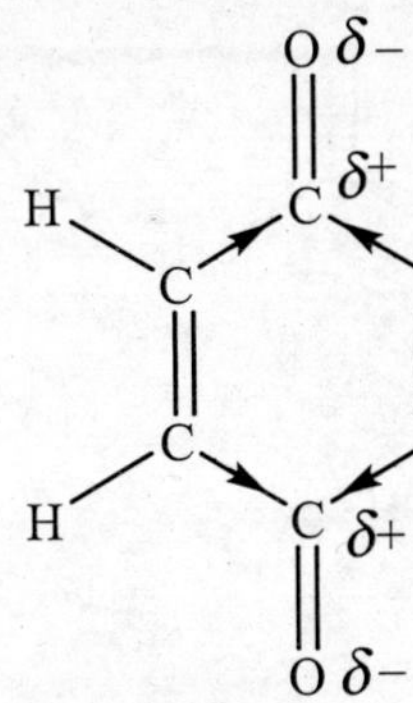

Benzoquinone

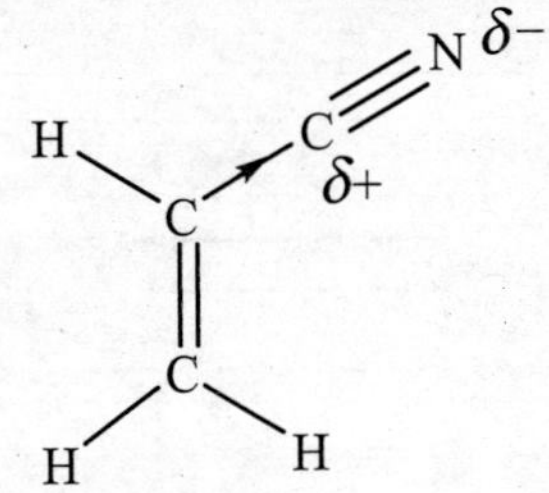

Propenenitrile
(Acrylonitrile)

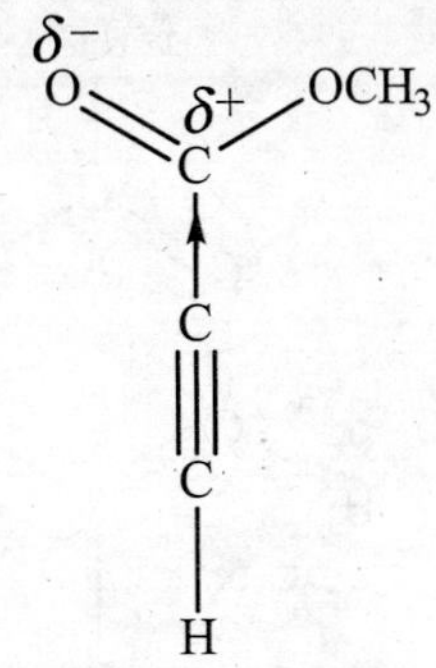

Methyl propynoate

Unreactive

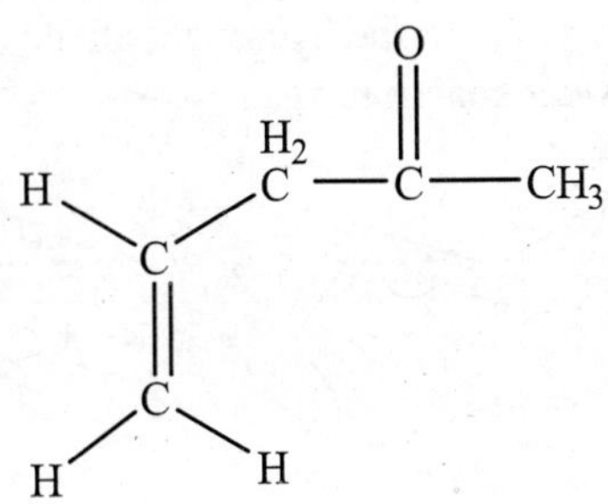

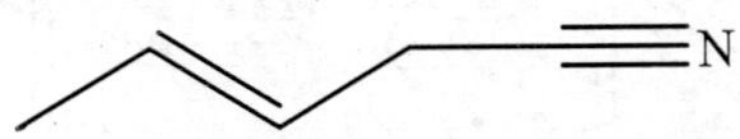

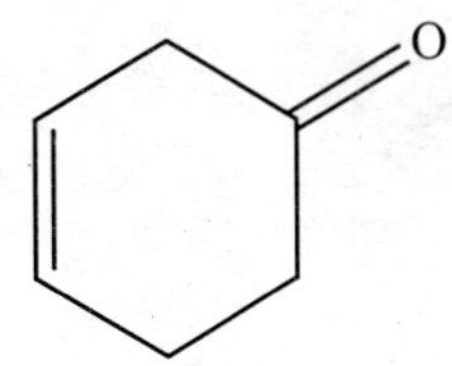

When 1, 3-butadiene and maleic anhydride are heated, a 6-membered ring product is produced 100% of the time.

16.3.5.2 Stereochemistry

The Diels-Alder reaction is stereospecific, meaning that the dienophile's conformation is preserved in the final product and the reaction is a syn addition.

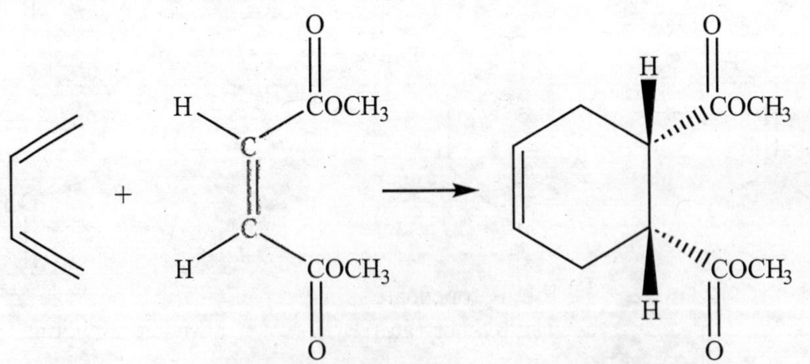

Dimethyl maleate
(a *cis*-dienophile)

Dimethyl cyclohex-4-ene-*cis*-1,2-dicarboxylate

16.3.5.3 Stereospecificity

One product stereoisomer is generated, and the original dienophile's stereochemistry is maintained throughout the process. For example:

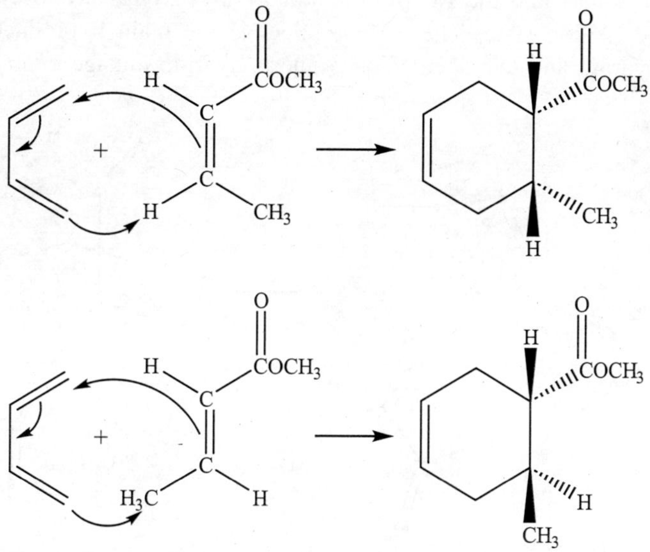

Woodward-Hoffmann Rules: The initial material orbitals must be converted into product orbitals of the same symmetry in any coordinated process.

Endo and Exo Product: The isomer with the substituent nearest to the longest bridge—also known as "*syn*"—is designated with the prefix endo. The isomer with the substituent positioned "anti"—farthest—from the longest bridge is designated with the prefix exo. The terms "longest" and "shortest" here denote how many atoms make up the bridge.

longest bridge

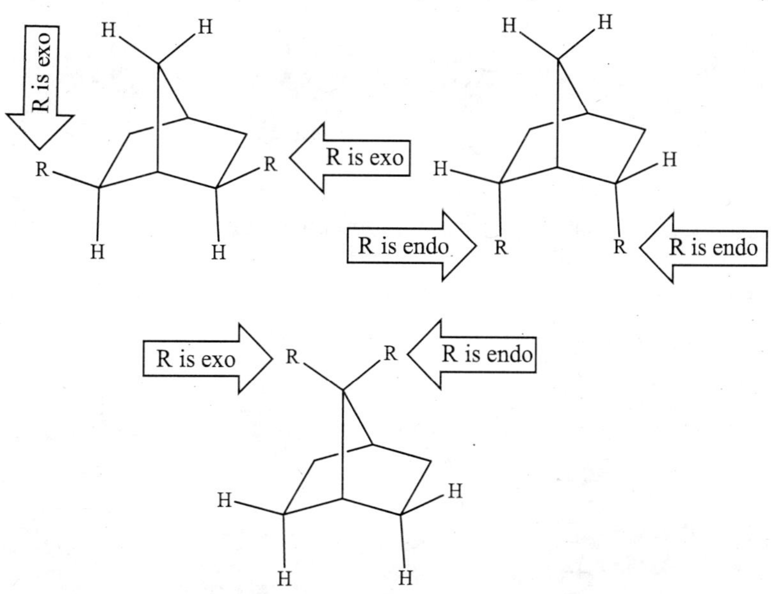

The LUMO of maleic anhydride and the HOMO of cyclopentadiene interact favourably when the molecules get close to one another. It is also possible for favourable secondary orbital contacts to arise in the endo direction between the HOMO of the cyclopentadiene carbons at the diene's C_2 and C_3 locations and the LUMO of the carbonyl groups. Molecular orbital factors endorsing an endo transition state. The endo product is the main byproduct of the reaction between maleic anhydride and cyclopentadiene. The anhydride linkage endo is present in the main product.

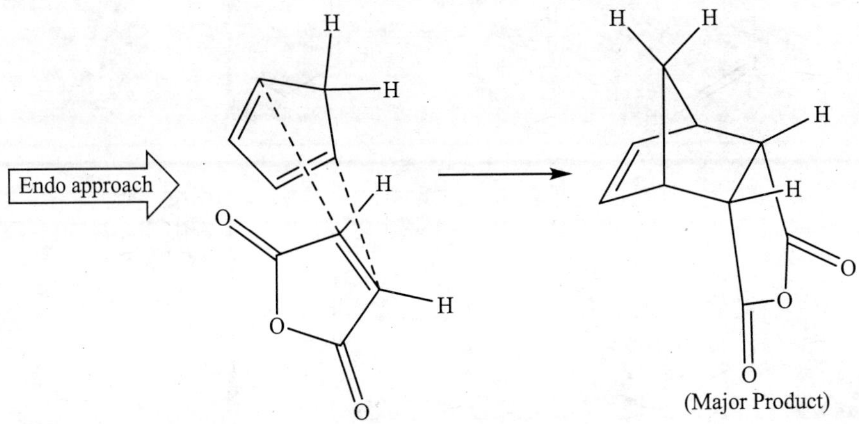

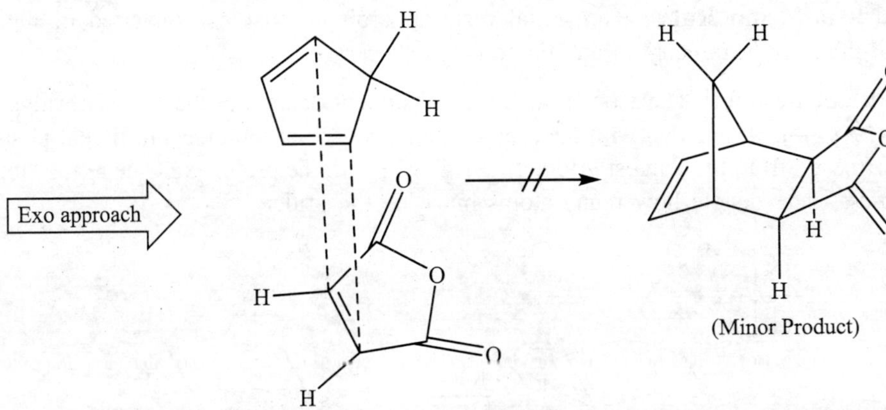

Secondary orbital overlap favors the endo product formation.

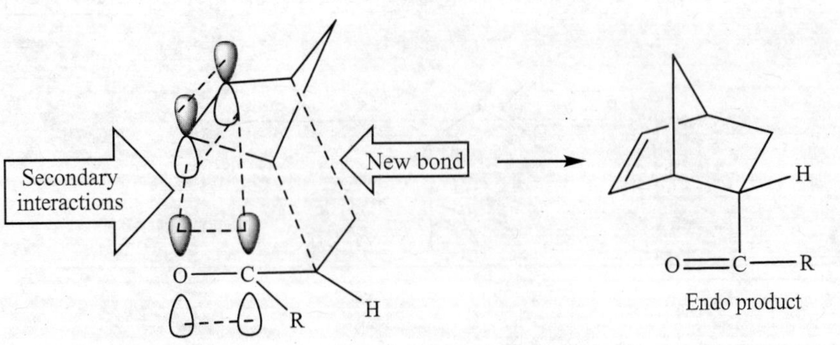

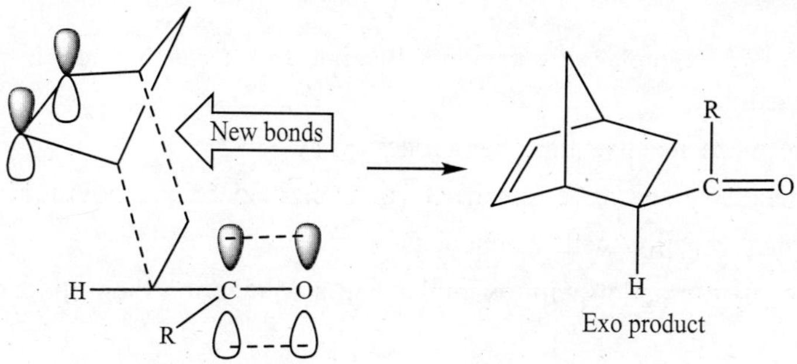

No secondary orbital overlap in the exo transition state

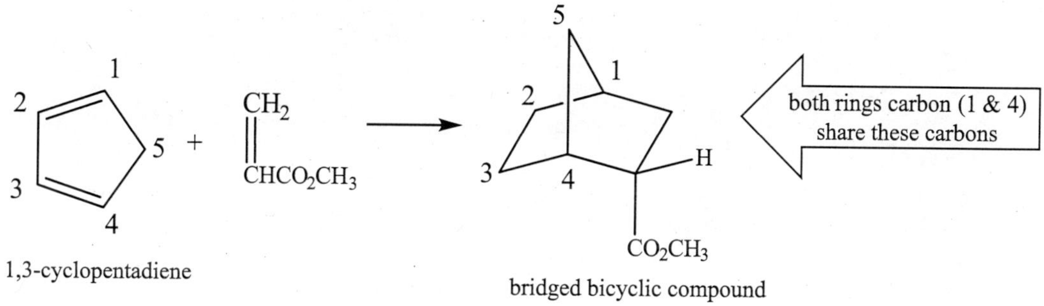

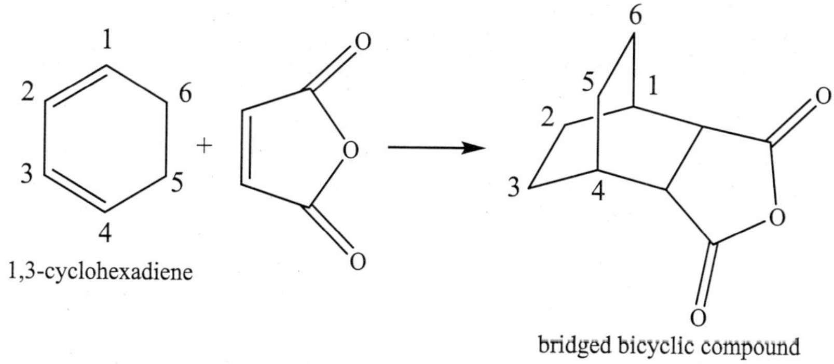

QUESTIONS

16.1 Explain with examples:

(a) How do bulky group affect S_N1 reaction.

(b) How does S_N2 reactions give rise to inverted product?

16.2 Write a short note on any two from the following:

(a) Nucleophilic addition reaction

(b) Nucleophilic substitution reaction

(c) Diels-Alder reaction

16.3 Addition of HBr to Propene-1 does not follow Markownikoffls rule in presence of peroxide. Why?

16.4 By using appropriate examples, discuss and illustrate the stereochemical implications of S_N1 and S_N2 reactions.

16.5 Halo alkanes are more reactive than haloarenes. Why?

16.6 Explain stereochemistry and stereospecificity of Diels-Alder Cycloaddition.

16.7 Explain free radical reaction with mechanism.

16.8 Explain rearrangement reaction with examples and also mention the impact of migrating group.

16.9 State the difference between S_N1 and E_1.

CHAPTER 17

Drugs: Synthesis and Applications

A drug refers to any substance that is ingested and subsequently alters or improves a physiological or psychological function within the human body. A proprietary name, also referred to as a trade name or brand name, serves the purpose of identifying a certain commercial product and differentiating it from other similar products in the market. Furthermore, every pharmaceutical product is assigned a generic name that can be universally used by any pharmaceutical company for identification purposes.

17.1 ASPIRIN

17.1.1 Introduction

The patent for aspirin was granted to Bayer in the year 1893, rendering it one of the oldest drug. One of the most widely utilized pharmaceutical substances. The substance possesses analgesic properties, serving as a pain-relieving agent, as well as antipyretic properties, which aid in reducing fever. The perception of pain is a fundamental protective mechanism that humans possess. The experience of pain enables individuals to engage in behaviors that mitigate the risk of sustaining additional harm to their physical well-being. As an example, the act of withdrawing one's hand from a heated surface or the inability to sustain running due to a muscular strain might serve as instances of physical limitations.

The brain perceives pain as a sensory experience when it receives neural signals transmitted by diverse pain receptors distributed throughout the body. The activation of these receptors is triggered by chemical compounds referred to as prostaglandins. Prostaglandins are bioactive lipid compounds synthesized at the sites of tissue injury or infection, playing a crucial role in the process of tissue repair and regeneration. Prostaglandins regulate various physiological processes, including inflammation, blood circulation, blood clot formation, and labor initiation. Upon their release, prostaglandins elicit the inflammatory response. The inflammatory response is an innate physiological reaction that promptly ensues subsequent to tissue injury. The primary roles of the immune system encompass safeguarding the body from detrimental substances, eliminating deceased or deteriorating tissue, and facilitating the regeneration of healthy tissue. The inflammatory response elicits manifestations such as edema, nociception, and pyrexia.

Aspirin and non-steroidal anti-inflammatory medicines (NSAIDs), such as ibuprofen, are classified as mild analgesics. Their mechanism of action involves the inhibition of prostaglandin formation at the site of injury, resulting in a reduction of inflammation, fever, and pain. Analgesics are classified as non-narcotics due to their lack of interference with brain function. The mechanism of action of aspirin involves the inhibition of prostaglandin formation. Upon entering the cell, aspirin molecules chemically change the cyclooxygenase enzyme to impede the manufacture of prostaglandins. Additionally, it exhibits anti-inflammatory effects by inhibiting the formation of prostaglandins.

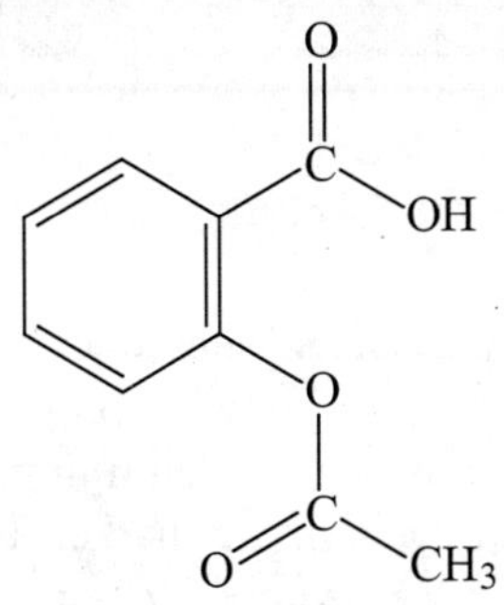

2-acetoxybenzoic acid
(aspirin)

17.1.2 Biological Activity

The emergence of novel analgesics, such as Tylenol, has been documented since approximately 400 B.C. that the act of chewing willow bark has been associated with the alleviation of pain and fever. During the early 19th century, empirical evidence demonstrated that salicin, the primary compound found in the bark, undergoes conversion into salicylic acid within the human body. Despite the efficacy of salicylic acid in pain management, its taste was found to be highly unpleasant and its administration was associated with the occurrence of vomiting. In the year 1890, the Bayer Company, situated in Germany, synthesized an ester derivative of salicylic acid that exhibited improved taste and reduced irritability to the human body. This newly developed compound was named **aspirin**. Presently, aspirin remains the most extensively utilized pharmaceutical compound on a global scale. Aspirin demonstrates efficacy in the reduction of fever (antipyretic) and the provision of analgesic effects.

17.1.3 Reaction Involved in Synthesis

The experimental procedure involves combining solid salicylic acid with liquid acetic anhydride, followed by the addition of a catalytic quantity of H_2SO_4. The resulting mixture is then subjected to heating at a temperature of 90°C for a duration of 10 minutes. The production of aspirin involves the conversion of salicylic acid (2-hydroxybenzoic acid) into aspirin through an esterification process. This process entails the transformation of the hydroxyl (–OH) group of salicylic acid sulphuric into an ester group. The procedure is conducted utilizing ethanoic anhydride in the presence of intense sulphuric acid or phosphoric acid, which serves as a catalyst. Furthermore, the solution is subjected to gentle heating. Subsequently, it is vital to separate and refine the aspirin compound from the amalgamation.

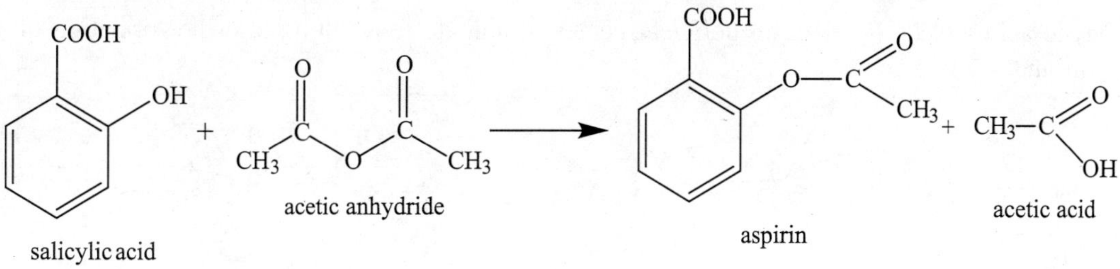

After the completion of the reaction, the resulting product undergoes a cooling process, which induces the formation of crystalline structures. Subsequently, the product undergoes suction filtration and is subsequently subjected to a washing process utilizing cooled water. Due to the limited solubility of aspirin in water under low temperature conditions, this procedure facilitates the selective removal of soluble acids without affecting the aspirin product. Subsequently, the product undergoes a purification process. The process of purifying employs a method referred to as recrystallization. The process of recrystallization entails the dissolution of impure crystals in heated ethanol, resulting in the dissolution of those impurities. This results in the formation of a saturated solution of aspirin.

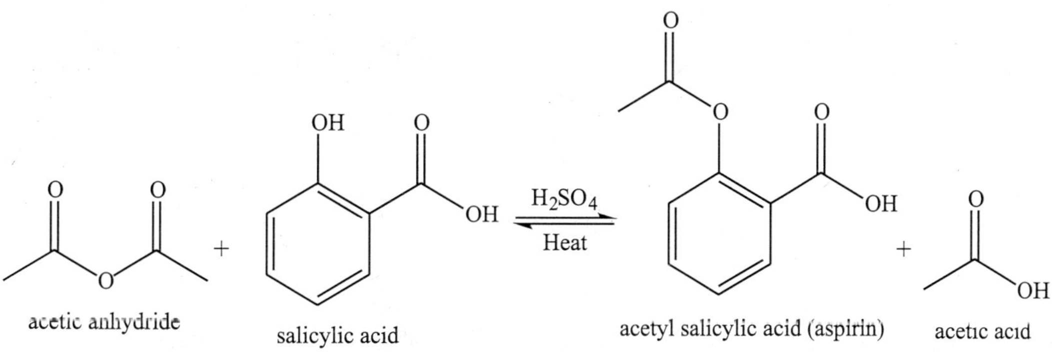

Subsequently, the solution is subjected to a gradual cooling process. The reduction in solubility of aspirin leads to its crystallization from the solution, facilitating its separation through the process of filtration. The confirmation of purity can be achieved through the determination of the melting point. Pure substances exhibit distinct melting points, which can be modified by the introduction of impurities.

The melting point of pure aspirin ranges from 138°C to 140°C, while salicylic acid exhibits a higher melting point of 159°C. The presence of a mixture is associated with a decrease in the melting point. The determination of the reaction yield can be achieved by evaluating the mass of the salicylic acid employed and the mass of the resulting product. Allow the substance to reach ambient temperature and introduce water as the solvent for crystallization. Facilitate the process of crystallization. The crystals should be subjected to a filtration process.

17.1.4 Reaction Mechanism

Salicylic acid attacks on electron deficient center of anhydride resulting into the formation of aspirin and acetic acid.

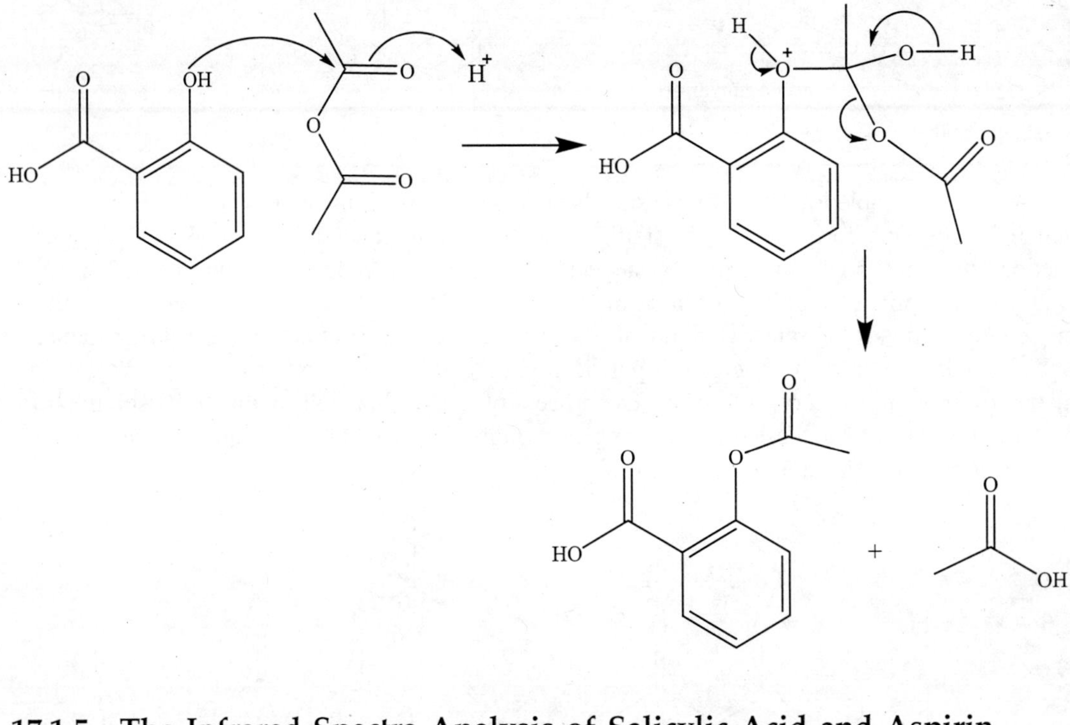

17.1.5 The Infrared Spectra Analysis of Salicylic Acid and Aspirin

The infrared spectra of salicylic acid and aspirin are depicted below. There exist certain similarities in the observed spectra, characterized by prominent peaks ranging from 1050 to 1410 cm^{-1}, which can be attributed to the presence of C–O bonds in alcohol and ester compounds. The presence of carboxylic acid is indicated by prominent peaks observed in the spectral range of 1700 to 1750 cm^{-1}, which can be attributed to the C=O functional group. Both spectra exhibit wide peaks ranging from 2500 to 3000 cm^{-1}, which can be attributed to the presence of the hydroxyl (OH) group in carboxylic acid. Both spectra exhibit prominent peaks within the wavenumber range of 2850 to 3090 cm^{-1}, which can be attributed to the presence of C–H bonds.

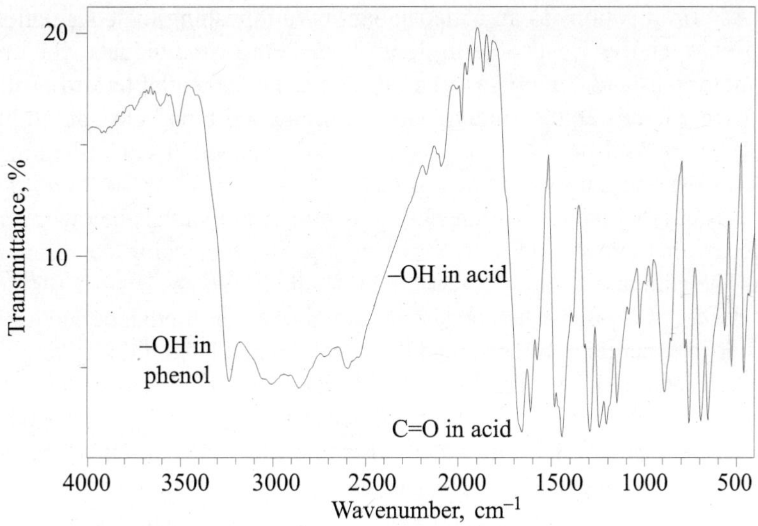

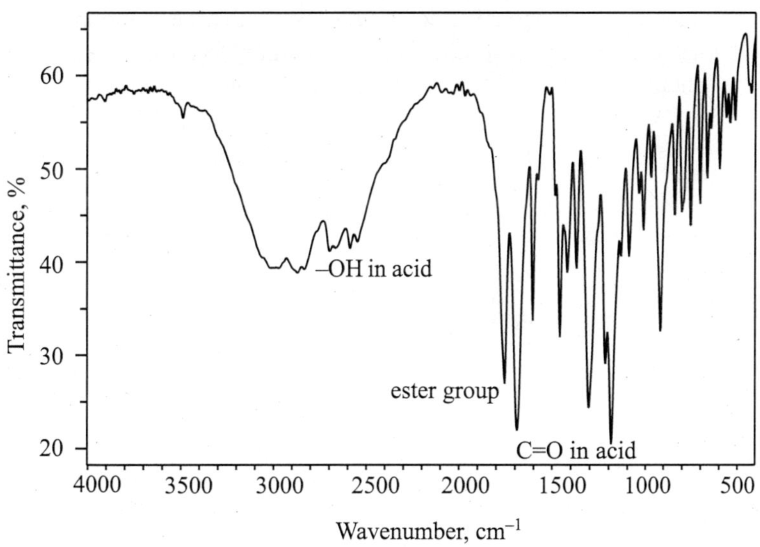

These peaks overlap with the broad –OH peak. Furthermore, it is worth noting that there are discernible distinctions, such as the occurrence of a secondary peak ranging from 1700 to 1750 cm^{-1}, which can be attributed to the existence of an ester group within the molecular structure of aspirin. The salicylic acid spectrum exhibits a discernible peak ranging from 3200 to 3600 cm^{-1}, which can be attributed to the existence of the hydroxyl (–OH) functional group. Conversely, this particular peak is absent in the spectrum of aspirin.

17.1.6 The Impact of Aspirin

The medication of aspirin is associated with adverse effects such as gastrointestinal discomfort and bleeding. The potential outcomes can encompass both advantageous and unfavorable consequences. Aspirin functions as an anticoagulant by diminishing the coagulation capacity of blood. This attribute renders it valuable in the management of individuals who are susceptible to myocardial infarctions and cerebrovascular accidents. The potential hazard of this side effect is heightened when administered to individuals with impaired blood clotting abilities. Adverse effects associated with aspirin usage encompass gastric and duodenal inflammation, perhaps resulting in ulceration and subsequent hemorrhaging.

Moreover, it is advised against administering aspirin to individuals below the age of 12 due to its association with Reye's syndrome, a potentially life-threatening condition affecting the liver and brain. The administration of ethanol in alcoholic beverages has been found to enhance the immediate effects of aspirin. The concurrent administration of ethanol and aspirin has been observed to result in synergistic effects, leading to heightened susceptibility to gastric mucosal bleeding and an elevated risk of ulceration. Aspirin has the potential to undergo modifications that can enhance its absorption and distribution, hence leading to improved bioavailability.

Aspirin is administered orally and is carried in the blood plasma as an aqueous solution. However, due to its predominantly non-polar nature, aspirin has limited solubility in water. To enhance its bioavailability, it is necessary to augment its water solubility. Chemical alteration can be employed to achieve this objective. This process entails the reaction of aspirin with an alkali, such as NaOH or $NaHCO_3$, resulting in the formation of an ionic salt. Aspirin exhibits limited solubility, whereas its sodium salt demonstrates enhanced solubility.

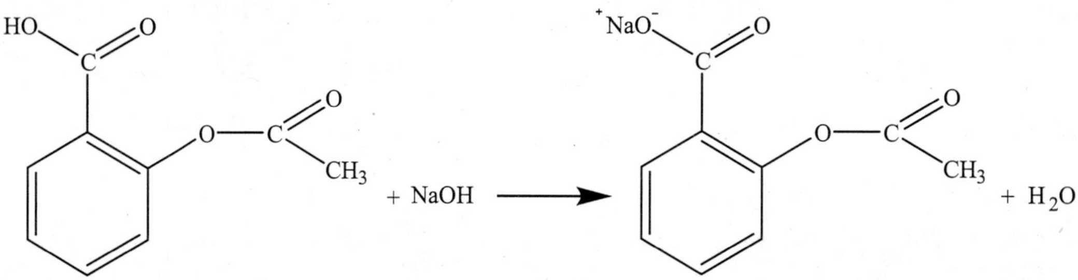

aspirin sodium 2-acetoxybenzoate

17.2 PHENACETIN

17.2.1 Introduction

Phenacetin, an analgesic agent, is recognized as the first synthetic medicinal compound in global medical history. This particular medication represented an early instance of a non-opioid analgesic, distinct from its lack of anti-inflammatory properties. Phenacetin, an analgesic and antipyretic compound, was synthesized in the year 1878 by Harmon Northrop Morse, a chemist hailing from the United States. The pharmaceutical market saw its introduction in the year 1887. Nevertheless, the drug was removed from the market in the United States in 1983 due to the presence of interstitial nephritis in patients at levels deemed unsatisfactory, as well as the possible hazards associated with tumorigenicity. Similar to the United States, the majority of Western nations did not implement a ban on the marketing of phenacetin until the year 1983. Phenacetin constitutes a constituent of the pharmaceutical compound APC, which comprises aspirin, phenacetin, and caffeine.

At this point, the biochemist David Lester and Leon Greenberg demonstrated that the human body not only undergoes metabolic processes to convert phenacetin into acetaminophen, but also established the comparable efficacy of acetaminophen to phenacetin. Acetaminophen possesses the notable advantage of lacking methemoglobin-inducing properties and exhibiting non-carcinogenic characteristics.

Consequently, the utilization of acetaminophen became widespread. In typical usage, paracetamol has a consistently favorable safety profile. The organization's commendable safety reputation was tarnished upon the revelation that a significant dosage could potentially lead to severe and occasionally lethal liver injury.

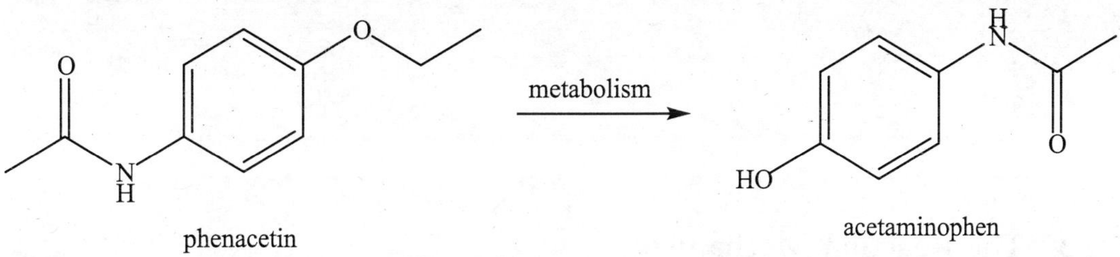

17.2.2 The Mechanisms of Action

The analgesic effects of phenacetin can be attributed to its impact on the sensory pathways inside the spinal cord. Furthermore, phenacetin exerts a depressive effect on cardiac function, functioning as a negative inotrope. This substance has antipyretic properties via modulating the temperature set point inside the central nervous system. Additionally, it is employed for the management of rheumatoid arthritis, which is a chronic inflammatory condition largely characterized by joint inflammation. The manifestation often includes joint inflammation characterized by elevated temperature, increased size, and discomfort. The substance undergoes metabolic processes within the human body, resulting in the formation of paracetamol (commonly known as acetaminophen), which is a noteworthy analgesic with clinical significance.

For a considerable period, Phenacetin, also known as acetophenetidin or N-(4-ethoxyphenyl) acetamide, was employed as a pharmaceutical agent in human and veterinary medicine to alleviate pain and reduce fever. The utilization of this substance in therapeutic practices commenced in 1887, and it was widely incorporated into analgesic formulations until its association with renal disease (nephropathy) resulting from the misuse of analgesic medications.

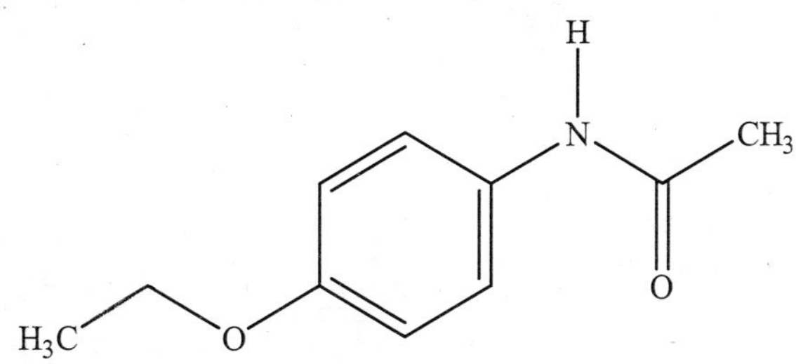

N-acetyl-p-aminophenol(phenacetin)

The synthesis of phenacetin involves a simple and efficient two-step process, commonly referred to as a "one-pot" organic synthesis. The experiment elucidates the fundamental process of amine-anhydride reaction leading to the formation of an amide. Additionally, it serves to strengthen the understanding of recrystallization principles and the calculation of melting points.

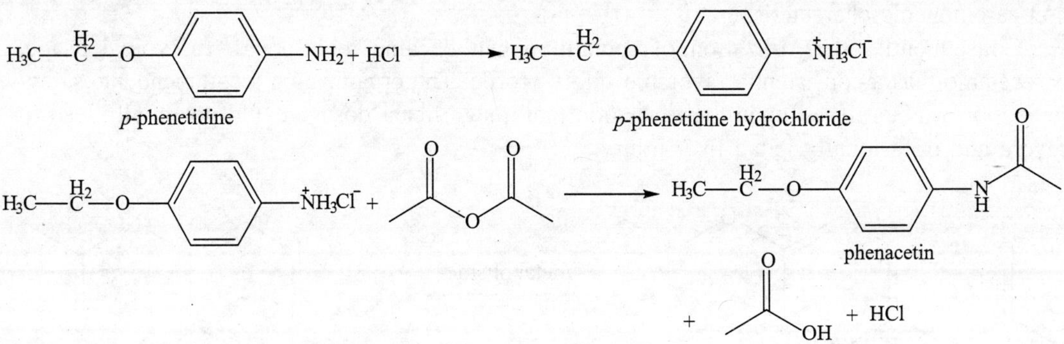

17.2.3 The Reaction Mechanism

The reaction proceeds through a mechanism known as nucleophilic acyl substitution.

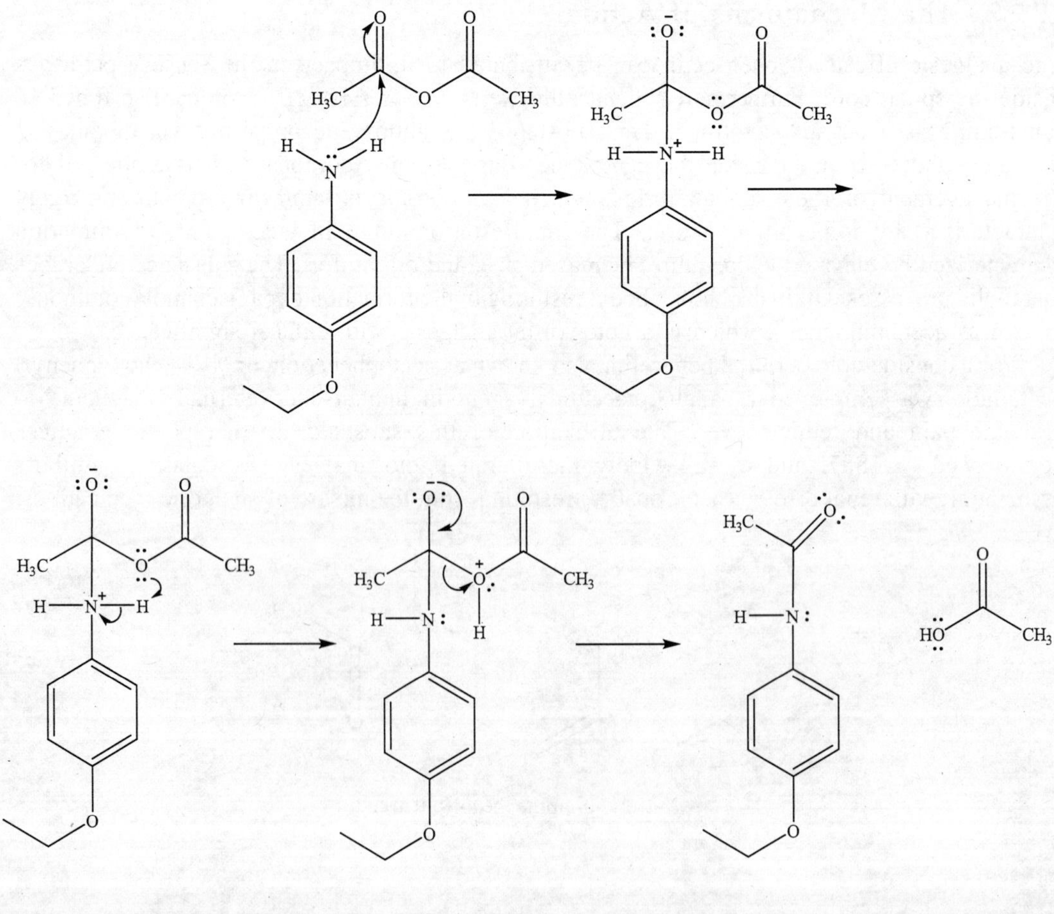

17.3 MELUBRIN

17.3.1 Introduction

The pharmaceutical company Sun Pharmaceutical Industries Ltd. manufactures the medication known as Melubrin. Chloroquine serves as the primary constituent in Melubrin products. Chloroquine is a pharmacological agent utilized for the treatment of malaria, exhibiting antiparasitic properties. Chloroquine is a pharmacological agent employed for the treatment of susceptible infections caused by *Plasmodium vivax, Plasmodium malariae, Plasmodium ovale*, and *Plasmodium falciparum*. Additionally, it is employed as a secondary therapeutic option for rheumatoid arthritis. The mechanism of action involves the elevation of haeme concentrations in the bloodstream, which has a toxic effect on the malarial parasite. The aforementioned action effectively eradicates the parasite and halts the propagation of the infection.

Chloroquine is widely utilized as a pharmaceutical agent for the management of malaria. This compound exhibits a high efficacy in rapidly eliminating erythrocytic schizonts across all Plasmodium species. The chemical structure under consideration is a derivative of quinoline, wherein a chlorine atom is located at the Meta position. The molecule consists of three nitrogen atoms.

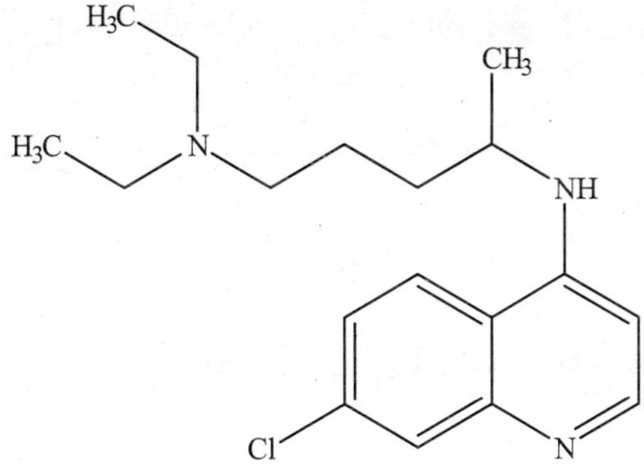

N[4]-(7-chloroquinolin-4-yl)-*N*[1], *N*[1]-diethylpentane-1,4-diamine (chloroquine)

The discovery of chloroquine can be attributed to Hans Andersag in the year 1934. Chloroquine, an aminoquinolone derivative, was initially synthesized in the 1940s with the primary purpose of addressing malaria. Historically, it served as the preferred pharmaceutical agent for the treatment of malaria until the emergence of more contemporary antimalarial drugs, including pyrimethamine, artemisinin, and mefloquine. Chloroquine and its derivative hydroxychloroquine have been repurposed for the treatment of several medical disorders, including as HIV, systemic lupus erythematosus, and rheumatoid arthritis. Chloroquine is commercially available under various brand names, including Melubrin, Resochin, Dawaquin, and Lariago.

The typical adverse effects associated with the administration of Melubrin include the occurrence of a rash, headache, disorientation, vomiting, nausea, convulsions, urticaria, and loss of vision, renal impairment, psychosis, and rheumatoid arthritis. The signs of this medication include malaria prophylaxis and rheumatoid arthritis, while its contraindications encompass Fabry disease, renal failure, hypotension, and dermatitis.

17.3.2 Physical and Chemical Properties of the Drug

The IUPAC name of the compound is N-(7-chloroquinolin-4-yl)-N, N-dimethylpentane-1, 4-diamine. The molecular formula of the compound under consideration is $C_{18}H_{26}N_3Cl$, with a corresponding molecular weight of 319.827 g/mol. The chemical compounds commonly known as aralen and nivaquin are the subject of discussion. The compound exhibits solubility in both water and methanol. The color of the substance is a white crystalline powder. The substance is present in a solid state, specifically in a crystalline form. The melting point of the substance is 87°C. The color of the object varies from white to a pale yellow.

17.3.3 Method of Preparation

The process of synthesis comprises three distinct stages that encompass all reactions.

Step 1: Synthesis of 4, 7-dichloroquinoline

The reaction involving the combination of *m*-chloroaniline (1) and diethyl 2-oxosuccinate (2) was conducted to get compound (3). Then cyclization of intermediate with NaOH/water and HCl to form compound (4). Upon exposure to $POCl_3$ under conditions of elevated temperature, the elimination of CO_2 takes place, resulting in the formation of 4, 7-dichloroquinoline (7).

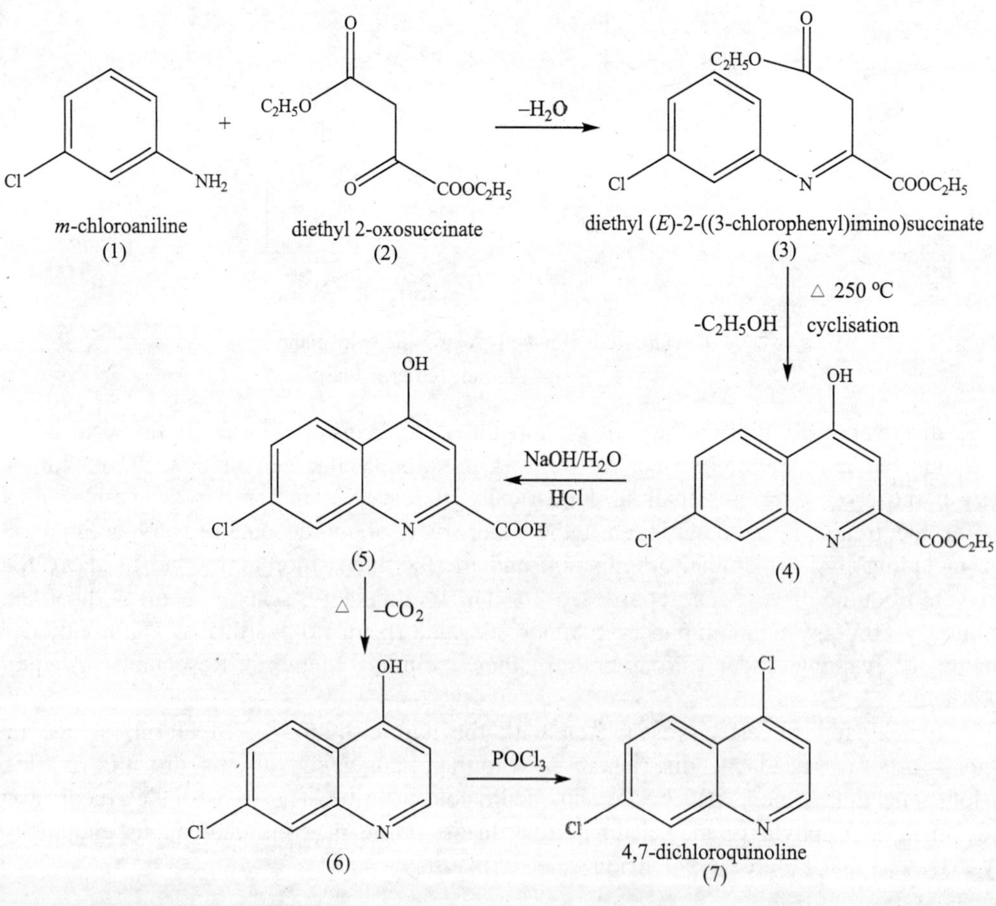

Step 2: Preparation of 1-diethyl amino-4-amino Pentane

The reaction between ethylene oxide (8) and diethylamine (9) results in the formation of chemical compound (10). The chemical compound (11) was obtained by subjecting it to treatment with acetone. Compound (6) undergoes further reduction using H_2/Ni and NH_3, resulting in the formation of 1-diethyl amino-4-aminopentane (12).

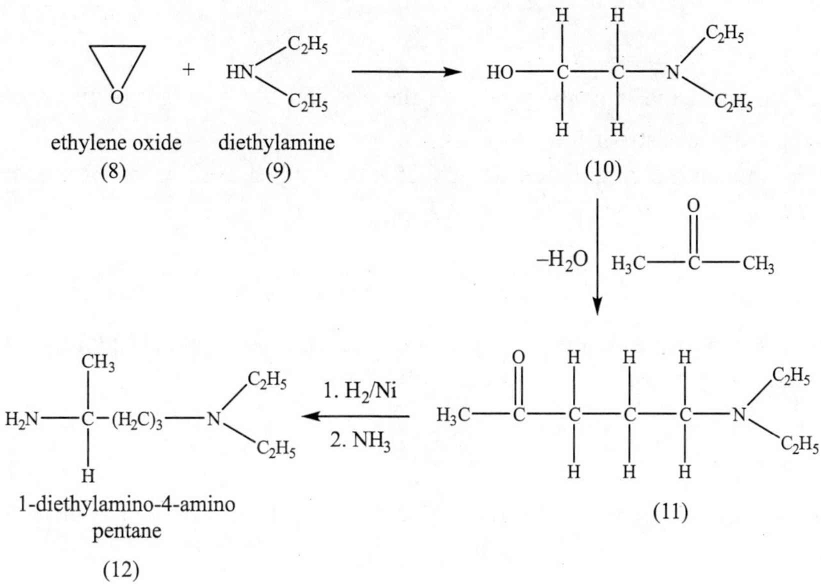

Step 3: Condensation

The compound chloroquine (13) can be synthesized by condensation of the products of Step 1 and Step 2.

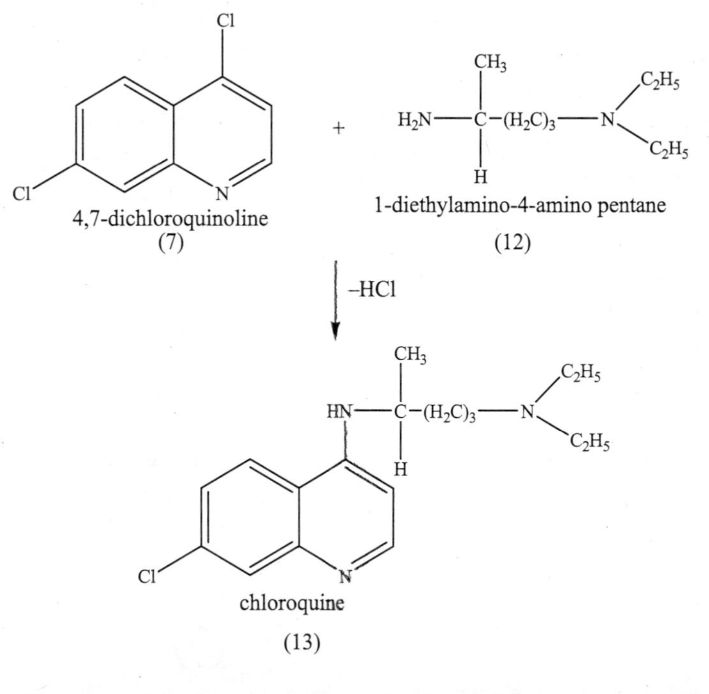

17.3.4 Applications

Primarily, it is employed as an antimalarial agent. This substance exhibits antihistaminic and anti-inflammatory properties. This medication is employed for the treatment of hepatic amoebiasis as well as rheumatoid arthritis.

17.3.5 Dose

The dosage of the medicine is contingent upon the age and body weight of the patients.

1. The regimen consists of four dosages.
2. The recommended cumulative dosage of chloroquine over a span of three days is 41.7mg.

17.3.6 pKa Value

The pKa value is a measure of the acidity or basicity of a compound. Chloroquine exhibits three distinct pKa values, specifically 4.0, 8.4, and 10.2.

17.3.7 Stability

The concept of stability refers to the state or condition of being stable, where an object. Chloroquine has thermal stability within the pH range of 4.0 to 6.5.

17.3.8 Storage Conditions

The conditions under which items are stored. Chloroquine exhibits stability under ambient temperatures.

The medication exhibits photosensitivity. Therefore, it is imperative to ensure that it is shielded from exposure to light.

The stability of the medication is increased in its powdered form. Therefore, it has the ability to be stored for a duration of approximately six months.

17.4 NOVALGIN

17.4.1 Introduction

Metamizole can be classified as a methyl derivative of melubrine and is additionally recognized as a prodrug of pyramidon with enhanced solubility. The initial introduction of Metamizole to the market occurred in Germany under the brand name "Novalgin" in the year 1922. Metamizole, also known as dipyrone, is a pharmaceutical compound utilized for its analgesic, antispasmodic, and antipyretic properties. Administration of the medication is often conducted orally or via intravenous infusion. While several countries permit over-the-counter access to it, other nations have implemented a prohibition on its usage owing to the probable occurrence of adverse consequences, such as agranulocytosis, which involves a perilous decrease in the count of white blood cells.

According to a study conducted by a pharmaceutical company, the incidence of agranulocytosis during the initial week of medication was found to be 1.1 cases per million individuals, in comparison to 5.9 cases per million individuals for diclofenac. This pharmaceutical compound is classified under the ampyrone sulfonate family and obtained its patent in the year 1922. For an extended period, this product was readily accessible without a prescription in numerous nations until its discontinuation as a result of the occurrence of significant negative consequences. Metamizole is commercially available under multiple brand names.

Metamizole is classified as a nonsteroidal anti-inflammatory drug (NSAID) belonging to the pyrazolone class. It has analgesic and antipyretic properties. The compound is identified by its chemical name as sodium N-(2, 3-dimethyl-5-oxo-1-phenyl-3-pyrazolin-4-yl)-N-methylaminomethanesulphonate. It has a molar mass of 334.38 g/mole and a molecular formula of $C_{13}H_{16}N_3NaO_4S.Na$. Metamizole is a crystalline powder that exhibits a white or almost white crystalline powder. It possesses a high solubility in water and a moderate solubility in alcohol.

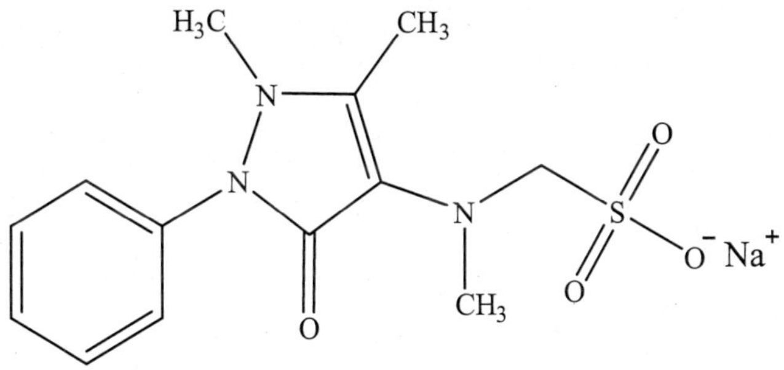

metamizole sodium

17.4.2 Medical Applications

The primary medical applications of this substance include managing pain during the perioperative period, treating acute injuries, alleviating colic symptoms, addressing cancer-related pain, managing various forms of acute and chronic pain, and reducing high fever that does not respond to other medications.

17.4.3 Special Populations

The administration of this substance during pregnancy is not recommended, although animal studies provide reassurance by demonstrating a low likelihood of causing birth abnormalities. Administration of this medication is contraindicated in the elderly and individuals with liver or kidney dysfunction. However, if treatment is deemed necessary for these populations, it is generally recommended to administer a reduced dosage and exercise caution. The utilization of this substance is not recommended during the period of lactation due to its excretion in breast milk.

17.4.4 The Potential Negative Consequences

Metamizole possesses the potential to induce blood-related toxicity, specifically blood dyscrasias. However, in comparison to non-steroidal anti-inflammatory medicines (NSAIDs), it exhibits a lower propensity for inducing kidney, cardiovascular, and gastrointestinal toxicity. It has the potential to induce bronchospasm or anaphylaxis, particularly in those with asthma. Significant adverse effects encompass agranulocytosis, aplastic anemia, hypersensitivity events such as anaphylaxis and bronchospasm, toxic epidermal necrolysis, and the potential to induce acute porphyria attacks due to its chemical similarity to sulfonamides. The relative risk of agranulocytosis exhibits significant variability depending on the country of estimation, leading to a pronounced divergence in opinions regarding the associated risk. The influence of genetics on the susceptibility to metamizole is potentially substantial.

17.4.5 Synthesis Mechanism

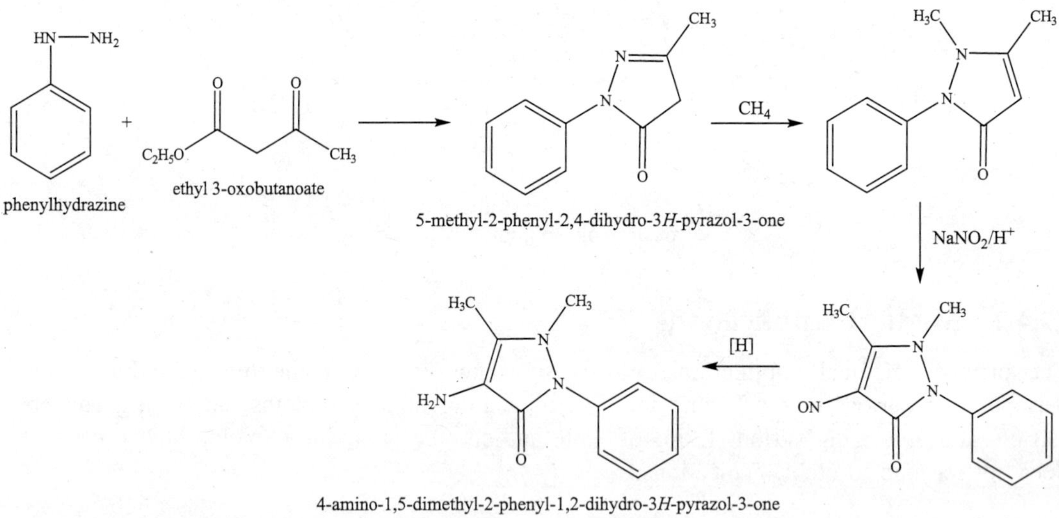

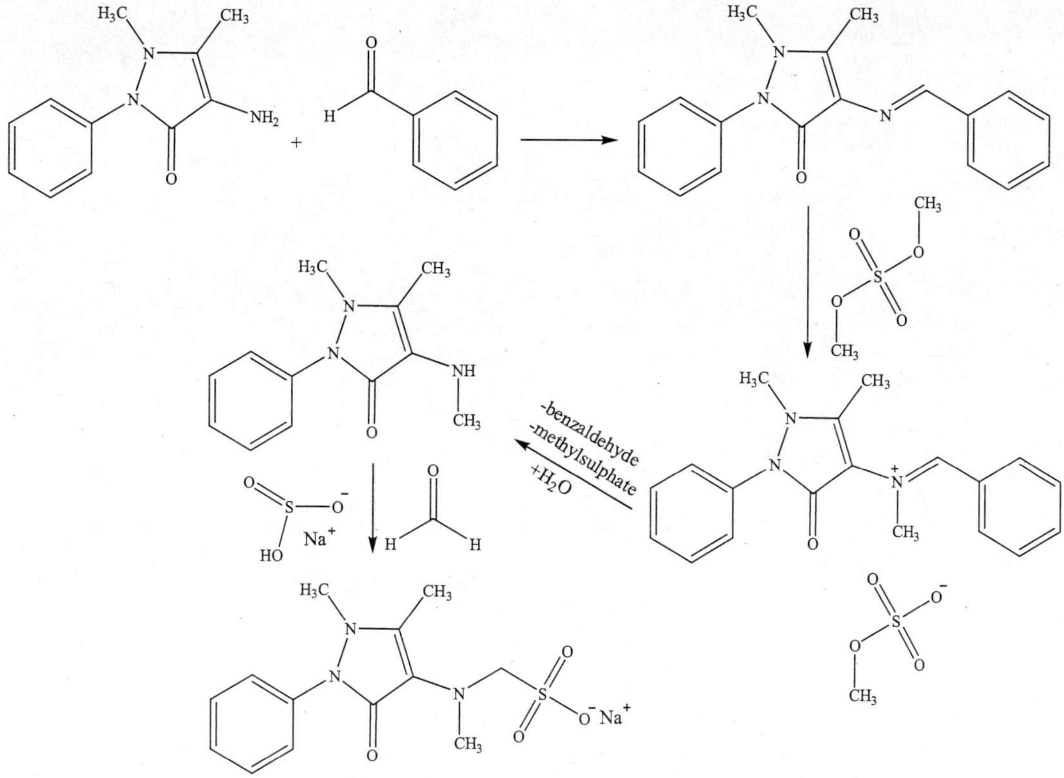

metamizole sodium

There is a suggestion that certain groups have a higher susceptibility to the development of metamizole-induced agranulocytosis compared to others.

QUESTIONS

17.1 Discuss the synthesis of a commonly used drug molecule by taking suitable example.

17.2 Write down the chemical reactions involved in the preparation of Novalgin.

17.3 Write the reactions involved in the synthesis of Aspirin. What are its impact on human health?

17.4 Write the reaction involved in the mechanism of Phenacetin.

17.5 What are the potential negative consequences of metamizole?

17.6 Write down the applications of Melubrin drug.

Index